Thermodynamics for Chemical Engineers

Thermodynamics for Chemical Engineers

Kenneth R. Hall
Gustavo A. Iglesias-Silva

Authors

Prof. Kenneth Richard Hall ret.
Texas A&M University
Bryan Research & Engineering
77845 College Station, TX
United States

Prof. Gustavo Arturo Iglesias-Silva
National Technological Institute of
Mexico - Technological Institute of
Celaya
Chemical Engineering
30810 Celaya, Guanajuato
Mexico

Library of Congress Card No.: applied for

British Library Cataloguing-in-Publication Data
A catalogue record for this book is available from the British Library.

Bibliographic information published by the Deutsche Nationalbibliothek
The Deutsche Nationalbibliothek lists this publication in the Deutsche Nationalbibliografie; detailed bibliographic data are available on the Internet at <http://dnb.d-nb.de>.

Print ISBN: 978-3-527-35030-8
ePDF ISBN: 978-3-527-83678-9
ePub ISBN: 978-3-527-83679-6

Typesetting Straive, Chennai, India
Printing and Binding CPI Group (UK) Ltd, Croydon, CR0 4YY

Printed on acid-free paper

C118890_270522

Contents

Preface

This book covers our experiences as instructors teaching thermodynamics in undergraduate- and graduate-level courses. Our years of teaching have indicated to us that many students have problems understanding many concepts of thermodynamics. Our intention with this book is to present the subjects in a manner that enables the reader to assimilate them as quickly as possible. This book is primarily a textbook for the usual two thermodynamic courses that appear in chemical engineering curricula, but it can also be useful in careers that require a sense of thermodynamics.

The first chapter contains introductory material to familiarize students with the notation, definitions, and variables used in thermodynamics The second chapter deals with the first law of thermodynamics by introducing a general equation applying to open and closed systems to enable students to analyze energy transfer problems. The third chapter introduces the PVT behavior of pure substances. The fourth and fifth chapters discuss the second law of thermodynamics and the necessary mathematical formality to calculate thermodynamic properties, respectively. Chapter 6 introduces practical applications of thermodynamics. Chapter 7 introduces to the students to all the concepts necessary to compute equilibrium problems. Chapter 8 introduces calculation of physical equilibrium among phases and includes new procedures for the calculation of equilibrium based upon the Gibbs energy. Finally, Chapter 9 discusses chemical equilibrium.

The authors acknowledge the many students that offered their comments and criticisms during their thermodynamic courses. These observations have encouraged the authors to write this textbook. In addition, the authors thank Prof. J. C. Holste for making available to us many problem statements that he has developed during his teaching career at Texas A&M University.

College Station Texas, USA
April 2022

Kenneth R. Hall
Gustavo A. Iglesias-Silva

1 Introduction

1.1 Definition

Welcome to a new world, one without time! This is the world of *thermodynamics*, and it is the world we shall study in this course. Thermodynamics is an engineering/science field of study that is an exceptionally powerful tool for solving difficult problems with relatively little effort. For example, consider a glass of water that over the course of a day is heated, cooled, stirred, boiled, and frozen. How could we find the final condition of the water as represented by its characteristics knowing only the initial condition? It would be necessary to know the position and momentum of each molecule of water as a function of time and integrate from the initial time to the final time. Given that the water may contain 10^{25} molecules, this problem poses a prohibitively difficult task. However, with thermodynamics, we can translate this problem from a time domain to a timeless domain. In that new domain, we can solve the problem by knowing only a few things (less than five) about the water. Surely, you would rather tackle the latter problem than the former one. The translational procedure to move between the real world and the thermodynamic world is the application of the laws of thermodynamics. These are observations for which we have found no contradictions. The means for translation is mathematics. This process/procedure will become obvious as we progress through the course.

What is thermodynamics? The original Greek words from which we derive the name are *thermos* = heat and *dynamos* = power. This analysis would imply that thermodynamics is a study of heat power. Indeed, early thermodynamisists studied extracting energy transferred as work from energy transferred as heat. However, thermodynamics is a much richer and broader topic than that. The topic has very few concepts combined with formal elegance. Therefore, thermodynamics in a broader sense is a mathematical description of the real world using its physical properties.

Thermodynamics comes in two "flavors": classical and statistical. Classical thermodynamics derives from macroscopic observations. Statistical thermodynamics derives from microscopic models. A modern bridge between the two is computer simulation that uses numerical techniques to apply principles of statistical thermodynamics to macroscopic problems to increase understanding. This course considers classical thermodynamics, but, when appropriate, introduces the concepts of statistical thermodynamics to provide deeper understanding.

Thermodynamics for Chemical Engineers, First Edition. Kenneth R. Hall and Gustavo A. Iglesias-Silva.

1.2 Dimensions, Fundamental Quantities, and Units

Thermodynamics has simple mathematics, but the concepts are usually more demanding for students. It is important to recall several important definitions before beginning a course in thermodynamics to emphasize the importance of concepts. Therefore, let us define some common concepts that are useful in thermodynamics.

The numbers are meaningless without the correct specification of what they represent. For example, it does not make sense to say *I reduced my weight by three* or *the distance that I walk every day is five*. We must state that we have reduced our weight by 3 lb or 3 kg, and we traveled 5 mi or 5 km. These specifications are *units*. The product of a number and a unit can express any physical quantity. The number multiplying the unit is the *numerical value* of the quantity expressed in that unit, that is

$$A = \{A\}[A] \tag{1.1}$$

where $\{A\}$ is the numerical value or magnitude of A when expressing the value of A with units $[A]$. For example, if you weigh 60 kg, then $\{A\} = 60$ and $[A]$ is kilograms. Generally, in figures and tables, we see the labels $A/[A]$, which indicate numerical values according to Eq. (1.1). Thus, the axis of a graph or the heading of a column in a table should be "m kg^{-1}," denoting the mass measured in kilograms instead of "m (kg)" or "Mass (kg)."

Many units exist for different physical quantities, so it is convenient to introduce a general designation for a single class of units. Then, all the units employed for a particular physical property have the same **dimension** and use a separate symbol for it. The concept dimension is the name of a class of units, and a particular unit is an individual member of this class. The use of dimensions requires establishing a scale of measure with specific units. The International System of Units (SI: Systeme International) sets these units by international agreement. The SI establishes seven base units for seven base physical quantities that do not depend upon any other physical property (such as the length of the King's foot, the mass of a 90% platinum and 10% iridium bar, or suchlike).

The SI is a system of units for which

- the unperturbed ground-state hyperfine transition frequency of the cesium atom, $\Delta\nu_{Cs}$, is 9 192 631 770 Hz
- the speed of light under vacuum, c, is 299 792 458 m/s
- the Planck constant, h, is $6.62607015 \times 10^{-34}$ J s
- the elementary charge, e, is $1.602176634 \times 10^{-19}$ C
- the Boltzmann constant, k, is 1.380649×10^{-23} J/K
- the Avogadro constant, N_A, is $6.02214076 \times 10^{23}$ mol^{-1}
- the luminous efficacy of monochromatic radiation of frequency 540×10^{12} Hz and K_{cd} is 683 lm/W

in which Hz (hertz), J (joule), C (coulomb), lm (lumen), and W (watt) have relationships to the following units: s (second), m (meter), kg (kilogram), A (ampere), K

Table 1.1 Units of the base units in the English Engineering System.

Physical property	Unit	Symbol
Length	Foot	ft
Mass	Pound-mass	lbm, #m
Time	Second	sec
Electric current	Ampere	amp
Temperature	Rankine	°R
Amount of substance	Pound-mole	lbmol, # mole
Luminous intensity	Foot-candles	ft Cd

(kelvin), mol (mole), and cd (candela): $\text{Hz} = \text{s}^{-1}$, $\text{J} = \text{kg}\,\text{m}^2/\text{s}^2$, $\text{C} = \text{A}\,\text{s}$, $\text{lm} = \text{cd}\,\text{m}^2$, $\text{m}^{-2} = \text{cd}\,\text{sr}$, and $\text{W} = \text{kg}\,\text{m}^2/\text{s}^3$.

The base units of the SI are as follows:

- Time (t) is a measure of the separation of events. The SI unit of time is second (s), and it comes directly from $\Delta\nu_{\text{Cs}}$ defined above in terms of Hz, which is s^{-1}.
- Length (l) is a measure of the separation of points in space. Meter, m, is the fundamental unit of length defined as the length of the path traveled by light under vacuum during a time interval of 1/299 792 458 of a second.
- Mass (m) is a measure of the amount of an object. Kilogram, kg, comes from Planck's constant expressed in J s, which equals $\text{kg}\,\text{m}^2/\text{s}$ with m and s defined under time and length.
- Ampere (A) is the SI unit for electric current, and it comes from the elementary charge, e, expressed in C (=A s) with s defined under time.
- Thermodynamic temperature (T) measures the "hotness" of an object and reflects the motion of its molecules. The unit of thermodynamic temperature is kelvin (K), and it comes from the Boltzmann constant expressed in J/K or $\text{kg}\,\text{m}^2/(\text{s}^2\ \text{K})$ (kg, m, and s already have definitions).
- The amount of substance is mole (n) in SI. It comes directly from N_A (Avogadro's number).
- The SI unit for luminous intensity in a given direction is candela (cd) defined by the luminous efficacy of monochromatic radiation of frequency expressed in lm/W.

Other systems of units exist, such as the English Engineering System (used sparingly throughout the world primarily in the United States and a few small political entities). The primary dimensions are force, mass, length, time, and temperature. The units for force and mass are independent. Table 1.1 contains the base units in this system.

A prefix added to any unit can produce an integer multiple of the base unit. For example, a kilometer denotes one thousand meters, and a millimeter denotes one thousandth of a meter: For example, $1\,\text{km} = 1 \times 10^3\,\text{m}$ and $1\,\text{mm} = 1 \times 10^{-3}\,\text{m}$. Table 1.2 contains the accepted prefixes for use with SI units.

Table 1.2 SI prefixes.

Name	Symbol	Factor	Name	Symbol	Factor
yotta	Y	10^{24}	deci	d	10^{-1}
zetta	Z	10^{21}	centi	c	10^{-2}
exa	E	10^{18}	milli	m	10^{-3}
peta	P	10^{15}	micro	μ	10^{-6}
tera	T	10^{12}	nano	n	10^{-9}
giga	G	10^{9}	pico	p	10^{-12}
mega	M	10^{6}	femto	f	10^{-15}
kilo	k	10^{3}	atto	a	10^{-18}
hecto	h	10^{2}	zepto	z	10^{-21}
deka	da	10^{1}	yocto	y	10^{-24}

1.3 Secondary or Derived Physical Quantities

All other quantities derive from base quantities by multiplication and division. These quantities have units that are a combination of the base units according to the algebraic relations of the corresponding physical quantities. Some derived physical properties are area, volume, force, pressure, etc.

Area is a measure of the surface of an object. It is two dimensional and in terms of dimensional symbols is L^2. The units employed to represent the area are square meters in the SI system or square feet in the English system.

Volume is the space occupied by an object and in terms of dimensions is the product of the lengths associated with the object: $L \times L \times L = L^3$. The SI unit for volume is the cubic meter. The volume of a substance depends upon the amount of the material, but a specific volume and molar volume (m^3/kg or m^3/mol) are independent of the amount of the material. The reciprocal of the specific volume is the density. Obviously, an object can have a total volume (the amount of space it occupies) and a specific or molar volume that applies to any sized object. **In this text, we use capital letters to denote the molar properties of a substance, so a total property would be the molar property multiplied by the number of moles.** For example, the total volume is

$$(nV) \equiv n \cdot V \tag{1.2}$$

The specific volume (volume per mass) is $(nV)/m = V/M$, where M is the molar mass (or molecular weight) and the molar volume is V (volume per mole).

Force induces motion or change. Newton's laws describe the action of forces in causing motion. Newton's second law states that force is the product of mass m and its acceleration a,

$$F = ma \tag{1.3}$$

Thus, force = (mass)(length)/(time)^2. The SI unit for force is newton (N), and according to the above equation, it is the force required to accelerate a mass of 1 kg at a rate of 1 m/s^2. Therefore,

$$1\ \mathrm{N} = 1\ \mathrm{kg\ m/s^2} \tag{1.4}$$

In the English Engineering System, the unit of force is pound force (lb$_f$), and 1 lb force imparts an acceleration of 32.1740 ft/s^2 to a mass of 1 lb. In this system, force is an independent dimension that requires a proportionality constant to be consistent with the definition Eq. (1.3)

$$F = \frac{ma}{g_c} \tag{1.5}$$

Thus,

$$1\ \mathrm{lb_f} = (1/g_c) \times 1\ \mathrm{lb_m} \times 32.1740\ \mathrm{ft/s^2} \rightarrow g_c = 32.1740\ \mathrm{lb_m\ ft\ /(lb_f\ s^2)}$$

and 1 lb$_f$ = 4.4482216 N. In this system of units, confusion exists between the terms mass and weight, and often, they are used synonymously. For example, when we buy meat by weight, we are interested in the amount of meat. Likewise, when we measure our body weight, we want to know the amount of fat or muscle present. Scales measure force. Weight refers to the force exerted upon an object by virtue of its position in a gravitational field. In most circumstances, this ambiguity is not a problem because the weight of an object is directly proportional to its mass (see Eq. (1.3)), and on Earth, the proportionality constant is the gravitational acceleration, which is essentially constant. We conclude that the mass of an object does not change, but its weight does depend on the gravitational field surrounding the object.

Example 1.1

An alien weighs 102 N on his planet and 700 N on Earth. The gravitational acceleration on Earth is approximately 9.802 m/s^2. What is the gravitational acceleration of the alien's planet? From the following table (moons are in italics), can you infer from which planet or moon this alien comes?

Planet/moons	Approximate gravitational acceleration (m/s^2)	Planets/moons	Approximate gravitational acceleration (m/s^2)
Mercury	3.70	Saturn	11.19
Venus	8.87	*Titan*	1.36
Moon	1.63	Uranus	9.01
Mars	3.73	*Titania*	0.38
Jupiter	25.93	*Oberon*	0.35
Io	1.79	Neptune	11.28
Europa	1.31	*Triton*	0.78
Ganymede	1.43	Pluto	0.61
Callisto	1.24		

Solution

First, we calculate the mass of the alien using the information from Earth:

$$m = \frac{F}{a} = \frac{700\ \text{N}}{9.802\ \text{m/s}^2} = \frac{700\ \text{kg m/s}^2}{9.802\ \text{m/s}^2} = 71.414\ \text{kg}$$

Because the mass is independent of the location, on its native planet or moon

$$a = \frac{F}{m} = \frac{102\ \text{N}}{71.414\ \text{kg}} = \frac{102\ \text{kg m/s}^2}{71.414\ \text{kg}} = 1.428\ \text{m/s}^2$$

Most likely, the alien comes from the moon of Jupiter: *Ganymede.*

Pressure is the force applied over an area. In the SI system, the unit of force is newton and that for area is square meters, so the unit of pressure is N/m^2. This unit is called pascal, and its symbol is Pa. In the English Engineering System, the usual pressure unit is pound force per square inch (psi). If an object rests on a surface, the force pressing on the surface is its weight. However, in different positions, the area in contact can be different, and it can exert different pressures. **Pressure is an observable and it is not a function of mass.**

$$P = \frac{F}{A} \tag{1.6}$$

We use various terms for pressure:

- Atmospheric pressure, P_{atm}, is the pressure caused by the weight of the Earth's atmosphere on an object. We might find this pressure called "barometric" pressure. The standard atmosphere (atm) is a unit of pressure equal to 101.325 kPa.
- Absolute pressure, P_{abs}, is the total pressure. An absolute pressure of 0 is perfect vacuum. All thermodynamic calculations must use absolute pressure.
- Gauge or manometric pressure, P_{man}, is the pressure relative to atmospheric pressure, $P_{man} = P_{abs} - P_{atm}$.
- Vacuum is a gauge pressure that is below the atmospheric pressure. It reports a positive number for vacuum.

We calculate the absolute pressure by adding the atmospheric pressure to the gauge pressure:

$$P_{abs} = P_{atm} + P_{man} \tag{1.7}$$

Many pressure measurement techniques are available. First, we must define primary and secondary measurement standards. A primary standard is a measurement device for which the theoretical relation between the measured physical property and the desired quantity (pressure in this case) is known exactly. A secondary standard is a measurement device for which the theoretical relation between the measured physical property and the desired quantity (pressure, temperature, etc.) is not known exactly. Secondary measurement devices must have calibrations against the primary standards.

First, let us consider manometers and dead-weight gauges. These are primary pressure devices commonly used in engineering. Figure 1.1 shows a manometer and

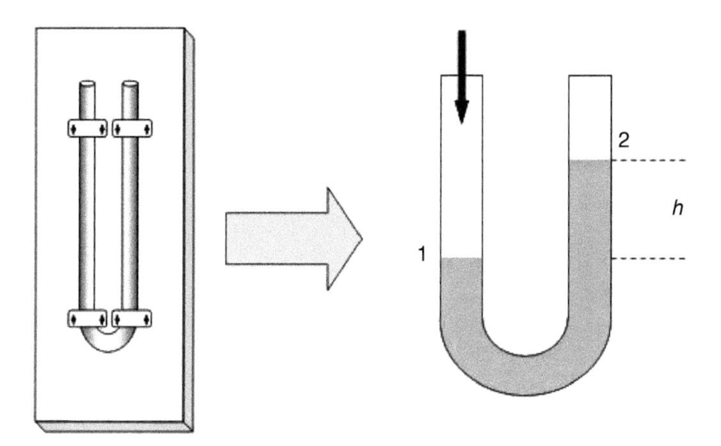

Figure 1.1 A U-tube glass manometer and its schematic diagram.

its schematic diagram. Manometers measure differential pressures. The differential equation that expresses pressure is

$$\frac{\partial P}{\partial z} = -\rho g \tag{1.8}$$

where ρ is the density of the fluid and g is the local acceleration of gravity. If the density is constant, the integration of the Eq. (1.8) is

$$P_2 - P_1 = -\rho g \Delta z = -\rho g(z_2 - z_1) = -\rho g h \tag{1.9}$$

In Figure 1.1, a force pushes on the fluid in position 1 and the atmospheric pressure acts against the fluid in position 2, then

$$P_1 = P_2 + \rho g h \tag{1.10}$$

This equation indicates that the hydraulic pressure (gauge pressure) is ρgh. This effect also is the pressure that a vertical column of fluid exerts at its base caused by gravity

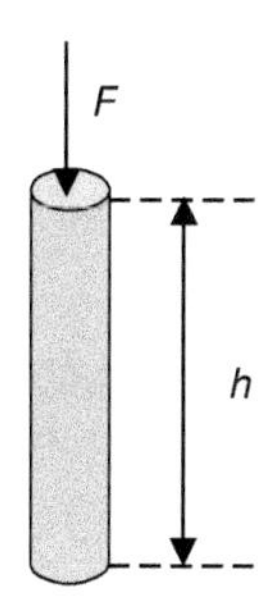

Applying Newton's law and the definition of pressure,

$$P = \frac{F}{A} = \frac{mg}{A} = \frac{\rho V^t g}{A} = \frac{\rho A h g}{A} = \rho h g \tag{1.11}$$

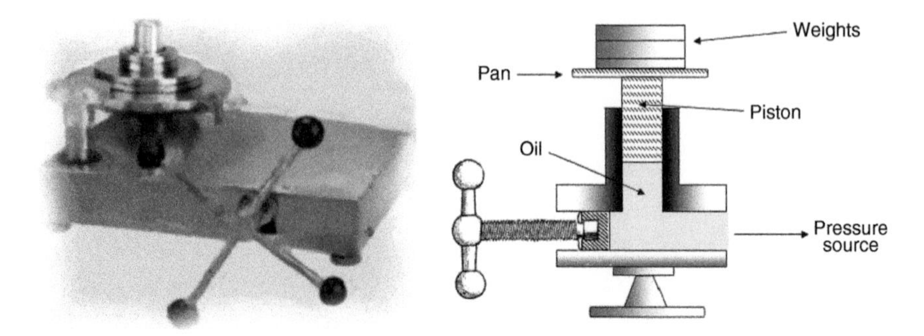

Figure 1.2 Dead-weight gauge.

in which ρ is the mass density (mass per volume). Equation (1.11) is the theoretical relation for pressure as a function of height and density of a fluid in a manometer. For very precise measurement of pressure with a manometer, the density of the measuring fluid must be corrected for its thermal expansion and for the height measuring device in a gravitational field.

Calibration of commercial mercury manometers uses the density at 0 °C, and the measuring scale (usually brass) has zero correction at either 70 or 77 °F. Because these devices measure distances, units of pressure exist such as mm Hg, in of H_2O, etc. Torr is the pressure equal to 1 mm of mercury at 0 °C in a standard gravitational field. One torr equals 133.322 Pa. The overall uncertainty in a manometer is approximately 8.1×10^{-5} Pa. Manometers are useful and practical for measurements from 10^{-3} to 2 atm.

Another primary standard is the dead-weight gauge. Figure 1.2 depicts such a gauge along with a schematic. In this case, masses impose the downward force on the plate and piston (with a known cross-sectional area).

The theoretical relationship for this device comes from a force balance between the working fluid (oil or air) and the weights:

$$\textit{Upward force} = \textit{downward force} \rightarrow PA = mg \tag{1.12}$$

in which P is the pressure; A is the cross-sectional area of the piston; m is the mass of the piston, pan, and load; and g is the local gravitational acceleration. Therefore,

$$P = \frac{mg}{A} \tag{1.13}$$

The ultimate accuracy for the best dead-weight gauges is about 1 part in 30 000 with a precision of 1 part in 20 000. Dead-weight gauges are the instrument of choice for the measurement of highest accuracy or for calibrating other pressure gauges. The secondary pressure devices are pressure transducers, bourdon tubes, and differential pressure indicators (DPIs) (Figure 1.3).

Example 1.2

Calculate the pressure that the atmosphere exerts on the head of a man who is 1.75 m tall and who is on top of a mountain that is 1000 m high. The temperature is 10 °C.

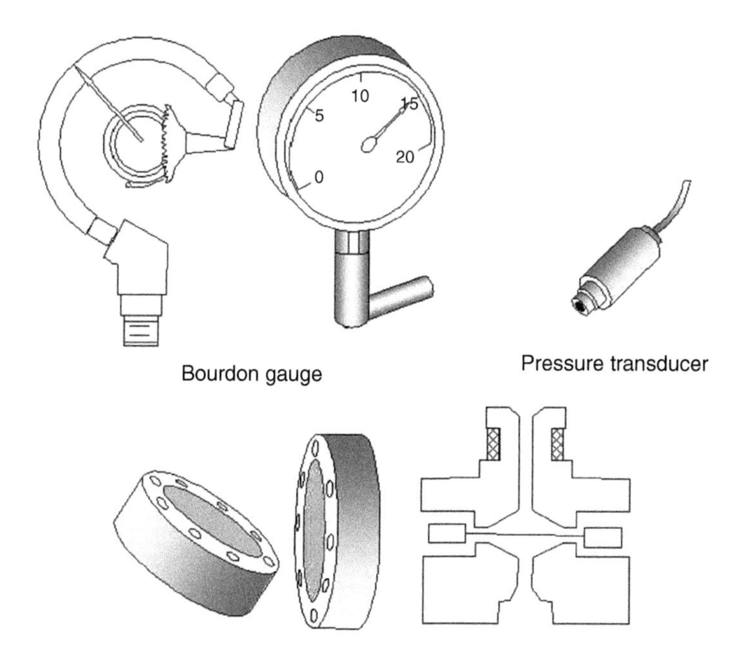

Figure 1.3 Secondary pressure measurement devices.

Assume that the pressure at the sea level is 101 kPa, $g = 9.80665\ \text{m/s}^2$, and the density of air in kg/m^3 is a function of temperature (but not pressure) given by

$$\rho_{air} = 1.29235 - 4.76457 \times 10^{-3}t + 1.66653 \times 10^{-5}t^2$$

Solution

Using Eq. (1.9).

$$P_{top} - P_{sea} = -\rho_{air} g \Delta z = -\rho_{air} g (z_{top} - z_{sea}) = -\rho_{air} g h_{mountain+man\ height}$$

$$\rho_{air} = 1.29235 - 4.76457 \times 10^{-3}(10) + 1.66653 \times 10^{-5}(10)^2 = 1.246371\ \text{kg/m}^3$$

$$\rho_{air} g h_{mountain+man\ height} = 1.246371\ \text{kg/m}^3 \times 9.89665\ \text{m/s}^2 \times (1000 + 1.75)\ \text{m}$$

$$\rho_{air} g h_{mountain+man\ height} = 12244.11\ \text{kg/(m s}^2) = 12244.11\ \text{Pa} = 12.244\ \text{kPa}$$

$$P_{top} = P_{sea} - \rho_{air} g h_{mountain+man\ height} = 101 - 12.244 = 88.766\ \text{kPa}$$

Example 1.3

A dead-weight gauge has the following specifications:

Piston diameter = 2.5 cm
Pan mass = 415 g
Cylinder mass = 500 g
Masses: #1: 100 g, #2: 500 g, #3: 1 kg, #4: 3 kg, and #5: 5 kg

An experimenter uses two #1, one #2, one #4, and 1 #5 masses to balance the gauge with the pressure of the system. What is the gauge pressure if the local acceleration of gravity is 9.805 m/s^2?

Solution

The total mass (pan plus weights) is

$$m = 500\ \text{g}\frac{1\ \text{kg}}{1000\ \text{g}} + 2\times 100\ \text{g}\frac{1\ \text{kg}}{1000\ \text{g}} + 500\ \text{g}\frac{1\ \text{kg}}{1000\ \text{g}} + 3\ \text{kg}$$
$$+ 5\ \text{kg} + 415\ \text{g}\frac{1\ \text{kg}}{1000\ \text{g}} = 9.615\ \text{kg}$$

$$A = \pi r^2 = 3.1416\times\left(\frac{2.5\ \text{cm}}{2}\times\frac{1\ \text{m}}{100\ \text{cm}}\right)^2 = 3.1416\times(0.0125)^2\ \text{m}^2$$
$$= 4.90875\times 10^{-4}\ \text{m}^2$$

Using Eq. (1.13)

$$P = \frac{mg}{A} = \frac{9.615\ \text{kg}\times 9.805\ \text{m/s}^2}{4.90875\times 10^{-4}\ \text{m}^2} = 193753\ \text{kg/(m s}^2) = 193753$$
$$\text{Pa} = 193.75\ \text{kPa} = 0.19375\ \text{MPa}$$

Temperature from statistical thermodynamics is a variable that describes how atoms or molecules distribute among the quantum energy levels available to them. In classical thermodynamics, the definition of temperature comes from the second law of thermodynamics, while its measurement comes from the zeroth law. **Temperature is an observable and not a function of mass.**

Now, imagine that you are a Neanderthal, and you have many hot stones. How can you tell a person of your tribe which stone is the hottest and then the next hottest? Probably, you can use different tones and sounds to solve the problem, but if you come upon a different tribe, and you want to explain the hotness, these sounds may not mean anything to them. Therefore, it is necessary to create a system for measuring the temperature that is the same for everyone. Temperature scales are the solution. We define these scales by using a primary thermometer to determine temperatures corresponding to the observed physical behavior, i.e. triple points, boiling points, and melting points. Like the primary pressure device, a primary thermometer is one for which a known relationship exists between some physical property and the absolute temperature. The examples of primary thermometers are gas thermometer, acoustic thermometer, noise thermometer, and total radiation thermometer. The relationship for the gas thermometer is

$$T = \lim_{P\to 0}\frac{PV}{R} \tag{1.14}$$

where V is the molar volume, P is the pressure, and R = 8.3144621 J/(mol K) is the universal gas constant (the gas constant is not actually a constant, rather it is an experimental number subject to change, but "constant" at any given time). It is convenient to use a reference point temperature

$$\frac{T}{T_{ref}} = \lim_{P\to 0}\frac{(PV)}{(PV)_{ref}} \tag{1.15}$$

The common reference temperature is the triple point of water (273.16 K). Primary thermometers are difficult to use and expensive, so their use is primarily to establish temperature scales and to calibrate secondary thermometers. The Kelvin and Celsius scales are by international agreement defined numerically by two points: absolute zero (0 K) and the triple point of water. At absolute zero, all kinetic motion of particles ceases, and the molecules and atoms are in their ground states. Gas thermometry establishes the Kelvin scale.

In the Celsius scale, the freezing point of water is 0 °C, and the boiling point of water at atmospheric pressure is 100 °C. Every unit is a degree Celsius. The name of this temperature scale comes from the Swedish astronomer Anders Celsius (1701–1744), who developed the temperature scale two years before his death. The mathematical relationship between the Kelvin and Celsius scales is

$$t/^{\circ}\mathrm{C} = T/K - 273.15 \tag{1.16}$$

In addition to the Kelvin and Celsius scales, two other scales exist in the English Engineering System: Rankine and Fahrenheit. The Rankine scale is an absolute temperature scale like the Kelvin scale. The relationship between them is

$$T/^{\circ}\mathrm{R} = 1.8T/K \tag{1.17}$$

A direct relation exists between the Fahrenheit and Rankine temperatures

$$t/^{\circ}\mathrm{F} = T/^{\circ}\mathrm{R} - 459.67 \tag{1.18}$$

and between the Fahrenheit and Celsius scales

$$t/^{\circ}\mathrm{F} = 1.8t/^{\circ}\mathrm{C} + 32 \tag{1.19}$$

Because primary thermometers are not useful for practical purposes, we use secondary thermometers for such applications. Examples of secondary thermometers are glass thermometers, thermistors, platinum resistance thermometers (PLTs), bimetallic strips, and thermocouples. These devices require calibration using an internationally recognized temperature scale based on primary thermometers and fixed points. PLTs also find use as transfer standards. Their calibration uses a primary thermometer to calibrate other temperature-measuring devices. The primary scale is the International Temperature Scale of 1990 (ITS-90) with its addendum the Provisional Low Temperature Scale of 2000 (PLTS-2000). ITS-90 contains seventeen fixed points and four temperature instruments. The temperatures range from 0.65 to 10000 K using a gas thermometer and a PRT. Below 0.65 K, the PLTS-2000 scale uses Johnson noise and nuclear orientation thermometers. Figure 1.4 depicts secondary thermometers.

Example 1.4

An experimentalist has problems with a PLT giving incorrect temperatures. He knows that the electrical resistance of a wire is a function of temperature

$$R = R_0[1 + \alpha(T - T_0)]$$

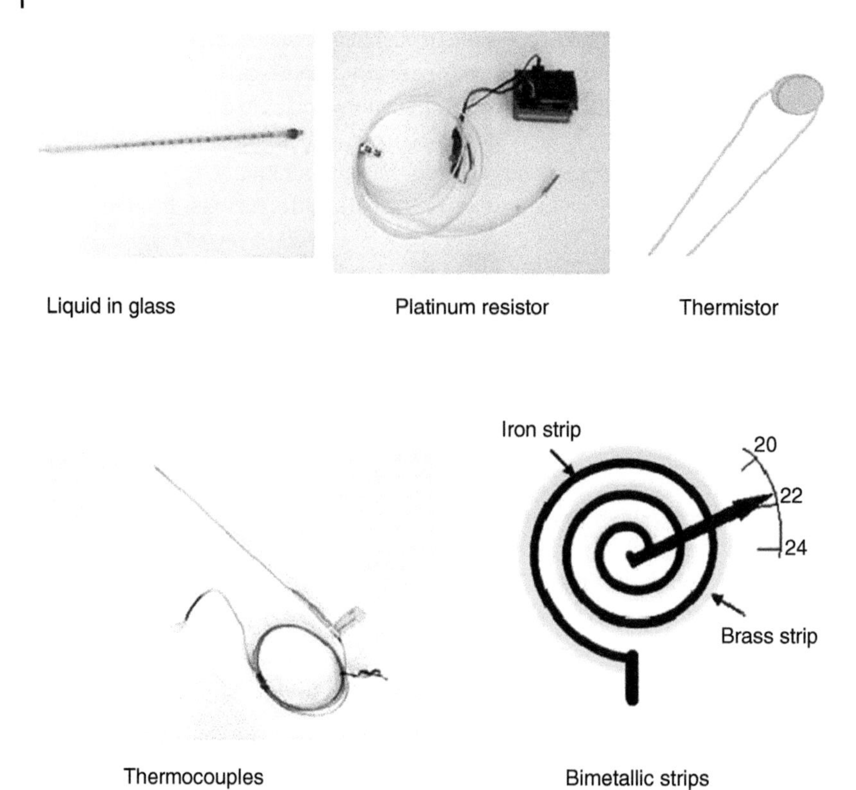

Figure 1.4 Secondary thermometers.

in which α is the temperature coefficient of resistance and R_0 and T_0 are reference resistance and reference temperature (triple point of water), respectively. The manufacturer says that the thermometer has a calibrated standard resistance of 10.5 Ω at 0.01 °C. The experimentalist measures a resistance of 12.94 Ω at 75 °C. Is the manufacturer correct? Assume $\alpha = 3.92 \times 10^{-3}\ °C^{-1}$.

Solution

Find the R_0:

$$R_0 = \frac{R}{[1 + \alpha(T - T_0)]} = \frac{12.94\ \Omega}{1 + 3.92 \times 10^{-3}\ °C^{-1} \times (75 - 0.01)\ °C} = 10.0003\ \Omega$$

Either the manufacturer is wrong, or the experimenter has a faulty resistance measurement.

It is possible to perform dimensional analysis in SI because any derived unit A can be expressed in terms of the SI base units by

$$A = l^{\alpha} m^{\beta} t^{\gamma} I^{\delta} T^{\epsilon} n^{\zeta} I_v^{\eta} \sum_{k=1} a_k$$

in which the exponents $\alpha, \beta, \gamma, \ldots$ and the factors a_k are numbers. The dimension of A is

$$\dim A = \mathrm{L}^{\alpha} \mathrm{M}^{\beta} \mathrm{T}^{\gamma} \mathrm{I}^{\delta} \Theta^{\epsilon} \mathrm{N}^{\zeta} \mathrm{J}^{\eta}$$

where L, M, T, I, θ, N, and J are the *dimensions* of the SI base units and α, β, γ, … are the dimensional exponents. The general SI-derived unit of A is $\mathrm{m}^{\alpha}\ \mathrm{kg}^{\beta}\ \mathrm{s}^{\gamma}\ \mathrm{A}^{\delta}\ \mathrm{K}^{\varepsilon}\ \mathrm{mol}^{\zeta}\ \mathrm{cd}^{\eta}$, which is obtained by replacing the dimensions of the SI base quantities with the symbols for the corresponding base units. For dimensionless quantities, the exponents are zero and $\dim A = 1$. This quantity does not have units (symbols) because its unit is 1 (the exponents are zero).

Example 1.5
Consider an object of mass m that moves uniformly a distance l in a time t. If the total kinetic energy is $(nE_k) = m\dot{z}^2/2$, express the velocity and the energy in terms of their symbols and express their dimensions.

Solution
Symbols mass = m, length = l, and time = t.
For the velocity: $\dot{z} = lt^{-1}$ and the kinetic energy is $(nE_k) = ml^2t^{-2}/2$.
For the dimensions:

$\dim \dot{z} = \mathrm{LT}^{-1}$ and the dimensional exponents are 1 and −1.
$\dim(nE_k) = \mathrm{ML}^2\mathrm{T}^{-2}$ and the dimensional exponents are 1, 2, and −2.

The SI-derived unit of the kinetic energy is then $\mathrm{kg\,m^2/s^2}$ named joule with the symbol J.

1.4 SI Usage of Units and Symbols

Several rules exist for using SI units and symbols. The most important are:

- Roman type (not italics or bold) denotes unit symbols regardless of the type used in the surrounding text.
- The unit symbols are lowercase except when the name of the symbol derives from a proper name. The unit name is in lowercase, but the symbol may be upper case, e.g. meter (m), kelvin (K), pascal (Pa), and second (s).
- Plurals used if indicated by normal grammar rules are henries and seconds. Lux, hertz, and siemens are the same for plural and singular. Unit symbols do not change when plural.
- A space appears between the numerical value and the unit or symbol, e.g. 10 Hz and 15 MPa. A period does not follow unit symbols unless at the end of a sentence.
- Raised dots or a space indicate products of symbols. Slashes, a horizontal line, or negative exponents denote quotients of symbols. No more than one slash should appear in any expression. For example,

$$1\ \text{newton} = 1\ \mathrm{N} = 1\ \mathrm{kg\ m/s^2} = 1\ \mathrm{kg\ m\ s^{-2}}$$

$$1\ \text{pascal} = 1\ \mathrm{Pa} = 1\ \mathrm{N/m^2} = 1\ \mathrm{kg\ m/(m^2\ s^2)} = 1\ \mathrm{kg/(m\ s^2)} = 1\ \mathrm{kg\ m^{-1}\ s^{-2}}$$

- Unit symbols and unit names do not appear together, e.g. meter per second, m/s, and $\mathrm{m\,s^{-1}}$ are correct while meter/s, meter per s, and meter $\mathrm{s^{-1}}$ are incorrect.

- Unit symbols or names do not have abbreviations, e.g. sec for s or seconds, sq mm for mm^2 or square millimeter, and cc for cm^3 or cubic centimeter. If the name of a unit appears, it must be written out in full.
- Conventions also exist for the use of the prefixes in Table 1.2. The following rules apply to prefix names, and symbols follow the first rule of the unit symbols (roman type not italics or bold) regardless of the type used in the surrounding text, the prefix name can be attached to a unit name without a space, which also applies to a prefix symbol attached to the unit symbol, e.g. mm (millimeter), TΩ (teraohm), and GHz (gigahertz).
- The union of a prefix and a unit symbol is a new, inseparable symbol that indicates a multiple or sub-multiple of the unit that follows all the rules for SI units. The prefix name is also inseparable from the unit name forming a single word when attached to each other, e.g. cm^3, μs^{-1}, megapascal, and microliter.
- Prefix names and symbols cannot appear more than once, e.g. nm (nanometer) is correct but not mμm ("millimicrometer").
- It is not desirable to use multiple prefixes in a derived unit formed by a division. It is preferable to reduce the prefixes to a minimum, e.g. it is correct to write 10 kW/ms, but it is preferable to write 10 MW/s because it contains only one prefix. This rule also applies to the product of units with prefixes, e.g. 10 kV s is preferable to 10 MV ms. When working with units that involve the kilogram, it is always preferable to use kg rather than g.
- Prefix symbols or names cannot appear alone, for example, 5 M.
- The kilogram is the only SI unit with a prefix as part of its name and symbol. The prefixes in Table 1.2 are applicable to the unit gram and the unit symbol g. For instance, 10^{-6} kg = 1 μg is acceptable but not 10^{-6} kg = 1 μkg (1 microkilogram).
- The prefix symbols and names are acceptable for use with the unit symbol °C and the unit name degree Celsius. The examples are 12 m °C (12 millidegrees Celsius).
- SI prefix symbols and prefix names may be used with the unit symbols and names: L (liter), t (metric ton), eV (electronvolt), u (unified atomic mass unit), and Da (dalton). However, although submultiples of liter, such as mL (milliliter) and dL (deciliter), are common, multiples of liter, such as kL (kiloliter) and ML (megaliter), are not. Similarly, although multiples of metric tonne such as kt (kilometric ton) are common, submultiples such as mt (millimetric ton) are not. The examples of the use of prefix symbols with eV and u are 80 MeV (80 megaelectronvolts) and 15 nu (15 nanounified atomic mass units), respectively.

1.5 Thermodynamic Systems and Variables

A system is a volume of space set aside to study. A boundary is a physical or imaginary surface (mathematical sense) that separates the system from the remainder of the universe. Everything outside the system is its surroundings. Three classes of systems exist depending on mass and energy transfer:

- *Isolated system* is one that has no mass or energy crossing its boundary
- *Closed system* is one that permits energy, but not mass, to cross its boundaries
- *Open system* is one that permits energy and mass to cross its boundaries

The state is the condition of the system. Its physical properties determine the state of a system. A property is a characteristic. An open system is in **steady state** if its properties vary with position but not with time. When the properties of any system vary with neither position nor time, the system is in **equilibrium**.

Application of thermodynamics is possible when the system consists of one or more parts with spatially uniform properties. Each of these parts is a phase. Many events can happen in a system. This sequence of events is a process. In thermodynamics, the process is **reversible** if the system can return from its final state to its initial state without finite changes in the surroundings. If the system requires finite changes in the surroundings to return from its final state to its initial state, the process is **irreversible**.

In classical thermodynamics, we utilize the physical properties mentioned in Sections 1.2 and 1.3. They fall into two groups:

Intensive properties do not depend on the amount of the material in the object. They are point functions, so they exist at each point within the system and can vary from point to point (if the system is not at equilibrium). They are not additive. The examples of these properties are pressure, temperature, density, and specific volume. Consider a pure component in a closed system that contains a single phase at equilibrium. For such a system, any intensive variable depends upon two other intensive variables,

$$I_j = f(I_1, I_2) \quad \text{for } j = 3, 4, \cdots n \tag{1.20}$$

For example, if we know the density and temperature of a pure substance, then we know other properties such as pressure and surface tension. In the case of mixtures, the intensive property depends upon the two intensive variables and the composition of the mixture. A pure component is a special case of a mixture with a composition of 100% of the component.

Extensive properties are those that depend upon the amount of the material in the object. They are additive; that is, the value of the property for the object is the sum of the values of all its constituent parts. The examples are total volume, mass, length, and area. Again, for a pure component in a closed system with a single equilibrium phase, any extensive property is a function of two intensive properties plus the amount of the material

$$E_j = mf(I_1, I_2) \quad \text{for } j = 1, 2, \cdots n \tag{1.21}$$

in which m is the mass of the system. Equation (1.21) enables the definition of a specific property

$$\frac{E_j}{m} = f(I_1, I_2) \quad \text{for } j = 1, 2, \cdots n \tag{1.22}$$

All specific properties are intensive properties.

1.6 Zeroth Law

The statement of the law is: The initial law of thermodynamics we shall investigate is the zeroth law (discovered after the first and second laws, hence the zeroth law). It has but one function: it enables construction of thermometers. The statement of the law is as follows:

Two systems separately in thermal equilibrium with a third system are in thermal equilibrium with each other.

Mathematically, this statement is

$$F_1(P_A, V_A, P_B, V_B) = 0 \quad A \text{ and } B \text{ are in thermal equilibrium}$$

$$F_2(P_C, V_C, P_B, V_B) = 0 \quad C \text{ and } B \text{ are in thermal equilibrium}$$

We can solve each equation for one variable, e.g. P_B

$$P_B = f_1(P_A, V_A, V_B)$$

$$P_B = f_2(P_C, V_C, V_B)$$

Thus, $f_1 = f_2$ and A and C are in equilibrium with B. If A and C are in equilibrium with B, then by the zeroth law, they are in equilibrium with each other and

$$F_3(P_A, V_A, P_C, V_C) = 0$$

This implies that f_1 and f_2 contain V_B in such a manner that it cancels exactly, e.g.

$$f_1 = \Theta_1(P_A, V_A)\xi(V_B) + \zeta(V_B)$$

$$f_2 = \Theta_2(P_C, V_C)\xi(V_B) + \zeta(V_B)$$

If this is the case, then

$$\Theta_1(P_A, V_A) = \Theta_2(P_C, V_C) = \Theta_3(P_B, V_B)$$

or

$$\theta = \Theta(P, V)$$

When we study the second law of thermodynamics, this final equation establishes the empirical temperature and the definition of the absolute temperature. Absolute thermodynamic temperature is a mathematical function that depends upon two variables, commonly pressure and molar volume.

Problems for Chapter 1

1.1 What thermodynamic variable results from the zeroth law of thermodynamics?

1.2 A European data sheet provides a pressure specification of 350 kg_f/cm^2. Convert this pressure to units of atmospheres. (This pressure unit was widely used in both research and practical applications for many years and persists today even though it is not an accepted SI unit.) The kg_f unit is defined analogously to the lb_f unit using the same value for the acceleration due to gravity but in appropriate SI units.

1.3 If the density of the gasoline fluctuates between 650 and 870 kg/m^3, what is the minimum and maximum volume that 10000 kg occupies in a cylindrical steel tank? What is the mass of gasoline? What is the minimum and maximum height of the tank if its diameter is 1 m. Calculate the mass of the two tanks if the density of steel is 7850 kg/m^3.

1.4 A black and white cat that weighs 5 kg kneads the stomach of its owner with its two front paws. What is the pressure that it exerts upon his owner's stomach? Consider that the paws are circular with a diameter of 1.8 cm.

1.5 A diver plunges vertically into the sea until reaching a depth of 100 m. To decompress, he must reduce his pressure by 122.625 kPa and rest for one minute; to do so, he must climb a certain distance. What pressure does the diver endure at the end? How many minutes should he rest? The density of water is 1 g/cm^3, $g = 9.81$ m/s^2.

1.6 Calculate the force caused by pressure acting on a horizontal hatch of 1 m diameter for a submarine submerged 600 m from the surface. What is the force if the submarine is on the surface of water?

1.7 The municipal water service fills a house cistern with a mass flow rate of 10 kg/s. The dimensions of the cistern are 7 m long by 1.2 m wide and 1.5 m high. Convert the mass flow rate to volumetric flow rate in l/min. How long does it take to fill the cistern? Consider the density of water equal to 1000 kg/m^3.

1.8 At which temperature is the Celsius scale three times the Kelvin scale? Also, at which temperature is the Celsius scale three times the Fahrenheit scale?

1.9 A pressurized tank filled with gas has a leak through a hole with a 0.4 mm diameter. An engineer finds a quick temporary solution to put a weight on the hole, so the gas does not escape. What is the mass of the weight that the engineer must put on the hole if the pressure inside of the tank is 2750 kPa?

1.10 An American travels by car to México. Suddenly, he notices that the speed limit signs say 90 km/h in the country and is 60 km/h in the city. Unfortunately, his car speedometer shows the speed in miles per hour. He is driving the car at 70 the speed limit in most highways in America and at 40 in the city. Is he violating the law? A mile is 1609 m.

1.11 What is the total mass and weight of an oak barrel containing 59 gal of Cabernet Sauvignon wine? The mass of the barrel is 110 kg. The density of the wine is 0.985 g/cm^3.

1.12 Shock-compression experiments on diamond have reported the melting temperature of carbon at a pressures of up to 1.1 TPa (6^{11} Mbar). Convert the pressure into atm, psia, MPa, and inches of Hg.

1.13 An equation of state presents the pressure as a function of temperature and molar volume. A simple equation is

$$P = \frac{RT}{V-b} - \frac{a}{V}$$

The unit of P is MPa, and the value of R is 8.314. What units should R, a, and b have if the volume is cm^3/mol and T is in kelvins?

1.14 What is the pressure in atm at point B in the following diagram if an open container contains oil with a density of $0.845\,g/cm^3$?

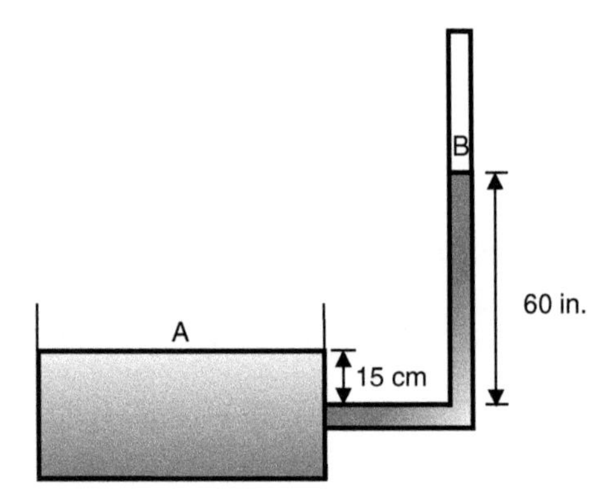

1.15 If a mercury barometer reads 27 in. of mercury, what is the pressure in atm? Is this device measuring absolute pressure? The density of mercury is $13\,590\,kg/m^3$.

1.16 Find the expression for the difference pressure between P_2 and P_1 for a manometer containing three different fluids:

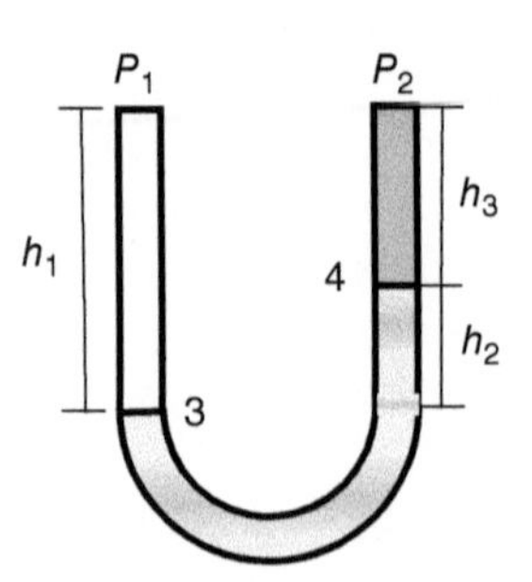

1.17 A manometer can measure pressure differences of a fluid flowing through an orifice as shown in the figure below. The expression for measuring the pressure differences is

$$P_2 - P_1 = (\rho_{fluid} - \rho_l)gh$$

Find the expression for the difference in pressure between P_2 and P_1. The density of the liquid in the manometer is ρ_l

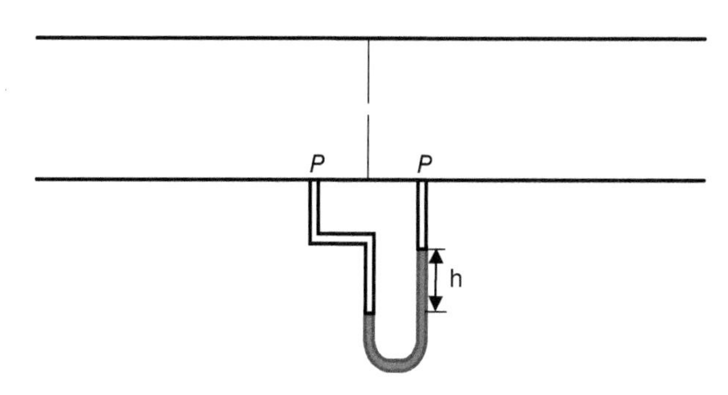

1.18 A pressure gauge is connected in the same tank as a lamp oil manometer. If the display of the gauge reads 5 kPa, what is the height of oil in the manometer? Consider the density of the oil to be 0.81 g/cm^3.

2

Energy and the First Law

2.1 Introduction

In Chapter 1, we described thermodynamics as a mathematical transformation of problems from the time domain (real world) into a timeless domain (thermodynamic world) in which problems are easier to solve. We must introduce two variables of *purely thermodynamic origin*, energy and entropy, to achieve this transformation. In this chapter, we investigate energy.

Because energy is a thermodynamic concept, we cannot assign it a rigorous definition. This indicates that our study of thermodynamics is an exercise in abstract thinking. This should not be frightening or difficult for you because all your mathematical studies have been abstract. The problem is that in science courses, you may have adjusted to definitions as substitutions for understanding. Thermodynamics is much easier and more satisfying if you abandon your familiar "crutches" at this point and work harder at understanding concepts.

To prove the point, take a minute or two and "define" energy to yourself. Remember, you cannot use the term energy in the definition. Well, do not waste too much time. Energy is beyond our sensory powers (we cannot see, feel, smell, hear, or otherwise experience it physically), yet we all "know" what it is. This is because we are *comfortable* with the word (rather than the concept) having heard the term all our lives (as children, our parents said we had too much of it; as adults, we hear that the world has too little of it or that it is too expensive). Of course, we can experience the *effects* of energy, e.g. watching an automobile drive by or touching a hot cup.

Let us accept energy as a mathematical concept that we invent to quantify our observations that **certain systems can influence their environments**. In other words, these systems possess some property (energy) whose *consequences* can be observed as interactions with their surroundings. We do not really require a formal definition, even pedagogically, if we accept the concept and learn its use in solving practical problems of engineering importance. The mathematical formalism we employ in solving these problems is a general statement of the first law of thermodynamics.

Thermodynamics for Chemical Engineers, First Edition. Kenneth R. Hall and Gustavo A. Iglesias-Silva.

2.2 Energy

Although we cannot experience energy, we can easily recall manifestations of its presence in our daily lives, for example, cooking, electricity, home heating, bicycling, explosions, etc. In each case, a system is energetic and, therefore, can influence its environment in some manner. However, these consequences of energy are difficult to use in a quantitative sense, and we must resort to mathematics, guided by physics, to generate equations that facilitate our study of energy and help us to visualize the concept of energy more clearly.

At this point, we evoke Newton's law of motion from his magnificent *Principia Mathematica* (1687). He stated that force is proportional to mass multiplied by acceleration. For our purposes, we consider a single particle and write

$$F(z) = m\frac{d^2z}{dt^2} = m\ddot{z} \tag{2.1}$$

where F is the force and is a function of the position of the particle, z is the vector of coordinates for the particle, m is the mass, $\dot{z}$ is the velocity vector, $\ddot{z}$ is the acceleration, and t is time. Now, using the chain rule and the definition of velocity,

$$F(z) = m\frac{d\dot{z}}{dz}\cdot\frac{dz}{dt} = m\dot{z}\cdot\frac{d\dot{z}}{dz} \tag{2.2}$$

in which velocity is the time derivative of position. Using separation of variables, we obtain

$$F(z)\cdot dz = m\dot{z}\cdot d\dot{z} \tag{2.3}$$

Upon integration, Eq. (2.3) becomes

$$\frac{m\dot{z}^2}{2} - \int F(z)\cdot dz = \text{constant} \tag{2.4}$$

The first term of Eq. (2.4) is the kinetic energy of the particle and the second term is its potential energy. For a single particle, the sum of the kinetic and potential energies is the total energy. Therefore, total energy must be the constant of integration. Of course, we seldom deal with a single particle, and we must generalize this result to a system of many particles. This exercise is not really part of a course in thermodynamics, but you can review the derivation in H. Goldstein, *Classical Mechanics*. Addison-Wesley Pub. Co., Cambridge, MA., 1950. In essence, Eqs. (2.3, 2.4) become

$$\left(\frac{m\dot{z}^2}{2}\right)_{cm} + \sum_i \left(\frac{m\dot{z}^2}{2}\right)_i - \left\{\left[\int F(z)\cdot dz\right]_{cm} + \sum_i \left[\int F(z)\cdot dz\right]_i\right\} = E \tag{2.5}$$

In this equation, the subscript *cm* denotes "center-of-mass," the summation is over all particles, and E is the total energy of the system (and the constant of integration). If we sum over all particles in the universe, it is apparent that **the energy of the**

Table 2.1 Equations to calculate energies.

Property	Total energy	Energy per mole	Energy per mass
Internal energy	nU	U	$nU/m = U/M$
Kinetic energy	nE_K	E_K	$nE_K/m = E_K/M$
Potential energy	nE_P	E_P	$nE_P/m = E_P/M$

Note: M is the molar mass.

universe is constant. We now invent symbols for the various terms in Eq. (2.5) for "shorthand" purposes:

$$nE_K = \left(\frac{m\dot{z}^2}{2}\right)_{cm} \tag{2.6}$$

$$nE_P \equiv -\left[\int F(z)\cdot dz\right]_{cm} \tag{2.7}$$

$$nU \equiv \sum_i \left(\frac{m\dot{z}^2}{2}\right)_i - \left\{\sum_i \left[\int F(z)\cdot dz\right]_i\right\} \tag{2.8}$$

Equations (2.6–2.8) refer to the properties of the system; E_K is the kinetic energy of the *system* (energy that the system possesses reflected by its motion with respect to some reference plane), E_P is the potential energy of the *system* (energy that the system possesses because of its interaction with some external field, e.g. gravity, electric field, and magnetic field), and U is the "catch-all" term, and we call it the internal energy (energy that the system possesses because it is composed of energetic particles). Because we have established our nomenclature in such a way that capital letters denote molar properties, we must multiply the terms by the number of moles, n. Therefore, the total energy of the system is

$$nE = nU + nE_K + nE_P \tag{2.9}$$

Of course, for a system with a constant number of moles, we can write

$$E = U + E_K + E_P \tag{2.10}$$

Table 2.1 illustrates the various equations to calculate these energy terms with different units. It is necessary to be consistent with units when solving problems. **Do not forget** Table 2.1, you do so at the risk of your grade in this course.

Example 2.1

A ball thrown with a velocity of 5 m/s reaches a height of 6 m. If the mass of the ball is 0.2 kg and the internal energy of the ball is 6 J, what is total energy of the system?

Solution

Consider the ball as the system:

Calculate the kinetic energy: $nE_K = \left(\frac{m\dot{z}^2}{2}\right) = \frac{0.2\ \text{kg} \times (5\ \text{m/s})^2}{2} = 2.5\ \text{J}$

Calculate the potential energy: $nE_P = mgz = 0.2\ \text{kg} \times 9.81\ \text{m/s} \times 6\ \text{m} = 11.8\ \text{J}$

Then, the total energy from Eq. (2.9) is

$$nE = n(U + E_K + E_P) = 6 + 2.5 + 11.8 = 20.3\ \text{J}$$

Thus, energy is a constant of integration composed of three identifiable parts – do you begin to see why it is difficult to define? Energy is a powerful concept we use in our study of thermodynamics. This concept also simplifies our task – historically, before we had the concept of energy, we employed the caloric theory with many problems of interpretation and application. Let us investigate energy in more detail and learn some of its practical applications.

2.3 First Law of Thermodynamics

Recall that if the system of particles we use in deriving Eq. (2.4) is the entire universe, energy would still be the constant of integration and **the energy of the universe is constant**. Another way of stating this observation is "**energy is conserved**," which is our word-statement of the first law of thermodynamics. Although this is a simple statement, it is a powerful concept that permits us to solve many complicated problems with relative ease. To quantify our statement, we introduce the general accounting equation (GAE).

The GAE applies to any countable quantity. This important equation has applications beyond the field of accounting (e.g. all differential equations):

$$ACCumulation = TRANSfer + GENeration + CONVersion \tag{2.11}$$

To apply the accounting equation, we define our system, select the countable property of interest, and apply Eq. (2.11). In the GAE, the *TRANS*, *GEN*, and *CONV* terms are net quantities – transfer represents input minus output *across the boundaries of the system*; generation represents production-less destruction *within the system*; and conversion represents appearance-less disappearance *within the system* (this is an interchange of the countable quantity among separately identifiable forms). Accumulation occurs over some stated period of time, the accounting period, during which the system passes from an initial state to a final state. We can think of many examples: money in a bank account, population of a country, and properties such as mass and energy. The GAE is the ultimate source of all differential equations.

Example: Apply (2.11) to the accumulation of US$ in a bank account.

First, we must select an accounting period, i.e. the time period over which we intend to observe the system, in this case, the bank account. Let us select one month. The accumulation of US$ in the account over a one-month period would be

$$ACC = TRANS + GEN + CONV$$

Or, in terms, we associate with bank accounts

$$Balance = (Deposits - Withdrawals) + (Interest - Fees)$$
$$+ (Conversion\ of\ Mexican\ Pesos\ into\ US\$)$$

When we apply the GAE to energy, we obtain a mathematical statement of the first law. Let us look at each term in this equation. The accumulation term is a simple application of Eq. (2.10): the energy accumulated equals the energy in the final state less energy in the initial state

$$ACC = E_f - E_i = \Delta(U + E_K + E_P)_{SYS} \tag{2.12}$$

This term applies to the entire system, and Δ denotes the final state values minus the initial state values. The values of U, E_K, and E_P refer to the system as a whole, not to any effect within the system, and not to any transfer across the system boundaries. If the system contains matter or energy in any form, U has some value. If the entire system moves relative to some reference plane, E_K has a value. If an external field acts upon the entire system, E_P has a value. At the risk of being boring, we emphasize again that this term and its components refer to the *entire system* viewed as an entity.

The *TRANS* term is the most common right-hand side (rhs) part of Eq. (2.11). We must deduce from our experiences how energy crosses the boundaries of a system. Two of these terms exist only at the boundaries of the system and they are *not properties of the system*. Rather, they are mechanisms for energy transfer across the boundaries. Three energy transfer mechanisms exist: energy transferred as heat, energy transferred as work, and energy transferred by mass (with mass being a property). It is common to use the terms heat transfer and work rather than energy transfer as heat and energy transfer as work; however, these are misnomers. They are energy transfer mechanisms called heat and work. In each case, the transfer can be *into* the system (which we denote as positive) or *out of* the system (which we denote as negative). We use summation notation to denote the net energy transfer across the boundaries and retain the Δ notation for *final–initial*. We also use the convention that **the energy entering the system is positive and the energy leaving the system is negative**:

$$\left(\sum x \equiv \text{input of } x - \text{output of } x\right)$$

One energy transfer term we experience frequently is **heat**. Actually, our senses tell us that temperature differences exist, and we can determine that one object is hotter or colder than another. However, we can also observe that when a hot object contacts a cold object for an extended period of time under reasonably isolated conditions, the hot body becomes cooler and the cold body becomes hotter.

Energy transfer as heat denotes all energy crossing the boundaries of the system caused by a temperature difference between the system and its surroundings.

Heat is also a *path function*. The amount of energy transferred as heat as a process passes from state 1 to state 2 depends upon the path followed when moving from 1 to 2. Three mechanisms exist to transfer energy as heat: conduction, convection, and radiation. In conduction, the energy flows from one atom or molecule to another, and no bulk flow occurs. Therefore, conduction is the most common form of energy transfer as heat in a solid. In convection, the energy flows because of bulk flow of a fluid contacting the system boundary. Convection is the most common form for energy transfer as heat in liquids and gases. Finally, in radiation, the energy is transferred as heat via electromagnetic radiation emitted by a body caused by a temperature difference with its surroundings. While conduction and convection require that atoms and molecules in the surroundings actually contact the system boundary, radiation can occur in the absence of any contact medium and can transfer energy as heat through vacuum.

The net energy transferred as heat across a boundary is

$$\text{Heat} \equiv \sum Q = \text{energy transferred in as heat} - \text{energy transferred out as heat} \tag{2.13}$$

Again, the summation sign refers to input minus output. A special system that we use frequently in thermodynamics is the one that does not exchange energy with its surroundings (it is insulated). We term such a system **adiabatic**.

Work is a "catch all" energy transfer mechanism.

Energy transfer as work refers to all energy crossing the boundaries of the system caused by any driving force other than temperature but excluding mass transfer.

Like heat, work is also a *path function*. The amount of energy transferred as work as a process passes from state 1 to state 2 depends upon the path followed moving from 1 to 2. Several forms exist for energy transfer as work, but thermodynamics, we use variations of the mechanical work defined as the force acting through distance. For example, if we pull or push an object, we can move it from point A to B. The push or pull is a force that acts upon the object over a physical distance performing work. Mathematically, this is

$$W_{rev} = \int F dz \tag{2.14}$$

in which F is the force vector acting over the object and z is the distance vector (length of displacement under the action of force). The notation W_{rev} denotes energy transfer as work in a reversible process. It is possible to define force and position only for reversible processes. In a volumetric system, we can expand or compress the system against an opposing pressure. Now, using the definition of pressure ($P = F/A$), Eq. (2.14) becomes

$$W_{rev} = \int PA dz = -\int P dV \tag{2.15}$$

in which P is the pressure and V is the volume. The minus sign is necessary because when the energy transferred as work enters a system, its volume decreases and vice versa. To visualize this, consider an ideal piston without friction with pressure acting over the system as shown in Figure 2.1. Equation (2.14) gives the work per unit mole or unit mass. To calculate the total work, $(nW)_{rev}$, it is necessary to multiply Eq. (2.14) or (2.15) by the number of moles.

Table 2.2 presents some forms for work. The net energy transfer across the boundaries of a system as work is

$$\text{Work} \equiv \sum W = \text{energy transferred in as work} - \text{energy transferred out as work} \tag{2.16}$$

Figure 2.1 Volumetric compression.

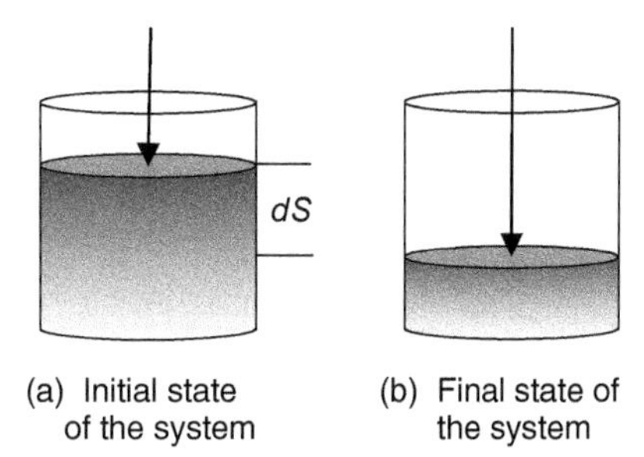

Table 2.2 Different forms for work.

Type	Formula	Observations
General	$W_{rev} = \int Fdz$	F, force
		S, distance
Volume changes	$W_{rev} = -\int PdV$	P, pressure
		V, volume
Shaft work	$W_{rev} = \int \tau d\theta$	τ, torque
		θ, angle of shaft rotation
Surface expansion	$W_{rev} = \int \chi dA$	χ, surface tension
		A, area
Electrical work	$W_{rev} = \int I^2 Rdt$	I, intensity
		R, resistance
		t, time
Reversible cell	$W_{rev} = \int edq$	e, emf
		q, charge of the cell
Stressed bar	$W_{rev} = \int \sigma Vd\varepsilon$	σ, stress
		V, volume
		ε, strain
Work of polarization	$W_{rev}/V = \int \mathbf{E}d\mathbf{P}$	$\mathbf{E}$, electric field
		$\mathbf{P}$, polarization
Magnetic work	$W_{rev}/V = \mu_0 \int \mathbf{H}d\mathbf{M}$	μ_0, $4\pi \times 10^{-7}$
		$\mathbf{H}$, magnetic field
		$\mathbf{M}$, magnetic moment per unit volume

Remember, heat and work are not properties of the system. They are mechanisms for energy transfer across the boundaries of the system and exist only at the boundaries of the system not within it. Therefore, it is not permissible to say heat or work in or out. Also, the notations ΔQ and ΔW are meaningless.

The **mass transfer** term represents the energy crossing the boundaries of the system because mass crosses the boundaries. The **heat** and **work** mechanisms are

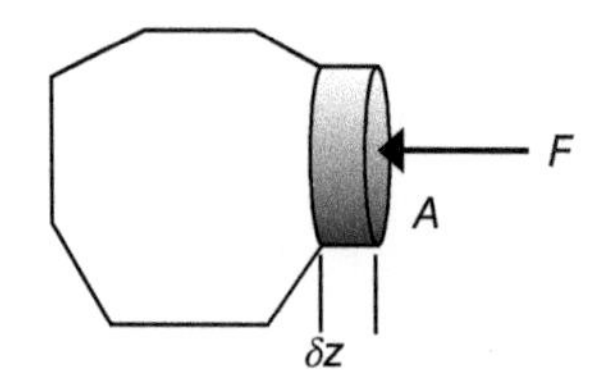

Figure 2.2 Flow of a constant volume of mass.

relatively straightforward, but the mass transfer term requires some scrutiny. Consider the flow of a constant volume of mass (a closed system), as shown in Figure 2.2, into an open system. If we consider the closed system flowing into the open system, its initial energy is $(U + E_K + E_P)_i$ and the final energy is $(U + E_K + E_P)_f$.

The energy required to "push" the mass into the system is

$$F\delta z = PA\delta z = PV \quad (2.17)$$

in which V is the molar volume of the closed system, F is the acting force, A is the cross-sectional area, and δz is the width of the system. Thus, the necessary energy transferred across the boundaries of the system to move the mass is the final energy that equals the initial energy of the system plus the energy transferred as work to move the mass from just outside the system to just inside the system:

$$E_{mt} = (U + E_K + E)_f = (U + E_K + E_P)_i + PV \quad (2.18)$$

and the net energy per mole caused by mass transfer is

$$Net\,E_{mt} = \sum(U + PV + E_K + E_P) \quad (2.19)$$

To obtain the total net energy, multiply Eq. (2.19) by the total number of moles. Finally, we can collect all our energy transfer terms:

$$TRANS = \sum Q + \sum W + \sum (U + PV + E_K + E_P)_{mt} \quad (2.20)$$

The *GEN* term is particularly simple for the first law. Because energy is conserved, it is neither created nor destroyed within the system, and

$$GEN = 0 \quad (2.21)$$

The *CONV* term arises from ignorance. A deity creating thermodynamics would have defined a mass–energy function, and we would not require the conversion term. However, when we have separate functions for mass and energy, we must ask: does energy convert into something else (negative term) or does anything else convert into energy (positive term) within the system? We know of no phenomenon that converts energy into something else (with the possible exception of black holes), but we must recognize the Einstein relationship between mass and energy. Clearly, if a nuclear reaction occurs within the system, we convert mass into energy. Thus, the *CONV* term in the first law is

$$CONV = -(\delta m)c^2 \quad (2.22)$$

in which (δm) is the change in mass and c is the velocity of light. However, this is the total energy released, so we must convert it into energy per mole by dividing by the number of moles. Of course, $m/n = M$, the molecular weight (molar mass). Because energy increases when mass decreases, we require a negative sign. In most problems of practical importance in chemical engineering, no nuclear reaction occurs within the system or, if one does occur, we can redefine the system such that the nuclear reaction lies outside of it. Therefore, in general, we can ignore this term. Now, we are ready to establish the final form of the GAE as applied to energy and formulate the general first law of thermodynamics:

$$\Delta(U + E_K + E_P)_{SYS} = \sum Q + \sum W + \sum (U + PV + E_K + E_P) - (\delta m/M)c^2 \tag{2.23}$$

While this is the general form, because normally, we intend to define systems to exclude nuclear reactions, we can use the reduced form:

$$\Delta(U + E_K + E_P)_{SYS} = \sum Q + \sum W + \sum (U + PV + E_K + E_P)_{mt} \tag{2.24}$$

We can also define a new energy function and call it enthalpy:

$$H \equiv U + PV \tag{2.25}$$

In chapter 5, we provide a more rigorous development of this property. Finally, the first law on a per mole basis becomes

$$\Delta(U + E_K + E_P)_{SYS} = \sum Q + \sum W + \sum (H + E_K + E_P)_{mt} \tag{2.26}$$

Interestingly, U, H, E_K, and E_P are properties of the system; therefore, they are point functions and their values are independent of the path followed as the system progresses from point 1 to point 2. As a result, we can rewrite Eq. (2.26):

$$\begin{aligned} ({}_1Q_2)_{rev} + ({}_1W_2)_{rev} &= ({}_1Q_2)_{irr} + ({}_1W_2)_{irr} \\ ({}_1Q_2)_{rev} &\neq ({}_1Q_2)_{irr} \\ ({}_1W_2)_{rev} &\neq ({}_1W_2)_{irr} \end{aligned} \tag{2.27}$$

in which subscript *irr* denotes an irreversible process. This is true because the terms on the left-hand side of the equation do not depend on the path (they are point functions), so it is possible to select a reversible path between states 1 and 2 or an irreversible one with no effect on the left-hand side. Therefore, the sum of the energy transferred as heat and the energy transferred as work for a reversible process equals the sum of the energy transferred as heat and the energy transferred as work for an irreversible process, but the energy transferred as heat and the energy transferred as work are different for the two processes.

Application of Eq. (2.26) is straightforward, and when applied in a systematic manner always provides a correct equation describing an energy balance problem. How do we use this expression in solving thermodynamics problems? It is simply a matter of asking questions about the problem and operating on the equation based on the answers. The basic development of this technique appeared in [1]. First, we define the system and draw a sketch of it. Then, we ask the following questions:

1. Do any nuclear reactions occur within the system?
   If NO, then $-(\delta m/M)c^2 = 0$ (note: the answer is almost always NO)
   then go to 2.

2. Does any mass cross the boundaries of the system?
   If NO, then $\sum(H + E_K + E_P) = 0$
   If YES, then
      Are E_K and/or E_P significant?
         If NO, then E_K and/or $E_P \cong 0$
         If YES, then retain E_K and/or E_P and go to 3

3. Does a mechanism for energy transfer as work exist?
   If NO, then $\sum W = 0$
   If YES, then retain $\sum W$ and go to 4

4. Can the system transfer energy as heat with its surroundings?
   If NO, it is adiabatic and $\sum Q = 0$
   If YES, retain $\sum Q$ and go to 5

5. Is the process in a steady state?
   If YES, then $\Delta(U + E_K + E_P)_{SYS} = 0$
   If NO, then is the process cyclic?
      If YES, then $\Delta(U + E_K + E_P)_{SYS} = 0$ for a complete cycle
      If NO, then are E_K and/or E_P significant?
         If NO, then E_K and/or $E_P \cong 0$
         If YES, then retain E_K and/or E_P

Note: in step 2, E_K and E_P refer to individual mass transfer streams crossing the system boundary; however, in step 5, E_K and E_P refer to the kinetic and potential energy of the system itself.

Figure 2.3 is a flow sheet of this procedure. After responding to the questions, we obtain the correct equation describing the process be it isolated, closed, open, steady state, or unsteady state. What remains is inserting numbers to attain the numerical solution. This may be a somewhat complicated process, but, in principle, it is possible if the problem has a solution. It is important to note that nothing has been said about how to select the system. This matter is more complicated, and only the information provided, and your own experience, provides guidance. Sometimes, improper selection of the system can lead to impossible solutions. Let us solve some simple problems with the procedure to become familiar with it.

2.4 Application of Solution Procedure to Simple Cases

When applying the first law to problems, it is important to follow an invariant procedure to assure arriving at the correct answer:

- Draw a sketch of the system
- Write the entire first law
- Ask the questions and cancel all terms that do not apply
- Do not skip any of the above steps

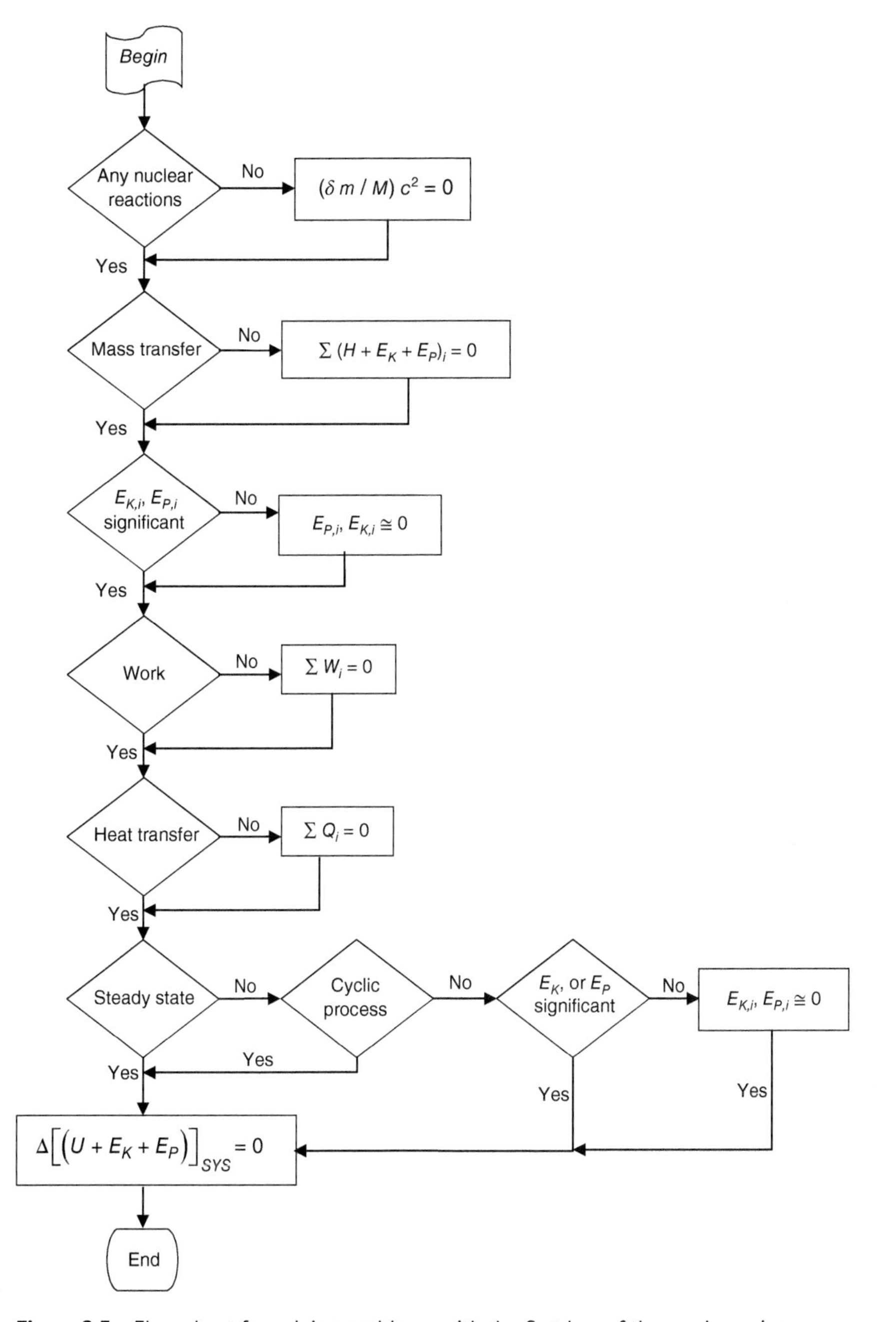

Figure 2.3 Flow sheet for solving problems with the first law of thermodynamics.

Example 2.2 ***"Free Expansion"***
A rigid, stationary, insulated box contains a membrane that divides the internal volume in half. One side of the volume contains a gas in which the membrane is soluble; the other side is a vacuum. When the membrane dissolves, the gas fills the entire volume. What does the first law say about this system?

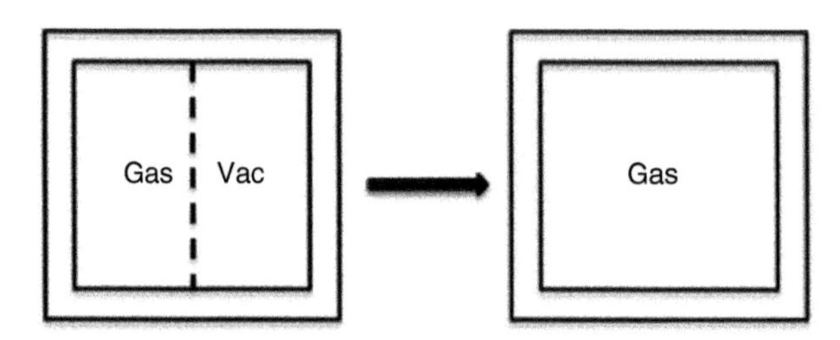

First law: $\Delta(U+E_K+E_P)_{SYS} = \sum Q + \sum W + \sum(U+PV+E_K+E_P)_{mt} - (\delta m/M)c^2$

Questions:

1. No, so $(\delta m/M)c^2 = 0$
2. No, so $\sum(H+E_K+E_P) = 0$
3. No, so $\sum W = 0$
4. No, so $\sum Q = 0$
5. No, no, no, so $\Delta(U)_{SYS} = 0$

In other words, the expansion occurred with no change in internal energy even though an irreversible process occurred within the system. We refer to this process as a "**free expansion**."

Example 2.3 ***Closed System***
If we remove the insulation from the system in case I, and the gas is initially at the temperature of the surroundings, what is the first law analysis?
First law: $\Delta(U+E_K+E_P)_{SYS} = \sum Q + \sum W + \sum(U+PV+E_K+E_P)_{mt} - (\delta m/M)c^2$

Questions:

1. No, so $(\delta m/M)c^2 = 0$
2. No, so $\sum(H+E_K+E_P) = 0$
3. No, so $\sum W = 0$
4. Yes (when the gas expands, its temperature changes), so $\sum Q$ remains
5. No, no, no, so $\Delta(U)_{SYS}$ remains

The first law becomes

$$\Delta U_{SYS} = \sum Q \tag{2.28}$$

The change in the internal energy of the system equals the energy transferred as heat across the boundary of the system; thus, we can measure the change in internal energy for this system.

Example 2.4 ***Piston and Cylinder***
If we compress a gas by pushing a frictionless piston into a rigid, uninsulated cylinder, what is the first law analysis?

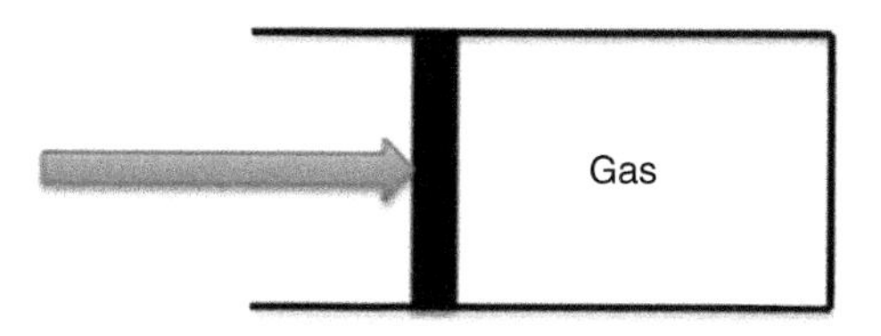

First law: $\Delta(U+E_K+E_P)_{SYS} = \sum Q + \sum W + \sum(U+PV+E_K+E_P)_{mt} - (\delta m/M)c^2$

Questions:

1. No, so $(\delta m/M)c^2 = 0$
2. No, so $\sum(H+E_K+E_P) = 0$
3. Yes (the volume of the cylinder changes), so $\sum W$ remains
4. Yes (compression changes the temperature of the gas), so $\sum Q$ remains
5. No, no, no, so $\Delta(U)_{SYS}$ remains

The first law becomes $\Delta U_{SYS} = \sum Q + \sum W$
This is the form you recall from chemistry classes.

Example 2.5 ***Flow System at Steady State***
If a fluid flows through a rigid, insulated, horizontal pipe that contains an obstruction, what is the first law analysis?

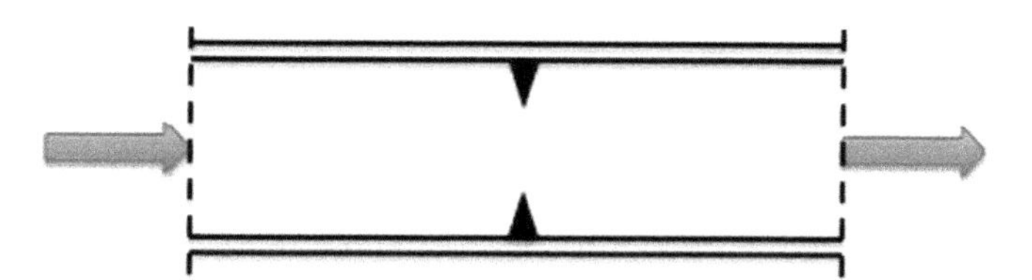

First law: $\Delta(U+E_K+E_P)_{SYS} = \sum Q + \sum W + \sum(U+PV+E_K+E_P)_{mt} - (\delta m/M)c^2$

Questions:

1. No, so $(\delta m/M)c^2 = 0$
2. Yes, but kinetic and potential energies are negligible (this is an assumption, but it often is valid), so $\sum H$ remains
3. No, so $\sum W = 0$
4. No (insulated pipe), so $\sum Q = 0$
5. Yes, so $\Delta(U+E_K+E_P)_{SYS} = 0$

The first law becomes $\sum H = H_{in} - H_{out} = -(H_{out} - H_{in}) = -\Delta H = 0 = \Delta H$.

We call this a "**throttling process**" – it is a constant enthalpy process (isenthalpic).

Example 2.6 ***Closed System at Constant Pressure***
Consider an uninsulated cylinder fitted with a frictionless piston upon which a constant mass rests. If we transfer some energy as heat into the cylinder, what is the first law analysis?
First law: $\Delta(U+E_K+E_P)_{SYS} = \sum Q + \sum W + \sum(U+PV+E_K+E_P)_{mt} - (\delta m/M)c^2$

Questions:

1. No, so $(\delta m/M)c^2 = 0$
2. No, so $\sum(H+E_K+E_P) = 0$
3. Yes, the volume changes, so $\sum W = -\int PdV = -P(V_2 - V_1)$
4. Yes, so $\sum Q$ remains
5. No, no, no, so $\Delta(U)_{SYS}$ remains

Then, the final energy equation is

$$\Delta U = \sum Q - P(V_2 - V_1) \text{ or } \sum Q = (U_2 + PV_2) - (U_1 + PV_1) = H_2 - H_1 = \Delta H$$

$$\Delta(H) = Q \tag{2.29}$$

This equation shows us how to measure enthalpy changes experimentally. This equation would also result from Example 2.5 if we remove the insulation from the pipe.

Obviously, these are simple applications of the first law, but they should have demonstrated to you how to use the formalism to address problems. Now, we can examine applications of the first law to more nearly practical examples in which the systems are units used in the chemical industry, such as condensers, pumps, and heat exchangers (note that these units should be called energy transfer as heat exchangers, but the convention is to call them heat exchangers).

2.5 Practical Application Examples

From this point on, we shall not include the nuclear reaction term in our first law analyses. Very few practical applications require it.

2.5.1 Compressors/Pumps

A compressor is a mechanical device that increases the pressure of a gas, usually for transportation through a pipe. We assume that a steady-state flow occurs in the compressor and that compressors are adiabatic (neither assumption is rigorously accurate, but both provide reasonably accurate solutions). Actually, compressors only approximate steady-state flow. Also, it is obvious that compressors transfer energy as heat because they become hot when operating. However, while the compressor becomes hot, the gas flowing through the compressor does not because it resides inside the compressor for a very short time, and energy transfer as heat is a function of time. Figure 2.4 is a schematic drawing of a compressor.

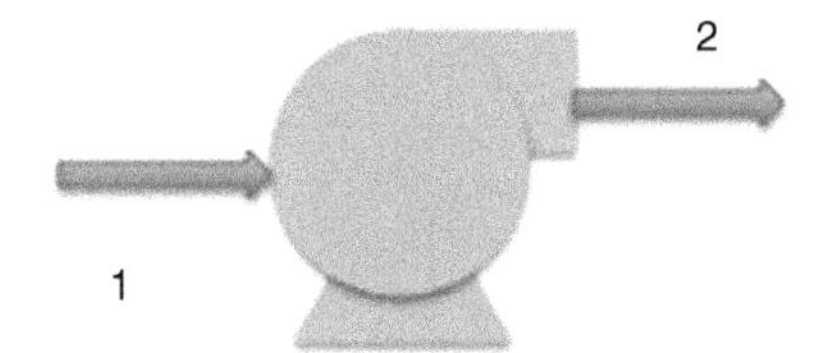

Figure 2.4 Schematic representation of a compressor.

If we change the words “compressor” to “pump” and “gas” to “liquid,” everything said in the above paragraph applies to pumps. Pumps are much less expensive to operate than compressors, and, because liquids are denser than gases, pumps are smaller than compressors when moving the same amount of mass.

First law: $\Delta(U + E_K + E_P)_{SYS} = \sum Q + \sum W + \sum(H + E_K + E_P)$

Note that we have eliminated the *CONV* term because in practical problems we do not need it.

Questions:

1. Yes, so $\sum(H + E_K + E_P)$ remains, but kinetic and potential energy changes are normally small compared to the change in enthalpy, so we only retain $\sum H$
2. Yes, it is necessary to transfer energy as shaft work to compress the gas, so $\sum W$ remains
3. No, we assume that the compressors are adiabatic (because the fluid flows through them so quickly that insufficient time exists for significant energy transfer as heat) and $\sum Q = 0$
4. We assume steady-state flow, so $\Delta(U + E_K + E_P)_{SYS} = 0$

The resulting equation applicable to compressors (or pumps) is

$$\sum W + \sum H = 0 = {}_1W_2 + H_{in} - H_{out} = {}_1W_2 + H_1 - H_2$$

$${}_1W_2 = H_2 - H_1 = \Delta H$$

Usually, compressors and pumps are horizontal, so the potential energy change is zero. The notation ${}_1W_2$ denotes energy transferred as work between points 1 and 2 in the system.

Remember, W_2, W_1, and ΔW are meaningless – work does not exist at points 1 or 2, enthalpy does exist at points 1 and 2 because enthalpy is a property. Work is not a property but only a mechanism to transfer energy across the boundaries of systems.

2.5.2 Turbines/Expanders

The analyses for turbines and expanders are identical to those for compressors and pumps. The difference is that turbines and expanders remove energy from the system while compressors and pumps add energy to the system. In other words, for compressors and pumps, $H_2 > H_1$, while for turbines and expanders, $H_2 < H_1$. Because enthalpy is a property, the temperature and pressure of the fluid at points 1 and 2

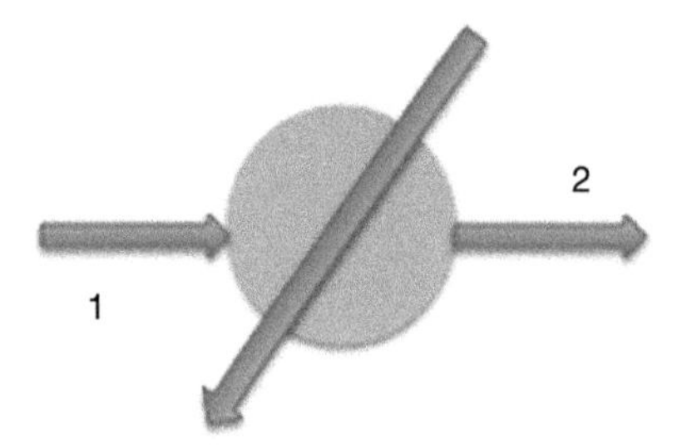

Figure 2.5 Condenser.

automatically provide the correct sign. Remember, the energy transfer as work is positive for compressors and pumps (energy in) but negative for turbines and expanders (energy out).

2.5.3 Condensers/Vaporizers/Reboilers

Condensers are heat exchangers – actually a misnomer, these are energy exchangers – that change vapor into liquid. We assume that they operate at steady state and that they are adiabatic with respect to their surroundings. Figure 2.5 is a drawing of a condenser. Vaporizers and reboilers are devices that change liquid into vapor. The analysis is identical for all three unit operations. Note that in Figure 2.5, the cross arrow for vaporizers and reboilers would have be drawn sloping upward.
First law: $\Delta(U + E_K + E_P)_{SYS} = \sum Q + \sum W + \sum (H + E_K + E_P)$

Questions:

1. Yes, so $\sum(H + E_K + E_P)$ remains, but kinetic and potential energy changes are normally very small compared to the change in enthalpy, so we only retain $\sum H$
2. No, so $\sum W = 0$
3. Yes, the coolant extracts energy as heat, so $\sum Q$ remains
4. We assume steady-state flow, so $\Delta(U + E_K + E_P)_{SYS} = 0$

The resulting equation applicable to condensers is

$$\sum Q + \sum H = 0 = {}_1Q_2 + H_{in} - H_{out} = {}_1Q_2 + H_1 - H_2$$

$${}_1Q_2 = H_2 - H_1 = \Delta H$$

The notation ${}_1Q_2$ denotes energy transferred as heat between points 1 and 2 in the system. **Remember, Q_2, Q_1, and ΔQ are meaningless – heat does not exist at points 1 and 2, enthalpy does exist at points 1 and 2 because enthalpy is a property. Heat is not a property but only a mechanism to transfer energy across the boundaries of systems.**

2.5.4 Heat Exchanger

Our next example is a "heat" exchanger. This is a device in which we transfer energy as heat from a hot stream to a cold stream without bringing them into direct contact.

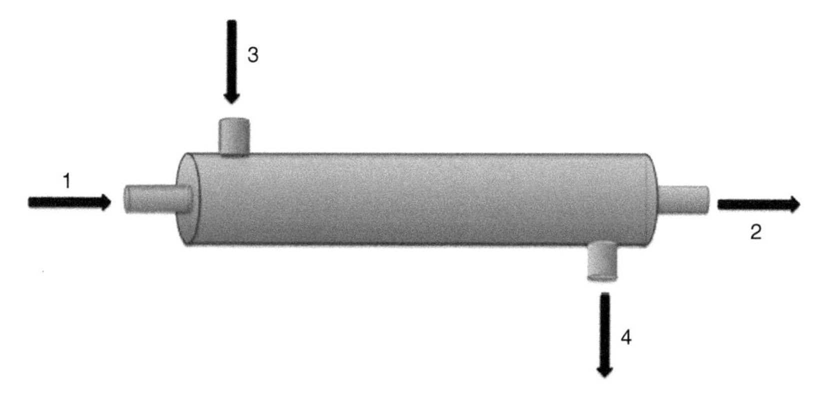

Figure 2.6 Heat exchanger.

Figure 2.6 is a schematic drawing of a simple heat exchanger. Here, we can consider two cases. The first one considers that the system has individual streams, so we have two systems: stream A and stream B. Next, we can consider the system as the entire heat exchanger.

Now, let us consider the system as only the tube(s) ($1 \rightarrow 2$)
First law: $\Delta(U + E_K + E_P)_{SYS} = \sum Q + \sum W + \sum (H + E_K + E_P)$

Questions:

1. Yes, so $\sum(H + E_K + E_P)$ remains, but kinetic and potential energy changes are normally small compared to the change in enthalpy, so we only retain $\sum H$
2. No, so $\sum W = 0$
3. Yes, the fluid in the tube exchanges energy as heat with the fluid in the shell, so $\sum Q$ remains
4. We assume steady-state flow, so $\Delta(U + E_K + E_P)_{SYS} = 0$

The resulting equation applicable to the fluid flowing through the tube is

$$_1Q_2 + H_1 - H_2 = 0$$
$$_1Q_2 = H_2 - H_1 = \Delta H$$

The properties of the tube fluid and the shell fluid determine the direction of energy flow (H_2 is either greater than or less than H_1). We note again that the enthalpies are the important information. This result is identical to that obtained for the condenser.

Now, we can consider the entire heat exchanger as the system:
First law: $\Delta(U + E_K + E_P)_{SYS} = \sum Q + \sum W + \sum (H + E_K + E_P)$

Questions:

1. Yes, so $\sum(H + E_K + E_P)$ remains, but kinetic and potential energy changes are normally small compared to the change in enthalpy, so we only retain $\sum H$
2. No, so $\sum W = 0$
3. No, energy does not cross the boundaries of the exchanger (they normally are insulated), so $\sum Q = 0$

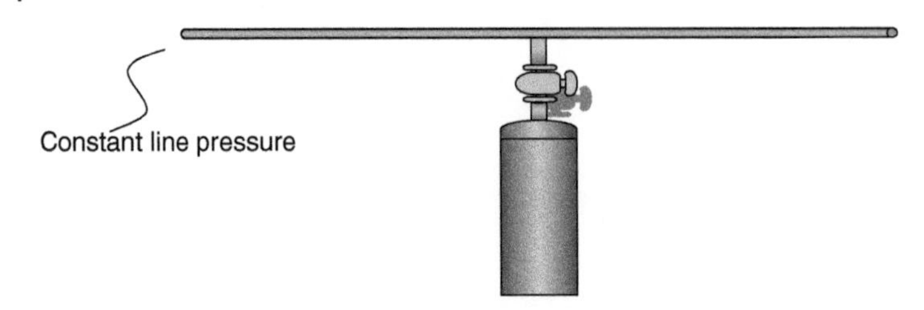

Figure 2.7 Sketch of the system.

4. We assume steady-state flow, so $\Delta(U+E_K+E_P)_{SYS}=0$

The resulting equation applicable to the entire exchanger is

$$0 = H_1 + H_3 - H_2 - H_4$$

or

$$H_2 - H_1 = -(H_4 - H_3)$$

These results show that the energy transferred between the shell fluid and the tube fluid is equal but with opposite signs.

2.5.5 Sample Cylinder

Assume that we want to pull a sample of a natural gas flowing through an insulated line, so we can perform an analysis. Figure 2.7 is a sketch of the system. The operation consists of opening the valve and allowing the evacuated cylinder to fill with gas until it reaches the line pressure. Assuming that the gas in the cylinder is the system, what is the first law analysis for this problem?

Solution
First law: $\Delta(U+E_K+E_P)_{SYS} = \sum Q + \sum W + \sum (H+E_K+E_P)$

Questions:

1. Yes, the system is open, so $\sum(H+E_K+E_P)$ remains, but kinetic and potential energy changes are normally small compared to the change in enthalpy, so we only retain $\sum H$
2. No, so $\sum W = 0$
3. Yes, energy can cross the boundaries of the cylinder as heat, so $\sum Q$ remains
4. The process is neither steady state nor cyclic, so $\Delta(U+E_K+E_P)_{SYS}$ remains, but E_K and E_P are probably negligible

The applicable equation is

$$\Delta U_{SYS} = H_{in}$$

However, if the initial pressure in the cylinder is zero (evacuated), then $U_i = 0$. Therefore, the equation reduces to

$$U_f = H_{in}$$

Congratulations – you have just solved an unsteady-state problem (that was not hard, was it).

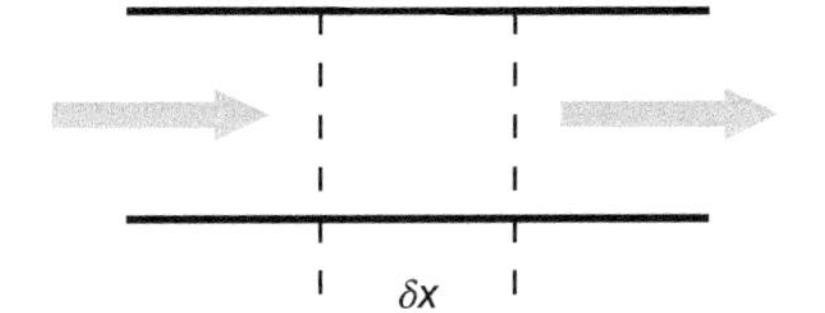

Figure 2.8 Diagram of an open system with unsteady flow.

2.6 Differential Form

Let us examine an element of a system, in particular an open system with unsteady-state flow. The mass contained in the element need not be constant. We can now apply our first law expression to the element shown in Figure 2.8

$$\Delta[n(U + E_K + E_P)]_{element} = \sum(nQ) + \sum(nW) + [n(H + E_K + E_P)]_x - [n(H + E_K + E_P)]_{x+\delta x} \tag{2.30}$$

Now, we can allow δx to shrink to infinitesimal size, and we get the differential expression for the first law

$$d[n(U + E_K + E_P)]_{element} = đq + đw - d[n(H + E_K + E_P)] \tag{2.31}$$

The expressions $đq$ and $đw$ are inexact differentials (they are infinitesimally small changes in system characteristics), but they exist only at the boundaries of the system and are not properties of the system. The negative sign in Eq. (2.31) is a result of the definition of a derivative (final–initial not input–output).

2.7 Inserting Time: Unsteady-State Flow Process

We can also insert time (artificially, we do not *need* time in thermodynamics, but mechanical engineers prefer to use this formulation in many situations) into the expression. Having gone to the trouble of conceiving a timeless world, it might seem silly to insert time into the equations; however, sometimes, it is convenient to do so.

$$\Delta[n(U + E_K + E_P)]_{SYS} = \sum(n\dot{Q})\Delta t + \sum(n\dot{W})\Delta t + \sum[\dot{n}(H + E_K + E_P)]\Delta t$$

$$\frac{\Delta[n(U + E_K + E_P)]_{SYS}}{\Delta t} = \sum(n\dot{Q}) + \sum(n\dot{W}) + \sum[\dot{n}(H + E_K + E_P)]$$

$$\lim_{\Delta t \to 0} \frac{\Delta[n(U + E_K + E_P)]_{SYS}}{\Delta t} = \frac{d[n(U + E_K + E_P)]_{SYS}}{dt}$$

$$\frac{d[n(U + E_K + E_P)]_{SYS}}{dt} = \sum(n\dot{Q}) + \sum(n\dot{W}) + \sum[\dot{n}(H + E_K + E_P)] \tag{2.32}$$

in which $\dot{Q}$ is the "heat transfer" rate, $\dot{W}$ is the power, and $\dot{n}$ is the molar flow rate: $\dot{n} = A\rho\dot{z}$ in which A is the area, ρ is the molar density, and $\dot{z}$ is the velocity.

2.8 Recap

In this chapter, we have used the GAE to develop a mathematical expression for the observation that "energy is conserved." The GAE says that the accumulation of energy in a system over an accounting period equals the net energy transferred across the boundaries of the system plus the net energy generated within the system plus the net conversion of energy within the system from one form to another. Because energy is conserved, the generation term is zero. The conversion term could arise in the case of a nuclear reaction, but we choose to ignore this term because we can always select the system to exclude the reactor.

We have examined several examples to become familiar with using the first law of thermodynamics. In these examples, we have worked with open, closed, and isolated systems and with both steady-state and unsteady-state processes.

The fool-proof procedure we have developed for first law analyses of problems is

- **Draw a sketch of the system**
- **Write the entire first law**
- **Ask all the questions about the system and cancel terms that do not apply**
- **The applicable equation remains**

Problems for Chapter 2

2.1 What thermodynamic variable results from the first law of thermodynamics?

2.2 The total energy of a closed system decreases by 55 kJ when the system performs 150 kJ of work. How much heat is added to or removed from the system? Neglect energy changes in kinetic and potential energy.

2.3 A closed container with 1 mol of gas increases temperature from 25 to 300 °C in a constant pressure process at 120 kPa. During the process, a change of volume of 0.01 m^3 occurs. What is the amount of energy transferred as heat added to the container during the process? What is the change in the internal energy during the process? The heat capacity of the gas is 30 kJ/(kmol K). Consider $1\,J = 1\,m^3\,Pa$.

2.4 After heating 5 kg of water at 150 °C and 10 kPa pressure at a constant volume, the final temperature of water is 400 °C. Calculate the amount of energy transferred as heat assuming a constant heat capacity of 4.184 kJ/(kg K).

2.5 A spring obeys Hook's law in which the force is proportional to the distance. The proportionality constant is −2500 N/m and the distance is in meters, so the force is in newton. Calculate the work when the spring elongation is 0.2 m.

2.6 Complete the following table for a cyclic process. Considered each step a closed system

W^t (kJ)	Q^t (kJ)	ΔU^t (kJ)	W^t (kJ)
a → b		75	−50
b → c		100	
c → a	200		100
Total			

2.7 Consider the following piping arrangement. The inlet stream specifications are A: 105 kg/h at 65 °C; B: 60 kg/h at 5 °C; and D: 20 kg/h at 110 °C. All streams contain the same fluid, which has $CP = 4.1868$ kJ/(kg K) (independent of temperature). Hint: Take $H = 0$ kJ/kg at 0 °C.

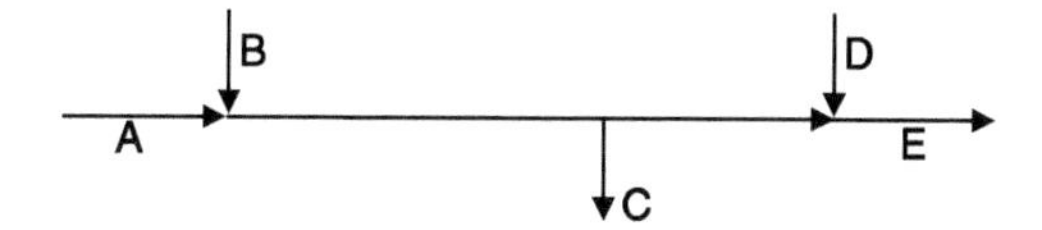

a. Determine the range of temperatures that can be achieved in streams C and E as the flow rate of stream C varies from 0 to 185 kg/h.

b. Determine the temperatures of streams C and E when the withdrawal rate through stream C is 40 kg/h.

2.8 Determine the form of the energy balance that is appropriate to describe a steady-state flow of gas through a valve, when the valve is closed sufficiently to maintain a significant pressure drop through the valve.

2.9 During a nonflow process in a rigid container, the internal energy of a gas decreases by 1400 J/mol of gas.

a. Determine the amount of energy transferred as heat during this process.

b. Determine the temperature change that occurs during this process. For this gas, $C_V = 20.8$ J/(mol K) and $C_P = 29.0$ J/(mol K).

2.10 Determine the form of the energy balance that is appropriate to describe a steady-state flow of gas through a pressure regulator.

2.11 During a nonflow process in an insulated container, the internal energy of a gas decreases by 1400 J/mol of gas.

a. Determine the amount of energy transferred as work during this process.

b. Determine the temperature change that would be required to accomplish an identical change in internal energy in a nonflow constant volume process. For this gas, $C_v = 20.8$ J/(mol K) and $C_P = 29.0$ J/(mol K).

2.12 Determine the form of the energy balance that is appropriate to describe the flow of m kg of gas from a line at P_1, T_1 through a valve into a piston cylinder arrangement in which the pressure and temperature remain constant at T_2 and P_2 by removing energy as heat if necessary. You may assume that the cylinder is originally empty. Clearly specify if internal energies and enthalpies used in the energy balance are at $T_1, P_1; T_2, P_2$ or some other conditions of temperature and pressure.

2.13 Determine a form of the energy balance that describes the pumping of oil from a depth of 4000 m to the surface. Assume that the flow is steady state and that no significant amount of energy is transferred as heat as the oil flows up the pipe.

2.14 Develop the energy balance that is appropriate for the system shown below. Develop the equations as far as possible. Assume that the piston operates without friction within an insulated cylinder that has an inside diameter D. During the process of interest, a gas at T_0 and P_0 enters the cylinder through the valve and the piston rises from its initial position to a final position at a distance h higher. The mass of the piston is m and an object with mass M rests on top of the piston.

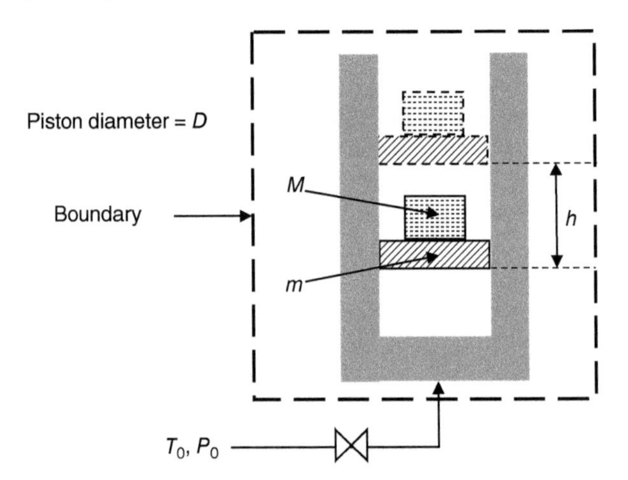

2.15 A cat with a mass of 4.5 kg climbs a 4 m tall tree (remember cats land upright). The fibula of a cat supports approximately a pressure of 1 485 440 Pa. The cat falls into sand and sinks 2.5 cm in the stand. During the fall, the cat does not change temperature nor is there any energy transfer as heat. Will the cat's bones endure the fall? Consider that its paws are circular and measure 2 cm in diameter. Also, you may assume that the cat is rigid, and it is a free fall.

2.16 A gas compressed into a cylinder passes through the following steps:

Step	Pressure (atm)	Volume (l)
1	4	7.6
2	4.1	5
3	12	2

Calculate the reversible work for the whole process.

2.17 880 kg/h of a substance enters a pipe that separates into two parts as shown in the following diagram:

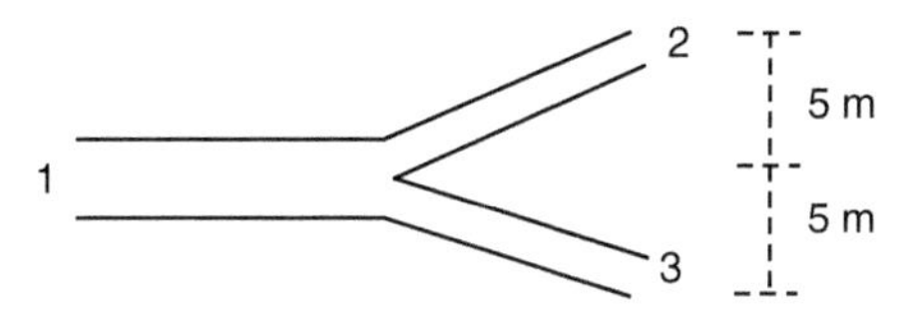

The pipes are insulated. The temperatures and velocities are $T_1 = 300$ K, $T_2 = 400$, $v_1 = 10$ m/h, $v_2 = 12$ m/h, $D_1 = 40$ cm, $D_2 = 15$ cm, and $D_3 = 35$ cm.
What is T_3? Assume $C_P = 2.5$ kJ/(kg K) and consider the density constant and the potential energy negligible.

2.18 Fluids flowing through two pipes mix in a chamber and exit through two other pipes. 1 kg/s of a fluid enters the first pipe at 30 °C, and 2 kg/s of the same fluid enters the second pipe at 100 °C. If the mixed fluid exits one pipe with a flow rate of 2.5 kg/s, what is the outlet temperature of the fourth pipe if the mixture receives 50 kJ/s as heat? Consider the fluid heat capacity constant to be 4.0 kJ/(kg K).

2.19 One mole of a gas expands until its volume increases to twice its initial volume. If the pressure at 300 K is

$$P = a/V + b/V^2$$

in which P is in bar and V is in cm^3/mol. What is the reversible work if $P = 10$ bar and $a = 21\,000$ cm^3 bar/mol and $b = 405\,200$ cm^6 bar/mol^2?

2.20 A car has a mass of 1300 kg and travels at 120 km/h. What is the kinetic energy? How much work must be done to stop it (justify your answer)?

2.21 In a quench process a 1 kg steel bar at 250 °C is submerged into an isolated container with cold water at a temperature of 30 °C. What is the mass of water if the final temperature is 60 °C? Consider the heat capacity of the steel and water to be 0.5024 and 4.182 kJ/(kg °C), respectively. In your calculations, consider $\Delta U = C_V \Delta T$.

2.22 A piston compresses 1.5 kg of a gas from 0.6 to 0.15 m^3 in a constant pressure process at 14.5 bar. If the enthalpy decreases by 2300 kJ/kg, what is the work, the amount of heat transferred to or from the gas, and the change of internal energy?

2.23 The heat exchanger illustrated below cools a liquid stream of 1200 kg/h that enters at 300 °C. The cooling medium is water at 25 °C. Calculate the amount of water needed to cool the liquid to 100 °C, if the water heats to 50 °C. What is the amount of energy transferred as heat received by the water?

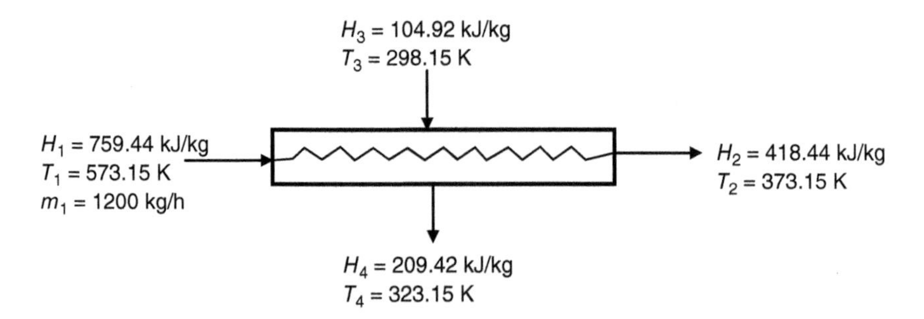

2.24 0.5 kg of a gas enters a compressor of 2 hp at 1 bar and leaves at 7 bar. The specific volume at the entrance of the compressor is 180 cm^3/g and its internal energy is 550 kJ/kg. When the gas leaves the compressor, its specific volume is 60 cm^3/g. What is the internal energy of the gas that exits the compressor if the compressor works for one hour?

2.25 In rural areas, it is common to use cisterns to store water. Pumps send groundwater to water tanks that are above the houses. Calculate the work required by the pump using the following diagram and data:

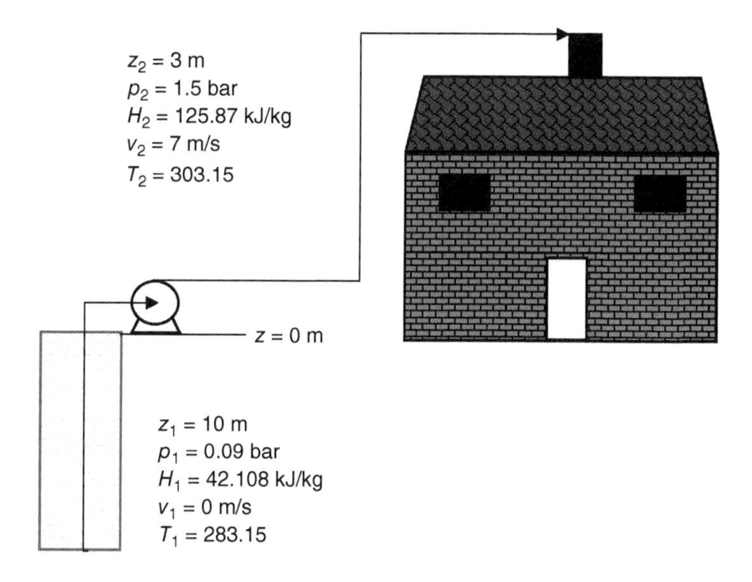

2.26 Carbon dioxide passes through an insulated pipe. Calculate the pipe diameter using the data from the process:

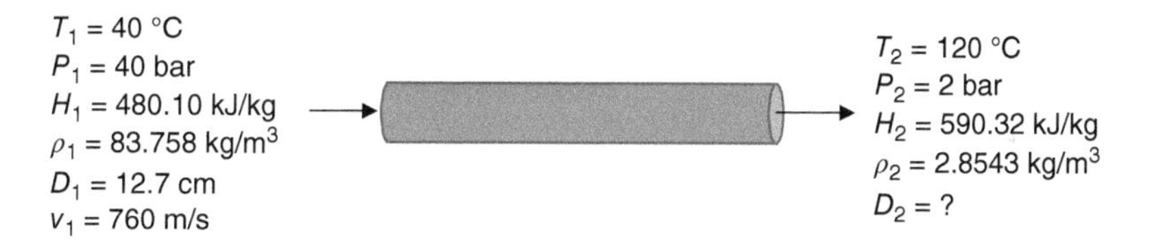

Reference

1 C.M. Sliepcevitch, J.E. Powers, W.J. Ewbank. *Foundations of Thermodynamic Analysis*. McGraw Hill, New York, 1971.

3

PVT Relations and Equations of State

3.1 Introduction

What is an equation of state (EOS)? Basically, it is a representation of the properties of a system in terms of other properties of the system or in terms of observables. Thus far, in this course, we have defined four properties of a system: U, the internal energy; H, the enthalpy; E_P, the potential energy; and E_K, the kinetic energy; and three observables: T, temperature; P, pressure; and V, volume (or ρ, density). The density is simply the reciprocal volume ($1/V$), so why do we use both of these concepts? Well, in many derivations, volume provides "simpler" results. On the other hand, for EOS density is definitely preferable. While volume/density is observable, it is more difficult to measure than temperature or pressure. The representation of an EOS can be a plot, a table, or a mathematical equation. We have occasions to use all three because an engineer must know how to obtain thermodynamic properties in terms of measurable quantities and to use such information as is available. Also, it is desirable to have usable expressions or correlations for the physical properties of the species involved in a process. In this chapter, we examine the pressure–volume/density–temperature representations for a pure substance. In Chapter 5, we focus upon the mathematical relationships to derive the thermodynamic properties from a *PVT* or $P\rho T$ relationship.

3.2 Graphical Representations

One of the most important diagrams for the prediction of thermodynamic properties is $P\rho T$. Figure 3.1 is a three-dimensional (3D) diagram for methane. The symbols are experimental measurements, and you can see different projections: $P\rho$, PT, and $T\rho$. One objective of thermodynamics is to represent this diagram with a single equation.

Figure 3.1 shows how complicated this diagram can be even for a simple substance such as methane. The pressure and density axes of the plot are logarithmic to present a clearer picture of the 3D diagram. It is also important to understand the P–T, P–ρ, and T–ρ projections of Figure 3.1. Unfortunately, these concepts evolved in the 1800s, and they suffer from what we believe are misconceptions associated with those early days of thermodynamics. The P–T diagram is the least complicated

Thermodynamics for Chemical Engineers, First Edition. Kenneth R. Hall and Gustavo A. Iglesias-Silva.

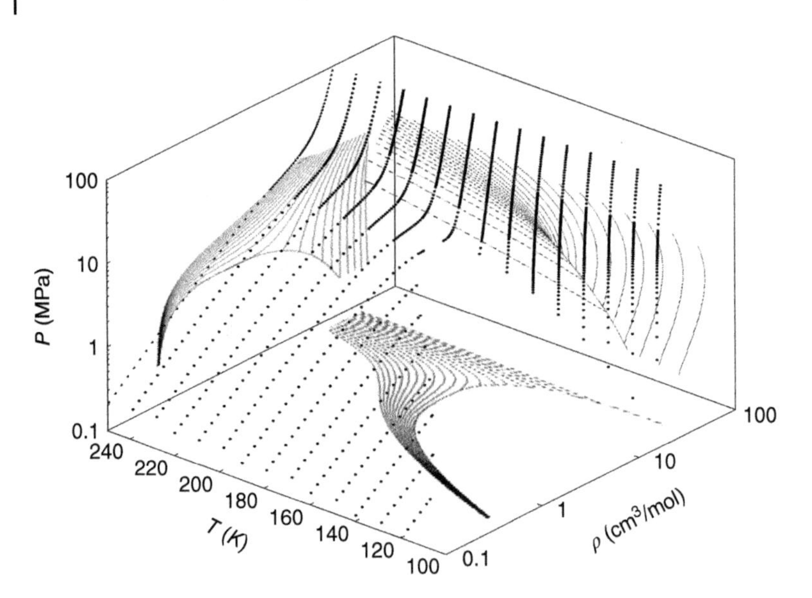

Figure 3.1 *PρT* diagram for methane.

to review and discuss. We can review the historical diagram, and the diagram we believe to be a better representation of the associated phenomena.

Figure 3.2a is the traditional representation of the *P*–*T* diagram. It divides the space into four regions: solid (S), liquid (L), vapor (V), and gas (G). The solid region is simply solid, and it persists to very high pressures (at least as long as the substance can exist physically under the exerted pressure). The line separating the solid region and the liquid region is the "freezing or melting curve" (when the temperature of the solid reaches the melting curve, the solid melts). On this curve, the solid and the liquid are in equilibrium. Then, there is a line at lower pressures separating the solid region from the vapor region, which is the sublimation curve (when the temperature of the solid reaches the sublimation curve, the solid vaporizes). Along this curve, the solid and the vapor are in equilibrium. The point on the diagram at which the solid, liquid, and vapor regions intersect is the triple point. At this point, the solid, liquid, and vapor are in equilibrium. Finally, the line emanating up and right from the triple point is the vaporization curve (or vapor pressure curve). On this curve, the vapor and liquid are in equilibrium. The curve ends at the critical point defined by the critical pressure (P_C), volume (V_C), or density (ρ_C), and temperature (T_C). Because it is not possible to have vapor and liquid coexisting above the critical temperature, the diagram uses a dashed line at T_C to separate the liquid region, L, and the vapor region, V, from the "gas" region, G.

However, an inconsistency exists with this representation. If we move along path 1 to 2 in Figure 3.2c, we can see the vapor and liquid coexisting when we touch the vaporization curve. The volume or density is significantly different on the vapor side of the curve than on the liquid side of the curve, and it is obvious that we have passed from the vapor region into the liquid region. However, if we take the path 1 to 3 to 4 to 2, we cannot *observe* any change in volume or density as we pass from V into G

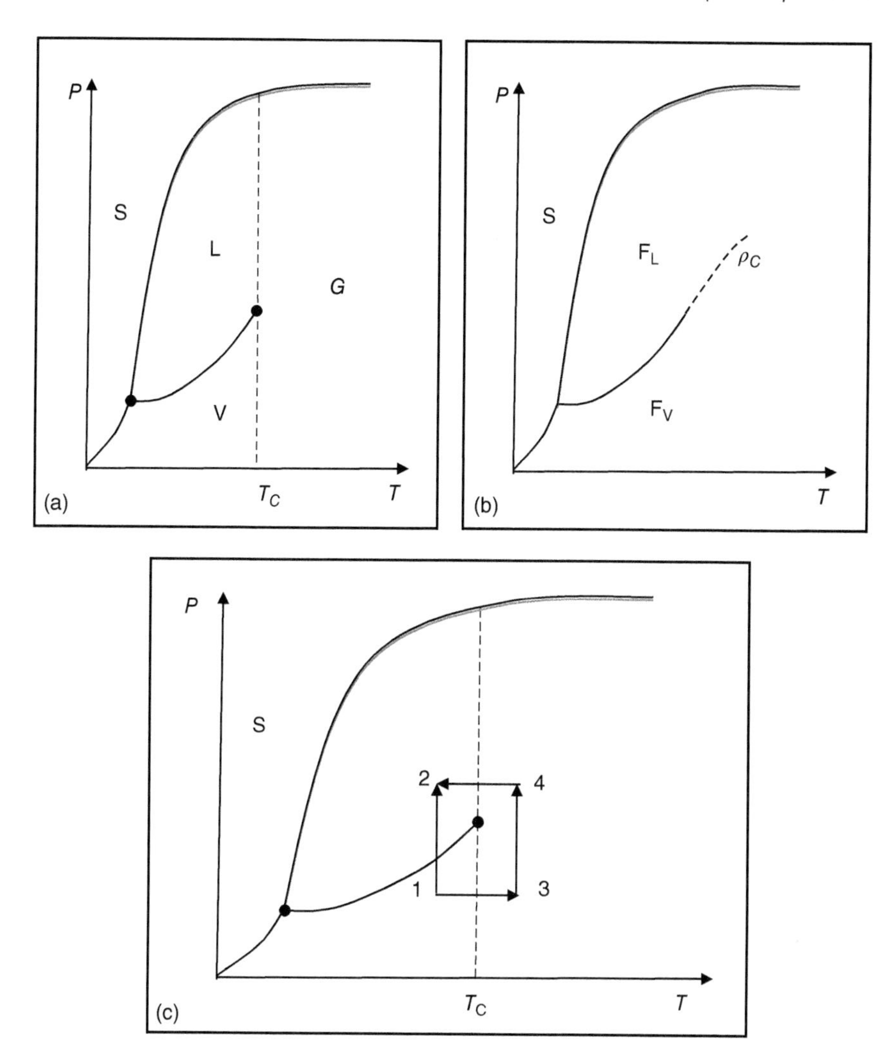

Figure 3.2 (a, b, c) Pressure–temperature diagrams.

or from G into L. The volume or density changes from point to point along this path are infinitesimally small, whereas along path 1 to 2, a finite change occurs at the vaporization curve. Therefore, L, V, and G are simply definitions for convenience.

Figure 3.2b seems to be a more rational approach. This diagram recognizes that Nature provides us with two phases: solid (S) and fluid (F). The fluid region has an arbitrary (and unnecessary) division into fluids with vapor-like volumes or densities (F_V) and those with liquid-like volumes or densities (F_L) separated by the dashed line that represents the critical volume or density locus. This definition is different from the one you learned in chemistry courses that created the diagram in Figure 3.2a.

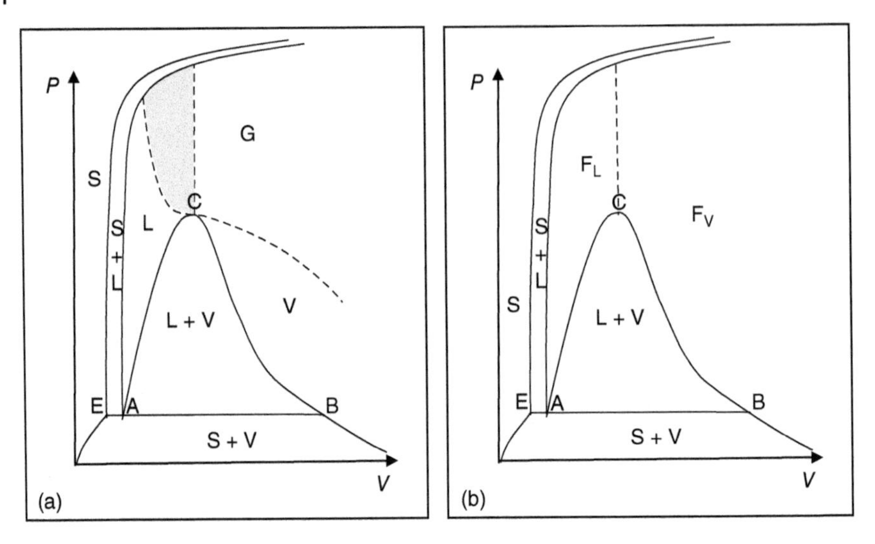

Figure 3.3 (a, b) Pressure–volume (*PV*) diagrams.

Why is this of any concern? In the petroleum industry, companies produce "oil" and "gas" from reservoirs. However, down hole petroleum is a single-phase fluid until the pressure drop caused by production crosses the saturated value. It is not uncommon for engineers employed in the petroleum industry to argue if the down hole fluid is "gas" or "oil." In fact, it is *fluid* until the pressure is below saturation when it becomes both. Figure 3.2b renders this conflict mute.

Figure 3.3a is a traditional pressure volume diagram, and Figure 3.3b incorporates the concepts introduced in Figure 3.2b. In Figure 3.3a, we see vapor (V), liquid (L), solid (S), L + V, S + L, and S + V regions and the gas (G) region separated into the vapor and liquid regions by the critical temperature (dashed line). Figure 3.3b has a solid region (S), a fluid with liquid-like density region (F_L), and a fluid with vapor-like density region (F_V). The critical volume or density line (dashed) separates these two regions (this division is totally unnecessary but can illustrate some previous misconceptions), and solid + liquid (S + L), solid + vapor (S + V), and vapor + liquid (V + L) regions.

Notice that some of the lines on the *P–T* diagram appear as regions on these diagrams. The vapor pressure curve becomes the V + L dome inside of which vapor and liquid appear as separate phases. The melting curve becomes the S + L zone in which the solid and fluid with liquid-like density appear as separate phases. The sublimation curve becomes the S + V zone in which the solid and fluid with vapor-like density appear as separate phases. Note also that the latter two zones do not have well-defined end points. It is not possible to observe two phases coexisting outside of these zones. The "triple point" on Figure 3.2a becomes a "triple line" on this diagram. The critical point is the point at which the vapor and liquid properties become identical (when approaching the critical point from inside the two-phase region), and the meniscus separating the vapor and liquid disappears. This point has considerable use in the development of correlations for properties of substances.

The plots also contain *isotherms* or curves at constant temperature. The isotherm that passes through the critical point is the critical isotherm, and isotherms below that temperature are "subcritical" isotherms, while those above the critical isotherm are "supercritical" isotherms. This dome is the *saturation curve*, and it consists of two sections, one from point A to point C, the saturated liquid curve, and one from point C to point B, the saturated vapor curve. Point C is the critical point. Saturation on this diagram indicates the onset of phase equilibrium. Subcritical isotherms touch the saturation curves. The line EAB is the triple point line where solid + liquid + vapor coexist in equilibrium (on a P–T plot, the triple point is actually a point, but on P–V,ρ or T–V,ρ plots, the triple "point" is a line).

Looking at the classical P–V plots illustrates an interesting concept. The region to the right of the critical isotherm but to the left of the critical volume is a gas. However, researchers "discovered" that fluids in this region behaved like liquids in extraction processes. They called it "supercritical extraction" and were amazed that a gas could behave in this manner. Of course, the new plots show clearly that the fluid left of the critical volume is the fluid with liquid-like density, so of course it behaves like a "liquid."

It is also possible to construct temperature volume or density diagrams. These diagrams are similar to pressure–volume diagrams, but they contain *isobars* (constant pressure curves) instead of isotherms. In this projection, the isobars have positive slopes, while isotherms on P–V plots have negative slopes in a pressure–volume projection. All the comments concerning F, F_L, and F_V are equally valid for these plots.

Figure 3.4 emphasizes the two-phase region, which contains different proportions of liquid and vapor. Point B contains more liquid than vapor, while point C has more vapor than liquid. Point A is a saturated liquid, and point D is a saturated vapor. If a container holds a fluid at temperature T_1 and is at equilibrium state B, the total volume in the container is the total volume of the liquid plus the total volume of the vapor

$$nV = n^l V^l + n^v V^v \tag{3.1}$$

in which n denotes moles, superscript l denotes the saturated liquid, and superscript v denotes the saturated vapor. Implicit in this equation is the expression

$$n = n^l + n^v \tag{3.2}$$

Using specific volume (per unit mass), the above equation becomes

$$m\,(nV/m) = m^l\left(n^l V^l/m^l\right) + m^v\left(n^v V^v/m^v\right) \tag{3.3}$$

or noting that $n/m = 1/M$ with M being the molar mass

$$m\,(V/M) = m^l\left(V^l/M\right) + m^v\left(V^v/M\right) \tag{3.4}$$

and

$$m = m^l + m^v \tag{3.5}$$

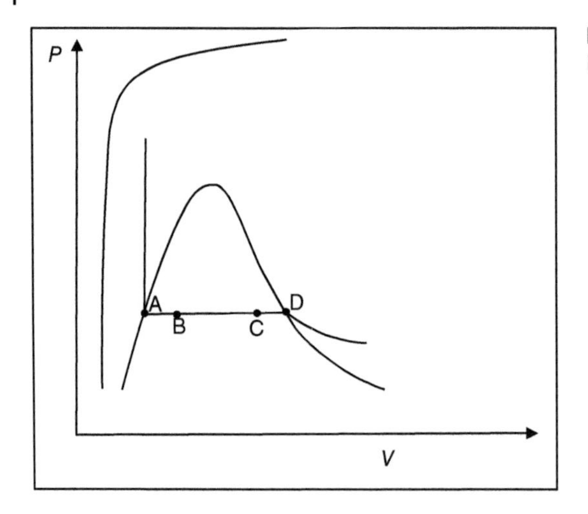

Figure 3.4 Constant quality lines in a PV diagram.

Dividing Eq. (3.4) by the total mass,

$$(V/M) = \frac{m^l}{m}\left(V^l/M\right) + \frac{m^v}{m}\left(V^v/M\right) \tag{3.6}$$

and substituting Eq. (3.5)

$$(V/M) = \frac{m - m^v}{m}\left(V^l/M\right) + \frac{m^v}{m}\left(V^v/M\right) \tag{3.7}$$

in which m^v/m is the vapor mass fraction, commonly called *quality* with the symbol χ

$$V = (1 - \chi)V^l + \chi V^v \tag{3.8}$$

In general, Eq. (3.8) applies to any specific thermodynamic property (mass or molar)

$$X = (1 - \chi)X^l + \chi X^v \tag{3.9}$$

in which X is the specific thermodynamic property, X^l is the saturated liquid property, and X^v is the saturated vapor property. Depending upon the use of mass or molar quantities, the quality of a pure component is

$$\chi = \frac{m^v}{m} = \frac{n^v}{n} \tag{3.10}$$

It is necessary to know either the saturation pressure or temperature to know X^l and X^v. If the value of X is available, then the quality is

$$\chi = \frac{X - X^l}{X^v - X^l} \tag{3.11}$$

Example 3.1

A container has vapor and liquid water in equilibrium. Calculate the quality and volume of the wet steam at 325 kPa, if the total enthalpy is 1000 kJ/kg.

Solution

The values for the enthalpy and the volume of saturated water at 325 kPa are

$$H^v = 2728.3\ \text{kJ/kg}$$

$$H^l = 573.197\ \text{kJ/kg}$$

$$V^v = 561.75\ \text{cm}^3/\text{g}$$

$$V^l = 1.076\ \text{cm}^3/\text{g}$$

Using Eq. (3.11) with $X = H$

$$\chi = \frac{H - H^l}{H^v - H^l} = \frac{1000 - 573.197}{2728.3 - 573.197} = \frac{426.803}{1958.3} = 0.218$$

Now, Eq. (3.8) provides the volume

$$V = (1 - \chi)V^l + \chi V^v = (1 - 0.218) \cdot 1.076 + 0.218 \cdot 561.75 = 123.3\ \text{cm}^3/\text{g}$$

3.3 Critical Region

The vicinity of the critical point is the critical region. Peculiar static and dynamic phenomena occur in this region. It is possible to approach the critical point from the single-phase region or from the two-phase region. In the critical region, some thermodynamic properties can have values of zero or infinity. From the classical point of view, at the critical point

$$\left(\frac{\partial P}{\partial V}\right)_T = \left(\frac{\partial^2 P}{\partial V^2}\right)_T = 0 \text{ at } T = T_C \tag{3.12}$$

This equation indicates a maximum (saturation curve) and an inflection (critical isotherm) at the critical point. It is conceivable that more than two derivatives can be zero at the critical point. If so, they must be zero by pairs; therefore, the first non-zero derivative must be odd (otherwise, the isotherms would exit the two-phase region with an incorrect slope). If only two derivatives equal to zero, then

$$\left(\frac{\partial^3 P}{\partial V^3}\right)_T > 0 \text{ at } T = T_C \tag{3.13}$$

In this text, we consider the function analytical (i.e. its derivatives are continuous functions) if the path is in the single-phase region and non-analytical if the path is in the two-phase region. In an attempt to quantify this behavior, scientists have formulated the *Scaling Hypothesis*, in which they express some thermodynamic functions proportional to temperature or density functions with specific exponents:

$$(\rho^l - \rho^v)\ \alpha \left(1 - T/T_C\right)^\beta \qquad \beta \approx 0.3285 \tag{3.14}$$

$$(H^l - H^v)\ \alpha \left(1 - T/T_C\right)^\beta \qquad \beta \approx 0.375 \tag{3.15}$$

$$\kappa_T \alpha \left(1 - T/T_C\right)^{-\gamma} \qquad \gamma = 1.125 \tag{3.16}$$

$$\left|P - P_C\right| \alpha \left(1 - T/T_C\right)^\delta \qquad \delta \approx 4.3 \tag{3.17}$$

$$\left(\frac{d^2 P}{dT^2}\right)_{sat} \alpha \left(1 - T/T_C\right)^{-\theta} \qquad \theta \approx 0.1 \tag{3.18}$$

$$\left[\frac{T_C}{P_C}\left(\frac{\partial P}{\partial T}\right)\right]^l - \left[\frac{T_C}{P_C}\left(\frac{\partial P}{\partial T}\right)\right]^v \alpha\left(1 - T/T_C\right)^\lambda \quad \lambda \approx 0.5 \tag{3.19}$$

An interesting inconsistency in this hypothesis is that Eqs. (3.14) and (3.15) have different "experimental" values for the same exponent. Although many physicists embrace the scaling hypothesis as being absolutely true, some excellent data seem to favor the classical behavior in the single-phase region. An important question is "What is the extent critical region?" Many physicists have worked to establish a dimensionless variable that represents the extent of the critical region. However, an accepted general variable is not yet available because of the complexity of the problem.

3.4 Tabular Representations

Although plots are very convenient, it is difficult to make them accurate. A table of $P\rho T$ data can be extremely accurate, but inconvenient. However, it is possible to make tables less inconvenient. You may be familiar with the Steam Tables. They are essentially a tabular representation of the EOS for water-containing $U/M, H/M$, and V/M (plus one more property) as functions of pressure and temperature. Construction of the tables is such that you may perform linear interpolations with considerable confidence. You may also make small linear extrapolations without too much worry. You may not yet have discovered much of the "excitement" associated with extrapolation in engineering and scientific calculations. Whenever possible, **avoid** extrapolation!

Table 3.1 is an excerpt from the Steam Tables we can use to illustrate interpolation and extrapolation within tables.

Example 3.2

What is the mass volume of steam at 0.075 MPa and 380 K?

Solution

We need to interpolate the values at 0.075 MPa between 373.15 and 398.15 K. A linear interpolation involves a truncated Taylor's series:

$$V = V_{ref} + (dV/dT)\left(T - T_{ref}\right) \tag{3.20}$$

$$dV/dT \approx \left(V_{ref2} - V_{ref1}\right) / \left(T_{ref2} - T_{ref1}\right) \tag{3.21}$$

Table 3.1 Steam properties.

P^{sat} (MPa)/T^{sat} (K)	$(V/M)^{sat}$ (m^3/kg)	373.15 K V/M (m^3/kg)	398.15 K V/M (m^3/kg)
0.075/364.94	2.2169	2.2698	2.4294
0.100/372.78	1.6937	1.6955	1.8167

Applying this procedure to our problem, we obtain

$$V_{380.00,0.75} = V_{373.15,0.075} + \frac{V_{398.15,0.075} - V_{373.15,0.075}}{T_{398.15} - T_{373.15}}(380.00 - 373.15)$$
$$= 2.2698 + \frac{2.4294 - 2.2698}{398.15 - 373.15}(380.00 - 373.15) = 2.3135\ \mathrm{m^3/kg}$$

This value should be very close to the actual value given a properly prepared table.

Example 3.3
What is the mass volume of saturated steam at 0.075 MPa (note: it is necessary only to specify the pressure or temperature for saturated steam because a saturation point is unique)?

Solution
This example involves an extrapolation. We know the derivative term in Eq. (3.20), but it is not necessarily a good value because the saturation point lies outside its range. Of course, it is not necessary to extrapolate to obtain the saturated volume because it appears explicitly in the table. We work this example to get an idea of extrapolation accuracy.

$$V_{364.94,0.075} = V_{373.15,0.075} + \frac{V_{398.15,0.075} - V_{373.15,0.075}}{T_{398.15} - T_{373.15}}(364.94 - 373.15)$$
$$= 2.2698 + \frac{2.4294 - 2.2698}{398.15 - 373.15}(364.94 - 373.15) = 2.2174\ \mathrm{m^3/kg}$$

Not too bad – the number is off by 0.022% in this case. However, we cannot count on being this lucky with every extrapolation even in the Steam Tables.

Example 3.4
What is the mass volume of steam at 0.090 MPa and 380 K?

Solution
Oops, this point lies between both the listed T and P values, and we need an additional Taylor's series:

$$V = V_{ref} + (dV/dP)\left(P - P_{ref}\right) \tag{3.22}$$

with

$$dV/dP \approx \left(V_{ref2} - V_{ref1}\right) / \left(P_{ref2} - P_{ref1}\right) \tag{3.23}$$

Applying both series to the data, we already know $V_{380.00} = 2.2135\ \mathrm{m^3/kg}$ at 0.075 MPa, so we need to calculate $V_{380.00}$ at 0.100 MPa:

$$V_{380.00,0.100} = V_{373.15,0.100} + \frac{V_{398.15,0.100} - V_{373.15,0.100}}{T_{398.15} - T_{373.15}}(380.00 - 373.15)$$
$$= 1.6955 + \frac{1.8167 - 1.6955}{398.15 - 373.15}(380.00 - 373.15) = 1.7287\ \mathrm{m^3/kg}$$

Now, using Eqs. (3.22) and (3.23)

$$V_{380.00,0.090} = V_{380.15,0.075} + \frac{V_{380.15,0.100} - V_{380.00,0.075}}{P_{0.100} - P_{0.075}}(0.090 - 0.075)$$
$$= 2.3135 + \frac{1.7287 - 2.3135}{0.100 - 0.075}(0.090 - 0.075) = 1.9626\ \text{m}^3/\text{kg}$$

These are all the procedures you need to use the Steam Tables.

Of course, the Steam Tables are not the only tabular representations of the EOS. Any table of experimental data is also a tabular representation. However, preparation of experimental data usually does not consider using interpolation or extrapolation.

3.5 Mathematical Representations

An EOS is a mathematical representation of the phase diagram. For example, we could have an equation to represent the *PVT* behavior of a system:

$$f(P, V, T) = 0 \tag{3.24}$$

This EOS is the most general mathematical form describing the *PVT* behavior of a system. Unfortunately, it is not very useful. However, given an equation such as Eq. (3.24), we can always expand it into:

$$P = P(T, V) \quad \text{or} \quad V = V(T, P) \quad \text{or} \quad T = T(P, V) \tag{3.25}$$

Each of these forms contains the same information, but the first form provides a better EOS in the sense that it has fewer terms that are easier to calculate than those in the other two. Of course, the *Principle of Maximum Inconvenience* dictates that the second form would be more useful because it is easier to measure P than it is to measure V.

We could write other EOS given the information we have in the course. For example, we could write

$$U = U(T, V) \quad \text{or} \quad H(T, P) \tag{3.26}$$

The problem with these two forms is U and H are not directly measurable as are T, P, and V. The forms in Eq. (3.26) require other thermodynamic properties for practical calculations, while those in Eq. (3.25) do not. However, none of these equations contain all the information available from thermodynamics. A **fundamental EOS** contains all the thermodynamic information about a system, and we shall learn about these equations later in this course.

We can classify EOS as *partial* or *universal*. Partial EOS represents a portion of the *PVT* phase space such as the vapor region or the liquid region, but they do not recognize phase separation. Universal EOS applies to any region of phase space including the two-phase region. An EOS can be universal without being a particularly good representation of the phase space.

Given a function that represents the *PVT* behavior, some of its derivatives are of practical interest:

$$\text{Thermal expansion: } \beta = \frac{1}{V}\left(\frac{\partial V}{\partial T}\right)_P \tag{3.27}$$

Isothermal compressibility: $\kappa_T = -\frac{1}{V}\left(\frac{\partial V}{\partial P}\right)_T$ (3.28)

Isothermal bulk modulus: $B_T = -V\left(\frac{\partial P}{\partial V}\right)_T = \frac{1}{\kappa_T}$ (3.29)

Adiabatic bulk modulus: $B_S = -V\left(\frac{\partial P}{\partial V}\right)_S$ (3.30)

Adiabatic compressibility: $\kappa_S = -\frac{1}{V}\left(\frac{\partial V}{\partial P}\right)_S$ (3.31)

Heat capacity at constant volume: $C_V = \left(\frac{\partial U}{\partial T}\right)_V$ (3.32)

Heat capacity at constant pressure: $C_P = \left(\frac{\partial H}{\partial T}\right)_P$ (3.33)

Thermal pressure coefficient: $\gamma_T = \left(\frac{\partial P}{\partial T}\right)_V$ (3.34)

Relationship between C_P and C_V: $C_P - C_V = TV\beta\gamma = TV\beta^2 B_T$ (3.35)

Recall the term "adiabatic" refers to a system that does not exchange energy as heat with its surroundings. Equations (3.32) and (3.33) serve as definitions of heat capacity (we shall revisit this concept later in the course). The subscript S represents another thermodynamic variable that we shall discuss later.

3.5.1 Perfect and Ideal Gas EOS

We can conceive a system of particles in which the particles have no volume and do not interact with each other. Using kinetic theory, it is possible to derive an EOS for such a system. Such a system of particles is an example of the perfect or ideal gas. A perfect gas is a special case of an ideal gas; however, we use the general term, ideal gas, throughout this text. The mathematical definition (as derived) of an ideal gas is

$$PV^{ig} = RT \tag{3.36}$$

with the boundary condition

$$U^{ig} = U(T) \tag{3.37}$$

in which the superscript *ig* indicates ideal gas, P is the absolute pressure, V^{ig} is the ideal gas molar volume, T is the absolute temperature, R is the universal gas constant, and U^{ig} is the internal energy of the ideal gas. The ideal gas is a hypothetical construct, and ideal gases do not exist in nature, but the ideal gas is a valuable reference system, and *real fluids would become ideal gases at zero pressure or density* (in essence, it would become a single molecule contained in an infinite volume).

Why do we write two equations, Eqs. (3.36) and (3.37), for the ideal gas? Equation (3.36) is the actual mathematical representation, but Eq. (3.37) is an essential boundary condition for an ideal gas: *the internal energy of an ideal gas is only a function of temperature*. Thus, both expressions are necessary to define completely the ideal gas.

If the internal energy is independent of pressure and volume, then the enthalpy of an ideal gas is also

$$H^{ig} = U^{ig} + PV^{ig} = U^{ig} + RT \Rightarrow H^{ig}(T) \tag{3.38}$$

Now, we should distinguish between perfect and ideal gas. From the kinetic theory of gases, molecules of a perfect gas do not have internal energy levels. Thus, for a perfect gas, Eq. (3.37) would become U^{ig} = constant; in other words, the internal energy of a perfect gas is not a function of temperature, while the internal energy of an ideal gas can be a function of T.

At this point, we should define a new property for the ideal gas: the heat capacity. This property essentially allows us to evaluate the change in internal energy or enthalpy for the ideal gas as temperature changes:

$$C_V^{ig} = \frac{dU^{ig}}{dT} \text{ along a path of constant volume} \tag{3.39}$$

$$C_P^{ig} = \frac{dH^{ig}}{dT} \text{ along a path of constant pressure} \tag{3.40}$$

Differentiation of Eq. (3.38) with respect to temperature provides a useful and interesting result:

$$\frac{dH^{ig}}{dT} = \frac{dU^{ig}}{dT} + R \text{ or } C_P^{ig} = C_V^{ig} + R \tag{3.41}$$

For an ideal gas, C_V^{ig} and C_P^{ig} are functions of temperature only, while for a perfect gas, they are constants.

3.5.2 Reversible Processes Involving Ideal Gases in Closed Systems

Because we shall use the ideal gas as a hypothetical reference state in the rigorous solution of many problems, it is necessary to develop a portfolio of ideal gas expressions. All these expressions derive from the first law of thermodynamics:

$$\Delta\left(U + E_P + E_K\right)_{sys} = \sum Q + \sum W + \sum\left(H + E_P + E_K\right)_{mt} - (\delta m/M)c^2 \tag{3.42}$$

Because we intend to deal with stationary, closed systems that do not contain nuclear reactions, Eq. (3.42) becomes

$$\Delta U_{sys} = \sum Q_{rev} + \sum W_{rev} \tag{3.43}$$

In addition, let us use systems that can exhibit only energy transfer as work via expansion:

$$W_{rev} = -\int PdV \tag{3.44}$$

This equation represents reversible energy transfer as work. Irreversible energy transfer as work requires more discussion, which comes later in the book.

3.5.2.1 Constant Volume (Isochoric) Process

Assume that a closed system is not insulated, and we transfer an amount of energy as heat across the boundaries. If the process is isochoric, no mechanism exists for energy transfer as work. Thus, Eq. (3.43) becomes

$$\Delta U^{ig} = \sum Q_{rev} = \left({}_1Q_2\right)_{rev} \tag{3.45}$$

The term $({}_1Q_2)_{rev}$ denotes the energy that crosses the boundaries of the system as heat as the system progresses reversibly from state 1 to state 2. Recall, the expression dQ is meaningless! So, in this case, the energy transferred as heat is the change in the internal energy of the ideal gas. If we reduce the energy transferred as heat to an amount comparable in magnitude to a differential amount, this equation becomes

$$dU^{ig} = đQ_{rev} \tag{3.46}$$

and the expression $đQ_{rev}$ represents the energy transferred as heat comparable in magnitude to a differential amount. This is not a differential, but it has an integral:

$$\int_1^2 đQ_{rev} = \left({}_1Q_2\right)_{rev} \neq \Delta Q_{rev} \tag{3.47}$$

If we substitute Eq. (3.47) into Eq. (3.46), we obtain

$$\left({}_1Q_2\right)_{rev} = \int_1^2 dU^{ig} = \int_1^2 C_V^{ig} dT = U_2^{ig} - U_1^{ig} \tag{3.48}$$

Also, we can substitute Eq. (3.41) into Eq. (3.48) to obtain the enthalpy change of the ideal gas in this process

$$\left({}_1Q_2\right)_{rev} = \int_1^2 dU^{ig} = \int_1^2 dH^{ig} - R\int_1^2 dT = H_2^{ig} - H_1^{ig} - R\left(T_2 - T_1\right) \tag{3.49}$$

This equation is valid for any ideal gas process because it considers the internal energy and the enthalpy only as functions of temperature and uses the general definition of heat capacity.

3.5.2.2 Constant Pressure (Isobaric) Process

In this case, Eq. (3.43) becomes

$$\Delta U_{sys}^{ig} = \sum Q_{rev} + \sum W_{rev} = \left({}_1Q_2\right)_{rev} + \left({}_1W_2\right)_{rev}$$

with

$$\left({}_1W_2\right)_{rev} = -\int_1^2 PdV^{ig} = -P\left(V_2^{ig} - V_1^{ig}\right) = -R\left(T_2 - T_1\right) \tag{3.50}$$

because the pressure is constant. Thus,

$$\left({}_1Q_2\right)_{rev} = U_2^{ig} - U_1^{ig} + R\left(T_2 - T_1\right) = H_2^{ig} - H_1^{ig} = \int_1^2 C_P^{ig} dT \tag{3.51}$$

3.5.2.3 Constant Temperature (Isothermal) Process

Equation (3.43) for a process that changes the state of the system from 1 to 2 is

$$\Delta U^{ig} = ({}_1Q_2)_{rev} + ({}_1W_2)_{rev} \tag{3.52}$$

Recalling that U^{ig} is a function only of temperature, if the temperature is constant, $\Delta U^{ig} = 0$ and

$$({}_1Q_2)_{rev} + ({}_1W_2)_{rev} = 0$$

then using the ideal gas equation with (3.52)

$$\begin{aligned}({}_1Q_2)_{rev} &= \int_1^2 PdV^{ig} = RT\int_1^2 \frac{dV^{ig}}{V^{ig}} = RT\ln\left[\frac{V_2^{ig}}{V_1^{ig}}\right] \\ &= \ln\left(\frac{RT/P_2}{RT/P_1}\right) = RT\ln\left(\frac{P_1}{P_2}\right) = -({}_1W_2)_{rev}\end{aligned} \tag{3.53}$$

Another way to obtain the pressure explicit form in Eq. (3.52) is to note

$$V^{ig} = RT/P$$

$$dV^{ig} = -RT\frac{dP}{P^2}$$

$$({}_1Q_2)_{rev} = \int_1^2 PdV^{ig} = -RT\int_1^2 P\frac{dP}{P^2} = RT\ln\left[\frac{P_1}{P_2}\right] = -({}_1W_2)_{rev} \tag{3.54}$$

Because the process is isothermal, the enthalpy change also is zero: $\Delta H^{ig} = 0$.

3.5.2.4 Adiabatic Process

In an adiabatic process, the energy transferred as heat is zero, $({}_1Q_2)_{rev} = 0$. Therefore, Eq. (3.43) reduces to

$$\Delta U^{ig} = ({}_1W_2)_{rev} \tag{3.55}$$

or in the differential form

$$dU^{ig} = đ({}_1W_2)_{rev} \tag{3.56}$$

Now, using Eq. (3.39) and the differential form for reversible work

$$C_V^{ig}dT = -PdV^{ig} \tag{3.57}$$

and inserting the ideal gas EOS

$$C_V^{ig}dT = -RT\frac{dV^{ig}}{V^{ig}} \tag{3.58}$$

which becomes upon rearrangement

$$\frac{dT}{T} = -\frac{R}{C_V^{ig}}\frac{dV^{ig}}{V^{ig}} \tag{3.59}$$

Substituting Eq. (3.41) into this equation provides

$$\frac{dT}{T} = -\frac{C_P^{ig} - C_V^{ig}}{C_V^{ig}}\frac{dV^{ig}}{V^{ig}} = -\left[\frac{C_P^{ig}}{C_V^{ig}} - 1\right]\frac{dV^{ig}}{V^{ig}} \tag{3.60}$$

and defining a heat capacity ratio

$$\gamma^{ig} \equiv \frac{C_P^{ig}}{C_V^{ig}} \tag{3.61}$$

Substituting this expression into Eq. (3.60) and integrating:

$$\int_1^2 \frac{dT}{T} = -\left[\gamma^{ig} - 1\right] \int_1^2 \frac{dV^{ig}}{V^{ig}}$$

$$\ln\left(T_2/T_1\right) = -\left[\gamma^{ig} - 1\right] \ln\left(V_2^{ig}/V_1^{ig}\right)$$

which produces a relationship between temperature and volume for an adiabatic ideal gas process:

$$\frac{T_2}{T_1} = \left(\frac{V_1^{ig}}{V_2^{ig}}\right)^{\gamma^{ig}-1} \tag{3.62}$$

Substituting the ideal gas EOS

$$\frac{T_2}{T_1} = \left(\frac{V_1^{ig}}{V_2^{ig}}\right)^{\gamma^{ig}-1} = \left(\frac{RT_1}{P_1}\frac{P_2}{RT_2}\right)^{\gamma^{ig}-1} = \left(\frac{T_2}{T_1}\right)^{1-\gamma^{ig}} \left(\frac{P_2}{P_1}\right)^{\gamma^{ig}-1} \tag{3.63}$$

which reduces to

$$\frac{T_2}{T_1} = \left(\frac{P_2}{P_1}\right)^{\frac{\gamma^{ig}-1}{\gamma^{ig}}} \tag{3.64}$$

Finally, we can equate Eqs. (3.62) and (3.64):

$$\left(\frac{V_1^{ig}}{V_2^{ig}}\right)^{\gamma^{ig}-1} = \left(\frac{P_2}{P_1}\right)^{\frac{\gamma^{ig}-1}{\gamma^{ig}}}$$

which reduces to

$$\left(\frac{V_1^{ig}}{V_2^{ig}}\right) = \left(\frac{P_2}{P_1}\right)^{\frac{1}{\gamma^{ig}}} \tag{3.65}$$

Another commonly used form of Eq. (3.65) is

$$P_1 V_1^{\gamma} = P_2 V_2^{\gamma} = PV^{\gamma} = \text{constant} \tag{3.66}$$

If the gas is perfect rather than ideal, the heat capacity at constant volume is only a function of γ^{ig}:

$$\gamma^{ig} \equiv \frac{C_P^{ig}}{C_V^{ig}} = \frac{R + C_V^{ig}}{C_V^{ig}} = \frac{R}{C_V^{ig}} + 1 \tag{3.67}$$

then,

$$\left({}_1W_2\right)_{rev} = \int_1^2 dU^{ig} = \int_1^2 C_V^{ig} dT = \frac{R}{\gamma^{ig} - 1} \int_1^2 dT = \frac{R\left(T_2 - T_1\right)}{\gamma^{ig} - 1} = \frac{RT_1}{\gamma^{ig} - 1}\left[\frac{T_2}{T_1} - 1\right] \tag{3.68}$$

but if we want Eq. (3.68) in terms of a P ratio rather than T ratio, then we can use Eq. (3.64) to obtain

$$\left({}_1W_2\right)_{rev} = \frac{RT_1}{\gamma^{ig}-1}\left[\left(\frac{P_2}{P_1}\right)^{\frac{\gamma^{ig}-1}{\gamma^{ig}}}-1\right] = \frac{P_1V_1^{ig}}{\gamma^{ig}-1}\left[\left(\frac{P_2}{P_1}\right)^{\frac{\gamma^{ig}-1}{\gamma^{ig}}}-1\right] \tag{3.69}$$

or in terms of volume ratios

$$\left({}_1W_2\right)_{rev} = \frac{RT_1}{\gamma^{ig}-1}\left[\left(\frac{V_1^{ig}}{V_2^{ig}}\right)^{\gamma^{ig}-1}-1\right] = \frac{P_1V_1^{ig}}{\gamma^{ig}-1}\left[\left(\frac{V_1^{ig}}{V_2^{ig}}\right)^{\gamma^{ig}-1}-1\right] \tag{3.70}$$

As a rule of thumb, we can approximate γ^{ig} as

$$\begin{aligned}\gamma^{ig} &= 1.67 \text{ monoatomic gases}\\ &= 1.40 \text{ diatomic gases}\\ &= 1.30 \text{ simple polyatomic gases}\end{aligned}$$

3.5.2.5 Polytropic Processes

Polytropic processes are idealized processes that "obey" the equation:

$$PV^{\delta} = \text{constant} \tag{3.71}$$

Such processes need not use ideal gases, although the similarity to Eq. (3.66) is obvious. In the case of ideal gas processes,

Process	δ
Isochoric	$-\infty$
Isobaric	0
Isothermal	1
Adiabatic	γ

All the processes treated so far are reversible. Working with an irreversible process requires knowing the efficiency of the process. For example, for the calculation of work, one should calculate the reversible work and then use a pattern efficiency to calculate the irreversible work.

Example 3.5

An ideal gas expands from an initial condition of 1.0 MPa and 300 K to a final state of 0.1 MPa and 300 K. This process could take any of the following three mechanically reversible paths (among others):

A. Isothermal expansion
B. Heating at a constant pressure, followed by cooling at a constant volume
C. Heating at a constant volume, followed by an adiabatic expansion

The heat capacities of this ideal gas are constant: $C_V^{ig} = (5/2)\,R$ and $C_P^{ig} = (7/2)\,R$. Calculate Q_{rev}, W_{rev}, ΔU^{ig}, and ΔH^{ig} for each path.

Solution

First, we draw a PV sketch of the different paths

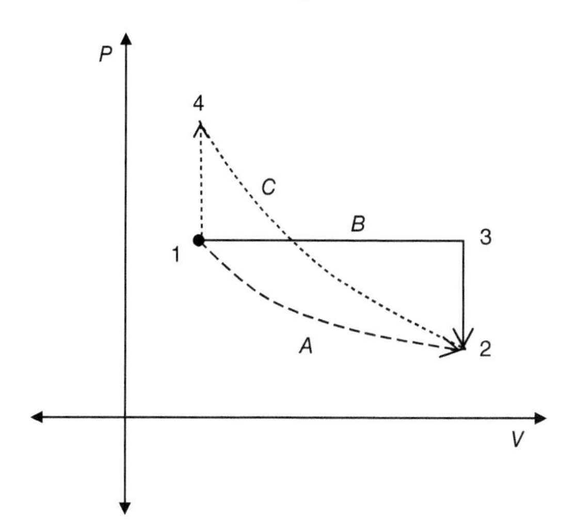

Path A (isothermal)

The working equation to solve this example is Eq. (3.52):

$$\Delta U^{ig} = U_2 - U_1 = \left({}_1Q_2\right)_{rev} + \left({}_1W_2\right)_{rev}$$

The process proceeds from 1 to 2 and $T_1 = T_2$. The internal energy and the enthalpy for an ideal gas depend only upon the temperature. Also, they are state functions; therefore, they are independent of the path. Thus, the change of the internal energy and enthalpy is zero for path A:

$$\Delta U^{ig} \equiv U_2^{ig} - U_1^{ig} = \Delta H^{ig} \equiv H_2^{ig} - H_1^{ig} = 0$$

These answers apply not only for Path A but also for Path B and Path C because all ultimately pass from point 1 to point 2, and the internal energy and enthalpy are state functions. Now, we can calculate the work and heat for Paths A, B, and C. First, we calculate the volumes at 1 and 2 (Note: rigorously, we should always use the most recent value of R, currently $R = 8.3144621(75)$ J/(mol K), but we can use 8.3145 in this course):

$$V_1^{ig} = RT/P_1 = \left[8.3415\ \left(\text{Pa m}^3/\,(\text{mol K})\right) \times 300\ \text{K}\right] / \left(10^6\ \text{Pa}\right)$$
$$= 0.0024944\ \text{m}^3/\text{mol} = V_4^{ig}$$

$$V_2^{ig} = RT/P_2 = \left[8.3415\left(\text{Pa m}^3/\,(\text{mol K})\right) \times 300\ \text{K}\right] / \left(10^5\ \text{Pa}\right)$$
$$= 0.024944\ \text{m}^3/\text{mol} = V_3^{ig}$$

Equation (3.51) applies to Path A:

$$\left({}_1Q_2\right)_{rev} = -\left({}_1W_2\right)_{rev} = RT\ln\left(P_1/P_2\right) = 8.3145\,(300)\ln\left(1/0.1\right) = 5743.5\ \text{J/mol}$$

Path B (isobaric heating followed by isochoric cooling)
The isobaric path is $1 \rightarrow 3$, so

$$\Delta U^{ig} = U_3^{ig} - U_1^{ig} = \int_1^3 C_V^{ig} dT = C_V^{ig}(T_3 - T_1) = ({}_1Q_3)_{rev} + ({}_1W_3)_{rev}$$
$$= ({}_1Q_3)_{rev} - \int_1^3 P dV = ({}_1Q_3)_{rev} - P_1(V_3 - V_1) = ({}_1Q_3)_{rev} - R(T_3 - T_1)$$

Rearranging

$$({}_1Q_3)_{rev} = C_V^{ig}(T_3 - T_1) + R(T_3 - T_1) = \left(C_V^{ig} + R\right)(T_3 - T_1) = C_P^{ig}(T_3 - T_1) = \Delta H^{ig}$$

We require T_3:

$$T_3 = P_3V_3/R = \left(10^6 \text{ Pa}\right)\left(0.024944 \text{ m}^3/\text{mol}\right) / \left(8.3145 \text{ Pa m}^3/(\text{mol K})\right) = 3000 \text{ K}$$

then

$$({}_1Q_3)_{rev} = C_P^{ig}(T_3 - T_1) = (7/2)(8.3145)(3000 - 300) = 78\,572 \text{ Pa m}^3/\text{mol}$$
$$= 78\,572 \left(\left(\text{N/m}^2\right) \text{m}^3/\text{mol}\right) = 78\,572 \text{ N m/mol} = 78\,572 \text{ J/mol} = \Delta H^{ig}$$

and

$$\Delta U^{ig} = U_3^{ig} - U_1^{ig} = \int_1^3 C_V^{ig} dT = (5/2)(8.3145)(3000 - 300) = 56\,123 \text{ J/mol}$$
$$({}_1W_3)_{rev} = U_3^{ig} - U_1^{ig} - ({}_1Q_3)_{rev} = 56\,123 - 78\,572 \text{ J/mol} = -22\,449 \text{ J/mol}$$

The isochoric path is $3 \rightarrow 2$, so

$$({}_3W_2)_{rev} = 0$$

And

$$\Delta U^{ig} = U_2^{ig} - U_3^{ig} = \int_3^2 C_V^{ig} dT = (5/2)(8.3145)(300 - 3000) = ({}_2Q_3) = -56\,123 \text{ J/mol}$$

and

$$\Delta H^{ig} = H_2^{ig} - H_3^{ig} = \int_3^2 C_P^{ig} dT = (7/2)(8.3145)(300 - 3000) = -78\,572 \text{ J/mol}$$

Thus, going from point 1 to point 2 via Path B

$$\Delta U^{ig} = 56\,123 - 56\,123 \text{ J/mol} = 0$$
$$\Delta H^{ig} = H_3^{ig} - H_1^{ig} + H_2^{ig} - H_3^{ig} = 78\,572 - 78\,572 \text{ J/mol} = 0 = H_2^{ig} - H_1^{ig}$$
$$Q_{rev} = ({}_1Q_3)_{rev} + ({}_3Q_2)_{rev} = 78\,572 - 56\,123 \text{ J/mol} = 22\,449 \text{ J/mol}$$
$$W_{rev} = ({}_1W_3)_{rev} + ({}_3W_2)_{rev} = -22\,449 + 0 \text{ J/mol} = -22\,449 \text{ J/mol}$$

Note that $Q_{rev} + W_{rev} = \Delta U^{ig} = 0$ because the overall path is still $1 \rightarrow 2$, which is isothermal.

Path C (isochoric heating followed by adiabatic cooling)
The isochoric path is $1 \rightarrow 4$, so

$$({}_1W_4)_{rev} = 0 \text{ J/mol}$$

We need T_4, and we can use Eq. (3.62)

$$\frac{T_4}{T_2} = \left(\frac{V_2^{ig}}{V_4^{ig}}\right)^{\gamma^{ig}-1}$$

$$T_4 = T_2\left[V_2^{ig}/\left(V_4^{ig} = V_1^{ig}\right)\right]^{\gamma^{ig}-1}$$
$$= 300\left[\left(0.024944\ \text{m}^3/\text{mol}\right)/\left(0.0024944\ \text{m}^3/\text{mol}\right)\right]^{7/5-1} = 753.6\ \text{K}$$

$$\Delta U^{ig} = U_4^{ig} - U_1^{ig} = \int_1^4 C_V^{ig} dT = (5/2)(8.3145)(753.6 - 300) = \left({}_1Q_4\right)_{rev} = 9429\ \text{J/mol}$$

$$\Delta H^{ig} = H_4^{ig} - H_1^{ig} = \int_1^4 C_P^{ig} dT = (7/2)(8.3145)(753.6 - 300) = 13\,200\ \text{J/mol}$$

The adiabatic path is $4 \to 2$, so

$$\left({}_4Q_2\right)_{rev} = 0\ \text{J/mol}$$

$$\left({}_4W_2\right)_{rev} = U_2^{ig} - U_4^{ig} = \int_4^2 C_V^{ig} dT = (5/2)(8.3145)(300 - 753.6) = -9429\ \text{J/mol}$$

$$\Delta H^{ig} = H_2^{ig} - H_4^{ig} = \int_4^2 C_P^{ig} dT = (7/2)(8.3145)(300 - 753.6) = -13\,200\ \text{J/mol}$$

Thus, going from point 1 to point 2 via Path C

$$\Delta U^{ig} = 9429 - 9429\ \text{J/mol} = 0$$

$$\Delta H^{ig} = 13\,200 - 13\,200\ \text{J/mol} = 0 = H_2^{ig} - H_1^{ig}$$

$$Q_{rev} = \left({}_1Q_4\right)_{rev} + \left({}_4Q_2\right)_{rev} = -150\,409 + 0\ \text{J/mol} = -150\,409\ \text{J/mol}$$

$$W_{rev} = \left({}_1W_4\right)_{rev} + \left({}_4W_2\right)_{rev} = 0 + 150\,409\ \text{J/mol} = 150\,409\ \text{J/mol}$$

Again, the heat and the work cancel because they must equal the change in internal energy.

3.5.3 Virial Equation of State

Using the techniques of statistical mechanics, it is possible to derive an EOS for real fluids. EOS that describes the behavior of real fluids generally employs a new variable called compressibility factor or compression factor or Z-factor or real-gas factor. Regardless of the name one prefers, the definition of this variable is

$$Z \equiv \frac{PV}{RT} = \frac{P}{RT\rho} \qquad (3.72)$$

For an ideal gas, the Z-factor is unity, $Z = 1$. The Z-factor is the ratio of the real fluid molar volume to the ideal gas molar volume because $V^{ig} = RT/P$, or it is the ratio of the ideal gas molar density to the real fluid molar density because $\rho^{ig} = P/RT$

$$Z \equiv \frac{V}{V^{ig}} = \frac{\rho^{ig}}{\rho} \qquad (3.73)$$

In the *PVT* region that contains most practical applications, the value of Z is less than unity. However, in regions of high pressure, volume, or temperature, Z can be greater than unity.

In 1885, Thiesen [1] proposed an empirical polynomial function of molar volume or density for the Z-factor at a given temperature. Then in 1901, Kamerlingh Onnes further extended the work of Thiesen and popularized the use of the equation. This equation became known as the Leiden equation because Kamerlingh Onnes [2] was a professor at the University of Leiden. The equation is only valid in the single-phase region, and it cannot represent the *PVT* behavior in multiphase regions

$$Z = 1 + B\rho + C\rho^2 + D\rho^3 + \cdots = 1 + B/V + C/V^2 + D/V^3 + \cdots \tag{3.74}$$

At about the same time, researchers at the University of Berlin proposed an empirical polynomial function of pressure for the Z-factor at a given temperature

$$Z = 1 + B'P + C'P^2 + D'P^3 + \cdots \tag{3.75}$$

In 1927, Ursell [3] provided the fundamental, statistical–mechanical basis for both these expansions. His work showed that Z should be an infinite series in either density or pressure, and the name for either series became the *virial* equation. The name *virial* comes from the Latin word *vis* that stands for *power* (as in a power series). It is also more convenient to use a different notation for the coefficients in the virial equation: B_k with $k = 2, 3, 4\ldots$. Thus, in the density series, B_2 is the second virial coefficient, B_3 is the third virial coefficient, etc. The density series is the more useful form (we shall see why later), so its coefficients are the virial coefficients, and we define the coefficients in the pressure series in terms of those in the density series. Virial coefficients are functions of temperature for pure substances and functions of temperature and composition for mixtures.

Thus, the virial equation in summation notation becomes

$$Z = 1 + \sum_{k=1}^{\infty} B_{k+1}\rho^k = 1 + \sum_{k=1}^{\infty} B'_{k+1}P^k \tag{3.76}$$

Note that when the density or pressure approaches zero, Z approaches unity, which is the ideal gas value. Also, when the density approaches zero

$$B_2 = \lim_{\rho\to 0}\left[\frac{\partial Z}{\partial \rho}\right]_T,\quad B_3 = \lim_{\rho\to 0}\frac{1}{2}\left[\frac{\partial^2 Z}{\partial \rho^2}\right]_T,\ \ldots,\ B_{k+1} = \lim_{\rho\to 0}\frac{1}{(k-1)!}\left[\frac{\partial^{k-1} Z}{\partial \rho^{k-1}}\right]_T \tag{3.77}$$

So, the virial coefficients are ideal gas properties, although they represent real fluids. Also, using Eq. (3.76), it is possible to write

$$B'_2 = B/RT,\ B'_3 = \left(B_3 - B_2^2\right)/(RT)^2,\ B'_4 = \left(B_4 - 3B_2B_3 + 2B_2^3\right)/(RT)^3,\ \text{etc.} \tag{3.78}$$

Example 3.6

Find the relationship between B_2 and B'_2.

Solution

Many ways exist to find these relationships. We illustrate a procedure using Eq. (3.77)

$$B_2 = \lim_{\rho \to 0} \left[\frac{\partial Z}{\partial \rho} \right]_T \quad \text{and} \quad B_2' = \lim_{P \to 0} \left[\frac{\partial Z}{\partial P} \right]_T$$

The relation between these derivatives is

$$\left(\frac{\partial Z}{\partial P} \right)_T = \left(\frac{\partial Z}{\partial \rho} \right)_T \Big/ \left(\frac{\partial P}{\partial \rho} \right)_T$$

but $P = ZRT\rho$, so

$$\left(\frac{\partial Z}{\partial P} \right)_T = \frac{\left(\frac{\partial Z}{\partial \rho} \right)_T}{ZRT + (P/Z) \left(\frac{\partial Z}{\partial \rho} \right)_T}$$

and in the limit as P and ρ approach zero, Z approaches unity, so

$$\left(\frac{\partial Z}{\partial P} \right)_T = \frac{1}{RT} \left(\frac{\partial Z}{\partial \rho} \right)_T \quad \text{or} \quad B_2' = \frac{B_2}{RT}$$

Similar derivations are possible for the rest of the virial coefficients.

Because the virial equation has a firm basis in statistical mechanics, many theories use this equation. Statistical mechanics provides physical significance for the virial coefficients. For example, the term $B_2\rho$ reflects interactions between pairs of molecules, and the term $B_3\rho^2$ reflects three-body interactions.

Figure 3.5 is a schematic representation of the second virial coefficient as a function of temperature. At low temperatures, B_2 is negative because the long-range attractive forces tend to reduce the pressure of a gas below the ideal gas pressure. As the temperature increases, the short-range repulsive forces become more important, and the second virial coefficient becomes less negative. At the *Boyle temperature* (approximately 2.5 times the critical temperature), the attractive and repulsive forces balance and the second virial coefficient becomes zero. At higher temperatures, the molecular collisions increase, and the repulsive forces dominate causing B_2 to be positive. The second virial coefficient increases slowly and passes through a maximum for substances that remain intact at such high temperatures. The maximum temperature is observable for substances with low critical temperatures such as hydrogen, helium, and neon.

The temperature dependence of the third virial coefficient is similar to B_2. The maximum occurs at a lower temperature around (0.90–0.95) times the critical temperature (Figure 3.6). At temperatures lower than $0.86T_C$, experimental determination of the third virial coefficient is subject to large uncertainties. For example, at $0.75T_C$, the contribution of $B_2\rho$ to the compressibility factor of saturated vapor is about −0.15, while the contribution of $B_3\rho^2$ is about 0.002, well within the error limit of $B_2\rho$.

In engineering practice, a truncated form of the virial equation is useful in limited regions of the *PVT* surface. For pressures up to one half of the critical pressure, a virial equation truncated at the second virial coefficient is often adequate. A virial

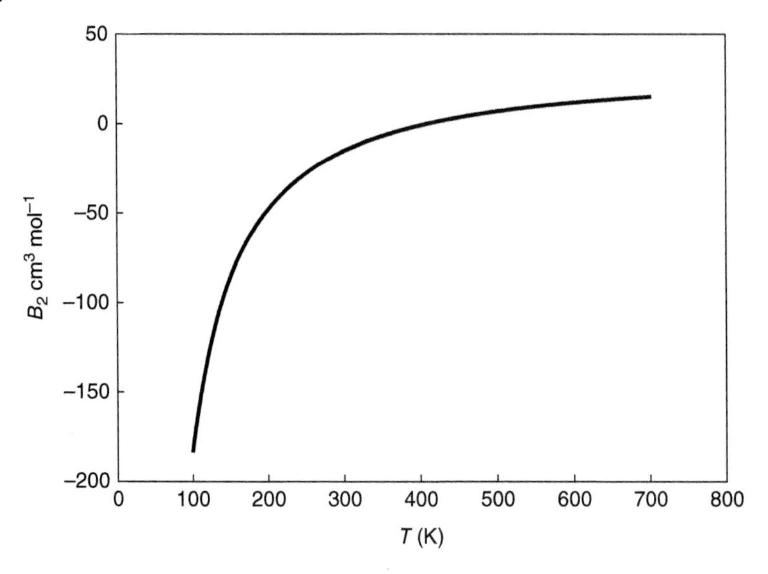

Figure 3.5 Second virial coefficient of argon.

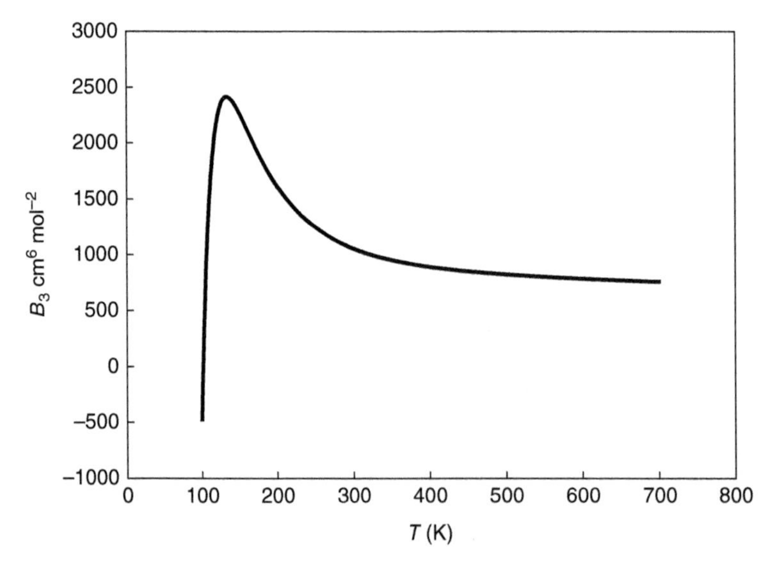

Figure 3.6 Third virial coefficient of argon.

equation truncated at the third virial coefficient may be sufficient for pressures up to two thirds of the critical pressure.

The virial equation is used in many engineering applications after truncation at the second virial coefficient. One should be aware that a truncated second virial equation in the Leiden form corresponds to an infinite series in the Berlin form, that is

$$1 + \frac{B_2}{V} = 1 + \frac{B_2 P}{RT} - \left(\frac{B_2 P}{RT}\right)^2 + \left(\frac{B_2 P}{RT}\right)^3 - \cdots + \cdots$$

Therefore, the Leiden form of the virial equation truncated at the second virial is usually preferable to the equivalent equation in the Berlin form for calculations at low pressures. At higher pressures, a curvature appears that requires additional terms in the equation. Note: at high temperatures, the Berlin form of the equation can have fewer terms because the isotherms become straighter than those of the Leiden form.

Figure 3.7 illustrates why the Z vs ρ form is preferable. The isotherms all have similar shapes, they do not cross (at least up to the temperatures contained on the plot), and the shapes are all similar. On both plots, the dotted line represents the highest temperature on the plot (higher temperatures would begin to fall back down on the plot), making it more difficult to interpret the plot, until they reach a horizontal line at $Z = 1$ the ideal gas limit. The dotted line is the Boyle temperature. This is obvious from Eq. (3.77), which says that the slope of a Z vs ρ plot is the second virial coefficient at the zero-density limit. The dotted line has the maximum slope of any isotherm at zero density, so it has the maximum value for the virial coefficient, which we have seen defines at the Boyle temperature.

3.5.3.1 Correlations for the Second and Third Virial Coefficient

Many correlations exist for the second and third virial coefficients. Pitzer and Curl [4] (1957) propose a dimensionless second virial coefficient as a linear function of the acentric factor (a term that reflects the non-spherical nature of a molecule)

$$\frac{B_2 P_C}{RT_C} = B_2^{(0)} + \omega B_2^{(1)} \tag{3.79}$$

with

$$B_2^{(0)} = 0.1445 - 0.33/T_R - 0.1385/T_R^2 - 0.0121/T_R^3 - 0.000607/T_R^8$$

$$B_2^{(1)} = 0.073 + 0.46/T_R - 0.50/T_R^2 - 0.097/T_R^3 - 0.0073/T_R^8$$

$$\omega = -\log\left(P^{sat}/P_C\right)_{T/T_C=0.7} - 1 \tag{3.80}$$

in which $B_2^{(0)}$ represents a reduced second virial coefficient for spherical molecules ($\omega = 0$), while $B_2^{(1)}$ is a correction term to account for non-spherical molecules. The variables in this correlation are the critical temperature, critical pressure, and a dimensionless temperature $T_R = T/T_C$ referred to as reduced temperature. The acentric factor appears in tables for many substances, or you can calculate it using a vapor pressure at $T_R = 0.7$ with Eq. (3.80).

Pitzer et al. [5] introduced the acentric factor in 1955 to characterize substances. For many substances, the acentric factor is close to 0.1. Abbot [6] suggests a simpler form for the Pitzer equation with

$$B_2^{(0)} = 0.083 - \frac{0.422}{T_R^{1.6}}$$

$$B_2^{(1)} = 0.139 - \frac{0.172}{T_R^{4.2}}$$

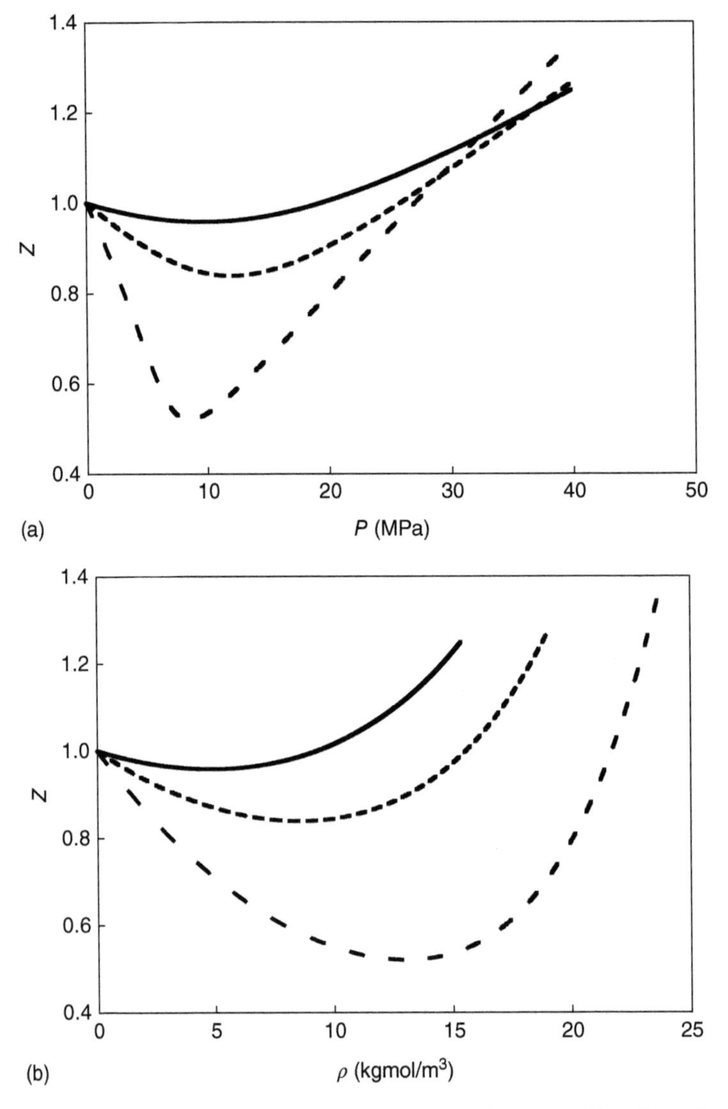

Figure 3.7 Compressibility factor as a function of pressure (a) and as a function of density (b).

An extension of the above equations is the Tsonopolous [7] correlation. He expresses the reduced second virial coefficient as

$$\frac{B_2 P_C}{RT_C} = B_2^{(0)} + \omega B_2^{(1)} + B_2^{(2)} \tag{3.81}$$

in which

$$B_2^{(0)} = 0.1445 - 0.33/T_R - 0.1385/T_R^2 - 0.0121/T_R^3 - 0.000607/T_R^8$$

$$B_2^{(1)} = 0.0637 + 0.331/T_R^2 - 0.423/T_R^3 - 0.008/T_R^8$$

$$B_2^{(2)} = a/T_R^6 - b/T_R^8$$

Table 3.2 Expressions for a and b in the Tsonopolous correlation.

Compound class	a	b
1-Alcohols	0.0878	$0.0098 + 0.0006957\mu_R$
Water	−0.0109	0
Ketones, aldehydes, and alkyl ethers	$-2.14 \times 10^{-4}\mu_R - 4.308 \times 10^{-21}\mu_R^8$	0
Nitriles and esters	$-2.112 \times 10^{-4}\mu_R - 3.877 \times 10^{-21}\mu_R^8$	0
Phenol	−0.0136	0
Haloalkanes	$-2.188 \times 10^{-11}\mu_R^4 - 7.831 \times 10^{-21}\mu_R^8$	

and ω is the acentric factor. In this equation, $B_2^{(2)}$ accounts for polarity in the molecules. Table 3.2 contains expressions for a and b as functions of the reduced dipole moment

$$\mu_R = \frac{10^5\mu^2 P_C}{T_C^2} \tag{3.82}$$

in which μ is the dipole moment in debye, P_C is the critical pressure in atmosphere, and T_C is the critical temperature in kelvin.

Iglesias-Silva and Hall [8] have proposed another correlation that uses the Boyle temperature:

$$B/b_0 = \left(\frac{T_{Boyle}}{T}\right)^{0.2}\left[1 - \left(\frac{T_{Boyle}}{T}\right)^{0.8}\right]\left[\frac{B_C}{b_0\left[(T_{Boyle}/T_C)^{0.2} - (T_{Boyle}/T_C)\right]}\right]^{(T_C/T)^n} \tag{3.83}$$

in which T_{Boyle} is the Boyle temperature and

$$B_C/V_C = -1.1747 - 0.3668\omega - 0.00061\mu_R$$

$$n = 1.4187 + 1.2058\omega$$

$$b_0/V_C = 0.1368 - 0.4791\omega + 13.81\left(T_{Boyle}/T_C\right)^2 \exp\left[-1.95\left(T_{Boyle}/T_C\right)\right]$$

Equation (3.80) provides ω and Eq. (3.82) provides μ_R.

Example 3.7

Abusleme and Vera [9] measured the second virial coefficient of ethanol at 363.15 K as $-863\ \text{cm}^3/\text{mol}$. Compare their result to the values predicted by the above correlations.

Data

$T_C = 513.9$ K, $P_C = 6.148$ MPa, $T_{Boyle} = 1086.9$ K, $V_C = 166.92\ \text{cm}^3/\text{mol}$, $\omega = 0.644$, and $\mu = 1.6909$ D

Solution

The reduced temperature is $T_R = 363.15/513.9 = 0.7067$

For the Abbot correlation

$$B_2^{(0)} = 0.083 - \frac{0.422}{T_R^{1.6}} = 0.083 - \frac{0.422}{(0.7067)^{1.6}} = -0.6524$$

$$B_2^{(1)} = 0.139 - \frac{0.172}{T_R^{4.2}} = 0.139 - \frac{0.172}{(0.7067)^{4.2}} = -0.6002$$

then

$$\begin{aligned} B_2 &= \left(\frac{RT_C}{P_C}\right)\left(B_2^{(0)} + \omega B_2^{(1)}\right) \\ &= \left[\left(8.3415\ \text{MPa cm}^3/(\text{mol K})\right)(513.9\ \text{K})/6.148\ \text{MPa}\right] \\ &\quad [0.6524 + 0.664 \times (-0.6002)] \\ &= -730.4\ \text{cm}^3/\text{mol} \end{aligned}$$

Now, using the Tsonopolous correlation

$$\mu_R = \frac{10^5 \mu^2 P_C}{T_C^2} = \frac{10^5 (1.6909\ \text{D})^2 (61.48\ \text{bar})}{(513.9\ \text{K})^2}\left(\frac{1\ \text{atm}}{1.01325\ \text{bar}}\right) = 65.69$$

$$\frac{B_2 P_C}{RT_C} = B_2^{(0)} + \omega B_2^{(1)} + B_2^{(2)}$$

with

$$\begin{aligned} B_2^{(0)} &= 0.1445 - 0.33/0.7067 - 0.1385/(0.7067)^2 \\ &\quad -0.0121/(0.7067)^3 - 0.000607/(0.7067)^8 = -0.6438 \end{aligned}$$

$$B_2^{(1)} = 0.0637 + 0.331/(0.7067)^2 - 0.423/(0.7067)^3 - 0.008/(0.7067)^8 = -0.6006$$

and for ethanol

$$a = 0.0878$$

$$b = 0.00908 + 0.0006957\mu_R = 0.00908 + 0.0006957 \times 65.69 = 0.0548$$

Then,

$$B_2^{(2)} = a/T_R^6 - b/T_R^8 = \frac{0.0878}{(0.7067)^6} - \frac{0.0549}{(0.7067)^8} = -0.1776$$

and

$$\begin{aligned} B_2 &= \left(RT_C/P_C\right)\left[B_2^{(0)} + \omega B_2^{(1)} + B_2^{(2)}\right] \\ &= 695.0\,[-0.6438 + 0.644 \times (-0.6006) - 0.1776] = -839.7\ \text{cm}^3/\text{mol} \end{aligned}$$

The Tsonopolous correlation provides a reasonably good prediction of the second virial coefficient for polar molecules. Now, using the Iglesias-Silva correlation

$$\begin{aligned} B_C/V_C &= -1.1747 - 0.3668\omega - 0.00061\mu_R \\ &= -1.1747 - 0.3668\,(0.644) - 0.00061\,(65.69) = -1.451 \end{aligned}$$

$$B_C = -1.451 V_C = -1.451\,(166.92) = -242.20092\ \text{cm}^3/\text{mol}$$

$$n = 1.4187 + 1.2058\omega = 1.4187 + 1.2058\,(0.644) = 2.1952$$

$$T_{Boyle}/T_C = 1086.9/513.9\text{ K} = 2.115$$

$$b_0/V_C = 0.1368 - 0.4791\omega + 13.81\left(T_{Boyle}/T_C\right)^2 \exp\left[-1.95\left(T_{Boyle}/T_C\right)\right]$$
$$= 0.1368 - 0.4791\,(0.644) + 13.81(2.115)^2 \exp\left[-1.95\,(2.115)\right] = 0.8275$$

$$T_{Boyle}/T = 1086.9/363.15\ \text{ K} = 2.993;\ T_C/T = 513.9/363.15\ \text{ K} = 1.451$$

$$B_2/b_0 = \left(\frac{T_{Boyle}}{T}\right)^{0.2}\left[1 - \left(\frac{T_{Boyle}}{T}\right)^{0.8}\right]\left[\frac{B_C}{b_0\left[\left(T_{Boyle}/T_C\right)^{0.2} - \left(T_{Boyle}/T_C\right)\right]}\right]^{(T_C/T)^n}$$

$$= (2.993)^{0.2}\left[1 - (2.993)^{0.8}\right]\left[\frac{-242.20092\text{ cm}^3/\text{mol}}{\left(138.1263\text{ cm}^3/\text{mol}\right)\left[(2.115)^{0.2} - (2.115)\right]}\right]^{(1.415)^{2.1952}}$$

$$= -6.449$$

$$B_2 = -6.449b_0 = -6.449\,(138.1263) = -890.777\text{ cm}^3/\text{mol}$$

The Iglesias-Silva–Hall correlation, like the Tsonopolous correlation, provides a good prediction for the second virial coefficient.

Correlations for the third virial coefficient are rare, but those that exist have forms similar to those for the second virial correlations. De Santis and Grande [10] developed one of the more successful correlations:

$$B_3/V_C^2 = B_3^{(0)} + dB_3^{(1)} + d^2B_3^{(2)} \tag{3.84}$$

with

$$B_3^{(0)} = 0.1961/T_R^{1/4} + 0.3972/T_R^5 + \left(0.06684T_R^4 - 0.5428/T_R^6\right)\exp\left(-T_R^2\right)$$
$$B_3^{(0)} = \left(64.5/T_R^9\right)\left[1 - 2.085\exp\left(-T_R^2\right)\right]$$
$$B_3^{(2)} = 801.7/T_R^7$$

and

$$d = \frac{\omega\alpha N}{b}$$

Equation (3.84) requires as input parameters the Avogadro number, N, the acentric factor, ω, the dipole polarizability, α, and the molecular volume, b, calculated from atomic radii and bond distances. Of course, the dipole polarizability and the bond distances for most substances are not readily available, so Orbey and Vera [11] developed a correlation that uses the acentric factor and the critical temperature and pressure

$$B_3\left(\frac{P_C}{RT_C}\right)^2 = B_3^{(0)} + \omega B_3^{(1)} \tag{3.85}$$

with

$$B_3^{(0)} = 0.01407 + 0.02432/T_R^{2.8} - 0.00313/T_R^{10.5}$$

and

$$B_3^{(1)} = -0.02676 + 0.0177/T_R^{2.8} + 0.040/T_R^3 - 0.003/T_R^6 - 0.00228/T_R^{10.5}$$

The main advantage of the virial equation is that the virial coefficients can come from intermolecular potentials, such as the Lennard–Jones potential

$$B_2 = 2\pi N_A \int_0^{\infty} [1 - \exp\left(u_{ij}\ /kT\right] r^2 dr \tag{3.86}$$

Here, N_A and k are the Avogadro number and the Boltzmann constant, respectively, and u_{ij} is the intermolecular potential of a pair of molecules i–j. The intermolecular potential is a function of the distance between molecules, r. Also, the third virial coefficient can result from using pairwise additivity (interaction only between two molecules) and an intermolecular potential

$$B_3 = -(8/3)\pi^2 N_A^2 \int\int\int [f_{12}f_{13}f_{23}]\, r_{12}r_{13}r_{23}dr_{12}dr_{13}dr_{23} \tag{3.87}$$

in which f_{ij} is the Mayer function defined as

$$f_{ij} = \exp\left(-u_{ij}/k_B T\right) - 1 \tag{3.88}$$

Obviously, it is desirable to have experimental measurements, but construction of equipment to make the measurements often is not economical. Several techniques exist to obtain experimental second and third virial coefficients. Among the most important are:

- PVT measurements
  Volumetric measurements at low pressures (generally differential).
  Volumetric measurements at high pressures (Burnett, isochoric-Burnett, magnetic suspension densimeter, and vibrating tube densimeter).
  Gas density determination (pycnometer).

The virial equation used directly with the PVT measurements provides the second and third virial coefficients. If one plots $(Z-1)/\rho$ vs the density, the intercept and the slope of the linear part of the curve are the second and third virial coefficients, respectively. Figure 3.8 illustrates this plot.

- Experiments that involve B_2 and its derivatives:
  Joule–Thomson coefficient as $P \rightarrow 0$: $\left(\frac{dT}{dP}\right)_{H,P\rightarrow 0} = \frac{1}{C_P^0}\left(T\frac{dB_2}{dT} - B_2\right)$
  Sound velocity, $\dot{z}_S$, and its pressure dependence are:

$$\dot{z}_S = RT\gamma^{ig}/M + \left\{\frac{\gamma^{ig}}{M}\left[2B_2 + 2\left(\gamma^{ig} - 1\right)T\frac{dB_2}{dT} + \left(\gamma^{ig} - 1\right)^2 \frac{T^2}{\gamma^{ig}}\frac{d^2B_2}{dT^2}\right]\right\}P + \cdots$$

  Pressure dependence of C_P: $\left(\frac{\partial C_P}{\partial P}\right)_{T,P\rightarrow 0} = -T\frac{d^2B_2}{dT^2}$

The limited experimental measurements of virial coefficients have led to developed prediction methods, as shown above. The methods based upon statistical mechanics expressions have not been as successful when applied in practical applications (although major advances in this endeavor may be imminent).

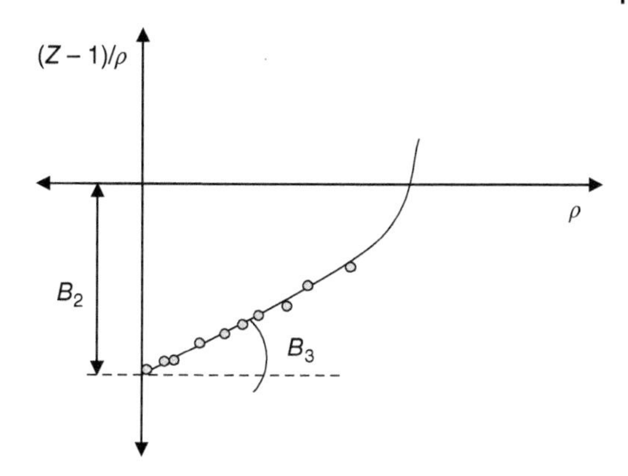

Figure 3.8 Experimental PVT measurements and calculation of the second virial coefficients.

3.5.4 Other Special Equations

3.5.4.1 Tait Equation

This equation [12] can correlate the density of liquids. It is an isothermal equation

$$1 - \rho_0/\rho = C \log \left[\frac{P + B}{P_0 + B} \right] \tag{3.89}$$

in which ρ_0 and P_0 are the reference density and pressure and B and C are constants (but not virial coefficients). Commonly the reference density is the atmospheric or saturation value. This equation can represent the volume or density.

3.5.4.2 Rackett Equation

One of the simplest equations to calculate the liquid density at saturation is the Rackett [13] equation:

$$V^{sat}/V_C = Z_C^{(1-T_R)^{2/7}} \tag{3.90}$$

in which V_C is the critical volume, Z_C is the Z-factor at the critical point, and T_R is the reduced temperature. If the critical volume is not available, one can substitute $RT_C Z_C/P_C$, then

$$V^{sat} = \left(\frac{RT_C}{P_C} \right) Z_C^{\left[1+(1-T/T_C)^{2/7}\right]} \tag{3.91}$$

If the critical compressibility factor is not available, it is possible to use

$$Z = \left[\left(0.3445 P_C/RT_C\right) V_C^{1.0135} \right]^{1/\left[1+(1-T_B/T_C)^{2/7}\right]} \tag{3.92}$$

in which T_B is the normal boiling point temperature.

3.5.4.3 Riedel Equation

This equation [14] uses the acentric factor when predicting saturated density

$$\rho^{sat} = \rho_C \left[1 + 0.85 \left(1 - T_R\right) + (1.6916 + 0.984\omega) \left(1 - T_R\right)^{1/3} \right] \tag{3.93}$$

3.5.4.4 Yen and Woods Equation

These authors [15] extend the Riedel equation without the acentric factor

$$\rho^{sat} = \rho_C \left[1 + \sum_{i=1}^{4} k_i \left(1 - T_R\right)^{i/3}\right] \tag{3.94}$$

in which

$$k_1 = 17.4425 - 214.578Z_C + 989.625Z_C^2 - 1522.06Z_C^3$$

$$k_2 = -3.28257 + 13.6377Z_C + 107.4844Z_C^2 - 384.211Z_C^3 \text{ for } Z_C < 0.26$$

$$k_2 = 60.2091 - 402.063Z_C + 501.0Z_C^2 + 641.0Z_C^3 \text{ for } Z_C > 0.26$$

$$k_3 = 0$$

$$k_4 = 0.93 - k_2$$

3.5.4.5 Chueh and Prausnitz Equation

For compressed liquids, Chueh and Prausnitz [16] suggest a modified Tait equation

$$\rho = \rho^{sat}\left[1 + 9\beta^{sat}\left(P - P^{sat}\right)\right]^{1/9} \tag{3.95}$$

with

$$\beta^{sat} RT_C/V_C = \left[1 - 0.89\omega^{0.5}\right]\phi \tag{3.96}$$

and

$$\ln\phi = 6.9547 - 76.2853T_R + 191.3060T_R^2 - 203.5472T_R^3 + 82.7631T_R^4 \tag{3.97}$$

This equation is useful for reduced temperatures between 0.4 and 0.98.

Chang and Zhao [17] propose the following equation for compressed liquid densities

$$V = V^{sat}\frac{A + 2.810^{\left(1.1-T_R\right)^B}\left(P_R - P_R^{sat}\right)}{A + 2.810\left(P_R - P_R^{sat}\right)} \tag{3.98}$$

in which $P_R = P/P_C$ is the reduced pressure and P_R^{sat} is the reduced pressure at saturation at the given temperature. Also,

$$A = 99.42 + 6.502T_R - 78.68T_R^2 - 75.18T_R^3 + 41.49T_R^4 + 7.257T_R^5$$

$$B = 0.381\,44 - 0.301\,44\omega - 0.084\,57\omega^2$$

3.5.4.6 Generalized Lee–Kesler Correlation

This correlation [18] can calculate Z-factors for gases and fluids. This Z-factor contains two terms such as the correlations for second virial coefficients,

$$Z\left(T_R, P_R\right) = Z^{(0)}\left(T_R, P_R\right) + \omega Z^{(1)}\left(T_R, P_R\right) \tag{3.99}$$

in which $Z^{(0)}$ is the Z-factor for spherical molecules and $Z^{(1)}$ is a shape-dependent Z-factor. Both Z-factors are functions of the reduced temperature and pressure. Tables for the values appear in the literature. The book includes an Excel® add-In, LK CALC.xlam, to calculate the compressibility values, and the instructions are in the Appendix section. This equation predicts values for liquids, but it does not replicate experimental measurements very accurately.

3.5.5 Cubic Equations of State

Many cubic EOS exist in the literature. They represent pressure as a function of temperature and a cubic polynomial in volume.

3.5.5.1 van der Waals (vdW) Equation of State

Johannes Diderik van der Waals [19] proposed the most famous cubic EOS in 1873. He proposed continuity between the vapor and the liquid states and represented it mathematically. He proposed a correction for the ideal gas EOS, $PV = RT$

$$P = \frac{RT}{V-b} - \frac{a}{V^2} \text{ or } \left(P + \frac{a}{V^2}\right)(V-b) = RT \tag{3.100}$$

This equation written in Z-factor form is

$$Z = \frac{V}{V-b} - \frac{a}{RTV} = \frac{1}{1-\rho b} - \frac{a\rho}{RT} \tag{3.101}$$

The first term reflects the effect of repulsive forces among molecules at short distances; b (co-volume) is non-zero because each molecule occupies a finite volume. The second term considers that attractive interactions exist among molecules. The main theoretical disadvantage of the equation is that it considers b to be constant when it should be volume dependent. This equation works qualitatively well when applied at constant temperature and the constants a and b come from experimental data. The major contribution of van der Waals (vdW) is the suggestion that

$$P = P^{rep} + P^{att}$$

in which P^{rep} is the repulsive contribution to pressure and P^{att} is the attractive contribution.

The vdW equation has three volume roots in the two-phase region, as shown in Figure 3.9. This curvature shows that the largest volume is the vapor volume; the smallest is the liquid volume; and the third one appears in the two-phase region and does not have any physical significance. All EOS exhibit a similar behavior in the two-phase region. We refer to the "loops" as *van der Waals loops.*

The two constants from the EOS come from solving the critical conditions:

$$\left(\frac{\partial P}{\partial V}\right)_{T_C} = 0 \text{ and } \left(\frac{\partial^2 P}{\partial V^2}\right)_{T_C} = 0$$

Then,

$$\left(\frac{\partial P}{\partial V}\right)_T = -\frac{RT}{(V-b)^2} + \frac{2a}{V^3} \tag{3.102}$$

and

$$\left(\frac{\partial^2 P}{\partial V^2}\right)_T = \frac{2RT}{(V-b)^3} - \frac{6a}{V^4} \tag{3.103}$$

At the critical point, these derivatives must equal to zero:

$$\frac{RT_C}{(V_C-b)^2} - \frac{2a}{V_C^3} = 0 \Rightarrow a = \frac{RTcV_C^3}{2(V_C-b)^2} \tag{3.104}$$

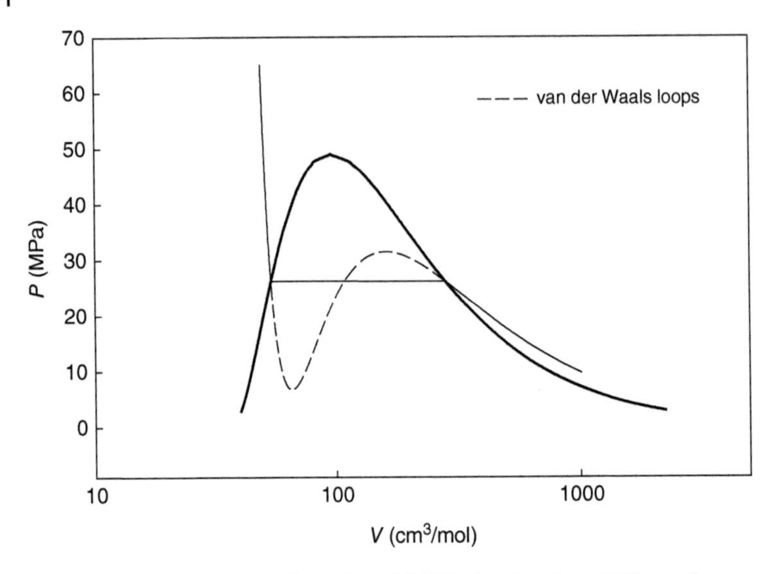

Figure 3.9 Phase diagram from the vdW EOS showing the vdW loops for argon.

$$\frac{2RT_C}{(V_C-b)^3}-\frac{6a}{V_C^4}=0 \Rightarrow a=\frac{2RT_CV_C^4}{6(V_C-b)^3}=\frac{RT_CV_C^4}{3(V_C-b)^3} \tag{3.105}$$

Equating the expressions,

$$\frac{RT_CV_C^3}{2(V_C-b)^2}=\frac{RT_CV_C^4}{3(V_C-b)^3} \Rightarrow \frac{1}{2}=\frac{V_C}{3\,(V_C-b)} \tag{3.106}$$

And solving for b,

$$b=\frac{V_C}{3} \tag{3.107}$$

Now, substituting this result into Eq. (3.104), we obtain a

$$a=\frac{9}{8}RT_CV_C \tag{3.108}$$

Thus, the vdW EOS becomes

$$Z=\frac{V}{V-V_C/3}-\frac{9RT_CV_C}{8RTV}=\frac{V/V_C}{V/V_C-1/3}-\frac{9T_CV_C}{8TV} \tag{3.109}$$

and the Z-factor at the critical point is

$$Z_C=\frac{V_C/V_C}{V_C/V_C-1/3}-\frac{9T_CV_C}{8T_CV_C}=\frac{3}{8} \tag{3.110}$$

The vdW EOS and all other cubic EOS give a single value of Z_C for all fluids. The value given by the vdW is high because most substances have values near $Z_C = 0.27$. It is also convenient to express the constant, a, in terms of critical properties

$$a=\frac{9}{8}RT_CV_C=\frac{9}{8}RT_C\left(\frac{3}{8}\frac{RT_C}{P_C}\right) \Rightarrow a=\frac{27}{64}\frac{R^2T_C^2}{P_C} \tag{3.111}$$

$$b = \frac{V_C}{3} = \frac{1}{3}\left(\frac{3}{8}\frac{RT_C}{P_C}\right) = \frac{1}{8}\frac{RT_C}{P_C}$$

Also, we can write the EOS using reduced properties

$$\frac{P}{P_C} = \frac{1}{P_C}\frac{RT}{(V-b)} - \frac{1}{P_C}\frac{a}{V^2} = \frac{RT}{P_C\left(V - \frac{1}{3}Vc\right)} - \frac{9RT_CV_C}{8P_CV^2} \tag{3.112}$$

Now, dividing and multiplying the first term by T_C

$$\frac{P}{P_C} = \frac{RT_C\frac{T}{T_C}}{P_CV_C\left(\frac{V}{V_C} - \frac{1}{3}\right)} - \frac{9RT_C}{8P_CV_C}\left(\frac{V_C}{V}\right)^2 \tag{3.113}$$

and collecting terms,

$$\frac{P}{P_C} = \frac{1}{Z_C}\left[\frac{T_R}{V_R - \frac{1}{3}} - \frac{9}{8V_R^2}\right] \tag{3.114}$$

Therefore,

$$P_R = \frac{1}{Z_C}\left[\frac{3T_R}{3V_R - 1} - \frac{9}{8V_R^2}\right] = \frac{8}{3}\left[\frac{3T_R}{3V_R - 1} - \frac{9}{8V_R^2}\right] = \frac{8T_R}{3V_R - 1} - \frac{3}{V_R^2} \tag{3.115}$$

This equation indicates that all substances obey a single EOS that describes the *PVT* behavior without any characteristic constants. Generalizing this concept, we can say that *the behavior of thermophysical properties for all substances is a unique function of reduced variables*. This is the **Corresponding of States Principle (CSP)**. This principle works well for simple molecules of similar shape. For example, the principle works reasonably well for properties of argon, neon, krypton, nitrogen, oxygen, etc. If we consider the shape of the molecule (size), we can include a third parameter such as the acentric factor. We already have used some correlations based on the CSP: for the second and third virial coefficients and the Z-factor.

3.5.5.2 Other Cubic EOS

Now, we can establish the general characteristics for two-parameter cubic EOS:

- All these equations have an incorrect repulsive term like the vdW EOS (co-volume independent of density)
- They predict thermodynamic properties with significant errors in some regions
- They are simple and approximate; therefore, they are popular in industrial applications and in computer programs that require many (thousands or millions) of calculations
- The attractive term must compensate for the incorrect repulsive term
- They satisfy the ideal gas limit when V tends to infinity
- They can represent the molar volume behavior reasonably well at low temperatures (this characteristic is necessary for calculation of vapor–liquid equilibria).

Table 3.3 Values for the parameters of different EOS.

EOS	θ	η	δ	ε
van der Waals (1873)	a	b	0	0
Clausius (1880) [35]	a/T	b	$2c$	c^2
Berthelot (1900) [36]	a/T	b	0	0
Redlich–Kwong (1949)	$a/T^{1/2}$	b	b	0
Lee–Erbar–Edmister (1976) [37]	$\theta_{LEE}(T)$	$\eta(b, \theta, T)$	b	0
Peng–Robinson (1976)	$\theta_{PR}(T)$	b	$2b$	$-b^2$
Knapp–Harmens (1980) [38]	$\theta_{KH}(T)$	b	cb	$-(c-1)b^2$

One way of expressing these cubic EOS in general form is

$$P = \frac{RT}{V-b} - \frac{\theta\,(V-\eta)}{(V-b)\left(V^2+\delta V+\varepsilon\right)} \tag{3.116}$$

where θ, η, δ, and ε are characteristic parameters that can be functions of temperature. For example, in the vdW EOS: $\theta = a$, $\eta = b$, $\delta = 0$, and $\varepsilon = 0$. Table 3.3 contains values for the parameters of various EOS.

3.5.5.3 Redlich–Kwong (RK) EOS

The Redlich–Kwong (RK) [20] equation appeared in 1949. This equation was the most successful attempt to correct the vdW EOS at that time. Researchers could not modify the parameter b of the EOS because it reflected molecules. This equation corrected the attractive term to include temperature

$$P = \frac{RT}{V-b} - \frac{a}{T^{0.5}\,V\,(V+b)} \tag{3.117}$$

The importance of this equation is that it describes experimental data better by correcting the attractive term. Although this concept is not completely correct because it only compensates for the attractive term, it has led to the development of more accurate EOS. If we apply the critical conditions to Eq. (3.117), we obtain

$$a = 0.427480\frac{R^2 T_C^{2.5}}{P_C} \tag{3.118}$$

and

$$b = 0.086640\frac{RT_C}{P_C} \tag{3.119}$$

This EOS improves upon the predictions of vdW, and because it is a cubic EOS, it has an expanded form of

$$V^3 - (RT/P)\,V^2 + (1/P)\left(a/T^{0.5} - bRT - Pb^2\right)V - ab/PT^{0.5} = 0 \tag{3.120}$$

Also, it is cubic in the Z-factor

$$Z^3 - Z^2 + \left(A - B - B^2\right)Z - AB = 0 \tag{3.121}$$

with

$$A = \frac{aP}{R^2T^{2.5}}$$

and

$$B = \frac{bP}{RT}$$

We can recast the RK as a two-parameter CSP expression by inserting (3.118) and (3.119) into (3.117)

$$P_R = \frac{T_R}{Z_C\left(V_R - 0.086640/Z_C\right)} - \frac{0.427480}{Z_C^2 T_R^{0.5} V_R\left(V_R + 0.086640/Z_C\right)} \tag{3.122}$$

in which the value of $Z_C = 1/3$ for the RK, then

$$P_R = \frac{3T_R}{\left(V_R - 0.25992\right)} - \frac{3.84732}{T_R^{0.5} V_R\left(V_R + 0.25992\right)} \tag{3.123}$$

The RK is not the only possible improvement to the vdW. Research continues, and there exist more than 100 forms similar to the RK. Figure 3.10 is a schematic diagram of the *PV* behavior for the RK.

This equation has three poles at which the pressure can go to $\pm\infty$: $V = 0$, $V = -b$, and $V = +b$. As shown in the figure, section A is the only one with physical significance. This region has the vdW loops for an isotherm below the critical temperature. Values of the molar volume less than b lead to unrealistic pressures. At temperatures greater than the critical temperature (supercritical temperatures), generally, a real root for the volume must exist, and the other roots must be imaginary. However, at very high pressures and temperatures, three real roots can occur. In this case, selection of the real volume should reflect the thermodynamic state of the substance.

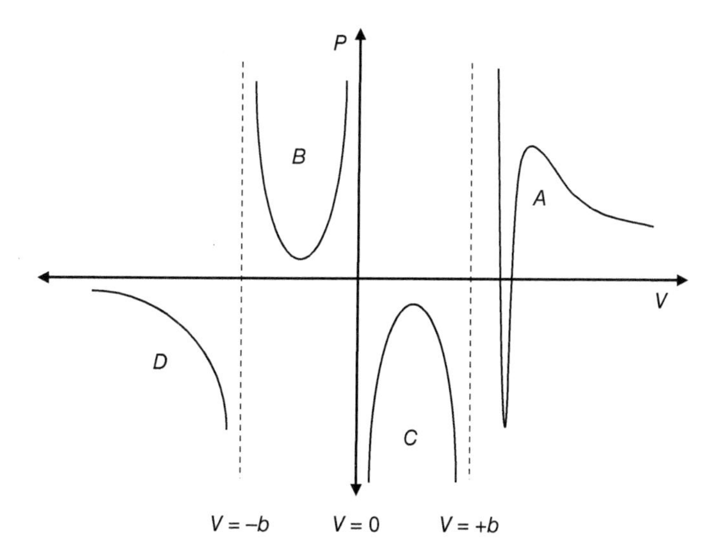

Figure 3.10 Schematic diagram of the *PV* behavior of a cubic EOS.

3.5.5.4 Soave–Redlich–Kwong (SRK) Equation of State

After more than two decades, a practical and good modification of the RK appeared: the Soave–Redlich–Kwong [21] (SRK) EOS.

$$P = \frac{RT}{V - b} - \frac{a(T)}{V(V + b)} \tag{3.124}$$

In this equation, the constant a of the RK EOS is temperature dependent while b is not

$$a(T) = a_C \alpha(T) = a_C \left[1 + m\left(1 - (T/T_C)^{1/2}\right)\right]^2 \tag{3.125}$$

$$m = 0.480 + 1.574\omega - 0.176\omega^2 \tag{3.126}$$

If we apply the critical conditions to Eq. (3.124)

$$a_C = 0.42747 \frac{R^2 T_C^2}{P_C} \tag{3.127}$$

with b the same as in RK

$$b = 0.086640 \frac{RT_C}{P_C}$$

and the critical compressibility factor is equal to 1/3 as in the RK.

3.5.5.5 Peng–Robinson (PR) Equation of State

Peng and Robinson [22] developed another popular cubic EOS. They sought to improve upon inaccuracies of the SRK. They succeeded in many senses, but, if one looks in detail, they did it at the expense of deteriorating the vapor prediction somewhat. The Peng–Robinson (PR) EOS is

$$P = \frac{RT}{V - b} - \frac{a(T)}{V(V + b) + (V - b)b} \tag{3.128}$$

As in the SRK, parameter a is temperature dependent, and parameter b is not

$$a(T) = a_C \alpha(T) = a_C \left[1 + m\left(1 - (T/T_C)^{1/2}\right)\right]^2 \tag{3.129}$$

$$m = 0.37464 + 1.54226\omega - 0.26992\omega^2 \tag{3.130}$$

The values of a_C and b from applying the critical conditions to Eq. (3.128) are

$$a_C = 0.45724 \frac{R^2 T_C^2}{P_C} \tag{3.131}$$

and

$$b = 0.07780 \frac{RT_C}{P_C} \tag{3.132}$$

The critical Z-factor for this equation is 0.307. The SRK and PR find wide use in the natural gas industry because actual computer programs require a large number of iterative calculations and speed is more important than accuracy in most cases. The accuracy of PR deteriorates badly at highly supercritical temperatures.

Kumar and Starling [23] have suggested general cubic EOS

$$\frac{P}{RT} = \frac{\lambda_4\rho + \lambda_5\rho^2 + \lambda_6\rho^3}{\lambda_0 + \lambda_1\rho + \lambda_2\rho^2 + \lambda_3\rho^3} = \frac{\rho + n_1\rho^2 + n_2\rho^3}{1 + d_1\rho + d_2\rho^2 + d_3\rho^3} \tag{3.133}$$

in which n and d are temperature-dependent parameters. For example, vdW has coefficients: $n_1 = -a/RT$, $n_2 = ab/RT$, $d_1 = -b$, and $d_2 = d_3 = 0$.

3.5.6 Multiparameter Equations of State

PVT equations can achieve greater accuracy by including more parameters. A multiparameter EOS has more parameters, but it is difficult to determine them from critical conditions. They result from curve-fitting experimental data. To describe multiple substances, it is necessary to calculate the parameters in terms of physical properties such as the critical temperature, pressure, boiling point, acentric factor, etc. Sometimes, this is not possible, and the EOS becomes a correlative model.

3.5.6.1 Benedict–Webb–Rubin (BWR) Equation of State

One of the first attempts at a multiparameter EOS was the Benedict–Webb–Rubin (BWR) [24] equation

$$P = \frac{RT}{V} + \frac{BoRT - Ao - \frac{Co}{T^2}}{V^2} + \frac{bRT - a}{V^3} + \frac{a\alpha}{V^6} + \frac{C}{V^3T^2}\left(1 + \frac{\gamma}{V^2}\right)\exp\left(-\frac{\gamma}{V^2}\right) \tag{3.134}$$

in which Bo, Ao, Co, b, a, α, C, and γ are characteristic parameters. The natural gas and petroleum industries use modifications of BWR widely in applications. The general form of these equations is a series of polynomials multiplied by exponentials. For calculation-intensive computer programs, the exponential terms in density require a great deal of time to calculate.

Another type of EOS, based on the augmented vdW theory, uses repulsive and attractive parts for the pressure or the compressibility factor. One of the most successful is the Boublik–Alder–Chen–Kreglewski (BACK) equation of state.

3.5.6.2 Boublik–Alder–Chen–Kreglewski

Chen and Kreglewski [25] suggested that the compressibility factor has the form

$$Z = Z^{rep} + Z^{att} \tag{3.135}$$

as in the original vdW. For the repulsive part, they proposed

$$Z^{rep} = \frac{1 + (3\alpha - 2)y + \left(3\alpha^2 - 3\alpha + 1\right)y^2 - \alpha^2y^3}{(1 - y)^3} \tag{3.136}$$

in which $y = 0.74048V^0/V$ and a is an anisotropic parameter that accounts the shape of the molecule. For spherical molecules, $\alpha = 1$. If it is unknown, an estimate is

$$\alpha = 1 + 0.3\omega \tag{3.137}$$

Table 3.4 Universal parameters for BACK.

D_{11}	−8.8043	D_{21}	2.9396	D_{31}	−2.8225	D_{41}	0.34
D_{12}	4.164627	D_{22}	−6.0865383	D_{32}	4.7600148	D_{42}	−3.1875014
D_{13}	−48.203555	D_{23}	40.137956	D_{33}	11.257177	D_{43}	12.231796
D_{14}	140.4362	D_{24}	−76.230797	D_{34}	−66.382743	D_{44}	−12.110681
D_{15}	−195.23339	D_{25}	−133.70055	D_{35}	69.248785		
D_{16}	113.515	D_{26}	860.25349				
		D_{27}	−1535.3224				
		D_{28}	1221.4261				
		D_{29}	−409.10539				

V^0 is the close-packed volume of the molecular hard core given by

$$V^0 = V^{00}\left[1 - 0.12\exp\left(-3u^0/kT\right)\right] \tag{3.138}$$

in which u^0 is a characteristic parameter and V^{00} is the value of V^0 at zero temperature. An estimation is $V^{00} = 0.21V_C$. The attractive term is a polynomial expansion

$$Z^{att} = \sum_{n=1}^{4}\sum_{m=1}^{9} mD_{nm}(u/kT)^n\left(V^0/V\right)^m \tag{3.139}$$

here u is the characteristic energy, which is a function of temperature:

$$\frac{u}{k} = \frac{u^0}{k}\left(1 + \frac{\eta}{kT}\right) \tag{3.140}$$

For spherical molecules, $\eta/k = 0$ and

$$\frac{\eta}{kT_C} = 0.505\omega + 0.702\omega^2 \tag{3.141}$$

in which T_C is the critical temperature in kelvin and ω is the acentric factor. Equation (3.139) has 24 universal constants, D_{nm}, shown in Table 3.4. BACK has three characteristic parameters as shown in Table 3.5.

Recently, a new EOS has become popular because it predicts the thermodynamic properties adequately for a wide range of substances. This equation is the Perturbed Chain-Statistical Associating Fluid Theory (PC-SAFT) [30],

$$Z = 1 + Z^{hs} + Z^{chain} + Z^{disp} + Z^{ass} + \cdots \tag{3.142}$$

Z^{hs}: is the hard sphere contribution for the compressibility factor
Z^{chain}: is the chain formation contribution to the compressibility factor
Z^{disp}: is the dispersion contribution to the compressibility factor. The residual part of the BACK EOS is a dispersion contribution.
Z^{ass}: is the association term of the compressibility factor.

Table 3.5 Characteristic parameters of the BACK EOS.

Substance	α	V^{00}	u^0/k	η/k
Argon	1.0000	16.290	150.86	0.00
Xenon	1.0231	25.499	294.38	1.20
Nitrogen	1.0000	19.457	123.53	3.00
Carbon monoxide	1.0153	19.820	130.46	4.20
Carbon dioxide	1.0571	19.703	284.28	40.00
Sulfur dioxide	1.0710	25.346	383.56	88.00
Hydrogen sulfide	1.0440	20.672	373.66	15.00
Methane	1.0000	21.576	190.29	1.00
Ethane	1.0370	31.118	298.03	19.00
Propane	1.0410	42.598	353.11	34.00
n-Butane	1.0510	53.855	399.56	51.00
iso-Butane	1.0482	54.682	383.11	47.00
n-Pentane	1.0566	65.751	435.83	70.72
iso-Pentane	1.0565	64.958	432.20	62.71
neo-Pentane	1.0498	65.518	409.59	51.28
n-Hexane	1.0720	77.228	468.33	90.11
n-Heptane	1.0626	90.404	465.99	130.00
n-Octane	1.0981	96.556	517.52	134.50
n-Decane	1.1349	110.720	558.07	181.57
2,2,4-Trimethylpentane	1.0588	101.270	468.62	125.00
Ethylene	1.0330	28.031	279.55	12.00
Propylene	1.0443	38.979	350.07	32.00
1-Butene	1.0539	50.224	393.38	52.00
Methylpropene	1.0544	50.140	394.54	49.00
Cyclohexane	1.0583	64.772	522.46	70.72
Benzene	1.0613	54.289	529.24	72.00
Toluene	1.0944	65.843	558.94	91.00

3.5.7 Reference Equation of State

A reference EOS is an equation with a high level of accuracy required for scientific standard and technical applications. These EOS optimize a modified BWR functional form based on the most complete data set existing in the literature of *PVT* data, virial coefficients, heat capacities, enthalpies, speeds of sound, Joule–Thompson coefficients, etc. When developing the EOS, the authors check every known derivative that involves P, V, and T. Most of the reference EOS have the form

$$Z = 1 + \delta\left(\frac{\partial \alpha^r}{\partial \delta}\right)_\tau = 1 + \sum_{i=1}^{N_1} n_i d_i \tau^{t_i} \delta^{d_i} + \sum_{i=N_1+1}^{N_1+N_2} n_i \tau^{t_i} \delta^{d_i} \left(d_i - p_i \delta^{p_i}\right) \exp\left(-\delta^{p_i}\right) \tag{3.143}$$

in which α^r is a residual part, R is the universal gas constant, T is the temperature, ρ is the density, τ is the dimensionless temperature equal to T_C/T or $1 - T_C/T$, and δ is the dimensionless density equal to ρ/ρ_C. N_1 and N_2 are the number of polynomial and polynomial plus exponential terms and n_i, t_i, d_i, and p_i are adjusted parameters selected from a data bank of hundreds of terms. National Institute of Standards and Technology (NIST) has developed software (REFPROP®) that calculates the thermodynamic properties of pure substances and mixtures using reference EOS. A web version is free of charge for calculation of thermodynamic properties of pure substances (http://webbook.nist.gov/chemistry/).

3.6 Calculation of Volumes from EOS

Several procedures exist to calculate volumes from an EOS. For example, if the EOS is a cubic, we can apply the general solution for finding the roots of cubic polynomials.

A second approach is to use a numerical method. For example, the Newton–Raphson method

$$V_{j+1} = V_j - \frac{f\left(V_j\right)}{f'\left(V_j\right)} \tag{3.144}$$

In this equation, we need to write the EOS as $f(V_j) = 0$. Several approaches are available for writing this function. For instance, two forms for the vdW can be

$$f\left(V\right) = P - \frac{RT}{V-b} + \frac{a}{V^2} = 0 \text{ or } f\left(V\right) = \left(P + \frac{a}{V^2}\right)\left(V+b\right) - RT = 0 \tag{3.145}$$

Let us work with the first one, its derivative with respect to volume is

$$f'\left(V\right) = \frac{RT}{\left(V-b\right)^2} - \frac{2a}{V^3} \tag{3.146}$$

Then, Eq. (3.144) becomes

$$V_{j+1} = V_j - \frac{P - \frac{RT}{V_j - b} + \frac{a}{V_j^2}}{\frac{RT}{\left(V_j - b\right)^2} - \frac{2a}{V_j^3}} \tag{3.147}$$

If we know the pressure and temperature, we can use an iterative procedure to find one of the volume roots. We should start with $j = 0$, so we need an initial value of the volume. If we seek a vapor volume, the initial value can be the ideal gas value. If we want a liquid volume, the initial value can be the value of b plus a small increment because a pole exists at $V = b$.

Another procedure is to use a software program like Excel. Here, we can introduce the equation $f(V) = 0$ to any cell (e.g. A1) and the value of volume in a different cell (e.g. B1). We can activate the Solver window by going to the Data Tab and in Analysis we find the Solver function. Figure 3.11 is a copy of the Solver® window. In the "Set Target Cell," use B1. In the "By Changing Cells," enter the cells that should change to make the function equal zero (in this case, cell A1). Next, check "Value of" and verify that it contains a zero. Now, we can solve the EOS by clicking "Solve." Another

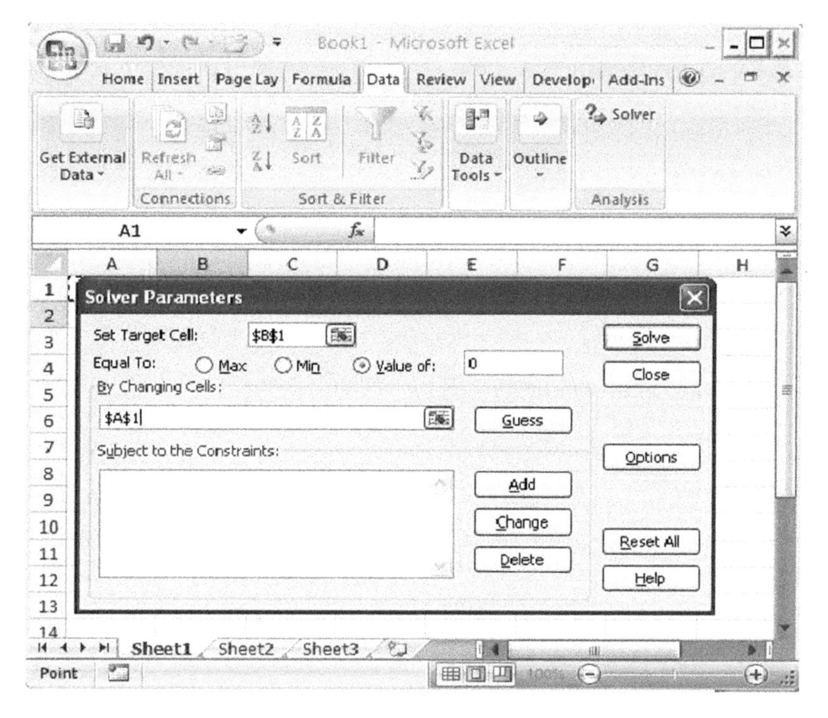

Figure 3.11 Solver window.

window appears, indicating that the solver has reached a solution. Cell A1 should contain the value of the volume and cell B1 should have a value close to zero.

Sometimes, it is desirable to solve for the Z-factor instead of the density or volume. For example, using the SRK EOS

$$P = \frac{RT}{V - b} - \frac{a(T)}{V(V + b)} \quad (3.124)$$

in terms of Z is

$$Z = \frac{V}{V - b} - \frac{a(T)}{RT(V + b)} \tag{3.148}$$

Substituting $V = ZRT/P$

$$Z = \frac{ZRT}{PV - Pb} - \frac{a(T)}{RT(ZRT/P + b)} = \frac{ZRT}{PV - Pb} - \frac{a(T)P}{RT(ZRT + bP)} \tag{3.149}$$

or

$$Z = \frac{Z}{Z - Pb/RT} - \frac{a(T)P}{(RT)^2(Z + bP/RT)} \tag{3.150}$$

Defining

$$A = \frac{a(T)P}{(RT)^2} \tag{3.151}$$

and

$$B = \frac{bP}{RT} \tag{3.152}$$

produces

$$Z = \frac{Z}{Z - B} - \frac{A}{Z + B} = \frac{Z^2 + BZ - AZ + AB}{Z^2 - B^2} \tag{3.153}$$

or

$$Z^3 - ZB^2 - Z^2 - BZ + AZ - AB = 0$$

which in cubic form is similar to Eq. (3.121)

$$Z^3 - Z^2 + \left(A - B - B^2\right) Z - AB = 0 \tag{3.154}$$

Now, changing to an iterative form

$$Z_{j+1} = 1 - \left(A - B - B^2\right) / Z_j + AB / Z_j^2 \tag{3.155}$$

This form works well to obtain a vapor-like Z-factor because the values of Z are relatively large (on the order of 1). The initial value in the above equation can be $Z = 1$. For the liquid-like Z-factor, the values of Z are small; therefore, an equation like (3.155) does not work. It is better to have the compressibility factor in the numerator, such as

$$Z_{j+1} = \frac{Z_j^2 - Z_j^3 + AB}{\left(A - B - B^2\right)} \tag{3.156}$$

Example 3.8

The vapor pressure of ethane at 250 K is 1.3 MPa. Calculate the liquid and vapor molar volume at this condition using the vdW EOS.

Data

$$T_C = 305.32\ \text{K}$$

$$P_C = 4.8722\ \text{MPa}$$

$$\omega = 0.0995$$

Solution

Calculate the a and b constants for vdW

$$a = \frac{27}{64}\frac{R^2 T_C^2}{P_C} = \frac{27}{64}\frac{(8.31451)^2\ \text{MPa}^2\ \text{cm}^6/(\text{mol K})^2 \times (305.32)^2\ \text{K}^2}{4.8722\ \text{MPa}}$$

$$= 558\,010.68\ \text{MPa cm}^6/\text{mol}^2$$

$$b = \frac{1}{8}\frac{RT_C}{P_C} = \frac{1}{8}\frac{8.31451\ \text{MPa cm}^3/(\text{mol K}) \times 305.32\ \text{K}}{4.8722\ \text{MPa}} = 65.129\ \text{cm}^3/\text{mol}$$

The initial value for the vapor volume is the ideal gas volume

$$V_0 = \frac{RT}{P} = \frac{8.31451\ \text{MPa cm}^3/(\text{mol K}) \times 250}{1.3\ \text{MPa}} = 1598.944\ \text{cm}^3/\text{mol}$$

j	V_j	$f(V_j)$	$f'(V_j)$
0	1598.94423	0.16305983	0.00061054
1	1331.8708	−0.02635412	0.00082302
2	1363.89217	−0.00049418	0.00079242
3	1364.51581	-1.808×10^{-7}	0.00079184
4	1364.51604	-2.4147×10^{-14}	0.00079184

For the liquid volume, the initial volume can be $b + 0.1b$,

$$V_0 = 65.129 + 6.5129 = 71.64\,\text{cm}^3/\text{mol}$$

j	V_j	$f(V_j)$	$f'(V_j)$
0	71.642297	−209.135162	45.9680038
1	76.191879	−90.475718	14.4619553
2	82.447998	−36.634090	4.93897664
3	89.865342	−13.635730	1.85939305
4	97.198773	−4.452775	0.80581338
5	102.724587	−1.109392	0.44110222
6	105.239632	−0.139842	0.33451817
7	105.657672	−0.003199	0.31932228
8	105.667689	-1.7848×10^{-6}	0.318966
9	105.667695	-5.5422×10^{-13}	0.3189658

3.7 Vapor Pressure and Enthalpy of Vaporization Correlations

Correlations for the vapor pressure are available and easy to use. The most common vapor pressure equation is the Antoine [27] equation

$$\ln P^v = A - \frac{B}{T + C} \tag{3.157}$$

Values of the constants are available for various substances [28, 29]. This equation works best below the boiling point. Above the boiling point, it is common to use an Antoine equation with different coefficients. Wagner [30] has developed a more accurate correlation for the vapor pressure based on a selection from a databank of temperature terms

$$\ln \frac{P^v}{P_C} = \left(\frac{T_C}{T}\right)\left(a\tau + b\tau^{1.5} + c\tau^3 + d\tau^6\right) \tag{3.158}$$

Generalizations of this equation have appeared using the acentric factor

$$a = -6.1559 - 4.0855\omega \tag{3.159}$$

$$b = 1.5737 - 1.0540\omega - 4.4365 \times 10^{-3} d \tag{3.160}$$

$$c = -0.8747 - 7.887\omega \tag{3.161}$$

$$d = \left(-0.4893 - 0.9912\omega + 3.1551\omega^2\right)^{-1} \tag{3.162}$$

This generalization does not work well for alcohols. Gómez-Nieto and Thodos [31] suggest a general vapor pressure equation

$$\ln \frac{P^v}{P_C} = \beta \left[\frac{1}{T_R^m} - 1\right] + \gamma \left[T_R^7 - 1\right] \tag{3.163}$$

$$m = 0.78425 \exp(0.089315s) - 8.5217 \exp(-0.74826s) \tag{3.164}$$

with

$$s = \frac{T_b \ln P_C}{T_C - T_b} \tag{3.165}$$

$$\beta = -4.26700 - \frac{221.79}{s^{2.5} \exp\left(0.03848s^{2.5}\right)} + \frac{3.8126}{\exp\left(2272.44s^{-3}\right)} + \Delta^* \tag{3.166}$$

in which T_b is the normal boiling point temperature and Δ^* is 0.41815, 0.19904, and 0.02319 for quantum gases (helium, hydrogen, and neon); otherwise, it is zero.

The Watson [32] equation provides a reasonably accurate estimate for enthalpy of vaporization

$$\Delta H^v(T) = \Delta H^v\left(T_0\right)\left[\frac{1 - T_R}{1 - T_{R,0}}\right]^{0.38} \tag{3.167}$$

in which T_0 is a reference temperature, usually the normal boiling point. Carruth and Kobayashi [33] propose a simple equation based on the Watson equation

$$\frac{\Delta H^v(T)}{RT_C} = 7.08\left(1 - T_R\right)^{0.354} + 10.95\omega\left(1 - T_R\right)^{0.456} \tag{3.168}$$

Morgan [34] proposed a simple equation using the acentric factor

$$\frac{\Delta H^v(T)}{RT_C} = d_1\left(1 - T_R\right)^{f(T_R)} \tag{3.169}$$

in which $f\left(T_R\right) = d_2 + d_3 T_R + d_4 T_R^2$ and

$$d_1 = 7.8149 + 11.409\omega + 2.167\omega^2 - 0.65342\omega^3 \tag{3.170}$$

$$d_2 = 0.81892 - 0.67637\omega + 1.2798\omega^2 - 0.47594\omega^3 \tag{3.171}$$

$$d_3 = -0.84408 + 1.8297\omega - 3.2435\omega^2 + 1.1449\omega^3 \tag{3.172}$$

$$d_4 = 0.41923 - 1.0892\omega + 1.9138\omega^2 - 0.65758\omega^3 \tag{3.173}$$

3.8 Ideal Gas Enthalpy Changes: Applications

We have calculated the enthalpy of an ideal gas. It depends on the temperature and there exist other practical applications of interest in the chemical industry. These applications occur in reactions in the gas phase at high temperature and generally at atmospheric pressure. Reactions can be treated as flow processes that occur over a given period of time.

3.8.1 Heat of Reaction

Consider a reaction that occurs at constant temperature and pressure in a flow process. The general energy balance is

$$\Delta\left[n\left(U+E_P+E_K\right)\right]_{SYS}=\sum n_iQ_i+\sum n_iW_i+\sum n_i\left(H+E_P+E_K\right)_i-(\delta m/M)\,c^2$$

Applying the questions from Chapter 2

1) No, so $(\delta m/M)c^2=0$
2) Yes, but kinetic and potential energies are negligible, so $\sum n_iH_i$ remains
3) No, so $\sum n_iW_i=0$
4) Yes, so $\sum m_iQ_i=q$
5) Yes, so $\Delta[m(U+E_P+E_K)]_{SYS}=0$

Then, the energy balance reduces to

$$\sum m_iQ_i+\sum m_iH_i=0 \tag{3.174}$$

In a reactor, the inlet is the reactants, and the outlet is the products, so

$$q=\sum_{products} n_iH_i-\sum_{reactants} n_iH_i\equiv\Delta H_r \tag{3.175}$$

in which ΔH_r is the heat of reaction. The amount of energy supplied as heat to the reaction maintains the medium at constant temperature and pressure because energy associated with the molecular bonds changes as the molecular and atomic arrangements alter during the reaction. The numerical values of heats of reaction defy organization in tabular form because they depend upon the reaction type and the temperature and pressure at which the reaction occurs.

3.8.1.1 Standard Heat of Reaction

It is possible to specify the initial and final states for all reactions at the same condition. This option avoids the problem of data representation discussed earlier. The standard heat of reaction has clearly defined the initial and final states

$$\Delta H_r^0\equiv\sum_{products} n_iH_i^0-\sum_{reactants} n_iH_i^0=q \tag{3.176}$$

in which ΔH_r^0 is the standard heat of reaction and H_i^0 is the enthalpy of a reactant or a product at standard conditions of temperature and pressure. The choice of standard states is arbitrary, but most tables use $T=25\,°\mathrm{C}$ (77 °F, 298 K) and $P=1$ atm. Under these conditions, the convention is to use the ideal gas for vapors and the liquid or solid state for liquids and solids. The pressure specification is not necessary to define ΔH_r^0 for vapors because it depends only on the temperature.

3.8.1.2 Standard Heat of Formation

Defining a standard heat of formation is useful and simplifies calculations. This is the standard heat of reaction when the reaction is the formation of a compound from its constituent elements, providing a single compound

$$\Delta H_f^0 \equiv H^0_{compound} - \sum_{\substack{constituent\\elements}} n_i H_i^0 \tag{3.177}$$

here ΔH_f^0 is the standard heat of formation, $H^0_{compound}$ is the enthalpy of the compound formed at the standard state, and H_i^0 is the enthalpy of a constituent element of the compound in the standard state. As a further simplification, the enthalpies of the elements in their standard states are zero by definition, implying that the enthalpy of the compound in its standard state equals the standard heat of formation and also the standard state

$$H_i^0 \equiv 0 \Rightarrow H^0_{compound} = \Delta H_f^0 \tag{3.178}$$

This enables tabulated values of the heat of formation in the standard state.

3.8.1.3 Standard Heat of Combustion

A special case of a standard heat of reaction is the standard heat of combustion. This reaction occurs when in a combustion a compound burns and oxidizes to form the most stable group of a product. For example, all the carbons present in the compound go to carbon dioxide:

all carbon → CO_2
all hydrogen → H_2O (must specify gas or liquid)
all sulfur → SO_2

The equation to represent this standard heat of reaction is

$$\Delta H_C^0 \equiv \sum_{\substack{oxidation\\products}} n_i H_i^0 - \sum_{O_2,\,Reactants} n_i H_i^0 = q \tag{3.179}$$

The heat of combustion is the easiest heat of reaction to measure experimentally; therefore, most tabulated values of ΔH_f^0 result from heat of combustion measurements.

3.8.2 Temperature Dependence of the Heat of Reaction

Let us start with the definition of heat capacity

$$C_P = \left(\frac{\partial H}{\partial T}\right)_P$$

By multiplying the number of moles and summing over all reactants and products

$$\sum_i n_i C_{P,i} = \left(\frac{\partial \sum_i n_i H_i}{\partial T}\right)_P \tag{3.180}$$

Then, because the standard state is an ideal gas

$$\sum_i n_i C_{P,i}^{ig} = \left(\frac{d\sum_i n_i H_i^0}{dT}\right)_P \Rightarrow d\left(\sum_i n_i H_i^0\right) = \left(\sum_i n_i C_{P,i}^{ig}\right) dT \tag{3.181}$$

and integrating from $T_0 = 298.15$ K to T

$$\sum_i n_i H_i^0(T) - \sum_i n_i H_i^0(T_0) = \int_{T_0}^{T} \left(\sum_i n_i C_{P,i}^{ig}\right) dT \tag{3.182}$$

Now, using the definition of heat of reaction

$$\Delta H_{r,T}^0 = \Delta H_{r,T_0}^0 + \int_{T_0}^{T} \left(\sum_i n_i C_{P,i}^{ig}\right) dT \tag{3.183}$$

The summation means products minus reactants, that is

$$\Delta H_{r,T}^0 = \Delta H_{r,T_0}^0 + \sum_{products} n_i \int_{T_0}^{T} {}_i C_{P,i}^{ig} dT - \sum_{reactants} n_i \int_{T_0}^{T} {}_i C_{P,i}^{ig} dT \tag{3.184}$$

For convenience, we define

$$\Delta C_P^{ig} \equiv \sum_{products} n_i C_{P,i}^{ig} - \sum_{reactants} n_i C_{P,i}^{ig} \tag{3.185}$$

Therefore, Eq. (3.184) becomes

$$\Delta H_{r,T}^0 = \Delta H_{r,T_0}^0 + \int_{T_0}^{T} \Delta C_P^{ig} dT \tag{3.186}$$

If the ideal heat capacity is

$$C_{P,i}^{ig}/R = \alpha + \beta T + \gamma T^2 + \delta T^{-2} \tag{3.187}$$

then

$$\Delta C_P^{ig}/R = \Delta\alpha + \Delta\beta T + \Delta\gamma T^2 + \Delta\delta T^{-2} \tag{3.188}$$

with

$$\Delta\alpha = \sum_{products} n_i \alpha_i - \sum_{reactants} n_i \alpha_i \tag{3.189}$$

$$\Delta\beta = \sum_{products} n_i \beta_i - \sum_{reactants} n_i \beta_i \tag{3.190}$$

$$\Delta\gamma = \sum_{products} n_i \gamma_i - \sum_{reactants} n_i \gamma_i \tag{3.191}$$

$$\Delta\delta = \sum_{products} n_i \delta_i - \sum_{reactants} n_i \delta_i \tag{3.192}$$

3.8.3 Practical Calculations

The general reaction equation is

$$\Delta H_r = \sum_{products} n_i H_i - \sum_{reactants} n_i H_i \equiv q \tag{3.193}$$

We must calculate the enthalpy for each component at the temperature of the reaction. Considering the change of enthalpy to be

$$H_i(T,P) = H_i\left(T_0,P\right) + \int_{T_0}^{T} C_P^{ig} dT + \Delta H^{ig}(T,P) \tag{3.194}$$

in which $\Delta H^{ig}(T, P)$ is a correction factor for the standard state and therefore for non-ideal considerations. Later, we can calculate this value in a systematic way by representing the difference between the enthalpy at T and P and the enthalpy of ideal gas at the same conditions. We can include the effects in a reaction systematically by considering that $H_i^0 = 0$ for all elements in their standard states (ideal gas). Thus,

$$H^0_{compound} = H_i\left(T_0\right) = \Delta H_f^0 \tag{3.195}$$

and

$$H^0_{elements} \equiv \Delta H_f^0 = 0 \tag{3.196}$$

and

$$H_i(T,P) = \Delta H^0_{f,i} + \int_{T_0}^{T} C_{P,i}^{ig} dT + \Delta H_i^{ig}(T,P) \tag{3.197}$$

Now, we apply this equation using the criteria

- $\Delta H_i^{ig}(T,P) \cong 0$ when the substances in the reaction are vapors, and the pressure is relatively low at reaction conditions. The standard state becomes the ideal gas for these substances.
- $\Delta H_i^{ig}(T,P) \cong \Delta H_i^{vap}\left(T_0\right)$ when the standard state of a substance is a liquid at 1 atm and exists as a vapor at reaction conditions. In effect, this converts the standard state into the ideal gas state.
- $\Delta H_i^{ig}(T,P) \cong -\Delta H_i^{vap}(T)$ when the standard state of a substance is an ideal gas, but in the reaction, the substance is liquid.
- $\Delta H^{ig}(T, P) \cong \Delta H^{vap}(T_0) - \Delta H^{vap}(T)$ when the substance exists as a liquid both at reaction conditions and the standard conditions.

Table 3.6 summarizes these values.

Sometimes, it is convenient to adjust thermodynamic tables to an appropriate reference at the standard state. For example, using the steam tables:

Find the standard heat of formation remembering that this is at 1 atm and 25 °C. Then, find the value of the enthalpy at those conditions (this should be zero if using the same reference state)

$$\Delta H^0_{f,298} - H_i^0\,(298.15\ \text{K},\ 1\ \text{atm}) \equiv 0 \tag{3.198}$$

Any difference from zero requires an enthalpy correction

$$\Delta H^0_{f,298} = H_i^0\,(298.15\ \text{K},\ 1\ \text{atm}) = H_{tables}\,(298.15\ \text{K},\ 1\ \text{atm}) + A \tag{3.199}$$

Table 3.6 Correction of the enthalpy for different conditions in a reaction.

Standard state	Reaction conditions	$\Delta H^{ig}(T, P)$
v	v	0
v	l	$-\Delta H_i^{vap}(T)$
l	v	$\Delta H_i^{vap}(T_0)$
l	l	$\Delta H^{vap}(T_0) - \Delta H^{vap}(T)$

in which A is the adjustment factor. For water,

$$\Delta H_{f,298}^{0} = -241\,818\ \text{J/mol} = \frac{-241\,818\ \text{J/mol}}{18.015\ \text{g/mol}} = -13\,423.15\ \text{kJ/kg}$$

and because at 298.15 K and 1 atm water is liquid, we use the saturated vapor enthalpy

$$H_{tables}(298.15\ \text{K},\ 1\ \text{atm}) \approx H^{v,sat}(298.15,\ 0.031\ \text{atm}) = 2442.5\ \text{kJ/kg}$$

The difference is minimal because we use an ideal gas; therefore,

$$A = -13\,423.15 - 2442.5 = -15\,865.6\ \text{kJ/kg}$$

Then, at any temperature and pressure,

$$H_i^0(T,\ P) = H_{tables}(T,\ P) - 15\,865 \tag{3.200}$$

3.8.3.1 Adiabatic Flame Temperature

The temperature that results from a complete combustion process in an adiabatic reaction without changes in kinetic or potential energy is the adiabatic flame temperature. This temperature occurs at chemical equilibrium. From Eq. (3.193),

$$\Delta H_r \equiv \sum_{products} n_i H_i - \sum_{reactants} n_i H_i = 0 \tag{3.201}$$

The unknown is the temperature, so we must calculate all enthalpies relative to elements at 25 °C. Given specified inlet conditions, all the enthalpies are available. The unknown is the temperature required to calculate the enthalpies, which makes the solution of the energy balance a trial-and-error procedure.

3.8.3.2 Reaction with Heat Transfer

Energy balance treatment is the same as in the adiabatic flame temperature section except that $q \neq 0$ in this case. The energy balance is Eq. (3.193)

$$\Delta H_r \equiv \sum_{products} n_i H_i - \sum_{reactants} n_i H_i = q \tag{3.193}$$

This case requires specifying two of the following:

- Inlet conditions
- Outlet conditions
- Energy transferred as heat

Solving the energy balance provides the third condition. Again, a trial-and-error procedure can utilize a root-solving method to find the unknown. The following example illustrates this method.

Example 3.9

Determine the flame temperature when burning ethane with 10% excess air. Assume that 95% of the ethane burns. Assume that the reaction is adiabatic. The inlet temperature of methane is 30 °C, the air enters at 35 °C, and the process occurs at 1 atm. Data

$$C_{P,i}^{ig}/R = a_i + b_i T + c_i T^2 + d_i T^{-2}$$

Substance	a	b	c	d	ΔH_f^0
Ethane	1.131	1.9225×10^{-2}	-2.164×10^{-6}	0	−83 820
Oxygen	3.639	5.0600×10^{-4}	0	-2.270×10^{4}	0
CO_2	5.457	1.0450×10^{-3}	0	-1.157×10^{5}	−393 509
Water	3.470	1.4500×10^{-3}	0	1.210×10^{4}	−241 818
Nitrogen	3.280	5.9300×10^{-4}	0	4.000×10^{3}	0

Solution

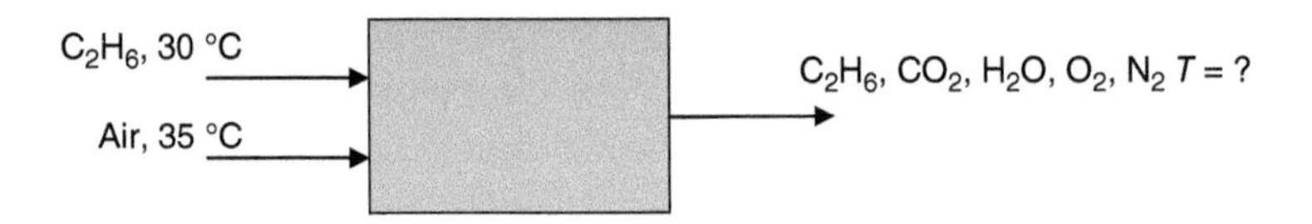

The reaction is

$$C_2H_6 + (7/2)\ O_2 \rightarrow 2CO_2 + 3H_2O$$

We must establish a basis, e.g. 1 mol of ethane. With this basis and 10% excess air, we have

moles of $O_2 = (7/2)\times1.10 = 3.85$
moles of $N_2 = (79/21)\times3.85 = 14.483$

The number of moles at the outlet is

moles of $C_2H_6 = 1-0.95\times1 = 0.05$
moles of $O_2 = 3.85-(7/2)\times0.95 = 0.525$
moles of $N_2 = 14.483$
moles of $CO_2 = 2\times0.95 = 1.9$
moles of $H_2O = 3\times0.95 = 2.85$

The energy balance is

$$\Delta H_r^0 \equiv \sum_{products} n_i H_i^0 - \sum_{reactants} n_i H_i^0 = 0 \quad (3.201)$$

Calculate the enthalpies using

$$H_i(T,P) = \Delta H^0_{f,i} + \int_{T_0}^{T} C^{ig}_{P,i} dT$$

The non-ideal term is not present because all the substances are gases, and the standard state is ideal gas. The next table provides the inlet and outlet mole values.

Substance	n_{inlet}	T_{inlet} (K)	n_{outlet}
Ethane	1	303.15	0.05
Oxygen	3.85	313.15	0.525
CO_2	0	0	1.9
Water	0	0	2.85
Nitrogen	14.483	313.15	14.483

Using the above equation in the energy balance produces

$$\begin{aligned}\Delta H^0_r &\equiv \sum_{products} n_i \Delta H^0_{f,i} - \sum_{reactants} n_i \Delta H^0_{f,i} - \sum_{reactants} n_i \int_{T_0}^{T} C^{ig}_{P,i} dT + \sum_{products} n_i \int_{T_0}^{T} C^{ig}_{P,i} dT \\ &= 0\end{aligned}$$

The first three terms are available because everything is known. The integral is

$$\begin{aligned}I_i(T) &= R \int_{T_0}^{T} \frac{C^{ig}_{P,i}}{R} dT \\ &= R\left[a_i\left(T - T_0\right) + \frac{b_i}{2}\left(T^2 - T_0^2\right) + \frac{c_i}{3}\left(T^3 - T_0^3\right) - d_i\left(\frac{1}{T} - \frac{1}{T_0}\right)\right]\end{aligned}$$

Therefore,

$$\Delta H^0_r \equiv \sum_{products} n_i \Delta H^0_{f,i} - \sum_{reactants} n_i \Delta H^0_{f,i} - \sum_{reactants} n_i I_i\left(T_{in}\right) + \sum_{products} n_i I_i\left(T_{out}\right) = 0$$

Calculating the known terms

$$\begin{aligned}\sum_{products} n_i \Delta H^0_{f,i} &= 0.05 \times (-83\,820) + 1.9 \times (-393\,509) + 2.85\,(-241\,818) \\ &= -1441\,039\ \text{J/mol}\end{aligned}$$

$$\sum_{reactants} n_i \Delta H^0_{f,i} = 1 \times (-83\,820) = -83\,820\ \text{J/mol}$$

$$\begin{aligned}&\sum_{reactants} n_i I_i\left(T_{in}\right) \\ &= R \sum_{reactants} n_i \left[a_i\left(T_{in} - T_0\right) + \frac{b_i}{2}\left(T_{in}^2 - T_0^2\right) + \frac{c_i}{3}\left(T_{in}^3 - T_0^3\right) - d_i\left(\frac{1}{T_{in}} - \frac{1}{T_0}\right)\right]\end{aligned}$$

$$= R\left[1.131\,(303.15-298.15)+\frac{1.9225\times10^{-2}}{2}\left(303.15^2-298.15^2\right)\right.$$

$$-\frac{2.164\times10^{6}}{3}\left(303.15^3-298.15^3\right)$$

$$+3.85\left\{3.639\,(313.15-298.15)+\frac{5.06\times10^{-4}}{2}\left(313.15^2-298.15^2\right)\right.$$

$$\left.+2.27\times10^{4}\left(\frac{1}{313.15}-\frac{1}{298.15}\right)\right\}$$

$$+14.483\left\{3.280\,(313.15-298.15)+\frac{5.93\times10^{-4}}{2}\left(313.15^2-298.15^2\right)\right.$$

$$\left.\left.-4\times10^{3}\left(\frac{1}{313.15}-\frac{1}{298.15}\right)\right)\right]$$

$$\sum_{reactants} n_i I_i\left(T_{in}\right)=8313.04\,\text{J/mol}$$

For the products,

$$\sum_{products} n_i I_i\left(T_{out}\right)=R\sum_{products} n_i\left[a_i\left(T_{out}-T_0\right)+\frac{b_i}{2}\left(T_{out}^2-T_0^2\right)+\frac{c_i}{3}\left(T_{out}^3-T_0^3\right)\right.$$

$$\left.-d_i\left(\frac{1}{T_{out}}-\frac{1}{T_0}\right)\right.$$

$$\sum_{products} n_i I_i\left(T_{out}\right)=R\left[0.05\left\{1.131\left(T_{out}-298.15\right)+\frac{1.9225\times10^{-2}}{2}\right.\right.$$

$$\left.\left(T_{out}^2-298.15^2\right)-\frac{2.164\times10^{6}}{3}\left(T_{out}^3-298.15^3\right)\right\}$$

$$+0.525\left\{3.639\left(T_{out}-298.15\right)+\frac{5.06\times10^{-4}}{2}\left(T_{out}^2-298.15^2\right)\right.$$

$$\left.+2.27\times10^{4}\left(\frac{1}{T_{out}}-\frac{1}{298.15}\right)\right\}$$

$$+1.9\left\{5.457\left(T_{out}-298.15\right)+\frac{1.045\times10^{-3}}{2}\left(T_{out}^2-298.15^2\right)\right.$$

$$\left.+1.157\times10^{5}\left(\frac{1}{T_{out}}-\frac{1}{298.15}\right)\right\}$$

$$+14.483\left\{3.280\left(T_{out}-298.15\right)+\frac{5.93\times10^{-4}}{2}\left(T_{out}^2-298.15^2\right)\right.$$

$$\left.\left.-4\times10^{3}\left(\frac{1}{T_{out}}-\frac{1}{298.15}\right)\right\}\right]$$

and the final equation is

$$-1\,441\,039+83\,820+\sum_{products} n_i I_i\left(T_{out}\right)-8313.036619=0$$

The above equation produces $T_{out} = 2147.65$ K using a root finding method. A simple way to solve the equation is to use a trial-and-error procedure. Provide two estimates of the temperature and calculate the function

$$f\left(T_{out}\right) = -1\,441\,039 + 83\,820 + \sum_{products} n_i I_i\left(T_{out}\right) - 8313.036619 = 0$$

Next, use linear extrapolations

$$T_{out}^{(j+2)} = \frac{T_{out}^{(j+1)} - T_{out}^{(j)}}{f\left(T_{out}^{(j+1)}\right) - f\left(T_{out}^{(j)}\right)}\left(0 - f\left(T_{out}^{(j+1)}\right)\right) + T_{out}^{(j+1)}$$

Using $T_{out}^{(1)} = 1000$ K and $T_{out}^{(2)} = 1500$ K as initial temperatures produces

$$T_{out} = 2147.65 \text{ K}$$

Problems for Chapter 3

3.1 Given the following conditions, determine if water is liquid or superheated vapor or at equilibrium. If it is at equilibrium, calculate the quality

a. At 300 °C and 5000 kPa
b. At 1000 kPa and a $H = 400$ kJ/kg
c. At 3500 kPa and $U = 1700$ kJ/kg

3.2 A container with 0.5 kg water is in equilibrium at 600 kPa. The volume of the container is 0.1 m^3. Calculate the temperature of the container and the masses of the vapor and liquid.

3.3 Calculate the reversible work in kJ during a steam expansion at a constant pressure of 700 kPa when the temperature changes from 200 to 500 °C.

3.4 Calculate the change of enthalpy and entropy when compressing steam in an isochoric process from 300 to 1000 kPa. Initially, the steam is at 150 °C.

3.5 Using the following data, estimate the specific volume, enthalpy, and entropy at 177 °C and 200 kPa. Compare your results to the values obtained from the Excel add-in.

At 150 °C and 200 kPa

$$V_1 = 0.9599 \text{ m}^3/\text{kg}\; H_1 = 2769.089 \text{ kJ/kg}\; S_1 = 7.28091 \text{ kJ/(kg K)}$$

At 300 °C and 200 kPa

$$V_2 = 1.3162 \text{ m}^3/\text{kg}\; H_2 = 3072.084 \text{ kJ/kg}\; S_2 = 7.89405 \text{ kJ/(kg K)}$$

3.6 A container has liquid water and vapor in equilibrium at 175 °C. The total mass is 2 kg. If the volume of the vapor is 25 times the volume of the liquid. What is its enthalpy?

3.7 After 13.131 kg of steam transfer from a line at 3.00 MPa and 321.64 °C to an insulated rigid container with a volume of 1.5 m^3, the temperature and pressure in the container are 450 °C and 3.00 MPa, respectively.

a. Determine the amount of material originally in the container.
b. Determine the temperature and pressure in the container before the filling process begins.
c. Determine the amount of energy that must be removed as heat to cool the steam to 25 °C if a heat transfer mechanism was available.

3.8 Superheated steam from a supply line operating at 30.0 MPa enters through a valve into a piston and cylinder arrangement that maintains a constant pressure of 5.0 MPa. The temperature of the steam in the cylinder is 300 °C immediately after 3.72 kg enters the cylinder, which initially was empty.

a. Determine the volume occupied by the steam in the cylinder and the temperature in the steam supply line, assuming the process is adiabatic.
b. Determine the amount of energy that transferred as heat and the final total volume of the fluid in the cylinder if the valve closes immediately after 3.72 kg enter, and the cylinder cools to 40 °C.

3.9 If a source of steam is available, one method for producing hot water is to mix steam with cold water as shown in the following diagram. Calculate the steam consumption rate required to produce 1250 kg/h of hot water at 85 °C and 101.325 psia, if cold water is available at 25 °C and 101.325 kPa, and the inlet steam is saturated vapor at 350 kPa.

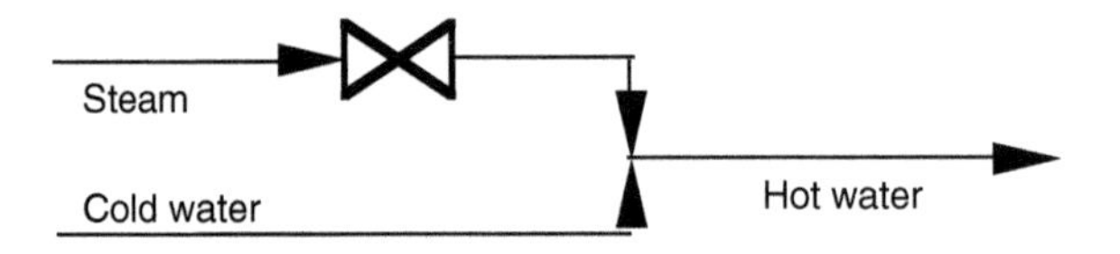

Note: You may assume that the properties of compressed liquid water are the same as those of saturated liquid water at the same temperature, i.e. that the properties of the compressed liquid are independent of pressure.

3.10 An ideal gas leaks from a container sufficiently slowly that the temperature of the gas remains constant. The volume of the container is 0.1 m^3 and the temperatures of the gas, cylinder, and surroundings are 300 K. How much energy transfers as heat when the pressure in the cylinder decreases from 100 to 1 bar?

3.11 An ideal gas with a molar mass of 29.0 and $C_P^{ig} = 3.5R$ leaks slowly (steady-state flow) from a container at 2.94 bar and 23 °C through a small hole into the surrounding atmosphere, which is at a pressure of 0.94 bar. Determine the temperature of the exit gas if the exit velocity from the small hole is 324 m/s. You may assume that the process is adiabatic.

3.12 An ideal gas with $C_P^{ig}/R = 3$ flows at steady state through a compressor and a heat exchanger. The temperature into the compressor is 300 K; the temperature out of the compressor is 500 K; the temperature out of the heat exchanger is 300 K.

a. How much energy (per mole of gas) transfers to or from the gas as work and as heat in this process? Express your answer in terms of R (e.g. $250R$).
b. If water enters the heat exchanger at 280 K, how many moles of water per mole of gas will maintain the water exit temperature at 340 K? Use $C_P^{ig}/R = 10$ for water.

3.13 A vertical piston and cylinder arrangement open to the atmosphere contains an ideal gas at 31.2 °C. The atmospheric pressure is 100.5 kPa. The piston has a mass of 4.62 kg and a diameter of 4.23 cm, and it initially is 12.8 cm above the bottom of the cylinder. The heat capacity at constant pressure of the ideal gas is $3.5R$. After 2.11 J of energy enters the gas as heat, the final temperature of the gas is 38.9 °C. (For this calculation, assume that no energy transfers as heat to the piston or the cylinder.)

a. Determine the pressure of the gas in the cylinder before and after the energy addition as heat.
b. Determine the height of the piston after the energy addition as heat.

3.14 50 mol of gas expand from a total volume of 1 m^3 to a total volume of 3 m^3 during a constant pressure non-flow process. The pressure is 100 kPa (1 kPa = 1 N/m^2 = 1 J/m^3), and the temperature of the gas changes from 300 to 365 K. For this gas, $C_V = 20.8$ J/(mol K) and $C_P = 29.0$ J/(mol K).

a. Determine the amount of energy transferred as heat during the expansion process.
b. Determine the amount of energy transferred as work during the process.
c. Determine the change in the internal energy of the gas during the process.

3.15 Air is drawn into an air supply system by a fan that raises the pressure of the air, and in the process, the temperature of the air. The process is adiabatic and reversible. The inlet conditions are 36 °C and 98.2 kPa, and the temperature at the outlet of the fan is 42 °C. For a steady-state flow of air through the system, determine

a. the pressure at the outlet of the fan;
b. the amount of work required by the fan per mole of air passing through the system.

3.16 Consider a perfect gas contained in an insulated piston and cylinder arrangement (assume negligible heat capacity for the piston and cylinder). For the gas, $C_P^{ig} = 41.868$ J/(mol °C) (independent of temperature). The initial temperature is 26.667 °C at 68.95 kPa. The gas is compressed to one half its original volume in an irreversible process that transfers 2326 kJ/(kg mol) of energy as work.

a. Determine the temperature and pressure of the gas after the compression process.
b. Determine the ratio of the work of this process relative to a reversible adiabatic compression process that accomplishes the same change in volume. Also, determine the final temperature and pressure resulting from the reversible process.

3.17 An ideal gas with $C_P^{ig} = 2.5R$ enters a turbine at a temperature of 700 K and a pressure of 4 bar. In the turbine, the gas expands in an adiabatic process to 1 bar and 442.5 K.
a. Determine the amount of work delivered by the turbine per mole of gas.
b. Is the expansion in the turbine reversible? (Justify your answer.)

3.18 The initial stage of a gas turbine compresses inlet air to three times the inlet pressure in a ***reversible adiabatic*** compression. Determine the temperature of the gas leaving the inlet stage, if the inlet conditions are 0.855 psia and 10 °C. You may treat air as in ideal gas with $C_P^{ig} = 3.5R$.

3.19 Consider an ideal gas with $C_V^{ig} = \frac{5}{2}R$ that undergoes the following three-step cycle in a closed system:
Step 1: ***Adiabatic reversible*** compression from 250 K and 1.2 to 4.6 bar;
Step 2: ***Reversible cooling*** at constant pressure (4.6 bar) to the original temperature;
Step 3: ***Isothermal reversible*** expansion to the original pressure.
a. Determine the temperature at the end of step 1.
b. Determine the mounts of energy transferred as heat and as work in each step. Indicate clearly in each case if the energy enters or leaves the gas.
c. Determine the net amounts of energy transferred as heat and as work for the entire cycle. Indicate clearly in each case if the energy enters or leaves the gas.

3.20 An ideal gas with $C_P^{ig}/R = 3$ passes through the following reversible cycle: isobaric expansion from 300 K and 2 bar until $V_2 = 2V_1$; adiabatic compression until $V_3 = V_1$; isochoric cooling to the original condition. What are ΔU^{ig}, ΔH^{ig}, Q, and W for each step and for the cycle?

3.21 An ideal gas with $C_P^{ig}/R = 3$ passes through the following reversible cycle: adiabatic compression from 300 K and 1 to 8 bar; isochoric cooling to 432.8 K; isothermal expansion to the original volume; and isochoric cooling to the original condition. What are Q and W for each step and for the cycle? Express your answers in terms of R.

3.22 An ideal gas passes through the following cycle:
- Isothermal expansion from 10 MPa and 300 K to 1 MPa;
- Isochoric (constant volume) heating until the pressure reaches 5 MPa;

- Isobaric cooling to the initial volume;
- Isochoric heating to the initial condition.

Find ΔU^{ig}, ΔH^{ig}, Q, and W for each step. Use $C_P^{ig} = 5R$.

3.23 A closed system contains 2.25 mol of an ideal gas that initially is at a temperature of 300 K and a pressure of 5.0 MPa. This gas has a heat capacity at a constant pressure of 3.5R. The gas then undergoes the following sequence of changes: reversible isothermal expansion to a pressure of 1.0 MPa; reversible adiabatic compression by a factor of 4 in volume; cooling at a constant volume to 5.0 MPa; and cooling at a constant pressure to 300 K.

a. Sketch this cycle on a PV diagram.
b. Determine W^{total}, Q^{total}, ΔU^{total}, and ΔH^{total} for each step and for the entire cycle.

3.24 An ideal gas with $C_P^{ig} = 31.987\ \text{J}/(\text{mol}\ °\text{C})$ that initially is at 30.7 °C and 1.469 bar undergoes the following cycle:

- Isothermal reversible compression to three times the original pressure;
- Isochoric transfer of energy as heat to the initial pressure;
- Isobaric transfer of energy as heat to the initial temperature.

Determine ΔU, ΔH, Q, and W for each step, and the net values of each for the entire cycle. For each of the last two steps, indicate whether the process involves heating or cooling.

3.25 An ideal gas with $C_P^{ig} = 3.5R$ undergoes the following sequence of three non-flow processes:

(1) adiabatic reversible compression from 26.7 °C and 1 bar to 102 °C and 2 bar
(2) heat transfer at a constant pressure raises the temperature to 454 °C; and
(3) adiabatic reversible expansion to 1 bar.

a. Sketch this sequence of processes on a PV diagram.
b. Calculate the energy per mole transferred as work to accomplish the first step. (Be sure to specify clearly in which direction the energy transfer occurs.)
c. Calculate the energy per mole required as heat to accomplish the second step.
d. Calculate the outlet temperature after the third step (expansion process).
e. Calculate the energy transferred as work per mole of gas during the expansion process.
f. Calculate the net energy transferred as work for the sequence of three processes.

Express your results in terms of R.

3.26 Consider an ideal gas with $C_V^{ig} = 1.5R$ that undergoes the following three-step cycle in a closed system

Step 1: *Isochoric heating* from 302 K and 1.30 bar to 604 K

Step 2: *Reversible isothermal expansion* to 1.30 bar
Step 3: *Reversible isobaric cooling* to the original state

a. Determine the pressure at the end of step 1.
b. Determine ΔU, ΔH, Q, and W for each step (in J/mol)
c. Determine the net values of ΔU, ΔH, Q, and W for the entire cycle (in J/mol) and state clearly whether the net transfer of energy as heat is into or out of the system.

3.27 Compare the ***isothermal reversible*** compression at 330 K of an ideal gas with a gas that obeys the virial equation with $B_2 = -125\ \text{cm}^3/\text{mol}$, $B_3 = 6472\ \text{cm}^6/\text{mol}^2$, and B_4, B_5, B_6,... = 0. The compression begins at a molar volume of 15 600 cm^3/mol and ends at 3423 cm^3/mol.

a. Determine the final pressure for each gas and the relative difference (percentage) between the two pressures.
b. Determine the relative difference (percentage) between the work required for the ideal gas and the work required for the gas that obeys the virial equation. Which gas requires the larger amount of work?

3.28 Saturated liquid isobutane at 53 °C completely fills a 3785 l tank.

a. Estimate the mass of isobutane in the tank (in kg).
b. Estimate the internal pressure in the tank if the fluid temperature warms to 61 °C at which the vapor pressure is 8.894 bar.

Use the Chang and Zhao equation to find the compressed volume. The critical parameters of isobutene are

$$T_C = 408.1\ \text{K}, P_C = 36.48\ \text{bar}, \omega = 0.181, \text{and } V_C = 262.7\ \text{cm}^3/\text{mol}$$

3.29 Determine the mass flow rate (in kilograms per hour) for normal pentane in the saturated liquid state at 375 K, if it flows with a linear velocity of 10 m/s through a pipe that has an inside diameter of 10 cm. For the calculation of the saturated liquid volume, use the Rackett equation. The critical parameters of pentane are

$$T_C = 469.7\ \text{K}, P_C = 33.7\ \text{bar}, \omega = 0.252, \text{and } V_C = 313\ \text{cm}^3/\text{mol}$$

3.30 Determine the mass flow rate (in kilograms per hour) for ethylene flowing in a pipeline at 310 K and 100 bar with a linear velocity of 40 m/s. The inside diameter of the pipe is 20 cm. Use a generalized correlation and assume the critical parameters of ethylene are

$$T_C = 282.35\ \text{K}, P_C = 50.418\ \text{bar}, \text{and } \omega = 0.087$$

3.31 Use the Lee–Kesler tables to determine the molar volume of isobutylene at 460 K and 44 bar. For isobutylene, $P_C = 40$ bar and $T_C = 417.9$ K.

3.32 Calculate the molar density of propane at 406.78 K and 63.72 bar. For propane, $P_C = 42.4766$ bar and $T_C = 369.825$ K.

3.33 $CClF_3$ (also known as Freon 13 or R-13) is a potential solvent for supercritical extraction purposes, and a useful refrigerant for cryogenic refrigerators. The material is supplied to customers in cylinders having an internal volume of 44 740 cm^3. When the cylinders contain the proper amount of R-13, the liquid phase occupies two thirds of the total volume at 21.1 °C. Determine the total mass of R-13 in the container and the fraction of the mass that is present in the vapor phase (quality). The vapor pressure at 21.1 °C is 33 bar
Useful information for R-13:

Critical properties				
T_C (K)	P_C (bar)	V_C (cm^3/mol)	Z_C	ω
301.88	38.79	180	0.2811	0.180

3.34 A container with a total volume of 1.92 l contains 33.3 mol of sulfur dioxide.

a. At what temperature is the container exactly filled with saturated liquid sulfur dioxide?
b. The vapor pressure of sulfur dioxide at 300 K is 4.1 bar. At 300 K, is the sulfur dioxide entirely in the liquid state, entirely in the vapor state or are both liquid and vapor phases present in the container? If both liquid and vapor phases are present, determine the fraction of the total volume occupied by the vapor phase. What is the pressure in the container?

The critical parameters for sulfur dioxide are

$$T_C = 430.8\text{ K}, P_C = 78.84\text{ bar}, \omega = 0.245, \text{and } V_C = 122\text{ cm}^3/\text{mol}$$

3.35 Derive an expression for the work required for a reversible isothermal compression from V_1 to V_2 of a fluid that obeys the vdW EOS.

3.36 Calculate the mass of carbon dioxide stored in a tank at 292 K with a total volume of 15 m^3 if 5/8 of the volume is liquid and the remainder is vapor. The pressure in the tank is 59.9 bar. Use the vdW EOS to find your estimate.

3.37 Liquid toluene at 26.7 °C enters an evaporator that operates at the normal boiling point of toluene (110.65 °C). The toluene leaves the unit as a saturated vapor at 1 atm. Estimate the evaporator duty (the energy supplied as heat to the toluene) required to vaporize 3 tons of toluene per hour. If you need an estimate of the enthalpy of vaporization, use the Carruth and Kobayashi correlation. Useful information

$$T_C = 591.8\text{ K}, \omega = 0.262, \text{ and } C_P^L = 83.703 + 0.517T - 1.49 \times 10^{-3}T^2 + 1.97 \times 10^{-6}T^3$$

3.38 A hexane stream at 450 K enters a condenser that operates at atmospheric pressure and provides a saturated liquid product stream. Determine the condenser duty (amount of energy transferred as heat) per mole of hexane condensed. The entire process occurs at a constant pressure. The normal boiling point temperature is 341.9 K. The Carruth and Kobayashi correlation can estimate the heat of vaporization. The critical temperature of hexane is 507.6 K and $\omega = 0.301$. The vapor heat capacity is

$$C_P^V/R = 3.025 + 53.722 \times 10^{-3}T - 16.971 \times 10^{-6}T^2$$

3.39 Before rural homes received electricity, gas lamps using acetylene as a fuel lit many such houses. Similar lamps are still in use in the mining industry. The fuel results from adding solid calcium carbide to liquid water to produce acetylene gas, and solid calcium hydroxide precipitates according to the following reaction:

$$CaC_{2(s)} + 2H_2O_{(l)} \rightarrow C_2H_{2(g)} + Ca(OH)_{2(s)}$$

The molar masses of calcium carbide and calcium hydroxide are 64.099 and 74.093, respectively.

a. Determine the mass of calcium carbide required to produce 0.0283 m^3 (at 15 °C, 1 atm) of acetylene gas.
b. Determine the amount of energy transferred to the surroundings per mole of calcium carbide reacted at 25 °C. Is the energy added to or removed from the surroundings?

Data

Substance	$\Delta H^0_{f,i}$
CaC_2	−59 800
H_2O	−285 830
C_2H_2	227 480
$Ca(OH)_2$	−986 090

3.40 A stream of pure benzene enters a condenser operating at a pressure of 3.35 bar as a vapor at 175 °C and leaves as a liquid at 34 °C. The vapor pressure of benzene is 3.35 bar at 124.7 °C. Determine the amount of energy removed as heat per mole of benzene passing through the condenser. You may treat the vapor as an ideal gas, and you may assume that the liquid heat capacity is independent of pressure. The ideal gas and the liquid heat capacities of benzene are

$$C_P^V/R = -0.206 + 39.064 \times 10^{-3}T - 13.301 \times 10^{-6}T^2$$

$$C_P^L = -31.663 + 1.3043T - 3.6078 \times 10^{-3}T^2 + 3.8243 \times 10^{-6}T^3$$

Use the Carruth and Kobayashi correlation to estimate the heat of vaporization.

3.41 Determine the standard heat of reaction for the synthesis of methanol from carbon monoxide at 500 °C

Data

$$C_{P,i}^{ig}/R = A_i + B_i T + C_i T^2 + D_i T^{-2}$$

Substance	*A*	*B*	*C*	*D*	$\Delta H_{f,i}^0$ (J/mol)
CH_3OH	2.211	1.22×10^{-2}	-3.45×10^{-6}	0	−200 660
CO	3.376	5.57×10^{-4}	0	-3.10×10^{3}	−110 525
H_2	3.249	4.22×10^{-4}	0	8.30×10^{3}	0

3.42 Determine the heat of combustion of methane at 500 K based upon water in the vapor state as a combustion product. The heat capacity is

$$C_P^{ig}/R = A + BT + CT^2 + DT^{-2}$$

Data

Substance	*A*	*B*	*C*	*D*	$\Delta H_{f,i}^0$ (J/mol)
CH_4	1.702	9.08×10^{-3}	-2.16×10^{-6}	0	−74 520
O_2	3.639	5.06×10^{-4}	0	-2.27×10^{4}	0
H_2O	3.47	1.45×10^{-3}	0	1.21×10^{4}	−241 818
CO_2	5.457	1.05×10^{-3}	0	-1.16×10^{5}	−393 509

3.43 Calculate the standard heat of combustion, based on SO_2 and H_2O both as vapors, of hydrogen sulfide at 25 °C and 1 atm

Data

Substance	$\Delta H_{f,i}^0$
H_2S	−20 630
O_2	0
SO_2	−296 830
H_2O	−241 818

3.44 Nitrogen and hydrogen in the proper stoichiometric ratio to make ammonia enter a constant pressure reactor at 25 °C. If the reaction goes to completion and the ammonia leaves the reactor at 600 K, determine the amount of energy transferred as heat to accomplish this reaction. Does the energy enter or leave the reactor as heat?

Data

$$C_{P,i}^{ig}/R = A_i + B_iT + C_iT^2 + D_iT^{-2}$$

Substance	A	B	C	D	ΔH_f^0 (J/mol)
Ammonia	3.580	3.02×10^{-3}	-2.164×10^{-6}	-1.86×10^{4}	−46 110
Nitrogen	3.280	5.93×10^{-4}	0	4.00×10^{3}	0
Hydrogen	3.250	4.22×10^{-4}	0	8.30×10^{3}	0

3.45 Determine the standard heat of formation of benzene at 200 °C based on the conventional choice of standard states for the various chemical species involved.

Data

$$C_P^{ig}/R = A + BT + CT^2 + DT^{-2}$$

Substance	A	B	C	D	ΔH_f^0 (J/mol)
C	1.771	7.710×10^{-4}	0	−86 700	0
H_2	3.249	4.220×10^{-4}	0	8 300	0
C_6H_6	−0.206	3.9064×10^{-2}	-1.3301×10^{-5}	0	82 930

3.46 Ethane converts into ethylene in a "cat-cracking" furnace in which the following chemical reaction occurs

$$C_2H_6 \rightarrow C_2H_4 + H_2$$

Consider a process in which a pure ethane stream enters the furnace at 25 °C and 1 bar. If 30% of the ethane converts into ethylene and hydrogen, and if the resulting mixture leaves the furnace at 450 °C and 1 bar, calculate the amount of energy transferred as heat per mole of ethane fed into the process. Does the energy enter or leave the furnace?

Data

$$C_{P,i}^{ig}/R = A_i + B_iT + C_iT^2 + D_iT^{-2}$$

Substance	A	B	C	D	ΔH_f^0 (J/mol)
Ethane	1.131	1.9225×10^{-2}	-5.561×10^{-6}	0	−83 820
Ethylene	1.424	1.4394×10^{-2}	-4.392×10^{-6}	0	52 510
Hydrogen	3.249	4.22×10^{-4}	0	8.30×10^{3}	0

3.47 Calculate the gross heating value of methane at 288.71 K. The gross heating value is the negative of the enthalpy of combustion when all water formed condenses to liquid.
Data

$$C_P^{ig}/R = A + BT + CT^2 + DT^{-2}$$

Substance	A	B	C	D	ΔH_f^0 (J/mol)
CH_4	1.702	9.08×10^{-3}	-2.16×10^{-6}	0	−74 520
O_2	3.639	5.06×10^{-4}	0	-2.27×10^{4}	0
H_2O	3.47	1.45×10^{-3}	0	$1.21\text{E} \times 10^{4}$	−241 818
CO_2	5.457	1.05×10^{-3}	0	$-1.16\text{E} \times 10^{5}$	−393 509

References

1 Thiesen, M. (1885). Untersuchungen über die Zustandsgleichung. *Ann. Phys.* 24: 467–492.

2 (a) Kamerlingh Onnes, H. (1901). Expression of the equation of state of gases and liquids by means of series. *Commun. Phys. Lab. Univ. Leiden* 71: 3–25. (b) Kamerlingh Onnes, H. (1902). Expression of the equation of state of gases and liquids by means of series. *Koninklijke Nederlandse Akademie van Wetenschappen Proceedings Series B, Physical Sciences*, 4: 125–147.

3 Ursell, H.D. (1927). The evaluation of Gibbs' phase-integral for imperfect gases. *Math. Proc. Cambridge Philos. Soc.* 23: 685–697.

4 Pitzer, K.S. and Curl, R.F. (1957). The volumetric and thermodynamic properties of fluids. III. Empirical equation for the second virial coefficient. *J. Am. Chem. Soc.* 79: 2369–2370.

5 Pitzer, K.S., Lippmann, D., Curl, R.F. et al. (1955). The volumetric and thermodynamic properties of fluids. II. Compressibility factor, vapor pressure and entropy of vaporization. *J. Am. Chem. Soc.* 77: 3433–3440.

6 (a) Abbot, M.M. (1973). Cubic equations of state. *AIChE J.* 19: 596–601; (b) Smith, J. and Van Ness, H.C. (1975). *Introduction to Chemical Engineering Thermodynamics*. McGraw Hill.

7 (a) Tsonopolous, C. (1974). An empirical correlation of second virial coefficients. *AIChE J.* 20: 263–272. (b) Tsonopolous, C. and Heidman, J.L. (1990). From the virial to the cubic equation of state. *Fluid Phase Equilib.* 57: 261–276.

8 Iglesias-Silva, G.A. and Hall, K.R. (2001). An equation for prediction and/or correlation of second virial coefficients. *Ind. Eng. Chem. Res.* 40: 1968–1974.

9 Abusleme, J.A. and Vera, J.H. (1989). Pure compound and cross second virial coefficients for *n*-amylamine – alcohol systems at 363.15 K. Measurements with a low-pressure Burnett-type apparatus. *Fluid Phase Equilib.* 45: 287–302.

10 De Santis, R. and Grande, B. (1979). An equation for predicting third virial coefficients of nonpolar gases. *AIChE J.* 25: 931–938.

11 Orbey, H. and Vera, J.H. (1983). Correlation for the third virial coefficient using T_C, P_C and as parameters. *AIChE J.* 29: 107–113.

12 Tait, P.G. (1888). Report on some of physical properties of fresh water and of sea water. In: *Physics and Chemistry of the Voyage of H.M.S. Challenger*, vol. 2, Part 4. London: HMSO, Printed by Morrison and Gibb for Her Majesty Stationary Office.

13 Rackett, H.G. (1970). Equation of state for saturated liquids. *J. Chem. Eng. Data* 15 (4): 514–517.

14 Riedel, L. (1954). Die Fiussigkeitsdichte im Sattigungszustand Untersuchungen uber eine Erweiterung des Theorems der ubereinstimmenden Zwstande. Teil II. *Chem. Ing. Tech.* 26: 259–264.

15 Yen, L.C. and Woods, S.S. (1966). A generalized equation for computer calculation of liquid densities. *AIChE J.* 12: 95–99.

16 Chueh, P.L. and Prausnitz, J.M. (1969). A generalized correlation for the compressibilities of normal liquids. *AIChE J.* 15: 471–472.

17 Chang, C.-H. and Zhao, X. (1990). A new generalized equation for predicting volumes of compressed liquids. *Fluid Phase Equilib.* 58: 231–238.

18 Lee, B.I. and Kesler, M.G. (1976). A generalized thermodynamic correlation based on three-parameter corresponding states. *AIChE J.* 21: 510–527.

19 van de Waals, J. D. (1873). Over de Continuiteit van den Gas-en Vloeistoftoestand. Doctoral Thesis, University of Leiden, Leiden.

20 Clausius, R. (1880). Über das Verhalten der Kohlensäure in Bezug auf Druck, Volumen und Temperatur. *Ann. Phys. Chem.* 9: 337–357.

21 Berthelot, D. (1900). Quelques Remarques sur l'Équation Charactéristique des Fluides. *Arch. Netherl. Sci. Exactes Naturelles*, II (5): 417–446.

22 Lee, B.I., Erbar, J.H., and Edmister, W.C. (1973). Prediction of thermodynamic properties for low temperature hydrocarbon process calculations. *AIChE J.* 19: 349–353.

23 Harmens, A. and Knapp, H. (1980). Three-parameter cubic equation of state for normal substances. *Ind. Eng. Fundam.* 19: 291–294.

24 Redlich, O. and Kwong, J.N.S. (1949). On the thermodynamics of solutions. V. An equation of state. Fugacities of gaseous solutions. *Chem. Rev.* 44 (1): 233–244.

25 Soave, G. (1972). Equilibrium constants from a modified Redkh–Kwong equation of state. *Chem. Eng. Sci.* 27: 1197–1203.

26 Peng, D.-Y. and Robinson, D.B. (1976). A new two-constant equation of state. *Ind. Eng. Chem. Fundam.* 15: 59–64.

27 (a) Kumar, K.H. and Starling, K.E. (1980). Comments on: "Cubic equations of state-which?". *Ind. Eng. Chem. Fundam.* 19 (1): 128–129. (b) Kumar, K.H. and Starling, K.E. (1982). The most general density-cubic equation of state: application to pure nonpolar fluids. *Ind. Eng. Chem. Fundam.* 21: 255–262.

28 Benedict, M., Webb, G.B., and Rubin, L.C. (1942). Mixtures of methane, ethane, propane and *n*-butane. *J. Chem. Phys.* 10: 747–758.

29 Chen, S.S. and Kreglewski, A. (1977). Applications of the augmented van der Waals theory of fluids.: I Pure fluids. *Ber. Bunsen Ges. Phys. Chem.* 81: 1048–1052.

30 Gross, J. and Sadowski, G. (2001). Perturbed-chain SAFT: an equation of state based on a perturbation theory for chain molecules. *Ind. Eng. Chem. Res.* 40: 1244–1260.

31 (a) Antoine, C. (1888). Tensions des Vapeurs: Nouvelle Relation entre les Tensions et les Températures. *C. R. Acad. Sci.* 107: 681–684. (b) Antoine, C. (1888). Calcul des Tensions des Diverses Vapeurs. *C. R. Acad. Sci.* 107: 778–780. (c) Antoine, C. (1888). Tensions des Diverses Vapeurs. *C. R. Acad. Sci.* 107: 836–8372.

32 Wichterle, J. and Linek, J. (1971). *Antoine Vapor Pressure Constants of Pure Compounds.* Praha: Academia.

33 Yaws, C.L. and Yang, H.-C. (1989). To estimate vapor pressure easily. A major compilation is: Antoine coefficients relate vapor pressure to temperature for almost 700 major organic compounds. *Hydrocarbon Process.* 68: 65–68.

34 Wagner, W. (1973). New vapour pressure measurements for argon and nitrogen and a new method for establishing rational vapour pressure equations. *Cryogenics* 13: 470–482.

35 Gómez-Nieto, M. and Thodos, G. (1977). A new vapor pressure equation and its application to normal alkanes. *Ind. Eng. Chem. Fundam.* 16: 254–259.

36 Watson, K.M. (1943). Thermodynamics of the liquid state-generalized prediction of properties. *Ind. Eng. Chem.* 35: 398–406.

37 Carruth, G.F. and Kobayashi, R. (1972). Extension to low reduced temperatures of three parameter corresponding states: vapor pressures, enthalpies and entropies of vaporization, and liquid fugacity coefficients. *Ind. Eng. Chem. Fundam.* 11: 509–516.

38 Morgan, D.L. (2007). Use of transformed correlations to help screen and populate properties within databanks. *Fluid Phase Equilib.* 256: 54–61.

4

Second Law of Thermodynamics

4.1 Introduction

We have learned that energy is a conserved variable and that transformations of energy follow the first law of thermodynamics. The second law, like the first law, is a part of normal human experience (except it is easier to observe its effects directly that those of the first law). Using the second law, we can determine the unique direction in which a spontaneous process proceeds, and how this direction relates to equilibrium (because spontaneous processes always proceed in the direction of equilibrium). For example, a stone can fall until it hits the ground, a perfume (concentrated solution) can disperse into the air (dilute solution) until it reaches a certain concentration, when two objects are in close contact excluding any interaction between them and their surroundings, the cold one always becomes hotter and the hot one becomes cooler. In other words, we must conserve energy, but we cannot do it just anyway we wish!

We know from experience that the above examples occur in a specified direction, but the second law tells us mathematically why they must proceed in those particular directions. Interestingly, we can apply the first law to a system moving in either direction, and formulate an expression for each path (the expression is the same but with opposite signs), but the first law cannot determine the correct direction. In general, natural processes (removed from external forces) proceed toward a state of macroscopic rest with uniform distributions of temperature and concentration. In other words, natural processes proceed only in one direction, and they cannot return spontaneously to their original states without assistance from external forces.

How can we relate the second law to our everyday experiences? We know that it can tell us if a process can proceed in certain directions, but we also can include *the efficiency* of the process and determine the thermodynamic feasibility and the direction of the process. In this chapter, we deal with new definitions, such as reversible and irreversible processes, reservoirs and heat engines, applications of the second law to ideal gases and second law calculations.

Thermodynamics for Chemical Engineers, First Edition. Kenneth R. Hall and Gustavo A. Iglesias-Silva.

4.2 General and Classical Statements of the Second Law

First, let us formulate a general statement of the second law that encompasses all its classical statements. A general statement of the second law would be

It is possible to conceive of processes that, while conserving energy, are impossible to construct.

This statement again describes observations. We observe that certain processes cannot occur, although it is conceivable that they could occur (but not in a stable manner). For example, consider the situation in which we have two, pure, miscible gas components A and B in a container separated by a membrane. Suddenly, the membrane ruptures, and the gasses form a mixture of A + B. An un-mixing process cannot occur spontaneously, although we can conceive of such a process and even write a first law expression for that process.

The second law, in any of its forms or statements, covers *all* processes and acts as a filter to prevent improper solutions to thermodynamics problems. Natural processes proceed in one direction, so we call them *irreversible*. Thus, an *irreversible process* is that one that cannot return it to its original state without a finite change in the surroundings of the system. Also, we can define its counterpart, a *reversible process*, that can return to its original state with an infinitesimal change in its surroundings. While reversible processes do not occur naturally, they can be powerful tools for solving thermodynamic problems.

Classical statements of the second law come from the nineteenth century. The most prominent among them are:

Kelvin Statement (KS): *It is impossible to extract heat from a reservoir and convert it wholly into work without causing other changes in the universe.*

Planck Statement (PS): *No cyclic process is possible whose result is the flow of heat from a single heat reservoir and the performance of an equivalent amount of work. This theorem is also called the Impossibility of perpetuum mobile (perpetual motion) of the second kind.*

Clausius Statement (CS): *Heat can never of itself flow from a lower to a higher temperature.*

Actually, all these statements are equivalent. KS and PS are obviously the same statement with British and German wording preferences. It is interesting to prove that CS is also the same statement. Before that, let us define reservoirs and heat engines.

Reservoir: This is a concept denoting an energy sink that can accept or deliver any amount of energy without changing temperature.

Heat Engine: This is a device that can transfer energy as heat to or from either a high temperature or a low temperature reservoir, and also it can transfer energy as work to or from its surroundings. Note that we use the same sign convention as earlier in this book: energy transfers are positive if they enter the system (the

Figure 4.1 Diagram of an engine.

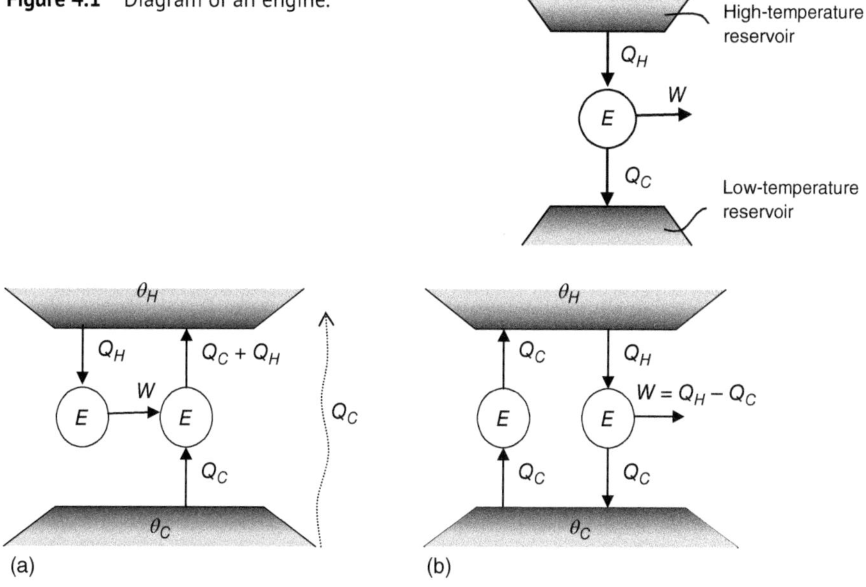

Figure 4.2 Schematic diagrams to prove equivalent between the statements of the second law (a) Engines exchanging work. (b) Engines not exchanging work.

engine in this case), and they are negative if they leave the system. Figure 4.1 illustrates a simple heat engine.

Now, we can prove that the statements are equivalent. Let us use a new notation: $+KS$ indicates that the object of our attention satisfies the Kelvin statement, and $-KS$ indicates that the object of our attention violates the Kelvin statement. It is obvious that

$$+KS \subset +CS \text{ and } +CS \subset +KS$$

in which the $\subset$ symbol denotes "implies." The statement reads as: "a process that satisfies the Kelvin statement also satisfies the Clausius statement." This is so because both Kelvin and Clausius statements describe observations that always occur as stated. If we can prove that the violation of the Kelvin statement is also the violation of the Clausius statement and vice versa, that is

$$-KS \subset -CS \text{ and } -CS \subset -KS$$

then the Kelvin statement is "equivalent to" the Clausius statement and $+KS \equiv +CS$.

We can conceive of two engines arranged as shown in Figure 4.2a. Clearly, the arrangement contains a violation, $-KS$. An energy balance reveals a net energy transfer as heat of Q_C from the cold reservoir to the hot reservoir. This violates CS, so we have $-KS \subset -CS$. Now, let us use another set of reservoirs connected by heat engines arranged differently as shown in Figure 4.2b. This arrangement contains a violation, $-CS$, and the energy balance shows that the net effect is to extract $Q_H - Q_C$ from the hot reservoir and convert it entirely into work; thus, $-CS \subset -KS$. This establishes the equivalence of KS (or PS) and CS.

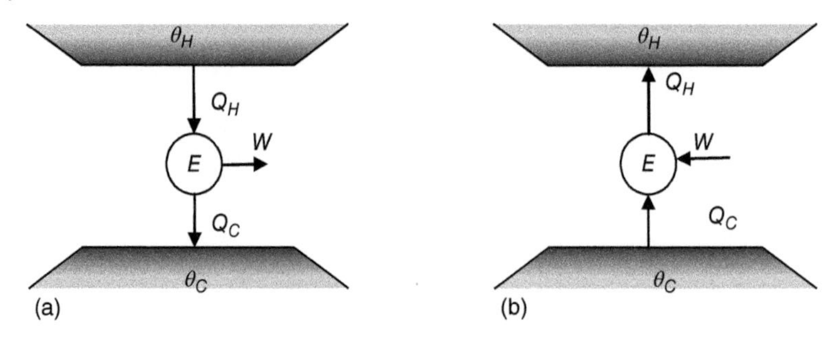

Figure 4.3 Schematic diagrams of (a) heat engine and (b) refrigerator.

4.3 Heat Engines, Refrigerators, and Cycles

In our daily lives, we often come into contact with engines. For example, our automobile might have a gasoline or diesel engine; a boat might use a steam engine; etc. In thermodynamics, we use two general types of engines:

A *heat engine* is a physical device that extracts energy as heat from a high-temperature reservoir, delivers energy as heat to a low-temperature reservoir, and transfers energy as work to the surroundings. In this case, $Q_H > 0$, $Q_C < 0$, and $W > 0$. Figure 4.3a shows the diagram of a heat engine. The engine is the system; thus, energy into the system is positive and energy out of the system is negative.

Another kind of engine is *a refrigerator or a heat pump*, which is a physical device that extracts energy as heat from a low-temperature reservoir and delivers energy as heat to a high-temperature reservoir, while energy transfer as work enters the system from the surroundings. In this case, $Q_H < 0$, $Q_C > 0$, and $W < 0$. One should note that the only difference between a refrigerator and a heat engine is the direction of the energy transfer as heat and work. Figure 4.3b shows a schematic diagram of a refrigerator with $Q_H < 0$, $Q_C > 0$, and $W > 0$.

Heat engines (namely, heat pumps) are physical devices that operate using thermodynamic cycles. A thermodynamic cycle contains multiple paths, and the system can reject or receive energy transfer as work or heat until the system returns to its initial state.

From the definition of heat engines and the statements of the second law, we can conclude that an engine cannot be 100% efficient. Then, what is the maximum efficiency a heat engine can achieve? Probably, the best guess would be the one that follows a reversible path through the process. Nicolas Léonard Sadi Carnot conceived a hypothetical engine in 1824 that operates along reversible paths. In 1834, Benoit Paul Émile Clapeyron represented Carnot's engine graphically on a pressure–volume diagram. The Carnot cycle consists of four paths:

1. Adiabatic reversible compression, followed by
2. Isothermal reversible expansion, followed by
3. Adiabatic reversible expansion, followed by
4. Isothermal reversible compression.

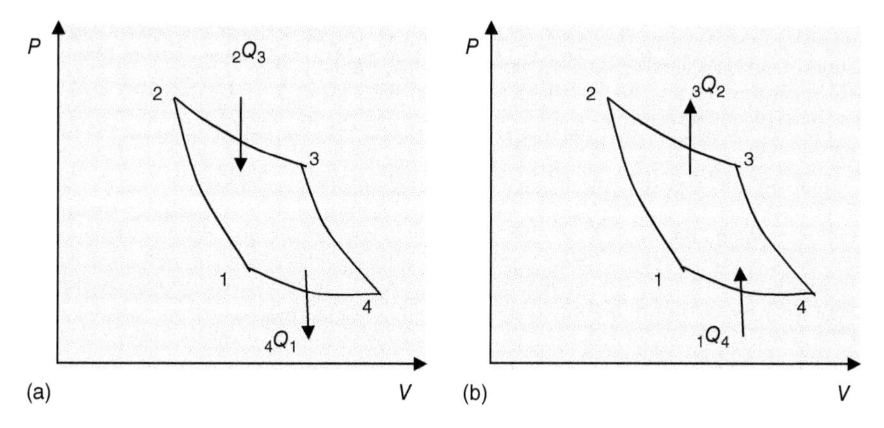

Figure 4.4 P–V diagram for (a) a Carnot engine and (b) a Carnot refrigerator.

Following the Carnot cycle in a heat engine:

1→2 Adiabatic reversible compression:	${}_1Q_2 = 0,\ {}_1W_2 > 0$	Heat engine
2→3 Isothermal reversible expansion:	${}_2Q_3 > 0,\ {}_2W_3 < 0$	
3→4 Adiabatic reversible expansion:	${}_3Q_4 = 0,\ {}_3W_4 < 0$	
4→1 Isothermal reversible compression:	${}_4Q_1 < 0,\ {}_4W_1 > 0$	

$W = {}_1W_2 + {}_2W_3 + {}_3W_4 + {}_4W_1 < 0$ for a heat engine. We often refer to ${}_1Q_2$ as Q_H because it comes from a hot reservoir and to ${}_4Q_1$ as Q_C because it flows to a cold reservoir.

In a refrigerator:

2→1 Adiabatic reversible expansion:	${}_2Q_1 = 0,\ {}_2W_1 < 0$	Refrigerator
1→4 Isothermal reversible expansion:	${}_1Q_4 > 0,\ {}_1W_4 < 0$	
4→3 Adiabatic reversible compression:	${}_4Q_3 = 0,\ {}_4W_3 > 0$	
3→2 Isothermal reversible compression:	${}_3Q_2 < 0,\ {}_3W_2 > 0$	

$W = {}_2W_1 + {}_1W_4 + {}_4W_3 + {}_3W_2 > 0$ for a refrigerator. Note that the only difference between the heat engine and the refrigerator is the direction in which the cycle operates (Figure 4.4).

4.4 Implications of the Second Law

To develop the mathematical expression for the second law, we investigate four propositions. Using these propositions, we can generate the fundamental formulas that describe the second law. The propositions are:

1. If a system undergoes a reversible cycle absorbing energy as heat, Q_H, from a hot reservoir with "hotness" described by θ_H and rejects energy as heat, Q_C, to a cold

reservoir with "coldness" described by θ_C, then the ratio of the energy transfer as heat is a function of θ_H and θ_C for the Carnot cycle

$$\frac{Q_H}{Q_C} = f(\theta_H, \theta_C) \tag{4.1}$$

2. The function in part 1 is the ratio of the reservoir temperatures.

$$f(\theta_C, \theta_H) = T_H/T_C \tag{4.2}$$

3. We can define a state function known as the entropy, which serves as a quantitative measure of the reversibility or irreversibility of a process

$$dS \equiv đQ_{rev}/T \tag{4.3}$$

in which $đ$ denotes an infinitesimally small quantity (on the order of a differential in size), but it is not a differential.

4. The total entropy change of the universe is zero for all reversible processes and positive for all irreversible processes. Processes cannot occur that would cause the total entropy of the universe to decrease.

$$\Delta S_{total} \geq 0 \tag{4.4}$$

To examine these propositions, we can use Carnot cycles.

Proposition 1 Let us utilize two Carnot cycles: one with steam as the working fluid and one with an ideal gas as the working fluid as shown in Figure 4.5.

Performing an energy balance on each cycle

$$-W = Q_H - Q_C \tag{4.5}$$

If we define the thermodynamic efficiency of the engine to be the energy output as work divided by the energy input as heat, we obtain

$$\eta = \frac{-W}{Q_H} = \frac{|W|}{Q_H} = \frac{Q_H - Q_C}{Q_H} = 1 - \frac{Q_C}{Q_H} \tag{4.6}$$

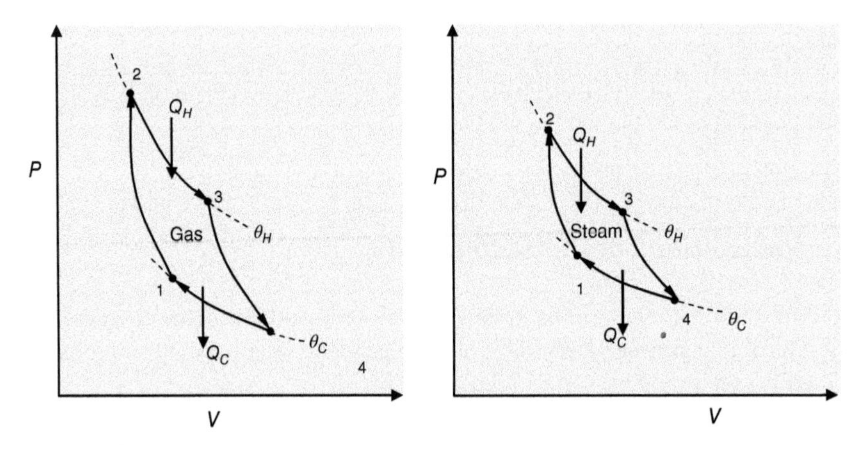

Figure 4.5 Carnot cycles using ideal gas and steam as a working fluid.

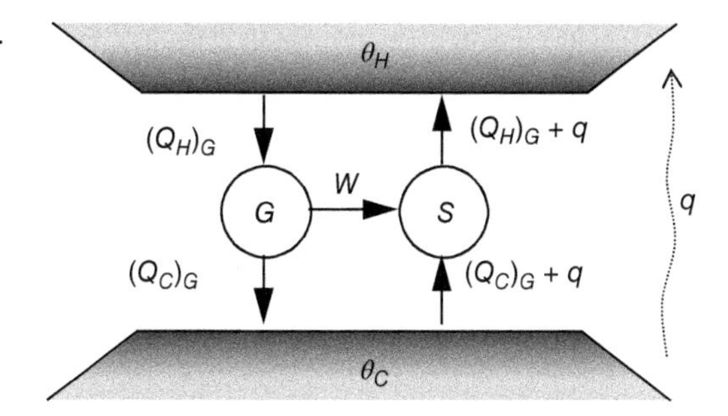

Figure 4.6 Cycles driving S with G.

Now, if both cycles transfer the same amount of energy as work

$$-W_S = -W_G = (Q_H - Q_C)_S = (Q_H - Q_C)_G \tag{4.7}$$

If Q_H is the same for both cycles, by definition, they have the same efficiencies from Eq. (4.6). Now, let us assume that Q_H are not the same for both cycles but that W is still the same. This indicates that

$$(Q_H)_S = (Q_H)_G + q \tag{4.8}$$

and using Eq. (4.7)

$$(Q_C)_S = (Q_C)_G + q \tag{4.9}$$

Now, we can create cycles operating between θ_H and θ_C driving S with G as shown in Figure 4.6. An energy balance demonstrates that the net effect is to transfer q from a lower temperature to a higher one, which violates the second law; therefore, $q = 0$ and $\eta_G = \eta_S$ or

$$(Q_C/Q_H)_S = (Q_C/Q_H)_G \tag{4.10}$$

This result would be the same for any working fluid, pressure, or volume because they are completely arbitrary in the definition of the cycles. Therefore, the only properties upon which Q_H/Q_C can depend are θ_H and θ_C. This observation along with Eq. (4.6) provides

$$\frac{Q_H}{Q_C} = f(\theta_H, \theta_C) = (1 - \eta)^{-1} \tag{Proposition 1}$$

Proposition 2 Now, let us see what the second law says about temperature. Using Carnot cycles in a new arrangement, as shown in Figure 4.7.

Using Proposition 1 for each cycle, we have

$$\frac{Q_H}{Q_I} = f(\theta_H, \theta_I) \tag{4.11}$$

$$\frac{Q_I}{Q_C} = f(\theta_I, \theta_C) \tag{4.12}$$

$$\frac{Q_H}{Q_C} = f(\theta_H, \theta_C) = \frac{Q_H}{Q_I}\frac{Q_I}{Q_C} = f(\theta_H, \theta_I) \cdot f(\theta_I, \theta_C) \tag{4.13}$$

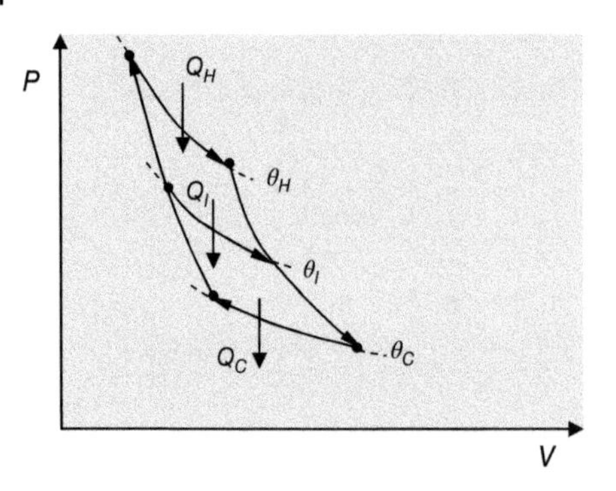

Figure 4.7 Arrangement of two cycles to prove Proposition 2.

but the only form of f that satisfies this expression is

$$f(\theta_i, \theta_j) = \frac{g(\theta_i)}{g(\theta_j)} \tag{4.14}$$

The function g is arbitrary. Kelvin chose $g(\theta) = \exp(\theta)$, which has certain advantages, but the modern selection is $g(\theta) = \theta = T$ in which T is the ideal gas temperature

$$g(\theta) = T = 273.16 \lim_{P \to 0} \left[\frac{(PV)_T}{(PV)_{273.16}} \right] \tag{4.15}$$

Then,

$$f(\theta_H, \theta_C) = \frac{\theta_H}{\theta_C} = \frac{T_H}{T_C} \qquad \text{(Proposition 2)}$$

Because Proposition 1 states that the efficiency is independent of the working fluid, we can use rigorously an ideal gas as a working fluid to prove Proposition 2. For an ideal gas, $PV = RT$ and $U^{ig} = f(T)$. For an ideal gas in an isothermal process, the first law says $\Delta U^{ig} = 0$ or $Q = -W$, using Eq. (3.5.2.12) for the two isotherms

$$_2Q_3 = RT_H \ln\left(\frac{V_3}{V_2}\right) \tag{4.16}$$

$$_4Q_1 = RT_C \ln\left(\frac{V_1}{V_4}\right) \tag{4.17}$$

Note that we have substituted $\theta_H = T_H$ and $\theta_C = T_C$. Now, considering the adiabatic paths and applying the first law to a closed system $Q = 0$ and $\Delta U = W$ and

$$C_V^{ig} dT = -PdV = -RT\frac{dV}{V} \tag{4.18}$$

or

$$C_V^{ig} \frac{dT}{T} = -RT\frac{dV}{V} \Rightarrow C_V^{ig} d(\ln T) = -Rd(\ln V) \tag{4.19}$$

Applying the above equation to the two adiabatic paths of the Carnot cycle produces

$$C_V^{ig} \ln\frac{T_H}{T_C} = R\ln\frac{V_1}{V_2} \tag{4.20}$$

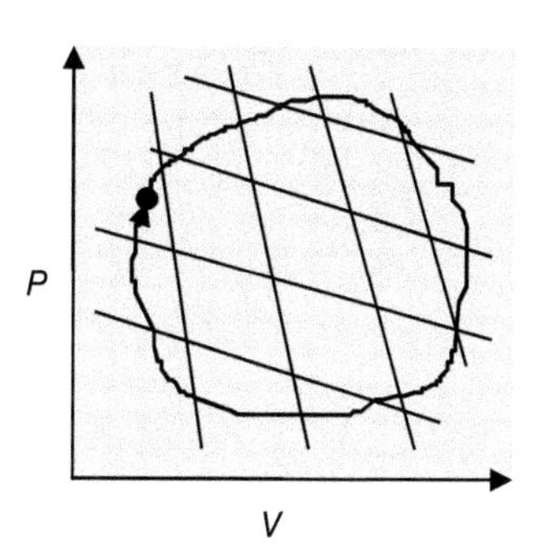

Figure 4.8 Arbitrary cycle separated into small Carnot cycles.

and

$$C_V^{ig} \ln \frac{T_C}{T_H} = -R \ln \frac{V_4}{V_3} \Rightarrow C_V^{ig} \ln \frac{T_H}{T_C} = R \ln \frac{V_4}{V_3} \tag{4.21}$$

Comparing Eqs. (4.20) and (4.21), it is apparent that

$$\frac{V_1}{V_2} = \frac{V_4}{V_3} \tag{4.22}$$

Dividing Eq. (4.16) by Eq. (4.17)

$$\frac{{}_2Q_3}{{}_4Q_1} = \frac{RT_H \ln\left(\frac{V_3}{V_2}\right)}{RT_C \ln\left(\frac{V_1}{V_4}\right)} \tag{4.23}$$

and using Eq. (4.22)

$$\frac{{}_2Q_3}{{}_4Q_1} = -\frac{T_H}{T_C} \tag{4.24}$$

Finally, we know that ${}_2Q_3 = Q_H$ and ${}_4Q_1 = -Q_C$, therefore

$$\frac{Q_H}{Q_C} = \frac{T_H}{T_C} \tag{Proposition 2}$$

and the efficiency for a Carnot cycle becomes

$$\eta = 1 - \frac{Q_C}{Q_H} = 1 - \frac{T_C}{T_H} \tag{4.25}$$

This is the maximum achievable efficiency in any heat engine.

Proposition 3 We can write Proposition 2 as

$$\frac{{}_2Q_3}{T_H} + \frac{{}_4Q_1}{T_C} = 0 \Rightarrow \sum \frac{(Q_i)_{rev}}{T_i} = 0 \tag{4.26}$$

It is possible to divide an arbitrary cycle into N Carnot cycles as shown in Figure 4.8. To provide an accurate representation, we can allow $N \to \infty$ and obtain

$$\oint \frac{đQ_{rev}}{T} = 0 \tag{4.27}$$

The efficiency of an irreversible cycle must be less than that of a reversible cycle (with equivalent work) because the irreversible cycle must overcome dissipative forces, thus

$$\eta_{rev} > \eta_{irrev} \tag{4.28}$$

Using the definition of Carnot efficiency

$$\left(\frac{Q_H}{Q_C}\right)_{rev} = \frac{T_H}{T_C} > \left(\frac{Q_H}{Q_C}\right)_{irrev} \tag{4.29}$$

and

$$\frac{(Q_C)_{irrev}}{T_C} > \frac{(Q_H)_{irrev}}{T_H} \tag{4.30}$$

Now, substituting ${}_2Q_3 = Q_H$ and ${}_4Q_1 = -Q_C$ into the irreversible cycle

$$0 > \frac{({}_2Q_3)_{irrev}}{T_H} + \frac{({}_4Q_1)_{irrev}}{T_C} \Rightarrow 0 > \sum \frac{(Q_i)_{irrev}}{T_i} \tag{4.31}$$

If infinitely many small cycles cover the irreversible cycle,

$$\oint \frac{dQ_{irrev}}{T} < 0 \tag{4.32}$$

By combining the two cyclic integrals (reversible and irreversible), we arrive at the Clausius inequality

$$\oint \frac{dQ}{T} \leq 0 \tag{4.33}$$

From a mathematical point of view, if $\oint dF = 0$ for any choice of paths, then F is a state function, and F depends on only the final and initial states. Returning to the case in which we have a cyclic integral equaling zero

$$dQ_{rev}/T \text{ is a state function} \tag{4.34}$$

Now, we can define this new state function as **Entropy**

$$dS \equiv dQ_{rev}/T \tag{Proposition 3}$$

S is the entropy, which derives from Greek έντροπία (entropia), which means *a turning toward.*

Proposition 4 For our final proposition, we use another arbitrary cycle with two paths. One is irreversible, the dashed line in Figure 4.9, and a reversible path, the solid line in the same figure.

The Clausius inequality states

$$\int_2^1 \frac{dQ_{rev}}{T} + \int_1^2 \frac{dQ_{irr}}{T} < 0 \tag{4.35}$$

Using Proposition 3, we have

$$S_1 - S_2 + \int_1^2 \frac{dQ_{irr}}{T} < 0 \tag{4.36}$$

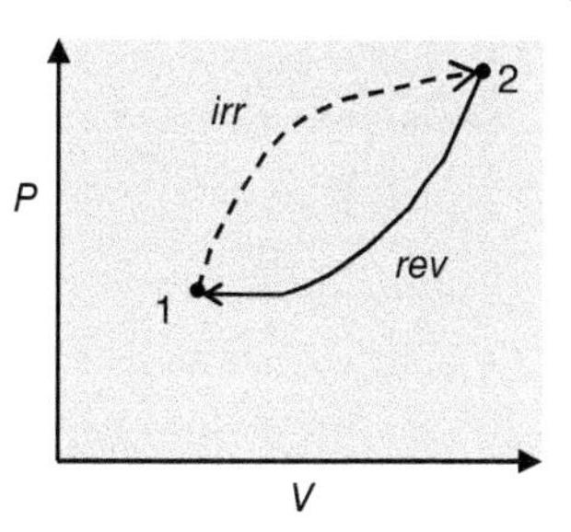

Figure 4.9 Arbitrary cycle with an irreversible path and a reversible path.

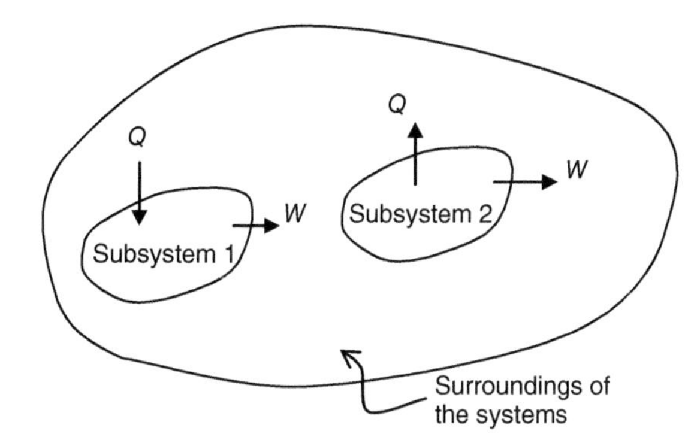

Figure 4.10 A general isolated system.

or

$$S_2 - S_1 = \Delta S > \int_1^2 \frac{đQ_{irrev}}{T} \tag{4.37}$$

Because an isolated system does not exchange energy with its surroundings, $đQ_{irrev} = 0$, thus

$$\Delta S \geq 0 \text{ for an isolated system} \qquad \text{(Proposition 4)}$$

The equal sign corresponds to a reversible process as shown in Proposition 3, so for an irreversible process,

$$\Delta S > 0 \tag{4.38}$$

This proposition reveals that the entropy of an isolated system can never decrease, but it remains constant when the process occurring is reversible. This is the *Principle of Increasing Entropy*. The universe is a system surrounded by nothing (because we define the universe as everything); therefore, the universe is an isolated system

$$\Delta S_{Universe} = \Delta S_{System} + \Delta S_{Surroundings} \geq 0 \tag{4.39}$$

An isolated system can consist of multiple systems and their individual surroundings within a single boundary, as shown in Figure 4.10. Then, the systems become part of the larger isolated system. If the entropy decreases in a region of this isolated system, this decrease must be offset by a larger increase of entropy in another part of the isolated system, making the entropy change greater than zero.

If we assume that the engine in Figure 4.11 does not transfer energy as work, $W = 0$. This case is a formal description of a situation that involves only energy transfer as heat.

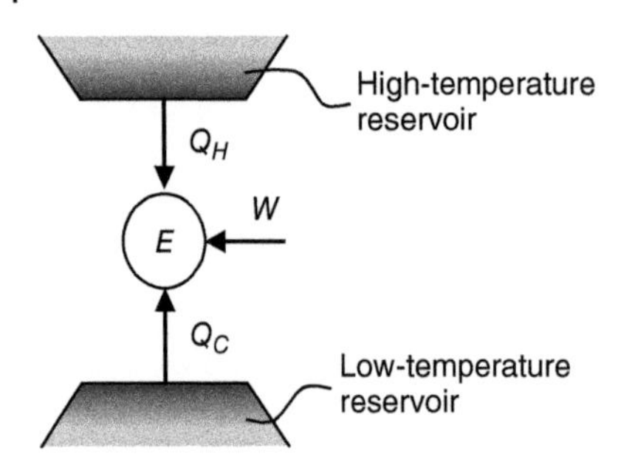

Figure 4.11 Engine.

If we apply the first law to the engine in Figure 4.11,

$$W + Q_C + Q_H = 0 \tag{4.40}$$

Because $W = 0$, $Q_C = -Q_H$. Thus, the total entropy change of the universe, considering the reservoirs as the surroundings, is

$$\Delta S_{Universe} = \Delta S_{Hot\ reservoir} + \Delta S_{Engine} + \Delta S_{Cold\ reservoir} \tag{4.41}$$

If the engine is a cyclic process

$$\Delta S_{Engine} = 0 \tag{4.42}$$

and for the reservoirs $dS \equiv đQ_{rev}/T$

$$\Delta S_{Hot\ reservoir} = -\frac{Q_H}{T_H} \tag{4.43}$$

$$\Delta S_{Cold\ reservoir} = -\frac{Q_C}{T_C} \tag{4.44}$$

Notice that the sign is negative because the systems now are the reservoirs, then

$$\Delta S_{Universe} = -\frac{Q_H}{T_H} - \frac{Q_C}{T_C} = -\frac{Q_H}{T_H} + \frac{Q_H}{T_C} = -Q_H\left[\frac{1}{T_H} - \frac{1}{T_C}\right] = Q_H\frac{(T_H - T_C)}{T_H T_C} \tag{4.45}$$

From the second law: $Q_H > 0$, $Q_C < 0$, and $T_H > T_C$

$$\Delta S_{Universe} \geq 0 \tag{4.46}$$

The equal sign applies to reversible processes. Therefore,

$\Delta S_{Universe} = 0$ for reversible process

$\Delta S_{Universe} > 0$ for irreversible process

Processes for which $\Delta S_{Universe} < 0$ cannot occur. Strictly speaking, energy transfer as heat cannot be a reversible process because $\Delta S_{Universe} = 0$ only when $T_H = T_C$. However, energy transfer as heat tends to zero ($Q \to 0$) when the temperatures of the reservoirs become equal.

The Clausius inequality shows that the change in entropy for the cycle is negative. That is, the entropy change in the environment during the cycle is larger than the entropy change transferred to the engine by heat from the hot reservoir. For a heat engine in which the energy transferred as heat, Q_H, from a reservoir at temperature T_H passes to the system, the entropy change of the reservoir (calculated as $\Delta S = Q_H/T_H$) passes to the system, and it must transfer to the environment to complete the cycle. In general, the engine temperature is less than T_H, implying an irreversible process.

An isentropic process need not be adiabatic or reversible, but, if it is reversible, it must be adiabatic. If the isentropic process is adiabatic, it must be reversible, but the adiabatic process is not necessarily isentropic.

This concludes our discussions of the mathematical foundations of the first three laws of thermodynamics. Now, we can examine more detailed applications of the second law.

Example 4.1

Calculate the remaining temperatures and energy requirements for the following arrangement of three Carnot cycles.

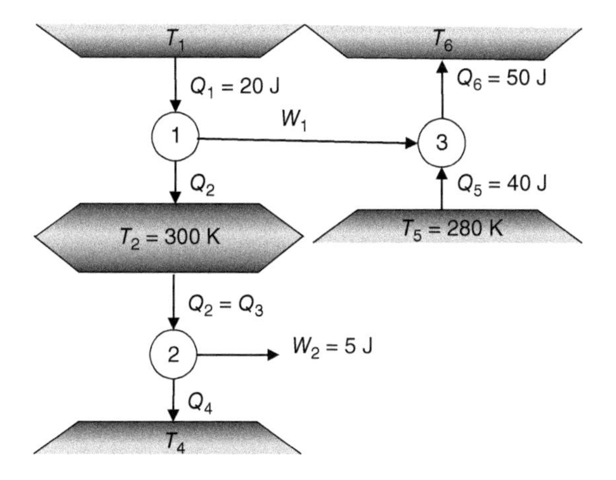

Solution

First, calculate temperature T_6 using Proposition 2

$$\frac{Q_6}{T_6} = \frac{Q_5}{T_5} \Rightarrow T_6 = T_5\frac{Q_6}{Q_5} = 280\frac{50}{40} = 350\ \text{K}$$

Applying the first law to engine 3 produces W_1

$$W_1 + Q_5 = Q_6 \Rightarrow W_1 = Q_6 - Q_5 = 50 - 40 = 10\ \text{J}$$

Q_2 results from applying the first law to engine 1

$$Q_1 = Q_2 + W_1 \Rightarrow Q_2 = Q_1 - W_1 = 20 - 10 = 10\ \text{J}$$

T_1 results from Proposition 2

$$\frac{Q_1}{T_1} = \frac{Q_2}{T_2} \Rightarrow T_1 = T_2\frac{Q_1}{Q_2} = 300\frac{20}{10} = 600\ \text{K}$$

then using the first law for engine 2 produces

$$Q_2 = Q_4 + W_2 \Rightarrow Q_4 = Q_2 - W_2 = 10 - 5 = 5\,\text{J}$$

Finally, using Proposition 2

$$\frac{Q_4}{T_4} = \frac{Q_2}{T_2} \Rightarrow T_1 = T_2\frac{Q_4}{Q_2} = 300\frac{5}{10} = 150\,\text{K}$$

Example 4.2

Two solid bodies with masses of $m_1 = 2$ kg and $m_2 = 1$ kg are in contact inside an adiabatic enclosure until they reach an equilibrium temperature, T_f. What is the entropy change of the process, if the initial temperatures are $T_1 = 25\,°\text{C}$ and $T_2 = 200\,°\text{C}$? What are the final temperatures and the entropy changes, if the two solid bodies have the same mass with heat capacities of $C_1 = 0.91$ kJ/(kg K) and $C_2 = 0.49$ kg/(kJ K). Consider the process to be reversible.

Solution

Applying the first law,

$$\Delta(mU/M)_{SYS} = 0$$

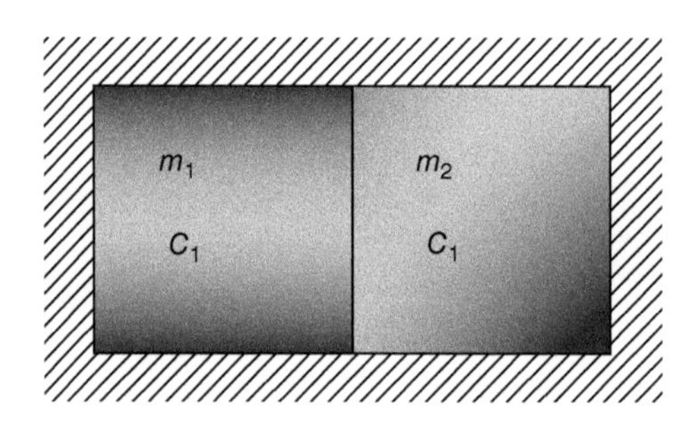

then

$$m_1C_1(T_f - T_1) + m_2C_2(T_f - T_2) = 0$$

or

$$T_f = \frac{m_1C_1T_1 + m_2C_2T_2}{m_1C_1 + m_2C_2} = \frac{2 \times 0.91 \times 298.15 + 2 \times 0.49 \times 373.15}{2 \times 0.91 + 2 \times 0.49} = 314.06\,\text{K}$$

The entropy changes for the bodies are

$$\begin{aligned}\Delta S_1 &= m_1 \int_{T_1}^{T_f} \frac{đQ_{rev}}{T} = m_1 \int_{T_1}^{T_f} C_1\frac{dT}{T} = m_1C_1 \ln\frac{T_f}{T_1} \\ &= 2 \times 0.91 \times \ln\frac{314.06}{298.15} = 0.095\,\text{kJ/K}\end{aligned}$$

$$\Delta S_2 = m_2C_2 \ln\frac{T_f}{T_2} = 2 \times 0.49 \times \ln\frac{314.06}{373.15} = -0.084\,\text{kJ/K}$$

The total change of entropy is

$$\Delta S_{Isolate\ \ system} = \Delta S_1 + \Delta S_2 = 0.095 - 0.084 = 0.011\,\text{kg/J} > 0$$

Now, **assume** $m_1 = m_2 = m$ and $C_1 = C_2 = C$, then

$$T_f = \frac{T_1 + T_2}{2}$$

and

$$\Delta S_{Isolated\ system} = m_1C_1 \ln \frac{T_f}{T_1} + m_2C_2 \ln \frac{T_f}{T_1} = mC \ln \frac{T_f^2}{T_1 \cdot T_2} = mC \ln \frac{[(T_1 + T_2)/2]^2}{T_1 \cdot T_2}$$

$$\Delta S_{Isolated\ system} = 2mC \ln \left[\frac{(T_1 + T_2)}{2\sqrt{T_1 \cdot T_2}} \right]$$

We can prove that the quantity in brackets is greater than 1, making $\Delta S_{Isolated\ system} > 0$. Clearly, the following statement is true

$$\frac{(\sqrt{T_1} - \sqrt{T_2})^2}{2} > 0$$

If we expand the square of the binomial

$$\frac{T_1 - 2\sqrt{T_1}\sqrt{T_2} + T_2}{2} = \frac{T_1 + T_2}{2} - \sqrt{T_1 \cdot T_2} > 0$$

Then,

$$\frac{T_1 + T_2}{2} > \sqrt{T_1 \cdot T_2}$$

and therefore

$$\Delta S_{Isolated\ system} > 0$$

4.5 Efficiency

We know that for reversible processes, the entropy change of the universe equals zero ($\Delta S_{Universe} = 0$) and that the entropy change is greater than 0 for irreversible processes. This inequality, $\Delta S_{Universe} > 0$, prevents direct calculation of practical irreversible processes and requires use of experimental results. Furthermore, all practical processes are irreversible to some degree. We have defined the thermodynamic efficiency for a heat engine as the ratio of the absolute value of the *net energy transferred as work* to the *net energy transferred as heat* in the machine in Eq. (4.6). Also, we have used this definition to obtain the Carnot efficiency in Eq. (4.25). Practical problems (irreversible) require use of a *pattern efficiency*, which is a comparison between the actual energy transferred as work and the reversible energy transferred as work. Two general cases exist: engines that produce energy transferred as work (such as heat engines) and engines that consume energy transferred as work (such as refrigerators)

$$\eta_P = \frac{W}{W_S} \text{ for a heat engine, } W < 0 \tag{4.47}$$

$$\eta_P = \frac{W_S}{W} \text{ for a refrigerator, } W > 0 \tag{4.48}$$

in which W is the energy transferred as work by the engine or the refrigerator operating irreversibly, and W_S is the energy transferred as work performed by a reversible process that arrives at the same final state. The subscript S denotes isentropic energy transferred as work because $\Delta S_{Universe} = 0$ for a reversible process. We can apply Eqs. (4.47) and (4.48) to practical problems following these steps:

1. Calculate the energy transferred as work if the process operates reversibly (knowing the initial and final conditions of the system). If the final temperature is unknown, use $\Delta S = 0$ to calculate the temperature.
2. Compare the energy transferred as work involved in the actual process to that obtained from the calculation for a reversible process. In this case, the initial and final conditions of the system are known, and we can calculate the pattern efficiency. If we desire, the final temperature using Eq. (4.47) or (4.48) provides the work and the ability to calculate the final temperature.

We always must remember that $0 < \eta_P < 1$ and that $|W| < |W_S|$ when energy transferred as work comes from an engine. The reverse is true when the energy transferred as work enters the engine.

Example 4.3
Establish the expression for the efficiency in a heat pump or a refrigerator for a Carnot cycle using T_H and T_C. Consider the schematic diagram in Figure 4.3b.

Solution
Applying the first law to a refrigerator

$$W_R - Q_H + Q_C = 0$$

then

$$W_R = Q_H - Q_C = -Q_C\left(1 - \frac{Q_H}{Q_C}\right)$$

or

$$W_R = -Q_C\left(1 - \frac{T_H}{T_C}\right)$$

In a refrigerator, the objective is to transfer energy as heat to the surroundings

$$\eta_R = \frac{Q_C}{W}$$

Substituting the expression for the energy transfer as work into the above equation

$$\eta_R = \frac{Q_C}{-Q_C\left(1 - \frac{T_H}{T_C}\right)} = \frac{1}{-\left(1 - \frac{T_H}{T_C}\right)} \Rightarrow COP \equiv \frac{T_C}{T_H - T_C}$$

This efficiency for a Carnot refrigerator is its coefficient of performance (COP).

4.6 Specific Heat/Heat Capacity

Let us return to the first law and combine it with the second law. The first law is

$$\Delta(U + E_K + E_P)_{SYS} = \sum Q + \sum W + \sum (U + PV + E_K + E_P) - (\delta m/M)c^2 \tag{2.23}$$

For our current purposes, we can look at a stationary, closed system without nuclear reactions, so

$$\Delta U = \sum Q + \sum W \tag{4.49}$$

or in differential form

$$dU = đQ_{rev} + đW_{rev} \tag{4.50}$$

In addition, let us assume that the work is expansion only, so

$$dU = đQ_{rev} - PdV \tag{4.51}$$

which becomes using Proposition 3

$$dU = TdS - PdV \tag{4.52}$$

or in terms of enthalpy

$$H = U + PV$$

$$dH = dU + PdV + VdP = TdS - PdV + PdV + VdP$$

$$dH = TdS + VdP \tag{4.53}$$

A useful function in thermodynamics is the specific heat or heat capacity (different names for the same function). For no particular reason, we use the term "heat capacity" in this book. The heat capacity is the amount of energy transferred as heat required to raise the temperature of a unit quantity of a substance by 1 K. The name comes from physicist Joseph Black, who conducted measurements of energy transferred as heat, in which he used the term *capacity for heat*. If the unit quantity of a substance is moles, then the property is the molar heat capacity. Often, the term specific heat refers to energy transfer per mass, and heat capacity heat refers to energy transfer per mass. However, either term may refer to either unit. For example, the energy transfer as heat required to raise the temperature of water by 1 K (or 1 °C) is approximately 75.31 J/mol or 4.184 J/g, which is the heat capacity of water. The SI units for the quantity are J/(mol K) or J/(g K), and the English units are Btu/(lbmol °F) or Btu/(lb °F). The symbol is C_X, where X indicates the path along which the energy flows as heat. Three paths are common: constant pressure, constant volume, and along the saturation curve

$$C_X \equiv \left[\frac{\delta Q_{rev}}{\delta T}\right]_X = T\left[\frac{\partial S}{\partial T}\right]_X \tag{4.54}$$

in which X denotes the path taken by the energy transfer as heat. The common paths used in thermodynamics are

Constant volume	V is constant
Constant pressure	P is constant
Along the saturation curve	σ denotes the saturation curve

thus

$$C_V = T\left[\frac{\partial S}{\partial T}\right]_V = \left[\frac{\partial U}{\partial T}\right]_V \tag{4.55}$$

$$C_P = T\left[\frac{\partial S}{\partial T}\right]_P = \left[\frac{\partial H}{\partial T}\right]_P \tag{4.56}$$

$$C_\sigma = T\left[\frac{\partial S}{\partial T}\right]_\sigma \tag{4.57}$$

4.6.1 Entropy Changes for Ideal Gases

An ideal gas satisfies

$$PV^{ig} = RT$$
$$U^{ig} = U(T) \tag{4.58}$$

of course, if U^{ig} is a function of T only so is H^{ig} because

$$H^{ig} = U^{ig} + PV^{ig} = U^{ig} + RT = H(T) \tag{4.59}$$

Therefore,

$$C_V^{ig} = T\left.\frac{dS^{ig}}{dT}\right|_V = \frac{dU^{ig}}{dT} \tag{4.60}$$

$$C_P^{ig} = T\left.\frac{dS^{ig}}{dT}\right|_P = \frac{dH^{ig}}{dT} \tag{4.61}$$

C_σ^{ig} cannot exist because an ideal gas has no saturation curve. From Eq. (4.59), we see that

$$H^{ig} = U^{ig} + RT$$

so

$$\frac{dH^{ig}}{dT} = \frac{dU^{ig}}{dT} + R \tag{4.62}$$

or in dimensionless form

$$C_P^{ig}/R = C_V^{ig}/R + 1 \tag{4.63}$$

We know from the second law that, for the paths specified, the heat capacities are

$$\frac{\Delta S^{ig}}{R} = \int_{T_1}^{T_2} \frac{C_V^{ig}}{R}\frac{dT}{T} = \int_{T_1}^{T_2} \frac{C_P^{ig}}{R}\frac{dT}{T} - \ln\frac{T_2}{T_1} \text{ at constant volume} \tag{4.64}$$

It is important to remember the entropy is a state function that depends upon the state conditions (variables specifying the state). In this case, one property is constant depending upon the path. Equations (4.55)–(4.57) apply to any substance for which the pressure or volume is constant or for calculations along the saturation curve. For an ideal gas, the heat capacity is a function only of temperature that can have a functional form such as

$$C_V^{ig}/R = A + BT + CT^2 + D/T^2 \tag{4.65}$$

from which the entropy becomes

$$\frac{\Delta S^{ig}}{R} = \int_{T_1}^{T_2} \frac{C_V^{ig}}{R}\frac{dT}{T} = A\ln\left(\frac{T_2}{T_1}\right) + B(T_2 - T_1) + \frac{C}{2}\left(T_2^{\ 2} - T_1^{\ 2}\right) - \frac{D}{2}\left(\frac{1}{T_2^{\ 2}} - \frac{1}{T_1^{\ 2}}\right) \tag{4.66}$$

or in terms of C_P^{ig}

$$\frac{\Delta S^{ig}}{R} = \int_{T_1}^{T_2} \frac{\left(C_P^{ig} - 1\right)}{R}\frac{dT}{T} = A\ln\left(\frac{T_2}{T_1}\right) + B(T_2 - T_1) + \frac{C}{2}\left(T_2^{\ 2} - T_1^{\ 2}\right) - \frac{D}{2}\left(\frac{1}{T_2^{\ 2}} - \frac{1}{T_1^{\ 2}}\right)$$

Rearranging Eq. (4.52) for an ideal gas

$$dS^{ig} = \frac{dU^{ig}}{T} + \frac{P}{T}dV^{ig} \tag{4.67}$$

and using the equation of state for an ideal gas and $dU^{ig} = C_V^{ig}dT$

$$dS^{ig} = C_V^{ig}\frac{dT}{T} + R\frac{dV^{ig}}{V^{ig}} \tag{4.68}$$

Integrating between two state conditions

$$\Delta S^{ig} = \int_{T_1}^{T_2} C_V^{ig}\frac{dT}{T} + R\int_{V_1}^{V_2}\frac{dV^{ig}}{V^{ig}} \tag{4.69}$$

and

$$\frac{\Delta S^{ig}}{R} = \int_{T_1}^{T_2} \frac{C_V^{ig}}{R}\frac{dT}{T} + \ln\frac{V_2^{ig}}{V_1^{ig}} \tag{4.70}$$

which at a constant volume is Eq. (4.52) for an ideal gas. Using $U^{ig} = H^{ig} - PV^{ig}$ in Eq. (4.67) produces

$$\frac{\Delta S^{ig}}{R} = \int_{T_1}^{T_2} \frac{C_P^{ig}}{R}\frac{dT}{T} - \ln\frac{P_2}{P_1} \tag{4.71}$$

Equations (4.70) and (4.71) apply to any process, reversible or irreversible, because ΔS is a state function and depends only on the state conditions.

Example 4.4

What is the final temperature if CO_2 expands from 7 to 1 bar in a closed system? The process is adiabatic and irreversible with an initial temperature of 700 K. The pattern efficiency of the process is 0.7.

$$\mathrm{CP/R} = 5.457 + 1.045 \times 10^{(-3)}T - 1.157 \times 10^5/T^2$$

Solution

For a reversible process, $\Delta S = 0$; then, we can calculate the final temperature of the reversible process using Eqs. (4.54) and (4.61)

$$\Delta S^{ig} = 0 = R\left[A\ln\left(\frac{T_2}{T_1}\right) + B(T_2 - T_1) + \frac{C}{2}\left(T_2^{\ 2} - T_1^{\ 2}\right) - \frac{D}{2}\left(\frac{1}{T_2^{\ 2}} - \frac{1}{T_1^{\ 2}}\right)\right] - R\ln\frac{P_2}{P_1}$$

then

$$0 = \left[5.457 \cdot \ln\left(\frac{T_2}{700}\right) + 1.045 \times 10^{-3} \cdot (T_2 - 700) + \frac{1.157 \times 10^5}{2}\left(\frac{1}{T_2^2} - \frac{1}{700^2}\right)\right] - \ln\frac{1}{7}$$

If we use a numerical method (such as Solver in Excel), we find that $T_2 = 498.72$ K. However, we can make our own iterative procedure by assuming temperatures and extrapolating using straight lines. For example, let us write the equation in this form

$$\ln\frac{P_2}{P_1} = \left[A\ln\left(\frac{T_2}{T_1}\right) + B(T_2 - T_1) + \frac{C}{2}\left(T_2^2 - T_1^2\right) - \frac{D}{2}\left(\frac{1}{T_2^2} - \frac{1}{T_1^2}\right)\right]$$

that is,

$$\left[A\ln\left(\frac{T_2}{T_1}\right) + B(T_2 - T_1) + \frac{C}{2}\left(T_2^2 - T_1^2\right) - \frac{D}{2}\left(\frac{1}{T_2^2} - \frac{1}{T_1^2}\right)\right] = -1.9459$$

Now, assume two temperatures for T_2, say 600 and 550. We know that it should be less than 700 K because it is an expansion. Now, calculate the *LHS* of the equation,

$$LHS(600) = -0.90306 \text{ and } LHS(550) = -1.3996$$

Because our objective is that $LHS = -1.9459$, we can use those two points to extrapolate and estimate a new temperature

$$T_2^{new} = \frac{600 - 550}{-0.90306 - (-1.3396)}[-1.9459 - (-0.90306)] + 600 = 494.987 \text{ K}$$

Use this temperature to obtain a better value

$$LHS(494.987) = -1.98731$$

Again, obtain a new temperature with this temperature and the closest to our objective

$$T_2^{new} = \frac{550 - 480.56}{-1.3396 - (-1.98731)}[-1.9459 - (-1.3396)] + 550 = 498.863 \text{ K}$$

Following this procedure, we have

$$LHS(498.863) = -1.94435 \text{ and } T_2^{new} = 498.72 \text{ K}$$

which is our result. Now returning to our problem, applying the first law for the reversible problem

$$\Delta U^{ig} = W_{rev} = \int_{T_1}^{T_2} C_V^{ig} dT = \int_{T_1}^{T_2} \left(C_P^{ig} - R\right) dT = R\int_{T_1}^{T_2} \left(C_P^{ig}/R - 1\right) dT$$

then

$$W_{rev} = R\left[(A-1)(T_2 - T_1) + \frac{B}{2}\left(T_2^2 - T_1^2\right) + \frac{C}{3}\left(T_2^3 - T_1^3\right) - D\left(\frac{1}{T_2} - \frac{1}{T_1}\right)\right]$$

$$W_{rev} = R\left[4.457(498.72 - 700) + \frac{1.045 \times 10^{-3}}{2}(498.72^2 - 700^2)\right.$$
$$\left. +1.157 \times 10^5 \left(\frac{1}{498.72} - \frac{1}{700}\right)\right]$$

$$W_{rev} = -956.46R$$

We can calculate the irreversible work using Eq. (4.47)

$$W_{irrev} = \eta_P W_S = 0.7(-956.46R) = -669.522R$$

Applying the first law to the irreversible process

$$\Delta U^{ig} = W_{irrev} = R\int_{T_1}^{T_2} \left(C_P^{ig}/R - 1\right) dT$$

thus

$$W_{irrev} = R\left[(A-1)(T_2 - T_1) + \frac{B}{2}\left(T_2^2 - T_1^2\right) + \frac{C}{3}\left(T_2^3 - T_1^3\right) - D\left(\frac{1}{T_2} - \frac{1}{T_1}\right)\right]$$

$$R\left[(A-1)(T_2 - T_1) + \frac{B}{2}\left(T_2^2 - T_1^2\right) + \frac{C}{3}\left(T_2^3 - T_1^3\right) - D\left(\frac{1}{T_2} - \frac{1}{T_1}\right)\right]$$
$$= -669.522R$$

$$R\left[4.457(T_2 - 700) + \frac{1.045\times10^{-3}}{2}\left(T_2^2 - 700^2\right) + 1.157 \times 10^5 \left(\frac{1}{T_2} - \frac{1}{700}\right)\right]$$
$$= -669.522R$$

Using an iterative procedure, we obtain

$$T_2 = 561.135 \text{ K}$$

4.7 Entropy Balance Equation for Open Systems

We can apply the general accounting equation (GAE) to entropy just as we did for energy, but we must consider that the entropy is always greater than or equal to zero, and the entropy is not conserved. Considering the system in Figure 4.12, we require entropy accumulation, transfer, and generation terms (no *CONVersion* term is necessary)

$$ACCumulation = TRANSfer + GENeration \tag{4.72}$$

These terms can depend on time. Because the second law establishes that the entropy of the universe must be greater or equal to zero if a process occurs, this inequality must appear in the generation term. The accumulation and transfer terms are similar to those of the energy balance. The *ACC* term is the time rate change of the entropy in the system

$$ACC = \left[\frac{\partial(nS)}{\partial t}\right]_{cv} \tag{4.73}$$

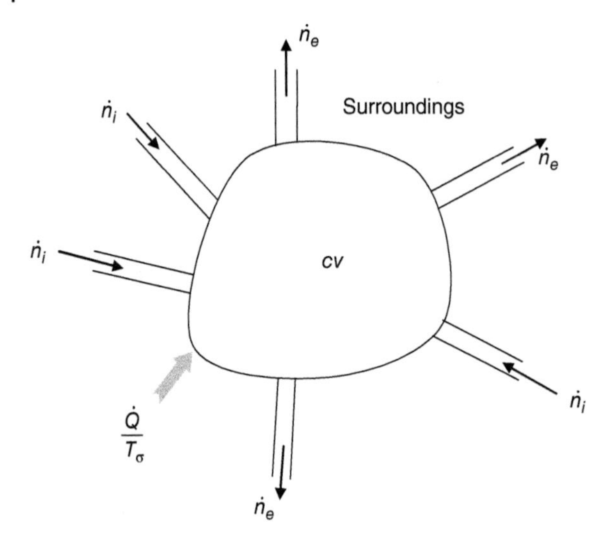

Figure 4.12 Diagram of an open system.

in which *cv* denotes control volume. The *TRANS* term considers the rate of entropy transfer caused by energy transfer as heat between the surroundings and the system (control volume) and the net rate of change in entropy of the flowing streams. The first transfer term is

$$\frac{\partial S^{total}}{\partial t} = \sum_j \frac{\dot{q}_j}{T_{\sigma,j}} \tag{4.74}$$

in which T_σ is the temperature of the surroundings and the dot above a quantity denotes its change with time. The net rate of change in entropy of the flowing streams considers the entropy change with respect to time of the inlet and outlet streams, that is

$$S_{mt} \equiv \sum_{inlets} \dot{n}_i S_i - \sum_{outlets} \dot{n}_i S_i = \sum_i \dot{n}_i S_i \tag{4.75}$$

Therefore, the general transfer term is

$$TRANS \equiv \sum_i \dot{n}_i S_i + \sum_j \frac{\dot{q}_j}{T_{\sigma j}} \tag{4.76}$$

The generation term is more complex. Considering a closed system, the entropy can change because of energy transfer as heat with the surroundings and internal irreversibility or dissipative effects (kinetic energy or work dissipated into an internal energy increase). If energy transfer as heat passes into the system through a boundary at the temperature of the surroundings, T_σ, the differential entropy of the system is

$$dS = (dS)_{ext} + (dS)_{in\text{-}irr} = \frac{đq}{T_\sigma} + (dS)_{in\text{-}irr} \tag{4.77}$$

in which *ext* and *in-irr* denote external energy transfer as heat and internal irreversibilities. Using Eq. (4.37) in differential form

$$\frac{đq}{T_\sigma} \leq \frac{đq}{T_\sigma} + (dS)_{in\text{-}irr} \tag{4.78}$$

and

$$(dS)_{in\text{-}irr} \geq 0 \tag{4.79}$$

This entropy increase caused by internal irreversibility is entropy production or entropy generation, S_{gen}.

Returning to the entropy balance

$$\left[\frac{\partial(nS)}{\partial t}\right]_{cv} = \sum_i \dot{n}_i S_i + \sum_j \frac{\dot{q}_j}{T_{\sigma,j}} + \dot{S}_{gen} \tag{4.80}$$

or

$$\dot{S}_{gen} = \left[\frac{\partial(nS)}{\partial t}\right]_{cv} - \sum_i \dot{n}_i S_i - \sum_j \frac{\dot{q}_j}{T_{\sigma,j}} \geq 0 \tag{4.81}$$

in which $\dot{S}_{gen}$ is the rate of entropy generation. Using Eq. (4.80) for simple cases such as a flow system at steady state

$$\left[\frac{\partial(nS)}{\partial t}\right]_{cv} = 0 \tag{4.82}$$

because the accumulation term is zero. Therefore,

$$0 = \sum_i \dot{n}_i S_i + \sum_j \frac{\dot{q}_j}{T_{\sigma,j}} + \dot{S}_{gen} \tag{4.83}$$

and for one inlet and one outlet,

$$0 = -\dot{n}\Delta S + \sum_j \frac{\dot{q}_j}{T_{\sigma,j}} + \dot{S}_{gen} \tag{4.84}$$

or

$$0 = -\Delta S + \sum_j \frac{q_j}{T_{\sigma,j}} + S_{gen} \rightarrow \Delta S = \sum_j \frac{q_j}{T_{\sigma,j}} + S_{gen} \tag{4.85}$$

in which $\Delta S = S_{out} - S_{in}$ and $\dot{n} = n_{in} = \dot{n}_{out}$ because of the continuity equation.

Example 4.5

Let us use a cross to split an ideal gas stream into three possible streams. If the cross has insulation, and the entering stream is at 3 bar and 320 K, the separated streams are at 1 bar and 250, 298, and 340 K. The flow rate of each exit stream is 1 kg/s. What is the entropy generation of the process? Assume that the process is at steady state and $C_P/R = 5/2$.

Solution

Using Eq. (4.83) with $\sum_j \frac{\dot{q}_j}{T_{\sigma,j}} = 0$,

$$0 = \sum_i \dot{n}_i S_i + \dot{S}_{gen}$$

or

$$\dot{n}_1 S_1 - \dot{n}_2 S_2 - \dot{n}_3 S_3 - \dot{n}_4 S_4 + \dot{S}_{gen} = 0$$

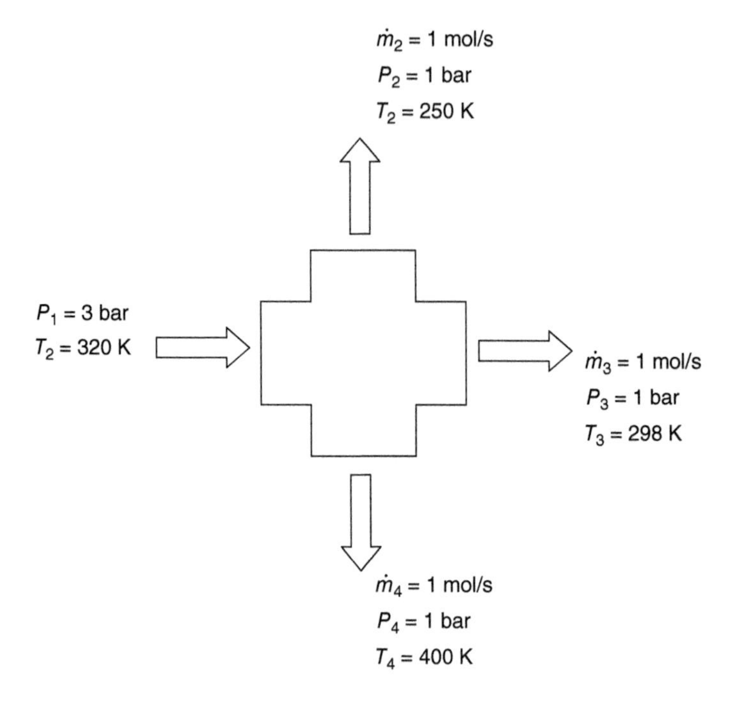

Then, using the continuity equation

$$\dot{S}_{gen} = \dot{n}_2 S_2 + \dot{n}_3 S_3 + \dot{n}_4 S_4 - (\dot{n}_2 + \dot{n}_3 + \dot{n}_4) S_1$$

or

$$\dot{S}_{gen} = \dot{n}_2 (S_2 - S_1) + \dot{n}_3 (S_3 - S_1) + \dot{n}_4 (S_4 - S_1)$$

Calculating the entropy differences provides

$$(S_2 - S_1) = C_P^{ig}(T_2 - T_1) - R \ln\left(\frac{P_2}{P_1}\right)$$
$$= \frac{5}{2} R \times (270 - 320) - R \times \ln\frac{1}{3} = -123.9R\,\text{J/(mol K)}$$

$$(S_3 - S_1) = C_P^{ig}(T_3 - T_1) - R \ln\left(\frac{P_3}{P_1}\right)$$
$$= \frac{5}{2} R \times (298 - 320) - R \times \ln\frac{1}{3} = -53.9R\,\text{J/(mol K)}$$

$$(S_4 - S_1) = C_P^{ig}(T_4 - T_1) - R \ln\left(\frac{P_4}{P_1}\right)$$
$$= \frac{5}{2} R \times (400 - 320) - R \times \ln\frac{1}{3} = 201.1R\,\text{J/(mol K)}$$

Then,

$$\dot{S}_{gen} = 1 \times (-123.9R) + 1 \times (-53.9R) + 1 \times (201R) = 23.3R\,\text{J/(s} \times \text{K)}$$

4.8 Availability and Maximum/Minimum Work

As energy becomes more expensive, we must examine processes with a goal to conserve it. For example, in highly exothermic reactions, we may wish to recover some of the energy transferred as heat. Likewise, we might want to transfer the minimum amount of energy as work to compressors. Of course, many other examples are conceivable. It would be convenient to calculate the maximum or minimum work without considering the details of the process. A quantity termed *availability* allows us to do this.

Let us examine a closed system and determine the maximum energy transfer as work available in the fluid within the system. We can extract the maximum amount of energy transfer as work if the system comes to equilibrium with its environment at T_0 and P_0. From the first law for a closed, stationary system

$$W = U_0 - U_1 - Q. \tag{4.86}$$

From the second law

$$\Delta S + \Delta S_0 \geq 0 \tag{4.87}$$

in which ΔS and ΔS_0 are the entropy changes of the system and the surroundings, respectively. If we multiply by the temperature of the surroundings

$$T_0 \Delta S + T_0 \Delta S_0 \geq 0 \tag{4.88}$$

However, the quantity $T_0 \Delta S_0$ equals Q_0, the energy transferred as heat to or from the surroundings, which equals $-Q$ (energy transferred as heat to or from the system), therefore

$$-Q \geq -T_0(S_0 - S_1) \tag{4.89}$$

and Eq. (4.86) becomes

$$W \geq U_0 - U_1 - T_0(S_0 - S_1) \tag{4.90}$$

Equation (4.90) is a maximum for an energy transfer as a work-producing process or a minimum for a process that accepts energy transfer as work when the process is reversible and the equality applies. Therefore, if the subscript m denotes minimum or maximum

$$W_m = U_0 - U_1 - T_0(S_0 - S_1) \tag{4.91}$$

The value of W_m is the total energy transfer as work; however; it is not the useful energy transfer as work. Consider the piston and cylinder of Figure 4.13. We can calculate the energy transfer as work when the piston moves a defined distance

$$W = \int_{z_0}^{z_1} F dz = \int_{z_0}^{z_1} \int [P_0 A_P + (F_{ext} - F_f)] dz \tag{4.92}$$

in which A_P is the cross-sectional area of the piston. Factoring A_P and using volume as a variable instead of distance

$$W = \int_{z_1}^{z_2} \left(P_0 + \frac{F_{ext} - F_f}{A_P} \right) A_P dz = \int_{V_0}^{V_1} \left(P_0 + \frac{F_{ext} - F_f}{A_P} \right) dV \tag{4.93}$$

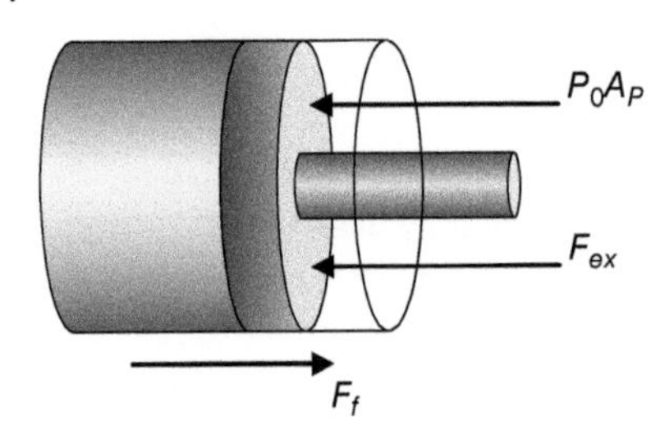

Figure 4.13 Schematic diagram of a piston–cylinder.

which becomes upon integration

$$W = P_0(V_1 - V_0) + \int_{z_0}^{z_1} (F_{ext} - F_f)dz \tag{4.94}$$

We now define the *useful work* for the process as

$$W_u = \int_{z_1}^{z_2} (F_{ext} - F_f)dz \tag{4.95}$$

because the work against the atmosphere (or surroundings) is unavailable for useful purposes. The maximum/minimum useful work in a process becomes

$$-A \equiv (W_u)_m = (U_0 + P_0V_0 - T_0S_0) - (U_1 + P_0V_1 - TS_1) \tag{4.96}$$

This is the *availability* of a closed system. If the process does not equilibrate with the environment, the maximum useful energy transfer as work is the difference in availability

$$(W_u)_m = A_1 - A_2 \tag{4.97}$$

In a steady-state, flow system (neglecting E_K and E_P of the flowing streams)

$$W = H_0 - H_1 - Q \tag{4.98}$$

and following the above analysis

$$W \geq H_0 - H_1 - T_0(S_0 - S_1) \tag{4.99}$$

or

$$W_m = (H_0 - T_0S_0) - (H_1 - T_0S_1) \tag{4.100}$$

In this case, W_m is the useful energy transfer as work, and the stream availability or *exergy* is

$$E \equiv (W_u)_m = (H_0 - T_0S_0) - (H_1 - T_0S_1) \tag{4.101}$$

Again, if the system is not in equilibrium with the surroundings

$$(W_u)_m = E_1 - E_2 \tag{4.102}$$

Problems for Chapter 4

4.1 If a heat pump that absorbs energy as heat from the surroundings at −3.889 °C and delivers energy as heat to a residence at 37.778 °C, determine the minimum fraction of the energy supplied as heat to the residence that must be purchased.

4.2 One concept proposed as a future source of electrical energy is to use the difference between the temperature at the surface of the ocean and the temperature at great depths to drive an electrical generation process. If the surface temperature is 30 °C and the deep water temperature is 4 °C, determine the maximum thermal efficiency possible and the minimum heat transfer rate from the warmer water required to generate 1 kW of electricity.

4.3 For the following Carnot engine, calculate the missing heats and temperatures. The efficiency in the engine is 0.6

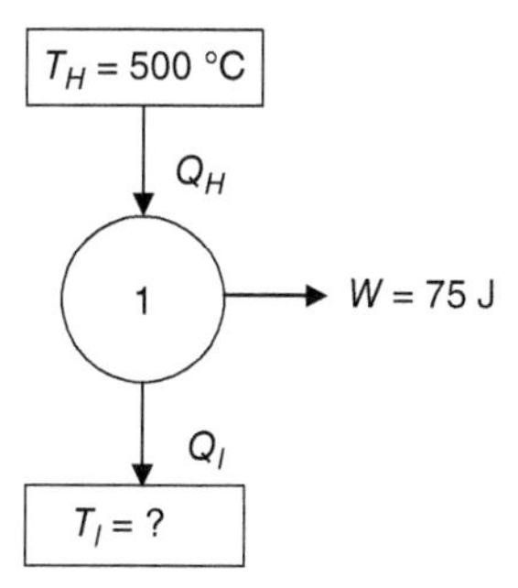

4.4 Consider a case in which natural gas is available at a cost of \$5.00 per 1×10^6 kJ energy content, and a Carnot engine can generate electricity using the energy provided by combustion of the gas. The engine rejects waste heat from the engine at 300 K. Determine the minimum temperature at which the energy transferred as heat from burning natural gas to the Carnot engine can produce electricity economically, if the sales price of the electricity is \$0.05/kWh.

4.5 A Carnot engine produces 100 kJ of work when it is operated between two reservoirs at 298 and 700 K. Calculate the change of entropy of the two reservoirs.

4.6 When heated at a constant pressure of 2 mol of an ideal gas at 30 kPa exit at 600 °C. The initial volume of the ideal gas is 0.1 m^3. Calculate the change of entropy during the process. Consider $C_P^{ig}/R = 7/2$.

4.7 An ideal gas compressed adiabatically in a cylinder passes from 1 to 15 bar. Estimate the final temperature if the process is reversible and the initial temperature is 20 °C. Calculate the changes of entropy of the process. Consider $C_P^{ig}/R = 7/2$.

4.8 Calculate the *COP* of a Carnot refrigerator in which the temperature is −5 °C. The refrigerator delivers heat to the surroundings at 27 °C.

4.9 A Carnot engine compresses steam adiabatically from 50 to 5000 kPa with saturated vapor at the end of the process. The work output is 1000 kJ/kg. Calculate the change of entropy of the process.

4.10 A reversible cyclic process performs work while exchanging heat with four reservoirs. The temperatures of tanks 1, 2, 3, and 4 are 1000, 300, 500, and 400 K, respectively. From tanks 1 and 4, 400 and 100 kJ transfer from the tank to the process, and the total work performed by the process is 100 kJ. For each reservoir, determine the amount and direction of heat transfer.

4.11 An apparatus compresses 1 mol of an ideal gas isothermally to 300 K and transfers heat by an amount of 4000 J. 3500 J of this heat passes to another ideal gas mole, which expands isothermally from 3 to 1 bar as a result. The compression and expansion processes are the same. The rest of the heat dissipates into the atmosphere at 25 °C. Assume that the processes are reversible. Is this overall process possible according to the second law of thermodynamics?

4.12 Steam, in a closed container initially at 500 kPa and 300 °C, cools to 30 °C (the temperature of the surroundings). Determine the final pressure, specific volume, quality, and the total entropy change of the universe for each of the following cases:

a. the cooling occurs at constant volume;
b. the cooling occurs at constant pressure.

4.13 A rigid vessel having a volume of 16.0 l contains 3.20 kg of water initially at 350 °C. If the vessel cools to 150 °C by exchanging energy as heat with the surroundings at 100 °C, determine the total entropy change of the universe during this process.

4.14 A simple method for determining the quality of steam in a supply line is to expand a portion of the stream through a valve to atmospheric pressure as shown below and to measure the temperature of the steam (now in the vapor phase only) at atmospheric pressure. The pressure in the main line is 2500 kPa, and all velocities may be considered negligible.

a. Determine the quality, temperature, enthalpy, and entropy of the steam in the line if the temperature of the expanded steam is 125 °C at 100 kPa.

b. The method fails if the expanded steam remains in the two-phase region. Determine the minimum quality in the line at 2500 kPa for which the method can work if the steam expands to 100 kPa.

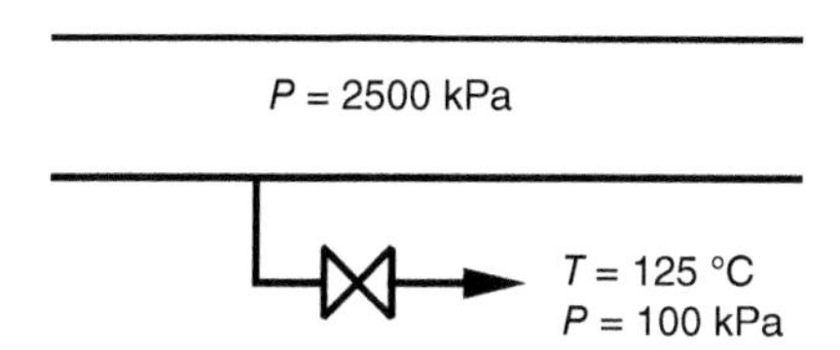

4.15 Determine the entropy change of oxygen in a change from 400 K and 0.4 bar if it is compressed to 800 K and 3 bar.
For oxygen: $C_P^{ig}/R = 3.639 + 0.506 \times 10^{-3}T - 0.227 \times 10^5/T^{-2}$

4.16 Liquid methanol at 62 °C is poured into a copper container that originally is at 0 °C. The methanol and the copper container reach thermal equilibrium at 33 °C. No energy leaves as heat to the surroundings during this process.

a. What mass of methanol was added, if the mass of the copper container is 3.72 kg?
b. Determine the total entropy change of the universe for this process.

Data

$$C_{P,CH_4O}/R = 13.431 - 0.051\,28T + 1.3113 \times 10^{-4}T^2$$

$$C_{P,Cu}/R = 2.677 + 8.15 \times 10^{-4}T + 3500T^{-2}$$

4.17 A piston/cylinder containing an ideal gas with $C_P^{ig} = 3.67R$ is capable of exchanging energy as heat with either of two reservoirs, one at 325 K and the other at 275 K. The gas initially is at a temperature of 300 K and a pressure of 3.43 bar.

a. Determine the entropy change (per mole of gas) of the gas and of the universe if the gas expands isothermally in a mechanically reversible process to $V_f = 2.70V_0$.
b. Determine the entropy change of the universe and the network (per mole of gas) for the entire two-step expansion/compression process if the gas is compressed to its original volume in a mechanically reversible isothermal process.
c. Determine the maximum thermal efficiency possible for a cyclic process operating between the two reservoirs.
d. List a series of steps that describe a process that would achieve the thermal efficiency calculated in part (c). The isothermal restriction does not apply, but the ratio of the maximum volume to the minimum volume in the process cannot exceed 2.70.

4.18 Calculate the entropy change of the universe when 1 mol of ethane gas

a. *cools at a constant volume* from 125 °C and 0.932 bar to 13 °C by exchanging energy as heat with a reservoir at 13 °C;
b. *heats at a constant volume* from 13 °C and 0.932 bar to 125 °C by exchanging energy as heat with a reservoir at 125 °C;
c. *heats at a constant pressure* from 13 °C and 0.932 bar to 125 °C by exchanging energy as heat with a reservoir at 125 °C.

For ethane: $C_P^{ig}/R = 1.131 + 1.9225 \times 10^{-2}T - 5.561 \times 10^{-6}T^2$

4.19 Steam passes through the following cycle in which all the steps are reversible: isothermal compression from 500 to 1900 kPa at 260 °C; isobaric heating from 260 to 450 °C at 1900 kPa; adiabatic expansion to the initial conditions. Determine the thermal efficiency of this cycle.

4.20 An isothermal, ***irreversible***, non-flow process requires 3150 J/mol to compress an ideal gas from 1 to 3 bar at 300 K. The ideal gas heat capacity at a constant pressure of 300 K is $C_P = 4.1R$.

a. Calculate the pattern efficiency for this process.
b. Calculate the total change in the entropy of the universe for this process.

4.21 Determine the final temperature for benzene at 400 K when compressed reversibly and adiabatically from 0.4 to 0.7 bar. (You may treat benzene as an ideal gas in this pressure range.)

For benzene: $C_P^{ig}/R = -0.206 + 3.9064 \times 10^{-2}T - 1.3301 \times 10^{-5}T^2$

4.22 Air at 26 °C and 1 atm enters a compressor that raises its pressure adiabatically to 5 atm. The air leaves the compressor at 218 °C and passes to an aftercooler where it cools at a constant pressure to 26 °C.

a. Determine the pattern efficiency of the compressor.
b. Determine the amount of energy removed as heat in the aftercooler per mole of air processed.
c. Determine the total entropy change of the universe (per mole of air compressed) for the entire process if the energy transferred as heat in the aftercooler enters the surroundings at 26 °C.

For air: $C_P^{ig}/R = 3.355 + 5.75 \times 10^{-4}T - 1.600 \times 10^{4}T^{-2}$

4.23 Calculate the total entropy change of the universe when 2 mol of an ideal gas at 320 K expands isothermally from 3 to 1 bar in a process that has a pattern efficiency of 75% (i.e. the amount of work actually done is 75% of what would have occurred in a reversible process) and the energy transferred as heat comes from a reservoir at 335 K.

4.24 Consider an ideal gas that undergoes the following cycle: adiabatic reversible compression from 300 K and 1 bar to one half the original volume; heating at constant a volume to 600 K by extracting heat from a reservoir at 600 K;

adiabatic reversible expansion to the original volume; cooling at a constant volume to the original temperature by delivering energy as heat to a reservoir at 300 K. The ideal gas heat capacity at a constant volume is 2.5R.

a. Determine the thermal efficiency of this cycle.
b. Determine the thermal efficiency of a Carnot cycle operating between the same two reservoirs.
c. Determine the entropy change of the universe per mole of gas for the complete cycle described above.
d. Determine the thermal efficiency of the cycle if, instead of being reversible, the adiabatic compression and expansion steps operate with pattern efficiencies of 90%.

4.25 Propane expands in a turbine from 2 bar and 700 to 500 K and 0.1 bar in a reversible process. If the process efficiency is 75%, what is the final temperature of the irreversible process?
For propane: $C_P^{ig}/R = 1.213 + 28.785 \times 10^{-3}T - 8.824 \times 10^{-6}T^2$

5

Thermodynamic Relations

5.1 Introduction

Thus far, we have learned three laws of thermodynamics. Also, we understand the *PVT* behavior of a pure substance, and we have calculated changes in the internal energy and the enthalpy for perfect and ideal gas systems. However, as engineers, we must deal with real systems when calculating the thermodynamic properties. A mathematical formality is necessary to calculate the thermodynamic properties and their changes. In this chapter, we define new thermodynamic properties, calculate the changes for these properties, and define new types of changes.

5.2 Mathematics Review

Before we establish the thermodynamic properties, we should review the mathematics required in thermodynamics. The definition of a function and its derivative is key to the development of thermodynamic properties.

5.2.1 Exact Differentials

If we consider a differential function of x and y variables

$$df = E(x, y)dx + F(x, y)dy \tag{5.1}$$

in which E and F are any functions of x and y, and we can apply the Euler test:

$$\left(\frac{\partial E}{\partial y}\right)_x = \left(\frac{\partial F}{\partial x}\right)_y \tag{5.2}$$

Then, if f exists and df is an exact differential, the integral is

$$\int_{x_i y_i}^{x_f y_f} df = f(x_f, y_f) - f(x_i, y_i) \tag{5.3}$$

The values of integrals of exact differentials depend only on the initial and final points and not on the path along which the integration proceeds. For example,

Thermodynamics for Chemical Engineers, First Edition. Kenneth R. Hall and Gustavo A. Iglesias-Silva.

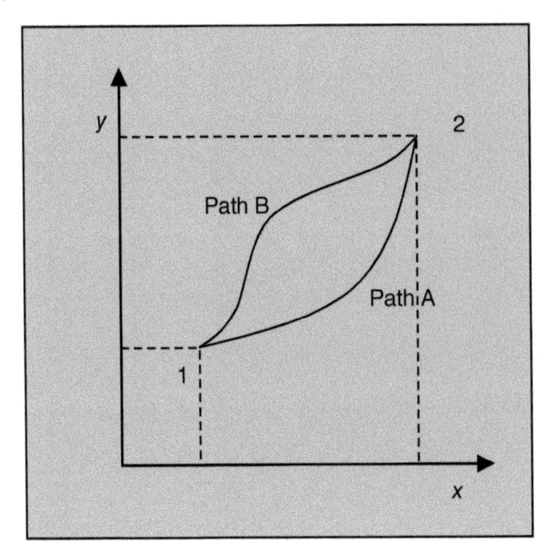

Figure 5.1 A function of exact differentials.

Figure 5.1 has two paths. If the functions have exact differentials, the integral could be

$$\int_A dx = x_2 - x_1 = \int_B dx$$

or

$$\int_A dy = y_2 - y_1 = \int_B dy$$

5.2.2 Inexact Differentials and Line Integration

Now let us consider the integral

$$\int_a^b F(x)\, dx \tag{5.4}$$

This integral is the area under a section of the curve bounded by $F(a)$ and $F(b)$. In Eq. (5.4), $F(x)$ is a function that depends upon the path of the function and the end points. Therefore, this function has an inexact differential, and the integral is a line integral. Returning to the thermodynamic properties, energy transferred as work and heat depends upon the path; therefore, they have inexact differentials defined as $đW$ and $đQ$, for which

$$\int_1^2 đW = {}_1W_2$$

and

$$\int_1^2 đQ = {}_1Q_2$$

State functions are exact differentials, and they depend only on the initial and final state.

5.2.3 Properties of Functions of Several Variables

A function is a mathematical relation in which each element (independent variables) of a given set (the domain of the function) is associated with an element (dependent variables) of another set (the range of the function). Then, $f = f(x_1, x_2, \ldots, x_m)$ indicates that the dependent variable f is a function of m independent variables x_i. For example, $f = f(x, y)$ means that the dependent variable f is a function of the independent variables x and y. The ***total differential of an exact function*** is

$$df = \sum_{i=1}^{m} \left(\frac{\partial f}{\partial x_i}\right)_{x_{i \neq j}} dx_i \Leftrightarrow f = f(x_1, x_2, \ldots, x_m) \tag{5.5}$$

The double arrow indicates equivalence. Often, functions in thermodynamics contain two independent variables

$$df = \left(\frac{\partial f}{\partial x}\right)_y dx + \left(\frac{\partial f}{\partial y}\right)_x dy \Leftrightarrow f = f(x, y) \quad \textbf{Math Relation 1} \tag{5.6}$$

Now, let us consider a general function

$$z = z(x, y) \Leftrightarrow dz = \left(\frac{\partial z}{\partial x}\right)_y dx + \left(\frac{\partial z}{\partial y}\right)_x dy \tag{5.7}$$

We can also express x as a function of the other two variables:

$$x = x(y, z) \Leftrightarrow dx = \left(\frac{\partial x}{\partial y}\right)_z dy + \left(\frac{\partial x}{\partial z}\right)_y dz \tag{5.8}$$

Substituting dx into Eq. (5.7)

$$dz = \left(\frac{\partial z}{\partial x}\right)_y \left[\left(\frac{\partial x}{\partial y}\right)_z dy + \left(\frac{\partial x}{\partial z}\right)_y dz\right] + \left(\frac{\partial z}{\partial y}\right)_x dy \tag{5.9}$$

and then collecting the differential terms

$$\left[1 - \left(\frac{\partial z}{\partial x}\right)_y \left(\frac{\partial x}{\partial z}\right)_y\right] dz = \left[\left(\frac{\partial z}{\partial x}\right)_y \left(\frac{\partial x}{\partial y}\right)_z + \left(\frac{\partial z}{\partial y}\right)_x\right] dy \tag{5.10}$$

Because dz and dy can have any value (+, 0, or −), the only way for Eq. (5.10) to be true in every case is for the terms in brackets to be zero. Therefore,

$$\left(\frac{\partial x}{\partial z}\right)_y \left(\frac{\partial z}{\partial x}\right)_y = 1 \quad \textbf{Math Relation 2} \tag{5.11}$$

and

$$\left(\frac{\partial z}{\partial y}\right)_x = -\left(\frac{\partial z}{\partial x}\right)_y \left(\frac{\partial x}{\partial y}\right)_z \tag{5.12}$$

A common way of expressing Eq. (5.12) using Eq. (5.11) is

$$\left(\frac{\partial z}{\partial x}\right)_y \left(\frac{\partial x}{\partial y}\right)_z \left(\frac{\partial y}{\partial z}\right)_x = -1 \quad \textbf{Math Relation 3} \tag{5.13}$$

Another useful math relation in thermodynamics is the *chain rule*

$$\left(\frac{\partial z}{\partial x}\right)_y = \left(\frac{\partial z}{\partial w}\right)_y \left(\frac{\partial w}{\partial x}\right)_y \quad \textbf{Math Relation 4} \tag{5.14}$$

Also, we can take the derivative of z with respect to y at a constant variable w. In this case, the derivation only acts on the differential terms of Eq. (5.7)

$$\left(\frac{\partial z}{\partial y}\right)_w = \left(\frac{\partial z}{\partial x}\right)_y \left(\frac{\partial x}{\partial y}\right)_w + \left(\frac{\partial z}{\partial y}\right)_x \quad \textbf{Math Relation 5} \tag{5.15}$$

Finally, we have a math relation that involves second derivatives:

$$\left[\frac{\partial}{\partial x}\left(\frac{\partial f}{\partial y}\right)_x\right]_y = \left[\frac{\partial}{\partial y}\left(\frac{\partial f}{\partial x}\right)_y\right]_x \quad \textbf{Math Relation 6} \tag{5.16}$$

Equation (5.16) is *Euler reciprocity relation*. All these relationships are useful in thermodynamics.

5.3 Fundamental Thermodynamics Equation

For a closed system, the first law in a differential form is

$$dU = đQ + đW \tag{5.17}$$

In a reversible process, the above equation becomes

$$(dU)_{rev} = đQ_{rev} + đW_{rev} \tag{5.18}$$

Using the second law

$$đQ_{rev} = TdS$$

and the definition for reversible energy transfer as (expansion) work

$$đW_{rev} = -PdV$$

then

$$(dU)_{rev} = TdS - PdV \tag{5.19}$$

Now, for an irreversible, closed system

$$(dU)_{irrev} = đQ_{irrev} + đW_{irrev} \tag{5.20}$$

Because the internal energy is a state function that depends only upon the final and initial state, we can write

$$(dU)_{irrev} = (dU)_{rev} = TdS - PdV \tag{5.21}$$

or

$$dU = TdS - PdV \tag{5.22}$$

Equation (5.22) is a fundamental equation because it contains the information from the zeroth, first, and second laws of thermodynamics. Applying Math Relation 1 to Eq. (5.22)

$$dU = TdS - PdV \Leftrightarrow U = U(S, V) \tag{5.23}$$

The implication of Eq. (5.23) is that entropy and specific volume are the natural and fundamental independent variables for internal energy, so $U = U(S,V)$ contains all thermodynamic information about a (pure component) system.

5.4 Legendre Transforms

Legendre transforms of Eq. (5.23) can produce fundamental equations for other thermodynamic properties that also contain all thermodynamic information about a system. The new thermodynamic properties are expressions of U minus conjugate pairs of intensive and extensive variables. Let us define a function of a single variable x, $y = y(x)$, for example, $y = 3x^2$ as shown in Figure 5.2.

The derivative is $dy/dx = 6x$ from which we can obtain x and substitute it into the original function to obtain $y = (dy/dx)^2/12$. Now, are these two functions identical? The answer is no because the function that contains the derivative is a differential equation

$$dy/dx = \sqrt{12y} \tag{5.24}$$

whose solution is a family of curves. Obtaining a particular y-function requires an initial condition, a boundary condition, or another kind of condition. One way to obtain the original function is to use lines tangent to the curve and obtain a family of y-intercepts as shown in Figure 5.3.

Knowing the y-intercepts as a function of the tangent is equivalent to knowing the original function $y = y(x)$. In mathematical form

$$\varphi = \varphi\left(\frac{dy}{dx}\right) \equiv y = y(x) \tag{5.25}$$

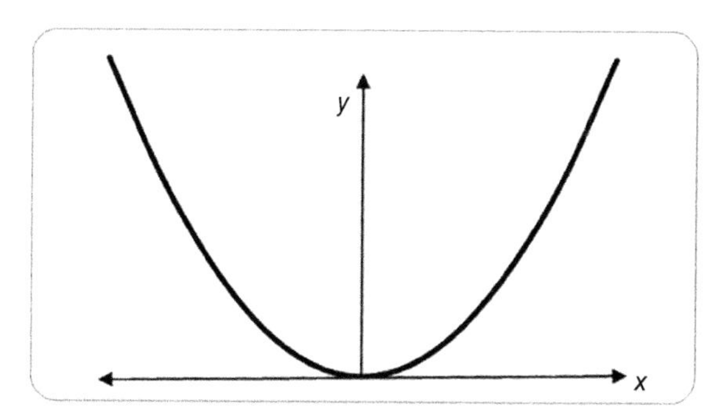

Figure 5.2 Function $y = 3x^2$.

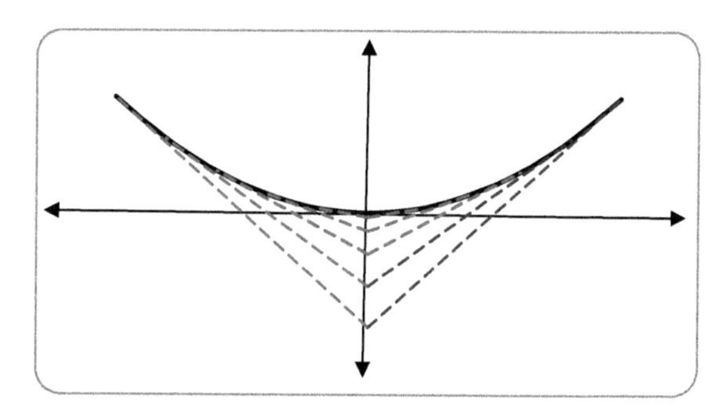

Figure 5.3 Tangents at each point in the curve.

An equation for the y-intercept that relates the function to its derivative results from calculating the tangent using two points:

$$\frac{dy}{dx} = \frac{y-\varphi}{x-0} \Rightarrow \varphi = y - x\frac{dy}{dx} \tag{5.26}$$

This equation defines the *Legendre transform*. With multiple independent variables, it becomes

$$\varphi = y - \sum_i x_i \left(\frac{\partial y}{\partial x_i}\right)_{x_{j\neq i}} \tag{5.27}$$

Consider f as a function of two variables, x_1 and x_2. Then, using Math Relation 1 gives

$$f = f(x,y) \Leftrightarrow df = \left(\frac{\partial f}{\partial x_1}\right)_{x_2} dx_1 + \left(\frac{\partial f}{\partial x_2}\right)_{x_1} dx_2 = D_1 dx_1 + D_2 dx_2 \tag{5.28}$$

in which $D_1 = \left(\frac{\partial f}{\partial x_1}\right)_{x_2}$ and $D_2 = \left(\frac{\partial f}{\partial x_2}\right)_{x_1}$. Now, we can create new functions using Eq. (5.27) and subtracting one or two derivatives because our function, f, depends upon two variables:

$$\varphi_1 = f - x_1\left(\frac{\partial f}{\partial x_1}\right)_{x_2} = f - D_1 x_1 \tag{5.29}$$

The differential is

$$d\varphi_1 = df - D_1 dx_1 - x_1 dD_1 \tag{5.30}$$

Substituting Eq. (5.28)

$$d\varphi_1 = (D_1 dx_1 + D_2 dx_2) - D_1 dx_1 - x_1 dD_1 = -x_1 dD_1 + D_2 dx_2 \tag{5.31}$$

and applying Math Relation 1 to this equation

$$d\varphi_1 = -x_1 dD_1 + D_2 dx_2 \Rightarrow \varphi_1 = \varphi_1(D_1, x_2) \tag{5.32}$$

This is a new function with independent variables D_1 and x_2. If we subtract the product of the other derivative times the other independent variable

$$\varphi_2 = f - x_2\left(\frac{\partial f}{\partial x_2}\right)_{x_1} = f - D_2 x_2 \tag{5.33}$$

and the differential is

$$d\varphi_2 = df - D_2 dx_2 - x_2 dD_2 = D_1 dx_1 - x_2 dD_2 \Leftrightarrow \varphi_2 = \varphi_2(x_1, D_2) \tag{5.34}$$

This Legendre transform is

$$\varphi_3 = f - x_1\left(\frac{\partial f}{\partial x_1}\right)_{x_2} - x_2\left(\frac{\partial f}{\partial x_2}\right)_{x_1} = f - D_1 x_1 - D_2 x_2 \tag{5.35}$$

and the differential is

$$\begin{aligned} d\varphi_3 &= df - D_1 dx_1 - x_1 dD_1 - D_2 dx_2 - x_2 dD_2 \\ &= (D_1 dx_1 + D_2 dx_2) - D_1 dx_1 - x_1 dD_1 - D_2 dx_2 - x_2 dD_2 \\ &= -x_1 dD_1 - x_2 dD_2 \end{aligned} \tag{5.36}$$

Thus

$$d\varphi_3 = -x_1 dD_1 - x_2 dD_2 \Leftrightarrow \varphi_3 = \varphi_3(D_1, D_2) \tag{5.37}$$

Equations (5.32), (5.34), and (5.37) are the Legendre transforms of a function with two independent variables. Now, using the fundamental equation for internal energy

$$dU = TdS - PdV \Leftrightarrow U = U(S, V) \tag{5.23}$$

Comparing this equation to Eq. (5.28): f is U, x_1 is S, x_2 is V, D_1 is T, and D_2 is $-P$, and the Legendre transforms are

$$\varphi_1 = U - TS \equiv A \Rightarrow A \equiv A(T, V) \text{ defined as } \textbf{Helmholtz energy} \tag{5.38}$$

$$\varphi_2 = U + PV \equiv H \Rightarrow H \equiv H(S, P) \text{ defined as } \textbf{Enthalpy} \tag{5.39}$$

$$\varphi_3 = U - TS + PV = H - TS \equiv G \Rightarrow G \equiv G(T, P) \text{ defined as } \textbf{Gibbs Energy} \tag{5.40}$$

Using Eqs. (5.32), (5.34), and (5.36) provides three additional fundamental equations for closed systems: enthalpy, Helmholtz energy, and Gibbs energy with

$$dU = TdS - PdV \tag{5.22}$$

$$dH = TdS + VdP \tag{5.41}$$

$$dA = -SdT - PdV \tag{5.42}$$

$$dG = -SdT + VdP \tag{5.43}$$

Applying Math Relation 1 to these forms of the Physics Summary Statement provides the ***thermodynamic definitions*** of T, V, P, and S. For the internal energy

$$dU = \left(\frac{\partial U}{\partial S}\right)_V dS + \left(\frac{\partial U}{\partial V}\right)_S dV \Leftrightarrow dU = TdS - PdV \tag{5.44}$$

and

$$T = \left(\frac{\partial U}{\partial S}\right)_V \tag{5.45}$$

$$-P = \left(\frac{\partial U}{\partial V}\right)_S \tag{5.46}$$

Table 5.1 contains all the thermodynamic definitions.

5.5 Maxwell Relations

If we apply the *Euler Reciprocity Relation* to Eq. (5.44)

$$\left[\frac{\partial}{\partial V}\left(\frac{\partial U}{\partial S}\right)_V\right]_S = \left[\frac{\partial}{\partial S}\left(\frac{\partial U}{\partial V}\right)_S\right]_V \tag{5.47}$$

or

$$\left[\frac{\partial T}{\partial V}\right]_S = -\left[\frac{\partial P}{\partial S}\right]_V \tag{5.48}$$

Table 5.1 Thermodynamic definitions of P, V, T, and S.

Variable	dU	dH	dA	dG
$-P=$	$\left(\frac{\partial U}{\partial V}\right)_S$		$\left(\frac{\partial A}{\partial V}\right)_T$	
$V=$		$\left(\frac{\partial H}{\partial P}\right)_S$		$\left(\frac{\partial G}{\partial P}\right)_T$
$T=$	$\left(\frac{\partial U}{\partial S}\right)_V$	$\left(\frac{\partial H}{\partial S}\right)_P$		
$-S=$			$\left(\frac{\partial A}{\partial T}\right)_V$	$\left(\frac{\partial G}{\partial T}\right)_P$

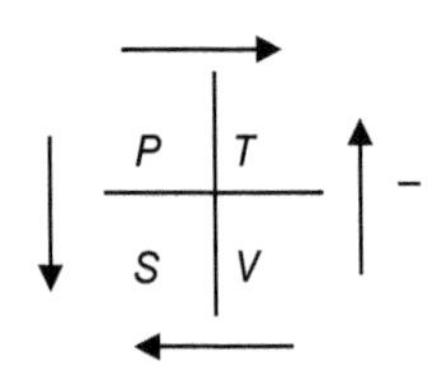

Figure 5.4 Nomograph to obtain the Maxwell relations.

This is the ***Maxwell Relation*** for the internal energy. A simple way of remembering the Maxwell relations is to use the nomograph in Figure 5.4. In this figure, we follow the direction of the arrows and the corresponding terms in the columns to get the derivative. If it is in a vertical direction, one of the derivatives is negative.

A mnemonic device for these relationships is

$$\frac{PV}{XY} \Leftrightarrow \frac{TS}{XY}$$

in which X and Y represent P, V, T, or S. Thus, Eq. (5.48) would be

$$\frac{PV}{VS} \Leftrightarrow \frac{TS}{VS}$$

This is not an equation. TS/VS represents $(\partial T/\partial V)_S$ with S (appearing in both numerator and denominator) representing the variable held constant. PV/VS represents $(\partial P/\partial S)_V$ with V (appearing in both numerator and denominator) as the variable held constant. Because the VS do not appear aligned above and below, we assign a negative sign to the derivative.

Applying the nomograph, the mnemonic device or *Euler Reciprocity Relation* to the fundamental relationships (Eqs. (5.22) and (5.41–5.43)) provides the Maxwell relations:

$$\left[\frac{\partial T}{\partial V}\right]_{S,n} = -\left[\frac{\partial P}{\partial S}\right]_{V,n} \tag{5.49}$$

$$\left[\frac{\partial S}{\partial V}\right]_{T,n} = \left[\frac{\partial P}{\partial T}\right]_{V,n} \tag{5.50}$$

$$\left[\frac{\partial T}{\partial P}\right]_{S,n} = \left[\frac{\partial V}{\partial S}\right]_{P,n} \tag{5.51}$$

$$\left[\frac{\partial V}{\partial T}\right]_{P,n} = -\left[\frac{\partial S}{\partial P}\right]_{T,n} \tag{5.52}$$

Holding the number of moles constant indicates that the system is closed. Open systems have different relationships (these appear later in the book).

5.6 Derivation of Thermodynamic Relationships

In this section, we formulate all the thermodynamic derivatives of intangibles (properties that are not measurable) in terms of observables (variables or properties that are measurable experimentally). All thermodynamic derivatives fall into six classes that reduce to simpler forms using a logical sequence of operations. These forms result in terms of observables such as pressure, volume, temperature, and heat capacity (either at a constant pressure or volume). In this procedure, the entropy is not reducible, but Section 5.9 demonstrates its calculation.

Let us denote the observables P, V, and T with the symbols $\boldsymbol{x}$, $\boldsymbol{y}$, and $\boldsymbol{z}$ and the intangibles U, H, A, G, and S with X, Y, and Z. Table 5.2 presents the classes and sequences to reduce them to observables.

After reading this statement, you can easily understand the concern of the scientific community upon the death of J.C. Maxwell. They lamented that only Maxwell could interpret Gibbs writing!

Table 5.2 Reduction of thermodynamic derivatives in terms of observables.

Class	General	Example	Reduction comments
1	$\left(\frac{\partial x}{\partial y}\right)_z$	$\left(\frac{\partial P}{\partial V}\right)_T$	Use only the Math Relation 4 $\left(\frac{\partial x}{\partial y}\right)_z = -\left(\frac{\partial x}{\partial z}\right)_y\left(\frac{\partial z}{\partial y}\right)_x \Leftrightarrow \left(\frac{\partial P}{\partial V}\right)_T = -\left(\frac{\partial P}{\partial T}\right)_V\left(\frac{\partial T}{\partial V}\right)_P$
2	$\left(\frac{\partial X}{\partial x}\right)_z$	$\left(\frac{\partial H}{\partial P}\right)_T$	Write the fundamental equation for dX and eliminate dS if necessary: $\left(\frac{dH}{dP}\right)_T = T\left(\frac{\partial S}{\partial P}\right)_T + V\left(\frac{\partial P}{\partial P}\right)_T$ Use a Maxwell relationship to eliminate dS $\left(\frac{\partial H}{\partial P}\right)_T = T\left(\frac{\partial S}{\partial P}\right)_T + V$ $\left(\frac{\partial H}{\partial P}\right)_T = V - T\left(\frac{\partial V}{\partial T}\right)_P$
3	$\left(\frac{\partial x}{\partial z}\right)_Z$	$\left(\frac{\partial P}{\partial T}\right)_H$	Expand the derivative into two derivatives of class 2: $\left(\frac{\partial P}{\partial T}\right)_H = -\left(\frac{\partial P}{\partial H}\right)_T\left(\frac{\partial H}{\partial T}\right)_P$ $= -\left(\frac{\partial H}{\partial T}\right)_P / \left(\frac{\partial H}{\partial P}\right)_T = \frac{-C_P}{V - T(\partial V/\partial T)_P}$
4	$\left(\frac{\partial X}{\partial Y}\right)_x$	$\left(\frac{\partial H}{\partial A}\right)_P$	Introduce a new variable different from z to reduce it to class 2 using the chain rule: $\left(\frac{\partial H}{\partial A}\right)_P = \left(\frac{\partial H}{\partial T}\right)_P\left(\frac{\partial T}{\partial A}\right)_P = \left(\frac{\partial H}{\partial T}\right)_P / \left(\frac{\partial A}{\partial T}\right)_P$

Table 5.2 (Continued)

Class	General	Example	Reduction comments
			$dA = -SdT - PdV$ $\left(\frac{\partial A}{\partial T}\right)_P = -S - P\left(\frac{\partial V}{\partial T}\right)_P$ $\left(\frac{\partial H}{\partial A}\right)_P = \frac{C_P}{-S - P(\partial V/\partial T)_P}$
5	$\left(\frac{\partial X}{\partial x}\right)_Z$	$\left(\frac{\partial H}{\partial P}\right)_A$	Write the fundamental equation for dX and reduce dS to a derivative of class 3: $\left(\frac{\partial H}{\partial P}\right)_A = T\left(\frac{\partial S}{\partial P}\right)_A + V$ $S = S(T, P)$ $dS = \left(\frac{\partial S}{\partial T}\right)_P dT + \left(\frac{\partial S}{\partial P}\right)_T dP$ Use observables and Maxwell relations $dS = \frac{C_P}{T} dT - \left(\frac{\partial V}{\partial T}\right)_P dP$ $T\left(\frac{\partial S}{\partial P}\right)_A = C_P\left(\frac{\partial T}{\partial P}\right)_A - T\left(\frac{\partial V}{\partial T}\right)_P$ Substituting into original equation $\left(\frac{\partial T}{\partial P}\right)_A = -\left(\frac{\partial T}{\partial A}\right)_P\left(\frac{\partial A}{\partial P}\right)_T = -\left(\frac{\partial A}{\partial P}\right)_T / \left(\frac{\partial A}{\partial T}\right)_P$ $\left(\frac{\partial A}{\partial P}\right)_T = -S\left(\frac{\partial T}{\partial P}\right)_T - P\left(\frac{\partial V}{\partial P}\right)_T = -P\left(\frac{\partial V}{\partial P}\right)_T$ $\left(\frac{\partial A}{\partial T}\right)_P = -S\left(\frac{\partial T}{\partial T}\right)_P - P\left(\frac{\partial V}{\partial T}\right)_P = -S - P\left(\frac{\partial V}{\partial T}\right)_P$ $\left(\frac{\partial H}{\partial P}\right)_A = V - T\left(\frac{\partial V}{\partial T}\right)_P - C_P\left[\frac{P\left(\frac{\partial V}{\partial P}\right)_T}{S + P\left(\frac{\partial V}{\partial T}\right)_P}\right]$
6	$\left(\frac{\partial X}{\partial Y}\right)_Z$	$\left(\frac{\partial H}{\partial G}\right)_A$	Introduce a new variable to reduce to derivatives of class 5: $\left(\frac{\partial H}{\partial G}\right)_A = \left(\frac{\partial H}{\partial P}\right)_A\left(\frac{\partial P}{\partial G}\right)_A = \left(\frac{\partial H}{\partial P}\right)_A / \left(\frac{\partial G}{\partial P}\right)_A$ $\left(\frac{\partial H}{\partial P}\right)_A = -S\left(\frac{\partial T}{\partial P}\right)_A + V$ $\left(\frac{\partial G}{\partial P}\right)_A = V + S\left[\frac{P(\partial V/\partial P)_T}{S + P(\partial V/\partial P)_P}\right]$ $\left(\frac{\partial H}{\partial G}\right)_A = \frac{V - T\left(\frac{\partial V}{\partial T}\right)_P - C_P\left[\frac{P\left(\frac{\partial V}{\partial P}\right)_T}{S + P\left(\frac{\partial V}{\partial T}\right)_P}\right]}{V + S\left[\frac{P(\partial V/\partial P)_T}{S + P(\partial V/\partial T)_P}\right]}$

5.7 Open Systems: Chemical Potential

Open systems require specifying the number of moles of each chemical species present to define the state of the system uniquely. Therefore, the total energy of the system is

$$nU = nU(nS, nV, n_1, n_2, \ldots, n_N) \tag{5.53}$$

Applying the energy balance to the ***total internal energy***, and using Math Relation 1, the **Summary Statement** pertinent to open systems is

$$\begin{aligned} d(nU) &= \left[\frac{\partial(nU)}{\partial(nS)}\right]_{nV,n} d(nS) + \left[\frac{\partial(nU)}{\partial(nV)}\right]_{nS,n} d(nV) + \sum_{i=1}^{N}\left[\frac{\partial(nU)}{\partial n_i}\right]_{nS,nV,n_{j\neq i}} dn_i \\ &= \left(\frac{\partial U}{\partial S}\right)_{nV,n} d(nS) + \left(\frac{\partial U}{\partial V}\right)_{nS,n} d(nV) + \sum_{i=1}^{N}\left[\frac{\partial(nU)}{\partial n_i}\right]_{nS,nV,n_{j\neq i}} dn_i \end{aligned} \tag{5.54}$$

or

$$d(nU) = Td(nS) - Pd(nV) + \sum_{i=1}^{N} \mu_i dn_i \tag{5.55}$$

in which μ_i is the chemical potential of specie i defined by Gibbs:

> If to any homogeneous mass we suppose an infinitesimal quantity of any substance to be added, the mass remaining homogeneous and its entropy and volume remaining unchanged, the increase of the energy of the mass divided by the quantity of the substance added is **the potential** for that substance in the mass considered. (For purpose of this definition, any chemical element or combination of elements in given proportions may be considered a substance, whether capable or not of existing by itself as homogeneous body.)

This intrinsic potential is a characteristic energy observed in Eq. (5.55) as the change in the total energy when adding n moles of a specie to the system at a constant total entropy and a constant total volume. Wilder Dwight Bancroft coined its name as "chemical potential." Some define this quantity as a measure of the tendency of particles to diffuse or simply as a rate of change. Comparing Eqs. (5.54) and (5.55), the chemical potential is

$$\mu_i \equiv \left[\frac{\partial(nU)}{\partial n_i}\right]_{nS,nV,n_{j\neq i}} \tag{5.56}$$

Using Eq. (5.30) for an open system, the Helmholtz energy becomes

$$d(nA) = d(nU) - Td(nS) - (nS)dT = -(nS)dT - Pd(nV) + \sum_{i=1}^{N} \mu_i dn_i \tag{5.57}$$

or

$$(nA) = (nA)\{T, (nV), n_1, n_2, \ldots\} \tag{5.58}$$

and its total differential is

$$\begin{aligned} d(nA) &= \left[\frac{\partial(nA)}{\partial T}\right]_{nV,n} dT + \left[\frac{\partial(nA)}{\partial(nV)}\right]_{T,n} d(nV) + \sum_{i=1}^{N}\left[\frac{\partial(nA)}{\partial n_i}\right]_{t,nV,n_{j\neq i}} dn_i \\ &= n\left(\frac{\partial A}{\partial T}\right)_{nV,n} dT + \left(\frac{\partial A}{\partial V}\right)_{T,n} d(nV) + \sum_{i=1}^{N}\left[\frac{\partial(nA)}{\partial n_i}\right]_{t,nV,n_{j\neq i}} dn_i \end{aligned} \tag{5.59}$$

The thermodynamic definitions of Table 5.1 remain unchanged. Using Eqs. (5.34) and (5.36), the thermodynamic relations for the enthalpy and the Gibbs energy for open systems become

$$d(nH) = Td(nS) + (nV)dP + \sum_{i=1}^{N}\mu_i dn_i \tag{5.60}$$

and

$$d(nG) = -(nS)dT + (nV)dP + \sum_{i=1}^{N}\mu_i dn_i \tag{5.61}$$

The total differentials of Eqs. (5.60) and (5.61) are

$$d(nH) = \left(\frac{\partial H}{\partial S}\right)_{P,n} d(nS) + n\left(\frac{\partial H}{\partial P}\right)_{nS,n} dP + \sum_{i=1}^{N}\left[\frac{\partial(nH)}{\partial n_i}\right]_{nS,P,n_{j\neq i}} dn_i \tag{5.62}$$

and

$$d(nG) = n\left(\frac{\partial G}{\partial T}\right)_{P,n} dT + n\left(\frac{\partial G}{\partial P}\right)_{T,n} dP + \sum_{i=1}^{N}\left[\frac{\partial(nG)}{\partial n_i}\right]_{P,T,n_{j\neq i}} dn_i \tag{5.63}$$

Therefore, the fundamental thermodynamic relations provide several definitions for the chemical potential

$$\mu_i \equiv \left[\frac{\partial(nU)}{\partial n_i}\right]_{nS,nV,n_{j\neq i}} \equiv \left[\frac{\partial(nA)}{\partial n_i}\right]_{t,nV,n_{j\neq i}} \equiv \left[\frac{\partial(nH)}{\partial n_i}\right]_{nS,P,n_{j\neq i}} \equiv \left[\frac{\partial(nG)}{\partial n_i}\right]_{P,T,n_{j\neq i}} = \overline{G}_i$$

The last of these is the most convenient for phase equilibrium calculations (which are open systems) because it involves only intensive variables other than the numbers of moles of each specie present. However, Eq. (5.57) allows the use of the equation of state easily, and it is the preferred form in process simulators. In summary, the fundamental relations are

$$d(nU) = Td(nS) - Pd(nV) + \sum_{i=1}^{N}\mu_i dn_i \tag{5.55}$$

$$d(nA) = -(nS)dT - Pd(nV) + \sum_{i=1}^{N}\mu_i dn_i \tag{5.57}$$

$$d(nH) = Td(nS) + (nV)dP + \sum_{i=1}^{N}\mu_i dn_i \tag{5.60}$$

$$d(nG) = -(nS)dT + (nV)dP + \sum_{i=1}^{N} \mu_i dn_i \quad (5.61)$$

with natural independent variables:

Internal energy	nU	nV	nS	$\{n_i\}$
Enthalpy	nH	nS	P	$\{n_i\}$
Helmholtz energy	nA	T	nV	$\{n_i\}$
Gibbs energy	nG	T	P	$\{n_i\}$

5.8 Property Change Calculations

To calculate changes in intangible properties, we exploit the knowledge that the state functions (properties) are ***exact differentials***, and therefore, the value of an integral between two states does not depend upon the choice of path. If we choose a path between state 1 and state 2 such that ***only one*** variable changes along each segment of the path

$$F_2 - F_1 = F(T_2, x_2) - F(T_1, x_1) = \Delta F = \int_1^2 dF = \int_{T_1}^{T_2} \left(\frac{\partial F}{\partial T}\right)_x dT + \int_{x_1}^{x_2} \left(\frac{\partial F}{\partial x}\right)_T dx \quad (5.64)$$

in which F denotes U, H, A, G, and S and x denotes either P or V. For these calculations, we need some combination of U, H, and S, and the derivatives of interest are with respect to temperature, volume, and pressure.

5.8.1 Temperature Derivatives

The derivatives with respect to temperature result from special cases of the first law and the definition of the heat capacity

$$đQ_{rev} = C_x dT \quad (4.54)$$

From the first law at a constant pressure and volume

$$\Delta H = Q_{rev} \quad (2.29)$$

and

$$\Delta U = Q_{rev}$$

In differential form,

$$dH = đQ_{rev} \quad (5.65)$$

and

$$dU = đQ_{rev} \quad (5.66)$$

Combining Eqs. (5.65) and (5.66) with Eq. (4.54)

$$C_P = \left(\frac{\partial H}{\partial T}\right)_P \tag{5.67}$$

$$C_V = \left(\frac{\partial U}{\partial T}\right)_V \tag{5.68}$$

The derivatives of entropy with respect to temperature come from the second law using Eq. (4.54)

$$dS \equiv \frac{đQ_{rev}}{T} \text{ or } dS \equiv \frac{C_X dT}{T}$$

In differential form

$$\left(\frac{\partial S}{\partial T}\right)_x \equiv \frac{C_x}{T} \tag{5.69}$$

in which x denotes a path of either constant P or constant V.

5.8.2 Volume Derivatives

Because volume derivatives are class 2, we begin with the fundamental equation for the internal energy, Eq. (5.22)

$$dU = TdS - PdV \tag{5.22}$$

then

$$\left(\frac{\partial U}{\partial V}\right)_T = T\left(\frac{\partial S}{\partial V}\right)_T - P$$

using a Maxwell relation to eliminate entropy

$$\left(\frac{\partial U}{\partial V}\right)_T = T\left(\frac{\partial P}{\partial T}\right)_V - P \tag{5.70}$$

The same procedure applies to enthalpy

$$dH = TdS + VdP \tag{5.41}$$

Then

$$\left(\frac{\partial H}{\partial V}\right)_T = T\left(\frac{\partial S}{\partial V}\right)_P + V\left(\frac{\partial P}{\partial V}\right)_T \tag{5.71}$$

Using a Maxwell relation to change the entropy derivative

$$\left(\frac{\partial H}{\partial V}\right)_T = T\left(\frac{\partial P}{\partial T}\right)_V + V\left(\frac{\partial P}{\partial V}\right)_T$$

For entropy, the derivative with respect to volume is a Maxwell relation

$$\left[\frac{\partial S}{\partial V}\right]_T = \left[\frac{\partial P}{\partial T}\right]_V \tag{5.50}$$

5.8.3 Pressure Derivatives

We need to calculate the derivative as a class 2, then follow the procedure to obtain

$$\left(\frac{\partial U}{\partial P}\right)_T = -T\left(\frac{\partial V}{\partial T}\right)_P - P\left(\frac{\partial V}{\partial P}\right)_T \tag{5.72}$$

and for enthalpy

$$\left(\frac{\partial H}{\partial P}\right)_T = V - T\left(\frac{\partial V}{\partial T}\right)_P \tag{5.73}$$

The derivative of the entropy with respect to pressure at a constant temperature is a Maxwell relation

$$\left[\frac{\partial S}{\partial P}\right]_{T,n} = -\left[\frac{\partial V}{\partial T}\right]_{P,n} \tag{5.52}$$

From the property change formulae

$$\Delta U = U(T_2, V_2) - U(T_1, V_1) = \int_1^2 dU = \int_{T_1}^{T_2} \left(\frac{\partial U}{\partial T}\right)_V dT + \int_{V_1}^{V_2} \left(\frac{\partial U}{\partial V}\right)_T dV$$

$$= \int_{T_1}^{T_2} C_V dT + \int_{V_1}^{V_2} \left[T\left(\frac{\partial P}{\partial T}\right)_V - P\right] dV \tag{5.74}$$

$$\Delta H = H(T_2, P_2) - H(T_1, P_1) = \int_1^2 dH = \int_{T_1}^{T_2} \left(\frac{\partial H}{\partial T}\right)_P dT + \int_{P_1}^{P_2} \left(\frac{\partial H}{\partial P}\right)_T dP$$

$$= \int_{T_1}^{T_2} C_P dT + \int_{P_1}^{P_2} \left[V - T\left(\frac{\partial V}{\partial T}\right)_P\right] dP \tag{5.75}$$

$$\Delta S = S(T_2, V_2) - S(T_1, V_1) = \int_1^2 dS = \int_{T_1}^{T_2} \left(\frac{\partial S}{\partial T}\right)_V dT + \int_{V_1}^{V_2} \left(\frac{\partial S}{\partial V}\right)_T dV$$

$$= \int_{T_1}^{T_2} \frac{C_V}{T} dT + \int_{V_1}^{V_2} \left(\frac{\partial P}{\partial T}\right)_V dV \tag{5.76}$$

$$\Delta S = S(T_2, P_2) - S(T_1, P_1) = \int_1^2 dS = \int_{T_1}^{T_2} \left(\frac{\partial S}{\partial T}\right)_P dT + \int_{P_1}^{P_2} \left(\frac{\partial S}{\partial P}\right)_T dP$$

$$= \int_{T_1}^{T_2} \frac{C_P}{T} dT - \int_{P_1}^{P_2} \left(\frac{\partial V}{\partial T}\right)_P dP \tag{5.77}$$

Example 5.1

Find the expressions for Eqs. (5.74)–(5.77) using an ideal gas.

Solution

We require the derivatives from the ideal gas equation of state: $P = RT/V^{ig}$

$$\left(\frac{\partial P}{\partial T}\right)_V^{ig} = \frac{R}{V^{ig}}$$

$$\left(\frac{\partial V}{\partial T}\right)_P^{ig} = \frac{R}{P}$$

Therefore,

$$\left[T\left(\frac{\partial P}{\partial T}\right)_V - P\right] = \frac{RT}{V^{ig}} - P = 0$$

$$\left[V^{ig} - T\left(\frac{\partial V}{\partial T}\right)_P\right] = V^{ig} - \frac{RT}{P} = 0$$

The internal energy and enthalpy become

$$\Delta U^{ig} = \int_{T_1}^{T_2} C_V^{ig} dT \tag{5.78}$$

$$\Delta H^{ig} = \int_{T_1}^{T_2} C_P^{ig} dT \tag{5.79}$$

We already knew these results because the internal energy and the enthalpy of an ideal gas are only functions of temperature. For the entropy (in dimensionless form)

$$\frac{\Delta S^{ig}}{R} = \frac{S^{ig}(T_2, V_2) - S^{ig}(T_1, V_1)}{R} = \int_{T_1}^{T_2} \frac{\left(C_V^{ig}/R\right)}{T} dT + \int_{V_1}^{V_2} \frac{dV}{V}$$

$$= \int_{T_1}^{T_2} \frac{\left(C_V^{ig}/R\right)}{T} dT + \ln \frac{V_2}{V_1} \tag{5.80}$$

$$\frac{\Delta S^{ig}}{R} = \frac{S^{ig}(T_2, P_2) - S^{ig}(T_1, P_1)}{R} = \int_{T_1}^{T_2} \frac{\left(C_P^{ig}/R\right)}{T} dT - \int_{P_1}^{P_2} \frac{dP}{P}$$

$$= \int_{T_1}^{T_2} \frac{\left(C_P^{ig}/R\right)}{T} dT - \ln \frac{P_2}{P_1} \tag{5.81}$$

These last two equations are equivalent using different variables.

5.9 Residual Properties

Although we can calculate property changes for U, H, and S from the above equations, we need to obtain changes for other properties. For example, if we want to calculate the change of Helmholtz energy, we must use its definition, $A \equiv U + TS$

$$\Delta A = \Delta U + \Delta(TS)$$

It is not clear how to calculate the second term. We need a reference term, and one convenient path is via *residual properties*. A residual property is the real property of the fluid minus the property of an ideal gas at the same temperature and volume or density or at the same temperature and pressure,

$$F^r = F(T, V \text{ or } \rho) - F^{ig}(T, V \text{ or } \rho) \tag{5.82}$$

$$F^R = F(T, P) - F^{ig}(T, P) \tag{5.83}$$

To develop these relationships, we use the T–V surface, which intersects the ideal gas plane at an infinite volume or zero density as shown in Figure 5.5.

Our path is isotherm from (T, V) to (T, ∞) to (T, V^{ig}), and our working equations are (in differential forms)

$$dU = C_V dT + \left[T\left(\frac{\partial P}{\partial T}\right)_V - P\right] dV \tag{5.74}$$

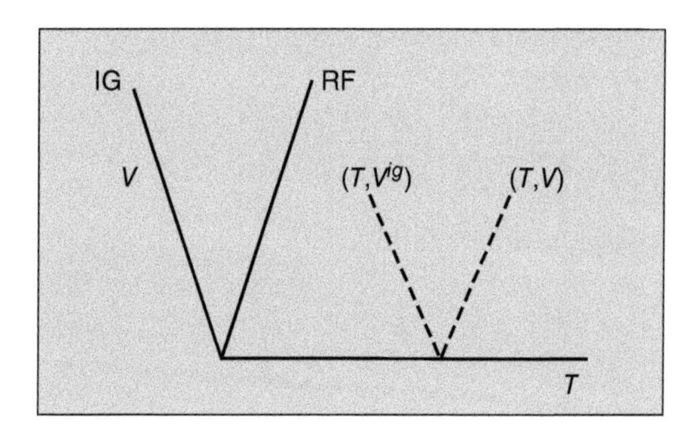

Figure 5.5 A path between the real fluid and ideal gas.

and the fundamental equation for the Helmholtz energy

$$dA = -SdT - PdV \tag{5.42}$$

Thus, U^r is

$$U(T,V) - U^{\text{ig}}(T,V) = \underbrace{U(T,V) - U^{\text{ig}}(T,\infty)}_{\int_\infty^V \left[T\left(\frac{\partial P}{\partial T}\right)_V - P\right] dV} + \underbrace{U^{\text{ig}}(T,\infty) - U^{\text{ig}}(T,V)}_{\int_V^\infty [0] dV} \tag{5.84}$$

but

$$dV = -d\rho/\rho^2 \tag{5.85}$$

and

$$P = ZRT\rho \tag{5.86}$$

so

$$\left(\frac{\partial P}{\partial T}\right)_\rho = \left(\frac{\partial Z}{\partial T}\right)_\rho RT\rho + ZR\rho \tag{5.87}$$

and

$$P - T\left(\frac{\partial P}{\partial T}\right)_\rho = ZRT\rho - \left(\frac{\partial Z}{\partial T}\right)_\rho RT^2\rho - ZRT\rho = -\left(\frac{\partial Z}{\partial T}\right)_\rho RT^2\rho = R\rho\left(\frac{\partial Z}{\partial(1/T)}\right)_\rho$$

thus

$$\frac{U^r}{RT} = \frac{1}{T}\int_0^\rho \left(\frac{\partial Z}{\partial(1/T)}\right)_\rho \frac{d\rho}{\rho} \tag{5.88}$$

For the Helmholtz free energy

$$A(T,V) - A^{\text{ig}}(T,V) = \underbrace{A(T,V) - A^{\text{ig}}(T,\infty)}_{-\int_\infty^V P dV} + \underbrace{A^{\text{ig}}(T,\infty) - A^{\text{ig}}(T,V)}_{RT\int_\infty^V \frac{dV}{V}} \tag{5.89}$$

Then

$$A(T,V) - A^{\text{ig}}(T,V) = \int_\infty^V \left[\frac{RT}{V} - P\right] dV \tag{5.90}$$

or in terms of the compressibility factor

$$\frac{A^r}{RT} = \int_0^\rho \left(\frac{Z-1}{\rho}\right) d\rho \tag{5.91}$$

Using the other identities

$$\frac{H^r}{RT} = \frac{U^r}{RT} + Z - 1 \tag{5.92}$$

$$\frac{G^r}{RT} = \frac{A^r}{RT} + Z - 1 \tag{5.93}$$

$$\frac{S^r}{R} = \frac{U^r}{RT} - \frac{A^r}{RT} \tag{5.94}$$

To use the *T–P* surface, we need to adjust only the ideal gas values

$$\begin{aligned} F(T,P) - F^{ig}(T,P) &= F(T,\rho) - F^{ig}(T,\rho) + [F^{ig}(T,\rho) - F^{ig}(T,P/RT)] \\ &= F^r + [F^{ig}(T,\rho) - F^{ig}(T,P/RT)] \end{aligned} \tag{5.95}$$

and

$$F^{ig}(T,\rho) - F^{ig}(T,P/RT) = 0 \quad \text{if } F = U, H \tag{5.96}$$

$$F^{ig}(T,\rho) - F^{ig}(T,P/RT) = -RT\ln Z \quad \text{if } F = A, G, -TS \tag{5.97}$$

In summary,

$$\frac{U^R}{RT} = \frac{U^r}{RT} \tag{5.98}$$

$$\frac{H^R}{RT} = \frac{U^r}{RT} + Z - 1 \tag{5.99}$$

$$\frac{A^R}{RT} = \frac{A^r}{RT} - \ln Z \tag{5.100}$$

$$\frac{S^R}{R} = \frac{S^r}{R} + \ln Z \tag{5.101}$$

$$\frac{G^R}{RT} = \frac{A^r}{RT} + Z - 1 - \ln Z \tag{5.102}$$

In the following graphic, the italicized equations are the more useful, although all are valid. It is also possible to derive the expression for the residual internal energy from the expression for the residual Helmholtz energy because A is a fundamental function of temperature and density.

$$\frac{U^r}{RT} = \frac{1}{T}\int_0^\rho \left(\frac{\partial Z}{\partial(1/T)}\right)_\rho \frac{d\rho}{\rho} \qquad \frac{\mathrm{U^R}}{\mathrm{RT}} = \frac{\mathrm{U^r}}{\mathrm{RT}}$$

$$\frac{A^r}{RT} = \int_0^\rho \left(\frac{Z-1}{\rho}\right) d\rho \qquad \frac{\mathrm{A^R}}{\mathrm{RT}} = \frac{\mathrm{A^r}}{\mathrm{RT}} - \ln \mathrm{Z}$$

$$\frac{S^r}{R} = \frac{U^r}{RT} - \frac{A^r}{RT} \qquad \frac{S^R}{R} = \frac{S^r}{R} + \ln Z$$

$$\frac{\mathrm{H^r}}{\mathrm{RT}} = \frac{\mathrm{U^r}}{\mathrm{RT}} + \mathrm{Z} - 1 \qquad \frac{H^R}{RT} = \frac{U^r}{RT} + Z - 1$$

$$\frac{\mathrm{G^r}}{\mathrm{RT}} = \frac{\mathrm{A^r}}{\mathrm{RT}} + \mathrm{Z} - 1 \qquad \frac{G^R}{RT} = \frac{A^r}{RT} + Z - 1 - \ln Z$$

$$\frac{\mathrm{PV^r}}{\mathrm{RT}} = 0 \qquad \frac{PV^R}{RT} = Z - 1$$

5.10 Property Changes Using Residual Functions

To calculate the change in a property as temperature and pressure change, we can add zero twice

$$F_2 - F_1 = F(T_2, P_2) - F(T_1, P_1) - F^{ig}(T_2, P_2) + F^{ig}(T_2, P_2) + F^{ig}(T_1, P_1) - F^{ig}(T_1, P_1) \tag{5.103}$$

Then, using the definition of residual property in pressure

$$F_2 - F_1 = \Delta F = F^R(T_2, P_2) + F^{ig}(T_2, P_2) - F^R(T_1, P_1) - F^{ig}(T_1, P_1) \tag{5.104}$$

or

$$F_2 - F_1 = \Delta F = F^R(T_2, P_2) - F^R(T_1, P_1) + \Delta F^{ig} \tag{5.105}$$

Likewise, in terms of volume

$$F_2 - F_1 = F(T_2, V_2) - F(T_1, V_1) - F^{ig}(T_2, V_2) + F^{ig}(T_2, V_2) + F^{ig}(T_1, V_1) - F^{ig}(T_1, V_1) \tag{5.106}$$

Now, using the definition of residual property in volume,

$$F_2 - F_1 = \Delta F = F^r(T_2, V_2) + F^{ig}(T_2, V_2) - F^r(T_1, V_1) - F^{ig}(T_1, V_1) \tag{5.107}$$

or

$$F_2 - F_1 = \Delta F = F^r(T_2, V_2) - F^r(T_1, V_1) + \Delta F^{ig} \tag{5.108}$$

For example, using the internal energy

$$\Delta U = U(T_2, V_2) - U(T_2, V_2) = U^r(T_2, V_2) - U^r(T_1, V_1) + \Delta U^{ig} = U^r(T_2, V_2) - U^r(T_1, V_1) + \int_{T_1}^{T_2} C_V^{ig} dT \tag{5.109}$$

or for enthalpy

$$\Delta H = H(T_2,P_2) - H(T_1,P_1) = H^R(T_2,P_2) - H^R(T_1,P_1) + \Delta H^{ig}$$
$$= H^R(T_2,P_2) - H^R(T_1,P_1) + \int_{T_1}^{T_2} C_P^{ig} dT \qquad (5.110)$$

For entropy, we have

$$\Delta S = S^R(T_2,P_2) - S^R(T_1,P_1) + \Delta S^{ig}$$

but

$$\Delta S^{ig} = \int_{T_1}^{T_2} \frac{C_P^{ig}}{T} dT - R\ln\frac{P_2}{P_1}$$

and

$$\Delta S = S^R(T_2,P_2) - S^R(T_1,P_1) + \int_{T_1}^{T_2} \frac{C_P^{ig}}{T} dT - R\ln\frac{P_2}{P_1} \qquad (5.111)$$

or in terms of volume

$$\Delta S = S^r(T_2,V_2) - S^r(T_1,V_1) + \int_{T_1}^{T_2} \frac{C_V^{ig}}{T} dT + R\ln\frac{V_2}{V_1} \qquad (5.112)$$

Finally, we might wish to calculate the change in a property relative to a standard state, usually at a selected temperature and pressure. Often, the selected temperature is 0 K. Let us use Eq. (5.100) to illustrate this calculation. The reason is that it contains A^r/RT, which has a direct and relatively simple connection to all good equations of state:

$$\frac{A(T,P) - A^{ig}(T_0,P_0)}{RT} = \frac{A(T,P) - A^{ig}(T,P)}{RT} + \frac{A^{ig}(T,P) - A^{ig}(T,P_0)}{RT}$$
$$+ \frac{A^{ig}(T,P_0) - A^{ig}(T_0,P_0)}{RT} \qquad (5.113)$$

Equation (5.100) allows us to replace the first term on the rhs with a more useful term

$$\frac{A(T,P) - A^{ig}(T_0,P_0)}{RT} = \left[\frac{A^r}{RT} - \ln Z\right] + \frac{A^{ig}(T,P) - A^{ig}(T,P_0)}{RT}$$
$$+ \frac{A^{ig}(T,P_0) - A^{ig}(T_0,P_0)}{RT} \qquad (5.114)$$

that is a straightforward function of the equation of state. The second term comes from the equation $dA^{ig} = -S^{ig}dT - PdV^{ig}$

$$dA^{ig} = -PdV^{ig} \text{ at constant } T$$
$$V^{ig} = RT/P$$
$$dV^{ig} = \frac{-RTdP}{P^2}$$
$$-PdV^{ig} = \frac{RTdP}{P}$$
$$\frac{A^{ig}(T,P) - A^{ig}(T,P_0)}{RT} = \int_{P_0}^{P} \frac{dP}{P} = \ln\left(\frac{P}{P_0}\right)$$

So, Eq. (5.114) becomes

$$\frac{A(T,P) - A^{ig}(T_0,P_0)}{RT} = \frac{A^r}{RT} - \ln Z + \ln\left(\frac{P}{P_0}\right) + \frac{A^{ig}(T,P_0) - A^{ig}(T_0,P_0)}{RT} \tag{5.115}$$

The final term requires more consideration. First, we rewrite it in terms of the fundamental variables for A

$$\frac{A^{ig}(T,P_0) - A^{ig}(T_0,P_0)}{RT} = \frac{U^{ig}(T,P_0) - U^{ig}(T_0,P_0)}{RT} + \frac{-TS^{ig}(T,P_0) + T_0S^{ig}(T_0,P_0)}{RT} \tag{5.116}$$

The internal energy of an ideal gas is a function only of temperature, and the first term on the rhs of Eq. (5.116) becomes

$$\frac{U^{ig}(T,P_0) - U^{ig}(T_0,P_0)}{RT} = \frac{1}{T}\int_{T_0}^{T} \frac{C_V^{ig}}{R} dT \tag{5.117}$$

and Eq. (5.115) becomes

$$\frac{A(T,P) - A^{ig}(T_0,P_0)}{RT} = \frac{A^r}{RT} - \ln Z + \ln\left(\frac{P}{P_0}\right) + \frac{1}{T}\int_{T_0}^{T} \frac{C_V^{ig}}{R} dT + \frac{-TS^{ig}(T,P_0) + T_0S^{ig}(T_0,P_0)}{RT}$$

Now, we expand the final term in Eq. (5.116)

$$\frac{-TS^{ig}(T,P_0) + T_0S^{ig}(T_0,P_0)}{RT} = \frac{\begin{array}{c}-TS^{ig}(T,P_0) + TS^{ig}(T_0,P_0)\\ -TS^{ig}(T_0,P_0) + T_0S^{ig}(T_0,P_0)\end{array}}{RT} \tag{5.118}$$

or rewriting

$$\frac{-TS^{ig}(T,P_0) + T_0S^{ig}(T_0,P_0)}{RT} = \frac{-T[S^{ig}(T,P_0) - S^{ig}(T_0,P_0)] - S^{ig}(T_0,P_0)(T - T_0)}{RT} \tag{5.119}$$

and employing Eq. (5.81)

$$\frac{A(T,P) - A^{ig}(T_0,P_0)}{RT} = \frac{A^r}{RT} - \ln Z + \ln\left(\frac{P}{P_0}\right) + \frac{1}{T}\int_{T_0}^{T} C_V^{ig} dT - \int_{T_0}^{T} \frac{C_P^{ig}}{R}\frac{dT}{T} - \frac{S^{ig}(T_0,P_0)}{R}\left(1 - \frac{T_0}{T}\right) \tag{5.120}$$

If T_0 is 0 K, it is common to assume that the entropy is zero, which would further simplify Eq. (5.120) to

$$\frac{A(T,P) - A^{ig}(T_0,P_0)}{RT} = \frac{A^r}{RT} - \ln Z + \ln\left(\frac{P}{P_0}\right) + \frac{1}{T}\int_{T_0}^{T} \frac{C_V^{ig}}{R} dT - \int_{T_0}^{T} \frac{C_P^{ig}}{R}\frac{dT}{T} \tag{5.121}$$

5.11 Generalized Correlations for Residual Functions

We saw before that the generalized correlation developed by Pitzer and Curl, which led to the Lee–Kesler tables to calculate the compressibility factor, is

$$Z(T_R, P_R) = Z^0(T_R, P_R) + \omega Z^1(T_R, P_R) \tag{5.122}$$

The fundamental equation of enthalpy as a function of pressure and temperature using Eq. (5.75) is

$$dH = C_P dT + \left[V - T\left(\frac{\partial V}{\partial T}\right)_P\right] dP \tag{5.123}$$

Also, the expression for the differential entropy according to Eq. (5.77) is

$$dS = \frac{C_P}{T} dT - \left(\frac{\partial V}{\partial T}\right)_P dP \tag{5.124}$$

Thus, the expression for the residual enthalpy is

$$H(T,P) - H^{ig}(T,P) = \underbrace{H(T,P) - H(T,0)}_{\int_0^P \left[V - T\left(\frac{\partial V}{\partial T}\right)_P\right] dP} + \underbrace{H^{ig}(T,0) - H^{ig}(T,P)}_{0} \tag{5.125}$$

but,

$$\left(\frac{\partial Z}{\partial T}\right)_P = -\frac{P}{RT^2}\left[V - T\left(\frac{\partial V}{\partial T}\right)_P\right] \tag{5.126}$$

Therefore,

$$H^R = H(T,P) - H^*(T,P) = -RT^2 \int_0^P \left(\frac{\partial Z}{\partial T}\right)_P \frac{dP}{P} \tag{5.127}$$

Also, using reduced properties

$$\frac{H^R}{RT_C} = -T_R^2 \int_0^{P_R} \left(\frac{\partial Z}{\partial T_R}\right)_P \frac{dP_R}{P_R} \tag{5.128}$$

For the residual entropy,

$$S(T,P) - S^{ig}(T,P) = \underbrace{S(T,P) - S(T,0)}_{-\int_0^P \left(\frac{\partial V}{\partial T}\right)_P dP} + \underbrace{S^{ig}(T,0) - S^{ig}(T,P)}_{\int_0^P \frac{R}{P} dP} \tag{5.129}$$

In terms of the compressibility factor, using Eq. (5.126),

$$\left(\frac{\partial V}{\partial T}\right)_P = \frac{R}{P}\left[T\left(\frac{\partial Z}{\partial T}\right)_P + Z\right] \tag{5.130}$$

Then,

$$S^R = S(T,P) - S^{ig}(T,P) = -R \int_0^P \left[T\left(\frac{\partial Z}{\partial T}\right)_P + Z - 1\right] \frac{dP}{P} \tag{5.131}$$

and using reduced variables

$$\frac{S^R}{R} = -T_R \int_0^{P_R} \left(\frac{\partial Z}{\partial T_R}\right)_P \frac{dP_R}{P_R} - \int_0^{P_R} (Z-1)\frac{dP_R}{P_R} \tag{5.132}$$

The forms of the integrals for the residual entropy and enthalpy are such that if Eq. (5.122) describes Z, then the residual enthalpy and entropy become

$$H^R(T_R, P_R) = [H^R(T_R, P_R)]^0 + \omega[H^R(T_R, P_R)]^1 \tag{5.133}$$

and

$$S^R(T_R, P_R) = [S^R(T_R, P_R)]^0 + \omega[S^R(T_R, P_R)]^1 \tag{5.134}$$

Therefore, tables exist for these two quantities and we can calculate them using the Excel® add-In LK CALC. Instructions are given in Appendix A.2.

Example 5.2

Find an expression for H^R and S^R using a pressure virial equation of state truncated after the second virial coefficient.

Solution

From

$$Z = 1 + \frac{B_2 P}{RT}$$

using Eq. (5.127)

$$H^R = -RT^2 \int_0^P \left(\frac{\partial Z}{\partial T}\right)_P \frac{dP}{P}$$

$$\left(\frac{\partial Z}{\partial T}\right)_P = \frac{P}{R}\left[\frac{1}{T}\left(\frac{dB_2}{dT}\right) - B_2\frac{1}{T^2}\right]$$

and

$$H^R = -RT \int_0^P \frac{P}{R}\left[\left(\frac{dB_2}{dT}\right) - B_2\frac{1}{T}\right]\frac{dP}{P} = -RT\left[\left(\frac{dB_2}{dT}\right) - B_2\frac{1}{T}\right]\frac{P}{R}$$

or in dimensionless form

$$\frac{H^R}{RT} = -\left[\left(\frac{dB_2}{dT}\right) - \frac{B_2}{T}\right]\frac{P}{R}$$

For the residual entropy using Eq. (5.131)

$$T\left(\frac{\partial Z}{\partial T}\right)_P + Z - 1 = \frac{P}{R}\left[\left(\frac{dB_2}{dT}\right) - B_2\frac{1}{T}\right] + \frac{B_2 P}{RT} = \frac{P}{R}\left(\frac{dB_2}{dT}\right)$$

thus

$$S^R = -R\int_0^P \left[\frac{P}{R}\left(\frac{dB_2}{dT}\right)\right]\frac{dP}{P} = -P\left(\frac{dB_2}{dT}\right)$$

or

$$\frac{S^R}{R} = -\frac{P}{R}\left(\frac{dB_2}{dT}\right)$$

Example 5.3

Calculate the enthalpy and entropy change of propane from $T_1 = 400\,\text{K}$ and $P_1 = 100\,\text{kPa}$ to $T_2 = 500\,\text{K}$ and $P_2 = 5000\,\text{kPa}$. Use the generalized correlation in Lee–Kesler tables.

Data

$$T_C = 369.89\ \text{K}$$

$$P_C = 4.2512\ \text{MPa}$$

$$\omega = 0.1521$$

$$C_P^*/R = 1.213 + 2.8785 \times 10^{-2}T - 8.824 \times 10^{-6}T^2$$

Solution

The reduced quantities at state 1 are

$$T_{R,1} = \frac{T_1}{T_C} = \frac{400}{369.89} = 1.0814$$

$$P_{R,1} = \frac{P_1}{P_C} = \frac{100}{4251.2} = 0.0235$$

then

$$\frac{[H^R]^0}{RT_C} = -0.0205$$

$$\frac{[H^R]^1}{RT_C} = -0.0154$$

and

$$H^R = \left\{ \frac{[H^R]^0}{RT_C} + \omega \frac{[H^R]^1}{RT_C} \right\} RT_C$$

$$H_1^R = -0.0205 + 0.1521 \times (-0.0154)[8.314 \times 369.85] = -70.2463$$

For entropy

$$\frac{[S^R]^0}{R} = -0.0127$$

$$\frac{[S^R]^1}{R} = -0.014$$

and

$$S^R = \left\{ \frac{[S^R]^0}{R} + \omega \frac{[S^R]^1}{R} \right\} R$$

$$S_1^R = -0.0127 + 0.1521 \times (-0.014)8.314 = -0.1233\ \text{J/(mol K)}$$

At state 2,

$$T_{R,2} = \frac{T_2}{T_C} = \frac{500}{369.89} = 1.3518$$

$$P_{R,2} = \frac{P_2}{P_C} = \frac{5000}{4251.2} = 1.7614$$

and using

$$\frac{[H^R]^0}{RT_C} = -0.7621$$

$$\frac{[H^R]^1}{RT_C} = -0.1591$$

then

$$H_2^R = [-0.7621 + 0.1521 \times (-0.1591)]\,[8.314 \times 369.85] = -2436.46\ \text{J/mol}$$

for entropy

$$\frac{[S^R]^0}{R} = -0.4079$$

$$\frac{[S^R]^1}{R} = -0.2270$$

and

$$S_2^R = -0.4079 + 0.1521 \times (-0.2270)8.314 = -0.4394\ \text{J/(mol K)}$$

using Eq. (5.110)

$$\Delta H = H^R(T_2, P_2) - H^R(T_1, P_1) + \int_{T_1}^{T_2} C_P^{ig}\, dT$$

$$\Delta H = 8.314\left[1.213(500 - 400) + \frac{2.8785 \times 10}{2}(500^2 - 400^2) - \frac{8.824 \times 10^{-6}}{3}(500^3 - 400^3)\right]$$
$$- 2436.46 + 70.2463 = 7920.4\ \text{J/mol}$$

using Eq. (5.111)

$$\Delta S = S^R(T_2, P_2) - S^R(T_1, P_1) + \int_{T_1}^{T_2} \frac{C_P^{ig}}{T}\, dT - R\ln\frac{P_2}{P_1}$$

$$\Delta S = 8.314\left[1.213\ln\frac{500}{400} + 2.8785 \times 10^{-2}(500 - 400) - \frac{8.824 \times 10^{-6}}{2}(500^2 - 400^2)\right]$$
$$- 8.314\ln\frac{5000}{100} - 0.4394 + 0.1233 = -13.173\ \text{J/(mol K)}$$

5.12 Two-Phase Systems – Clapeyron Equation

For two phases of a pure species that coexist at equilibrium, the stability criteria derived from the second law of thermodynamics require that

$$G^\alpha = G^\beta \tag{5.135}$$

In which, the superscripts α and β denote the individual coexisting phases.

As a system in equilibrium moves along its saturation curve, the differential changes in the coexisting phases must be equal

$$dG^\alpha = dG^\beta \tag{5.136}$$

Using the fundamental equation for the Gibbs energy (5.43) for at each phase in equilibrium

$$-S^{\alpha}dT + V^{\alpha}dP^{\sigma} = -S^{\beta}dT + V^{\beta}dP^{\sigma} \tag{5.137}$$

In which, P^{σ} is the equilibrium pressure. Regrouping the temperature and pressure terms

$$(S^{\beta} - S^{\alpha})dT = (V^{\beta} - V^{\alpha})dP^{\sigma} \tag{5.138}$$

thus

$$\frac{dP^{\sigma}}{dT} = \frac{S^{\beta} - S^{\alpha}}{V^{\beta} - V^{\alpha}} \tag{5.139}$$

Applying the definition of Gibbs energy, $G \equiv H - TS$, to both phases

$$G^{\beta} - G^{\alpha} = 0 = (H^{\beta} - H^{\alpha}) - T(S^{\beta} - S^{\alpha}) \tag{5.140}$$

and Eq. (5.139) becomes

$$\frac{dP^{\sigma}}{dT} = \frac{H^{\beta} - H^{\alpha}}{T(V^{\beta} - V^{\alpha})} \tag{5.141}$$

This is the Clausius–Clapeyron equation. For the special case of vapor–liquid equilibrium

$$\frac{dP^{sat}}{dT} = \frac{H^{v} - H^{l}}{T(V^{v} - V^{l})} = \frac{\Delta H^{vl}}{T\Delta V^{vl}} \tag{5.142}$$

This equation relates the slope of the vapor pressure curve to the enthalpy of vaporization, ΔH^{vl}, and the change of volume between vapor and liquid, ΔV^{vl}. At low to moderate pressures, the vapor volume is much greater than the liquid volume, so

$$\Delta V^{vl} = V^{v} - V^{l} \approx V^{v} = \frac{RT}{P^{sat}}$$

and

$$\frac{dP^{sat}}{dT} = \frac{\Delta H^{vl}}{T \cdot RT/P^{sat}} \tag{5.143}$$

If we rearrange this equation and separate the variables,

$$\frac{dP^{sat}}{P^{sat}} = \frac{\Delta H^{vl}}{R}\frac{dT}{T^2} \tag{5.144}$$

If the enthalpy of vaporization is constant, we can integrate to obtain the Clausius vapor pressure equation

$$\ln P^{sat} = -\frac{\Delta H^{vl}}{R}\left(\frac{1}{T}\right) + C = A + B/T \tag{5.145}$$

In which C is an integration constant. Obviously, this equation is valid only at low pressures and over a limited temperature range near the triple point temperature.

Example 5.4

Estimate the vapor pressure of ammonia at 300 K.

Data

$$T_C = 405.7\text{ K},\ P_C = 112.8\text{ bar},\ \text{and } T_b = 239.7\text{ K}$$

Solution

To estimate the vapor pressure, we can use the Clausius vapor pressure equation

$$\ln P^{sat} = A + B/T$$

We need two data points to determine two constants, e.g. the critical point and the normal boiling point,

$$\ln P_C = A + B/T_C \tag{A}$$

and

$$\ln P_b = \ln(1) = 0 = A + B/T_b \tag{B}$$

Solving for A in Eq. (5.II)

$$A = -\frac{B}{T_b}$$

and substituting this equation into Eq. (5.I),

$$\ln P_C = B/T_C - B/T_b = B\left(\frac{1}{T_C} - \frac{1}{T_b}\right)$$

Solving for B

$$B = \ln P_C/\left(\frac{1}{T_C} - \frac{1}{T_b}\right)$$

provides the value of the constants in terms of tabulated quantities

$$\ln P^{sat} = -\frac{B}{T_b} + \frac{B}{T} = B\left(\frac{1}{T} - \frac{1}{T_b}\right)$$

and

$$\ln P^{sat} = \ln P_C\left(\frac{1}{T} - \frac{1}{T_b}\right)\bigg/\left(\frac{1}{T_C} - \frac{1}{T_b}\right)$$

Multiplying and dividing the RHS of the equation by T_C

$$\ln P^{sat} = \ln P_C \frac{\left(\frac{T_C}{T} - \frac{T_C}{T_b}\right)}{\left(1 - \frac{T_C}{T_b}\right)} = \ln P_C \frac{\left(\frac{1}{T_R} - \frac{1}{T_{R,b}}\right)}{\left(1 - \frac{1}{T_{R,b}}\right)}$$

Now

$$T_R = \frac{T}{T_C} = \frac{300}{405.7} = 0.7395$$

$$T_{R,b} = \frac{T}{T_C} = \frac{239.7}{405.7} = 0.5908$$

$$\ln P^{sat} = \ln(112.8)l\frac{\left(\frac{1}{0.7395} - \frac{1}{0.5908}\right)}{\left(1 - \frac{1}{0.5908}\right)} = 2.3214$$

$$P^{sat} = \exp(2.3214) = 10.19 \text{ bar}$$

Problems for Chapter 5

5.1 A 5 kg metal block is compressed from 1 to 700 bar at a constant temperature. Estimate the change of entropy if the thermal expansion is constant and $\beta = 7.2\times10^{-6}$ K^{-1} and $\rho = 8000$ kg/m^3.

5.2 The math relation for P, V, and T is $\left(\frac{\partial P}{\partial T}\right)_V\left(\frac{\partial T}{\partial V}\right)_P\left(\frac{\partial V}{\partial P}\right)_T = -1$. Check the expression for the following EOS $P = RT/(V-b)$ in which b is a constant.

5.3 The value of a numerical derivative can be approximated numerically using forward differences

$$\left(\frac{\partial f}{\partial x}\right)_z = \left(\frac{f_2 - f_1}{x_2 - x_1}\right)_z = \left(\frac{f(x_2) - f(x_1)}{x_2 - x_1}\right)_z$$

in which $x_2 = x_1 + h$. Check the Maxwell relation $\left(\frac{\partial S}{\partial P}\right)_T = -\left(\frac{\partial V}{\partial T}\right)_P$ for steam at 1000 kPa and 350 °C using $h = 10$.

5.4 The natural variables for enthalpy are S and P. From the total differential of H, $V = \left(\frac{\partial H}{\partial P}\right)_S$. From tables at a constant entropy, an estimate of the volume is possible. Find the volume using the above derivative and the steam tables at 500 kPa and S = 7.3 kJ/(kg K). Use forward differences (with h = 10) to approximate the derivative and compare your approximation with the real volume.

5.5 Calculate the enthalpy change of an incompressible fluid for which $\beta = 5\times10^{-5}$ K^{-1} and $\rho = 2$ g/cm^3 if there is a change in the pressure of 1000 kPa at 300 K. Consider $M = 20$ g/gmol.

5.6 Show that $\left(\frac{\partial C_V}{\partial V}\right)_T = T\left(\frac{\partial^2 P}{\partial T^2}\right)_V$.

5.7 Show that $\left(\frac{\partial T}{\partial V}\right)_U > \left(\frac{\partial T}{\partial V}\right)_S$. Is this always true?

5.8 What is the value of $\left(\frac{\partial H}{\partial U}\right)_V$ at 55 °C if a substance exerts a pressure of 40 bar and the pressure behavior is

$$P = \frac{RT}{V-b} \text{ in which } b = 20\ \text{cm}^3/\text{mol and } C_V = R(3.5 + 2.14\times10^{-2}T)$$

5.9 Express $\left(\frac{\partial U}{\partial T}\right)_S$ in terms of observables variables.

5.10 A liquid is compressed from 250 K and 800 bar to 300 K and 900 bar. Use the following data to estimate the change of enthalpy. The volume and enthalpy derivatives can be approximated by forward differences.

T (K)	P (bar)	Volume (cm^3/mol)	Enthalpy (J/mol)
150	800	30.26	−755.35
200	800	34.629	1036.6
250	800	39.727	2749.2
300	800	45.367	4374.6
350	800	51.303	5910.1
150	900	29.75	−569.84
200	900	33.709	1198.5
250	900	38.226	2885.2
300	900	43.168	4490.2
350	900	48.359	6014.2

5.11 Methane gas passes through a valve. The inlet temperature is 400 K and 1 MPa. What is the outlet temperature if the pressure changes to 0.1 MPa? Data

Temperature (K)	Density (mol/l)	Internal energy (kJ/mol)	Enthalpy (kJ/mol)	Entropy (J/(mol K))
300	0.40776	12.072	14.524	87.922
350	0.34675	13.52	16.404	93.713
400	0.30205	15.083	18.393	99.023
450	0.26776	16.78	20.515	104.02
500	0.24058	18.623	22.78	108.79
550	0.21847	20.615	25.192	113.38
600	0.20012	22.755	27.752	117.84

5.12 Find the expression to calculate H^R/RT if a substance has the equation of state

$$P = \frac{RT}{V-b} - \frac{a}{T^{0.5}V(V+b)}$$

in which a and b are constants.

5.13 A gas with a critical temperature of 450 K, a critical pressure of 6.0 MPa, and $C_p^{ig} = 4.2R$ initially is at 300 K and 20 bar. Use the van der Waals equation of state to determine the final temperature and pressure after 300 J/mol is added as heat during a constant volume process.

5.14 Calculate the residual enthalpy of ethane at $T_2 = 500$ K and $P_2 = 1000$ kPa. Use the Lee–Kesler EOS and the virial EOS explicit in volume.

Data

$$\omega = 0.1,\ T_C = 305.3\,\text{K},\ P_C = 48.72\,\text{bar},\ Z_C = 0.279,\ \text{and}\ V_C = 145.5$$

5.15 Calculate the residual internal energy of ethane at 400 K and 70 bar. Using the appropriate generalized correlations.

5.16 How would you find S^R from steam tables?

5.17 Determine U^R, S^R, H^R, G^R, and A^R for *n*-octane at 580 K and 200 bar. Use a generalized correlation.
Data

$$T_C = 569.32\,\text{K},\ P_C = 24.97\,\text{bar},\ \text{and}\ \omega = 0.393$$

5.18 Methane is compressed from 1 bar at 200 K to 100 bar at 353 K. What is the entropy change using the Lee–Kesler EOS?
Data

$$T_C = 190.564\,\text{K},\ P_C = 45.992\,\text{bar},\ \text{and}\ \omega = 0.011$$

$$C_P^{ig}/R = 1.7025 + 9.0861 \times 10^{-3}T - 21.6528 \times 10^{-7}T^2$$

5.19 Determine the change in the entropy for a mole of ethane when it changes from 280 K and 1.5 bar to 366 K and 146 bar.
Data

$$T_C = 305.3\,\text{K},\ P_C = 48.72\,\text{bar},\ \text{and}\ \omega = 0.1$$

$$C_P/R = 1.1315 + 19.2363 \times 10^{-3}T - 55.6377 \times 10^{-7}T^2$$

5.20 Determine the molar change in entropy as ethylene changes from state 1 at 310.6 K and 5 bar to state 2 at 367.1 K and 20 bar. Is it possible to accomplish this change of state in a single adiabatic step? Justify your answer.
Data

$$T_C = 282.35\,\text{K},\ P_C = 50.418\,\text{bar},\ \text{and}\ \omega = 0.086$$

$$C_P^{ig}/R = 1.4241 + 14.3925 \times 10^{-3}T - 43.9107 \times 10^{-7}T^2$$

5.21 Estimate the amount of energy that must be removed as heat to cool butane from 395 to 320 K at a constant pressure of 370 bar.
Data

$$T_C = 425.125\,\text{K},\ P_C = 37.96\,\text{bar},\ \text{and}\ \omega = 0.201$$

$$C_P/R = 1.1315 + 19.2363 \times 10^{-3}T - 55.6377 \times 10^{-7}T^2$$

5.22 Nitrogen from a process line at 138 bar and 26.85 °C exhausts into the atmosphere through a valve. Determine the temperature of the exhausted gas, assuming that this is a steady-state flow process in which no energy is transferred as work or heat. You may also assume that the ideal gas heat capacity of nitrogen does not depend on temperature and that the value is $C_P = 3.50R$. Note that nitrogen may behave as an ideal gas at atmospheric pressure but not at 138 bar.
Data

$$T_C = 126.2\ \text{K},\ P_C = 34\ \text{bar},\ \text{and}\ \omega = 0.038$$

5.23 Propylene initially at 255 K and 1 bar is compressed reversibly and adiabatically. Estimate the final pressure if the temperature at the end of compression process is 550 K.
Data

$$T_C = 364.21\ \text{K},\ P_C = 45.55\ \text{bar},\ \text{and}\ \omega = 0.146$$

$$C_P^{ig}/R = 1.6370 + 22.7032 \times 10^{-3}T - 69.1421 \times 10^{-7}T^2$$

5.24 Methane initially at 50 °C and 20 bar expands to 7 bar in a turboexpander (turbine) during a reversible, steady-state flow process. Determine the outlet temperature of the methane.
Data

$$T_C = 190.564\ \text{K},\ P_C = 45.992\ \text{bar},\ \text{and}\ \omega = 0.01142$$

$$C_P^{ig}/R = 1.7025 + 9.0861 \times 10^{-3}T - 21.6528 \times 10^{-7}T^2$$

5.25 A gaseous stream at 220 °C and 40 bar expands in a turbine at 1 bar in an isentropic process. Determine the temperature of the expanded gas and the energy transferred as work. Use the Lee–Kesler EOS
Data

$$C_P^{ig}/R = 1.131 + 19.225 \times 10^{-3}T - 5.561 \times 10^{-6}T^2$$

$$T_C = 305.3\ \text{K},\ P_C = 48.72\ \text{bar},\ Z_C = 0.279,\ V_C = 145.5,\ \omega = 0.1$$

5.26 Oxygen at 140 bar and 25 °C passes through a valve and emerges at a pressure of 1 atm. Assuming that this is a steady-state flow process where no energy is transferred as heat or work in the valve, determine the temperature of the gas at the outlet. Use a valid generalized correlation at the T and P conditions.
Data

$$T_C = 154.6\ \text{K},\ P_C = 154.6\ \text{bar},\ \text{and}\ \omega = 0.022$$

$$C_P/R = 3.6390 + 5.06 \times 10^{-2}T - 2.27 \times 10^{4}T^{-2}$$

5.27 A stream of methane initially at 288 K and 27 bar is compressed in a single-stage compressor to 70 bar. The compressor has an operating efficiency of 0.85.

a. Determine the temperature of the gas leaving the compressor if the process is adiabatic and reversible.
b. Determine the ratio of densities of the gas entering and leaving the compressor if the process is adiabatic and reversible.
c. Determine the work actually required to compress the gas.
d. Determine the entropy change of the methane during the compression process.

Data

$$T_C = 190.564\ \text{K},\ P_C = 45.992\ \text{bar},\ \text{and}\ \omega = 0.011$$

$$C_P^{ig}/R = 1.7025 + 9.0861 \times 10^{-3}T - 21.6528 \times 10^{-7}T^2$$

5.28 Determine ΔH, ΔV, and T when saturated carbon dioxide vapor condenses to saturated carbon dioxide liquid at 59 bar. The vapor pressure of carbon dioxide is

$$\ln P^{sat}\ (\text{bar}) = 15.25 - \frac{3313.3}{T\ (\text{K})}$$

This notation means that P^{sat} has units of bars and T has units of kelvin.

Data

$$T_C = 304.2\ \text{K}\ \text{and}\ \omega = 0.224$$

5.29 A substance has a normal boiling point of 350 K, a critical temperature of 525 K, and a critical pressure of 37 bar. Estimate the acentric factor, ω, for this substance.

5.30 Use the Clausius equation to estimate the enthalpy of vaporization for propane at 350 K. The vapor pressure of propane at 350 K is 29.514 bar and dP/dT = 0.563 bar/K. The second and third virial coefficients at this temperature are $B_2 = -277\ \text{cm}^3/\text{mol}$ and $B_3 = 21\,338\ \text{cm}^6/\text{mol}^2$.

Data

$$T_C = 369.89\ \text{K},\ P_C = 42.512\ \text{bar},\ V_C = 200\ \text{cm}^3/\text{mol},\ \text{and}\ \omega = 0.152$$

5.31 Estimate the enthalpy of vaporization of a substance at its boiling temperature if the vapor pressure equation is

$$\ln P = 31.117 - \frac{6175}{T} - 0.0175T$$

The pressure is in kPa and the temperature in kelvin.

5.32 Estimate the vapor pressure of a substance at 300 K whose critical properties are $T_C = 500$ K and $P_C = 40$ bar and the acentric factor is 0.4. Use the Clausius vapor pressure equation.

5.33 The *PVT* behavior of a gas follows the equation of state:

$$Z = 1 + B_2P/RT$$

Estimate the value of the second virial coefficient at its normal boiling temperature if the vapor pressure is

$$\ln P^{sat}\ (\text{kPa}) = 13.86 - \frac{2774}{T\ (\text{K}) - 53.15}$$

and the enthalpy of vaporization is 30.77 kJ/mol.

Data

$$T_C = 420\ \text{K},\ P_C = 45\ \text{bar, and } Z_C = 0.27$$

5.34 A new substance has been discovered with an acentric factor of 0.2. Its critical pressure is 37 bar, and its critical temperature is 525 K. What is its boiling temperature?

6

Practical Applications for Thermodynamics

6.1 Fluid Flow

Almost all chemical process industries involve fluid flow. The examples include cooling water circulation through engines, cooling unit operations, transportation of fluids through a pipe, fluids passing through pumps, expanders, etc. Three basic concepts involved in fluid flow are the principle of momentum, the conservation of energy (first law of thermodynamics), and the conservation of mass. The problems in plant design (such as sizing) require using the principle of momentum. Thermodynamics provides relationships among pressure, velocity, density, and energy functions in such applications.

6.1.1 Flow Through Ducts

Let us consider the differential form of the first law, Eq. (2.31) (in which we do not consider nuclear reactions):

$$d[n(U+E_K+E_P)]_{SYS} = đ(nQ) + đ(nW) - d[n(H+E_K+E_P)] \tag{6.1}$$

$$\begin{aligned}[d(nU) + d(nE_K) + d(nE_P)]_{SYS} &= đ(nQ) + đ(nW) - d(nH) \\ &\quad - d(nE_K) - d(nE_P)\end{aligned} \tag{6.2}$$

If the number of moles is constant, $nE_K = m\dot{z}^2/2$ and $nE_P = mgz$, so

$$[dU + M\dot{z}d\dot{z} + Mgdz]_{SYS} = đQ + đW - dH - M\dot{z}d\dot{z} - Mgdz \tag{6.3}$$

For steady-state conditions,

$$0 = đQ + đW - dH - M\dot{z}d\dot{z} - Mgdz \tag{6.4}$$

or

$$dH + M\dot{z}d\dot{z} + Mgdz = đQ + đW \tag{6.5}$$

This is our basic fluid flow equation. If the flow rate is constant, then from the definition of molar flow rate,

$$\dot{n} = A\rho\dot{z} = constant \Rightarrow d\dot{n} = 0 = \rho\dot{z}dA + A\dot{z}d\rho + A\rho d\dot{z} \tag{6.6}$$

Thermodynamics for Chemical Engineers, First Edition. Kenneth R. Hall and Gustavo A. Iglesias-Silva.

This is the continuity equation in which the inlet mass equals the outlet mass. Dividing Eq. (6.6) by $\dot{n}$

$$\frac{dA}{A} + \frac{d\rho}{\rho} + \frac{d\dot{z}}{\dot{z}} = 0$$

Next, we consider several cases:

Case I. Horizontal and adiabatic flow in a constant cross-sectional pipe

The equation is

$$-dH = M\dot{z}d\dot{z} \tag{6.7}$$

and we know that $dH = TdS + VdP$; therefore,

$$TdS + VdP + M\dot{z}d\dot{z} = 0 \tag{6.8}$$

If we consider a constant cross section, then $dA = 0$, and

$$\frac{d\dot{z}}{\dot{z}} = -\frac{d\rho}{\rho} \tag{6.9}$$

Substituting into Eq. (6.8)

$$TdS + \frac{dP}{\rho} - M\dot{z}^2\frac{d\rho}{\rho} = 0$$

Maximum velocity occurs when the flow is frictionless, i.e. $dS = 0$ (constant entropy)

$$\frac{dP}{\rho} = M\dot{z}_{\max}^2\frac{d\rho}{\rho} \tag{6.10}$$

or

$$M\dot{z}_{\max}^2 = \left(\frac{dP}{d\rho}\right)_S \tag{6.11}$$

Using math relation (5.12),

$$M\dot{z}_{\max}^2 = -\left(\frac{\partial P}{\partial S}\right)_\rho\left(\frac{\partial S}{\partial \rho}\right)_P \tag{6.12}$$

and imposing the chain rule

$$\begin{aligned}\frac{\dot{z}_{\max}^2}{M} &= -\left(\frac{\partial P}{\partial T}\right)_\rho\left(\frac{\partial T}{\partial S}\right)_\rho\left(\frac{\partial S}{\partial T}\right)_P\left(\frac{\partial T}{\partial \rho}\right)_P = -\frac{C_P}{C_V}\left(\frac{\partial P}{\partial T}\right)_\rho\left(\frac{\partial T}{\partial \rho}\right)_P \\ &= \frac{C_P}{C_V}\left(\frac{\partial P}{\partial \rho}\right)_T\end{aligned} \tag{6.13}$$

This is the definition of sonic velocity or speed of sound.

Case II. Horizontal incompressible flow at steady state

$$dH + M\dot{z}d\dot{z} = đQ \tag{6.14}$$

Integrating between two points on the pipe,

$$H_2 - H_1 + M\frac{\dot{z}_2^2 - \dot{z}_1^2}{2} = {}_1Q_2 \tag{6.15}$$

and using the fundamental equation for enthalpy

$$TdS + VdP + M\dot{z}d\dot{z} = đQ \tag{6.16}$$

Again, integrating between two points on the pipe

$$\int_{S_1}^{S_2} TdS + \frac{\Delta P}{\rho} + M\frac{\dot{z}_2^2 - z_1^2}{2} = {}_1Q_2 \tag{6.17}$$

Considering that the irreversibility results in a friction term

$${}_1f_2 = \int_{S_1}^{S_2} TdS - {}_1Q_2 \tag{6.18}$$

Then, Eq. (6.17) is

$${}_1f_2 + \frac{\Delta P}{\rho} + M\frac{\dot{z}_2^2 - z_1^2}{2} = 0 \tag{6.19}$$

Using the definitions of flow rate and area $\dot{n} = A\rho\dot{z}$ and $A = \pi D^2/4$, then

$${}_1f_2 + \frac{\Delta P}{\rho} + \frac{M}{2}\left\{\frac{\dot{n}^2}{\left(\pi^2 D_2^4/16\right)\rho^2} - \frac{\dot{n}^2}{\left(\pi^2 D_1^4/16\right)\rho^2}\right\} = 0 \tag{6.20}$$

Rearranging the above equation

$${}_1f_2 + \frac{\Delta P}{\rho} + \frac{8\dot{n}^2 M}{\rho^2\pi^2}\frac{1}{D_2^4}\left\{1 - \left(\frac{D_2}{D_1}\right)^4\right\} = 0 \tag{6.21}$$

Generally, the friction term is a function of the Reynolds number and the ratio of diameters, $\beta = D_2/D_1$,

$${}_1f_2 = z_1^2 M\varphi\left(N_R, \beta, \ldots\right)/2 = \frac{8\dot{n}^2 M\varphi}{\rho^2\pi^2 D_1^4} \tag{6.22}$$

Now,

$$\frac{\Delta P}{\rho} + \frac{8\dot{n}^2 M}{\rho^2\pi^2}\frac{1}{D_2^4}\left\{1 - \beta^4(1-\varphi)\right\} = 0 \tag{6.23}$$

and the flow rate is

$$\dot{n}^2 = \frac{\pi^2 D_2^4}{8M\left\{\beta^4(1-\varphi) - 1\right\}}\rho\Delta P \Rightarrow \dot{n} = \frac{\pi D_2^2}{M2\sqrt{2}}\left\{\frac{1}{\beta^4(1-\varphi) - 1}\right\}^{0.5}\sqrt{\rho\Delta P} \tag{6.24}$$

Case III. Horizontal gas flow at steady state passing through a circular pipe containing an orifice

Using Eq. (6.19) for a compressible fluid

$${}_1f_2 + \int_{P_1}^{P_2} VdP + M\frac{\dot{z}_2^2 - \dot{z}_1^2}{2} = 0 \tag{6.25}$$

then from the continuity equation,

$$\frac{\pi D_1^2\dot{z}_1\rho_1}{4} = \frac{\pi D_2^2\dot{z}_2\rho_2}{4} \Rightarrow \frac{\dot{z}_1}{\dot{z}_2} = \frac{D_2^2\rho_2}{D_1^2\rho_1} = \beta^2\left(\frac{\rho_2}{\rho_1}\right) \tag{6.26}$$

and Eq. (6.25) becomes

$$_1f_2 + \int_{P_1}^{P_2} VdP = M\frac{\dot{z}_2^2}{2}\left\{\left(\frac{\dot{z}_1}{\dot{z}_2}\right)^2 - 1\right\} = 0 \tag{6.27}$$

or

$$_1f_2 + \int_{P_1}^{P_2} VdP = M\frac{\dot{z}_2^2}{2}\left\{\beta^4\left(\frac{\rho_2}{\rho_1}\right)^2 - 1\right\} \tag{6.28}$$

Solving for the velocity

$$\dot{z}_2^2 = \frac{_1f_2 + \int_{P_1}^{P_2} VdP}{\frac{M}{2}\left[\beta^4\left(\frac{\rho_2}{\rho_1}\right)^2 - 1\right]} \tag{6.29}$$

Using $d(PV) = PdV + VdP$

$$\dot{z}_2^2 = \frac{_1f_2 + P_2/\rho_2 - P_1/\rho_1 - \int_{V_1}^{V_2} PdV}{\frac{M}{2}\left[\beta^4\left(\frac{\rho_2}{\rho_1}\right)^2 - 1\right]} \tag{6.30}$$

This is a fundamental formulation for the velocity of a fluid passing through an orifice. The flow rate results from

$$\dot{n} = \dot{n}_1 = \dot{n}_2 = \dot{z}_2\rho_2\pi\frac{D_2^2}{4} \tag{6.31}$$

In this derivation, the terms accounting for energy loss and expansion factor effects are additive. We can consider two different scenarios. If it is an isothermal process and using the definition of compressibility factor, $Z = P/\rho RT$, we obtain a density ratio

$$\frac{\rho_2}{\rho_1} = \frac{Z_1P_2}{Z_2P_1} = \frac{Z_1}{Z_2}\left(1 - \frac{\Delta P}{P_1}\right) \tag{6.32}$$

in which $\Delta P = P_1 - P_2$. Then if

$$\frac{P_2}{\rho_2} - \frac{P_1}{\rho_1} = \frac{P_1Z_2}{\rho_1Z_1} - \frac{P_1}{\rho_1} = \frac{P_1}{\rho_1}\left(\frac{Z_2}{Z_1} - 1\right) = \frac{P_1}{\rho_1}\left(\frac{Z_2}{Z_1} - 1\right) \tag{6.33}$$

substituting into Eq. (6.30) provides

$$\dot{z}_2^2 = \frac{_1f_2 + \frac{P_1}{\rho_1}\left(\frac{Z_2}{Z_1} - 1\right) - \int_{V_1}^{V_2} PdV}{\frac{M}{2}\left[\beta^4\left\{\frac{Z_1}{Z_2}\left(1 - \frac{\Delta P}{P_1}\right)\right\}^2 - 1\right]} \tag{6.34}$$

It is well known that the passage of a fluid through an orifice is not isothermal, so we consider an adiabatic process for which

$$\dot{z}_2^2 = \frac{H_2 - H_1}{\frac{M}{2}\left[\beta^4\left(\frac{\rho_2}{\rho_1}\right)^2 - 1\right]} \tag{6.35}$$

or

$$\dot{z}_2^2 = \frac{\int_T^{T_2} C_P^{ig} dT + H_2^R - H_1^R}{\frac{M}{2}\left[\beta^4\left(\frac{\rho_2}{\rho_1}\right)^2 - 1\right]} \tag{6.36}$$

Here, any equation of state (EOS) can calculate the residual properties. The flow rate comes from

$$n = n_1 = n_2 = \dot{z}_2 \rho_2 \pi \frac{D_2^2}{4} \tag{6.37}$$

Case IV. Nonhorizontal nonadiabatic pipe
We start with the equation

$$dH + M\dot{z}d\dot{z} + Mgdz = đQ + đW \tag{6.38}$$

or

$${}_1f_2 + VdP + M\dot{z}d\dot{z} + Mgdz = đW \tag{6.39}$$

This is the basic fluid flow equation (sometimes called the mechanical energy balance equation). For the special case of an incompressible fluid with $W = 0$ and ${}_1f_2 = 0$,

$$VdP + M\dot{z}d\dot{z} + Mgdz = 0 \tag{6.40}$$

which becomes upon integration

$$VP + M\frac{\dot{z}^2}{2} + Mgz = constant \tag{6.41}$$

Case V. Flow through an isentropic nozzle
From Eq. (6.16),

$$VdP + M\dot{z}d\dot{z} = 0 \tag{6.42}$$

which becomes upon integration

$$M\frac{\left(\dot{z}_2^2 - \dot{z}_1^2\right)}{2} = -\int_{P_1}^{P_2} VdP \tag{6.43}$$

For simplicity, let us consider an ideal gas with constant heat capacities

$$PV^{\gamma^{ig}} = c \Rightarrow V = \left(\frac{c}{P}\right)^{1/\gamma^{ig}} \tag{6.44}$$

Then,

$$\begin{aligned} M\frac{\left(\dot{z}_2^2 - \dot{z}_1^2\right)}{2} &= -\frac{\gamma^{ig} c^{1/\gamma^{ig}}}{\gamma^{ig} - 1}\left(P_2^{-1/\gamma^{ig}+1} - P_1^{-1/\gamma^{ig}+1}\right) \\ &= \frac{\gamma^{ig} c^{1/\gamma^{ig}} P_2^{(\gamma^{ig}-1)/\gamma^{ig}}}{\gamma^{ig} - 1}\left[1 - \left(\frac{P_1}{P_2}\right)^{|\gamma^{ig}-1|/\gamma^{ig}}\right] \end{aligned} \tag{6.45}$$

but

$$V_2 = \left(\frac{c}{P_2}\right)^{1/\gamma^{ig}} \tag{6.46}$$

so

$$M\frac{(\dot{z}_2^2 - \dot{z}_1^2)}{2} == \frac{\gamma^{ig} P_2 V_2}{\gamma^{ig} - 1}\left[1 - \left(\frac{P_1}{P_2}\right)^{|\gamma^{ig}-1|/\gamma^{ig}}\right] \tag{6.47}$$

Taking the derivative of P with respect to V at constant S with $PV^{\gamma^{ig}} = c$

$$\left(\frac{\partial P}{\partial V}\right)_S V^{\gamma^{ig}} + P\gamma^{ig} V^{\gamma^{ig}-1} = 0 \Rightarrow \left(\frac{\partial P}{\partial V}\right)_S = -\frac{P\gamma^{ig}}{V} \tag{6.48}$$

and using the definition of sound speed

$$M\dot{z}_{max}^2 = \left(\frac{\partial P}{\partial \rho}\right)_S = -V^2\left(\frac{\partial P}{\partial V}\right)_S \tag{6.49}$$

$$M\dot{z}_{max}^2 = -VP\gamma^{ig} \tag{6.50}$$

At condition 2, $M\dot{z}_{2,max}^2 = -V_2 P_2 \gamma^{ig}$ and if $\dot{z}_1^2 = 0$, then the pressure ratio at the throat is

$$\frac{P_2}{P_1} = \left(\frac{2}{\gamma^{ig} + 1}\right)^{\frac{\gamma^{ig}}{\gamma^{ig}-1}} \tag{6.51}$$

Case VI. Flow through a valve

Example 2.5 in Chapter 2 described flow through a pipe containing an obstruction. *Cases III* and *IV* above describe common situations for such flow. However, a valve also presents an obstruction (as do rocks, clothing, and dead animals accumulated during pipeline construction). The common assumption for flow through a valve is that the kinetic and potential energy terms are negligible, which results in the flow being isenthalpic. However, the flow through a relief valve presents a different situation, one that often causes confusion among practicing engineers.

Flow through a pipe usually occurs at *relatively* low velocities, and pipes are usually horizontal. If the velocity of the fluid through a valve is low enough that the kinetic energy term is negligible, isenthalpic flow is a good approximation. If the velocity increases sufficiently that the kinetic energy is noticeable, isenthalpic flow becomes an assumption. In the case of a relief valve, the intention is to relieve the pressure in the pipe as fast as possible. This leads to essentially sonic flow through the relief valve, and as we have seen above that flow is isentropic! The moral to this story is let the first law guide you rather than relying on "everyone knows that…" Many very skilled and intelligent engineers do not know that the approximation for relief valves is isentropic flow, but now you do, and you can guide them in the discussion.

6.1.2 Properties of Sub-cooled Liquids (Compressed Liquid)

Before we develop the equations for compressors and pumps, let us determine how to calculate changes in enthalpy in a compressed liquid. Figure 6.1 contains three different paths to pass from a single-phase point to the saturated liquid.

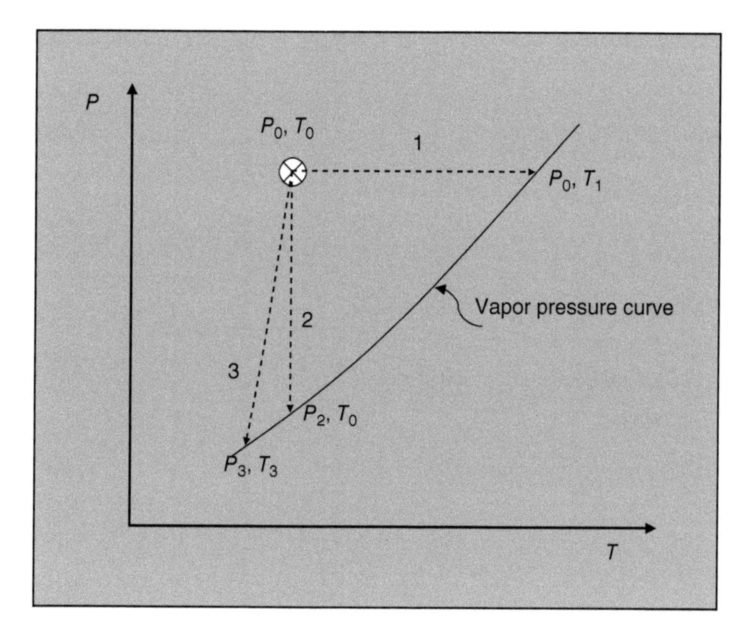

Figure 6.1 Different paths from a compressed liquid to the saturation liquid.

Path 1: Heating (cooling) at constant pressure from the compressed liquid to the saturation curve. Considering H and S as functions of T and P,

$$dH = \int_{T_0}^{T_1} \left(\frac{\partial H}{\partial T}\right)_P dT \Rightarrow H(P_0, T_1) - H(P_0, T_0) = \int_{T_0}^{T_1} C_P dT \tag{6.52}$$

$$dS = \int_{T_0}^{T_1} \left(\frac{\partial S}{\partial T}\right)_P dT \Rightarrow S(P_0, T_1) - S(P_0, T_0) = \int_{T_0}^{T_1} C_P \frac{dT}{T} \tag{6.53}$$

Path 2: Decreasing pressure at constant temperature.

$$dH = \int_{P_2}^{P} \left(\frac{\partial H}{\partial P}\right)_T dP \tag{6.54}$$

but

$$dH = TdS + VdP \Rightarrow \left(\frac{\partial H}{\partial P}\right)_T = T\left(\frac{\partial S}{\partial P}\right)_T + V = -T\left(\frac{\partial V}{\partial T}\right)_P + V = V(1 - \beta T) \tag{6.55}$$

and

$$H(P_2, T_0) - H(P_0, T_0) = \int_{P_0}^{P_2} [V(1 - \beta T)]\, dP \tag{6.56}$$

For the entropy,

$$\begin{aligned} dS = \int_{P_0}^{P_2} \left(\frac{\partial S}{\partial P}\right)_T dP &\Rightarrow S(P_2, T_0) - S(P_0, T_0) \\ &= -\int_{P_1}^{P_2} \left(\frac{\partial V}{\partial T}\right)_P dP = -\int_{P_1}^{P_2} V\beta dP \end{aligned} \tag{6.57}$$

Path 3: Decreasing pressure at a constant entropy. From the fundamental equation,

$$dH = VdP \Rightarrow H(P_3, T_3) - H(P_0, T_0) = \int_{P_0}^{P_3} VdP \tag{6.58}$$

Because $S(P_3, T_3) = S(P_0, T_0)$, we can calculate T_3 from

$$dT = \left(\frac{\partial T}{\partial P}\right)_S dP \tag{6.59}$$

This is a type 3 derivative, so in terms of observables,

$$\left(\frac{\partial T}{\partial P}\right)_S = -\left(\frac{\partial T}{\partial S}\right)_P \left(\frac{\partial S}{\partial P}\right)_T = \frac{T}{C_P}\left(\frac{\partial V}{\partial T}\right)_P \tag{6.60}$$

Substituting into Eq. (6.59)

$$dT = \frac{T}{C_P}\left(\frac{\partial V}{\partial T}\right)_P dP \tag{6.61}$$

and integrating

$$T_3 - T_0 = \int_{P_0}^{P_3} \frac{T}{C_P}\left(\frac{\partial V}{\partial T}\right)_P dP = \int_{P_0}^{P_3} \frac{TV\beta}{C_P} dP \tag{6.62}$$

With this path, we can obtain the temperature when we know P_3. All the paths should give the same state value for point (T_0, P_0) if that is the objective. The last path is useful in calculating the energy transferred as work using an isentropic pump as we will see later.

6.1.3 Pumps, Compressors, and Expanders

In this section, we follow the procedures established in Chapter 3 and consider irreversibility. To calculate the energy transferred as work or the temperature, we must follow a simple procedure previously discussed, but for the sake of clarity, we remind the reader of it. We have defined two pattern efficiencies:

$$\eta_P = \frac{{}_1W_2}{({}_1W_2)_S} \Rightarrow \text{Energy transfer as work in work}$$

-producing devices (expanders, turbines, etc.), $W < 0$

$$\eta = \frac{({}_1W_2)_S}{{}_1W_2} \Rightarrow \text{Energy transfer as work in work}$$

-consuming devices (pumps, compressors, etc.), $W > 0$

Subscript S denotes the energy transferred as work under isentropic conditions. From the first law for an adiabatic flow device in which the kinetic and potential energies are negligible,

$$W = \sum n_i H_i \tag{6.63}$$

With one inlet and one output, the above equation becomes

$$_1W_2 = H_2 - H_1 \tag{6.64}$$

Knowing all the inlet and outlet conditions, we can calculate the work using residual functions. Generally, the temperature of either the inlet or exit condition is unknown; therefore, H_1 or H_2 is unknown. One way to solve this problem is to use pattern efficiencies and the following procedure:

1. Find the inlet or outlet temperature assuming that the process is isentropic:

$$S_1 = S_2 \tag{6.65}$$

2. Here, we know either S_1 or S_2. Assuming that we know S_1, we can calculate the temperature from

$$\Delta S = 0 = S\left(T_2, P_2\right) - S\left(T_1, P_1\right) \tag{6.66}$$

In the above equation, the only unknown is T_2. We can calculate the enthalpy from generalized correlations or tables. In the case of water, one must check if the exit condition is in the two-phase region. If so, one should calculate the quality using the equilibrium conditions at the exit from the steam tables

$$\chi = \frac{S_1 - S_2{}^l}{S_2{}^v - S_2{}^l} \tag{6.67}$$

in which v and l stand for saturated vapor and liquid, respectively.

3. Next, we can calculate the enthalpy change for the isentropic process at the exit from generalized correlations or in the case of water

$$H\left(T_{2S}, P_2\right) = H_2{}^l + \chi\left(H_2{}^v - H_2{}^l\right) \tag{6.68}$$

4. Now, we can calculate the isentropic work using Eq. (6.64)

$$\left({}_1W_2\right)_S = H_{2S} - H_1 \tag{6.69}$$

5. To calculate the true temperature condition, we must calculate the true enthalpy at the exit using the pattern efficiency, Eqs. (6.64) and (6.69),

$${}_1W_2 = \eta_P\left({}_1W_2\right)_S \Rightarrow {}_1W_2 = H_2 - H_1 \Rightarrow H_2 = \eta_P\left({}_1W_2\right)_S + H_1 \tag{6.70}$$

or

$${}_1W_2 = \left({}_1W_2\right)_S/\eta_C \Rightarrow {}_1W_2 = H_2 - H_1 \Rightarrow H_2 = \left({}_1W_2\right)_S/\eta_C + H_1 \tag{6.71}$$

Condition 1 does not depend upon the process because it is the inlet condition.

6. With the true enthalpy, we know the change of enthalpy or the true energy transferred as work

$${}_1W_2 = H_2 - H_1 = \Delta H = f\left(T_1, P_1, T_2, P_2\right) \tag{6.72}$$

The unknown is the exit temperature. Several cases can exist

- Using the steam tables with the pressure and H_2 interpolation produces the temperature.
- Using the Lee–Kesler tables,

$$\Delta H = \int_{T_1}^{T_2} C_P^{ig} dT + H_2^R - H_1^R \tag{6.73}$$

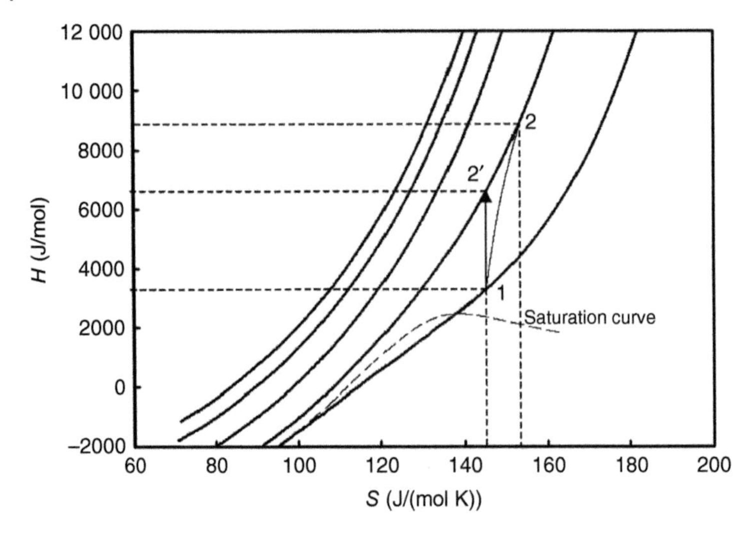

Figure 6.2 Enthalpy changes in an actual and an isentropic process.

We know ΔH, H_1^R and must employ an iterative procedure to get an answer: choose a T, and find the value of the integral and H_1^R. Propose a second T to find another enthalpy change. Linear extrapolation using two successive proposed temperatures produces the final T

$$T_2^{j+2} = \frac{T_2^{j+1} - T_2^j}{\Delta H_{j+1} - \Delta H_j}\left(\Delta H - \Delta H_{j+1}\right) + T_2^{j+1} \tag{6.74}$$

in which ΔH_{j+1} and ΔH_j are the enthalpy changes using the proposed exit temperatures T_2^{j+1}, T_2^j. When the enthalpy change equals the value given by Eq. (6.72), the result is the true exit temperature.

Figure 6.2 shows the true and isentropic changes on an H–S diagram for nitrogen. For pumps, the process occurs in the liquid region, and the enthalpy change comes from using the previously described paths.

Example 6.1

Methane expands adiabatically from 200 °C and 30 to 2 bar in a turbine. Calculate the final temperature if the efficiency is 0.75. Assume that a truncated second virial coefficient EOS is valid. The second virial of methane is

$$B_2 = 39 - 1.47 \times 10^4/T - 2.78 \times 10^6/T^2 - 6.1 \times 10^4/T^{2.5}$$

in which B is in cm^3/mol and T is in kelvin and

$$C_P^{ig}/R = 1.702 + 9.081 \times 10^{-3}T - 2.164 \times 10^{-6}T^2$$

Solution

From Eq. (5.111) for an isentropic process,

$$0 = S^R\left(T_2, P_2\right) - S^R\left(T_1, P_1\right) + \int_{T_1}^{T_2} \frac{C_P^{ig}}{T} dT - R \ln \frac{P_2}{P_1}$$

dividing by R

$$0 = \frac{S^R\left(T_2, P_2\right)}{R} - \frac{S^R\left(T_1, P_1\right)}{R} + \int_{T_1}^{T_2} \frac{C_P^{ig}}{R}\frac{dT}{T} - \ln\frac{P_2}{P_1}$$

Calculating the residual entropy, the derivative with respect to temperature of the second virial coefficient is

$$\left(\frac{dB_2}{dT}\right) = 1.47 \times 10^4/T^2 - 2.78 \times 10^6\,(-2)/T^3 - (-2.5)\,6.1 \times 10^4/T^{3.5}$$

The residual entropy comes from Eq. (5.131)

$$\frac{S^R}{R} = -\frac{P}{R}\left(\frac{dB_2}{dT}\right)$$

Therefore,

$$\frac{S^R}{R} = -\frac{P}{R}\left(1.47 \times 10^4/T^2 + 5.56 \times 10^6/T^3 + 15.25 \times 10^4/T^{3.5}\right)$$

At State 1,

$$\frac{S^R}{R} = -\frac{30}{83.14}\{1.47 \times 10^4/(473.15)^2 + 5.56 \times 10^6/(473.15)^3$$
$$+15.25 \times 10^4/(473.15)^{3.5}\} = -0.0539$$

and

$$0 = \frac{S^R\left(T_2, P_2\right)}{R} + 0.0539 + \int_{T_1}^{T_2} \frac{C_P^{ig}}{R}\frac{dT}{T} - \ln\frac{2}{30} = \frac{S^R\left(T_2, P_2\right)}{R} + I + 1.1222$$

with

$$I = \int_{T_1}^{T_2} \frac{C_P^*}{R}\frac{dT}{T} = 1.702\ln\left(\frac{T_2}{T_1}\right) + 9.081 \times 10^{-3}\left(T_2 - T_1\right)$$
$$-\frac{2.164 \times 10^{-6}}{2}\left(T_2^2 - T_1^2\right)$$

and

$$\frac{S^R\left(T_2, P_2\right)}{R} = -\frac{2}{83.14}\left(1.47 \times 10^4/T_2^2 + 5.56 \times 10^6/T_2^3 + 15.25 \times 10^4/T_2^{3.5}\right)$$

Finally, solving

$$-\frac{2}{83.14}\left(1.47 \times 10^4/T_2^2 + 5.56 \times 10^6/T_2^3 + 15.25 \times 10^4/T_2^{3.5}\right) + 1.702\ln\left(\frac{T_2}{T_1}\right)$$
$$+9.081 \times 10^{-3}\left(T_2 - T_1\right) - \frac{2.164 \times 10^{-6}}{2}\left(T_2^2 - T_1^2\right) + 1.1222 = 0$$

for T_2 gives

$$T_2 = 350.864\ \mathrm{K}$$

We can calculate the isentropic energy transferred as work from the change in enthalpy

$$\Delta H = H^R\left(T_2, P_2\right) - H^R\left(T_1, P_1\right) + R\int_{T_1}^{T_2} \frac{C_P^{ig}}{R}dT$$

with

$$\frac{H^R}{RT} = -\left[\left(\frac{dB_2}{dT}\right) - \frac{B_2}{T}\right]\frac{P}{R}$$

Then,

$$\int_{T_1}^{T_2} \frac{C_P^{ig}}{R} dT = 1.702\left(T_2 - T_1\right) + \frac{9.081 \times 10^{-3}}{2}\left(T_2^2 - T_1^2\right) - \frac{2.164 \times 10^{-6}}{3}\left(T_2^3 - T_1^3\right)$$

$$= 1.702\,(350.864 - 473.15) + \frac{9.081 \times 10^{-3}}{2}\left(350.864^2 - 473.15^2\right)$$

$$- \frac{2.164 \times 10^{-6}}{3}\left(350.864^3 - 473.15^3\right)$$

$$= -620.403$$

and

$$\frac{H^R}{RT} = -\left[\frac{1.47 \times 10^4}{T^2} + \frac{5.56 \times 10^6}{T^3} + \frac{15.25 \times 10^4}{T^{3.5}} - \left(39 - \frac{1.47 \times 10^4}{T} - \frac{2.78 \times 10^6}{T^2} - \frac{6.1 \times 10^4}{T^{2.5}}\right)\frac{1}{T}\right]\frac{P}{R}$$

so

$$\frac{H^R}{RT} = -\left[-\frac{39}{T} + \frac{2.94 \times 10^4}{T^2} + \frac{8.34 \times 10^6}{T^3} + \frac{21.35 \times 10^4}{T^{3.5}}\right]\frac{P}{R}$$

and at state 1,

$$\frac{H^R\left(T_1, P_1\right)}{RT_1} = -\left[-\frac{39}{473.15} + \frac{2.94 \times 10^4}{473.15^2} + \frac{8.34 \times 10^6}{473.15^3} + \frac{21.35 \times 10^4}{473.15^{3.5}}\right]\frac{30}{83.14}$$

$$= -0.04609$$

At State 2

$$\frac{H^R\left(T_2, P_2\right)}{RT_2} = -0.00772$$

then

$$\left({}_1W_2\right)_S = \Delta H = [(-0.00772)\,350.865 + (0.04609)\,473.15 - 620.403]\,8.314$$

$$= -4999.25\ \text{J/mol}$$

and the actual work is

$${}_1W_2 = \eta_P\left({}_1W_2\right)_S = -4999.25 \times 0.75 = -3749.44\ \text{J/mol} = \Delta H$$

Now, the final true temperature comes from

$$-3749.44 = \left\{-\left[-\frac{39}{T_2} + \frac{2.94 \times 10^4}{T_2^2} + \frac{8.34 \times 10^6}{T_2^2} + \frac{21.35 \times 10^4}{T_2^{2.5}}\right]\frac{2}{83.14}\right\} 8.314 + 181.303$$

$$+ 8.314\left\{1.702\left(T_2 - 473.15\right) + \frac{9.081 \times 10^{-3}}{2}\left(T_2^2 - 473.15^2\right)\right.$$

$$-\frac{2.164 \times 10^{-6}}{3}\left(T_2^3 - 473.15^3\right)\Bigg\}$$

The initial value is the same and solving for T_2 using SOLVER® in Excel® gives

$$T_2 = 382.485\ \text{K}$$

Example 6.2

A steam compressor operates at 200 kPa with a wet vapor quality of 0.9. Compression occurs adiabatically up to 1800 kPa with an efficiency of 0.8. What is the mass flow rate if the capacity of the compressor is 50 000 kW (1 kW = kJ/s)? Find the exit conditions of the compressor.

Solution

Data

$$P_1 = 200\ \text{kPa (at equilibrium)}$$

$$P_2 = 500\ \text{kPa}$$

The inlet temperature from steam tables is

$$T_1 = 120.23\,°\text{C} = 393.38\ \text{K}$$

At saturated conditions,

$$S^l = 1.5301\ \text{kJ}/(\text{kg K}),\ S^v = 7.1268\ \text{kJ}/(\text{kg K}),\ H^l = 504.701\ \text{kJ/kg},$$
$$\text{and } H^v = 2706.3\ \text{kJ/kg}$$

Calculate the entropy at the inlet from

$$S = S^l + \chi\left(S^v - S^l\right)$$

$$S_1 = 1.5301 + 0.9\,(7.1268 - 1.5301) = 6.5671\ \text{kJ}/(\text{kg K})$$

and the enthalpy is

$$H_1 = H^l + \chi\left(H^v - H^l\right) = 504.701 + 0.9\,(2706.3 - 504.701) = 2486.14\ \text{kJ/kg}$$

Because $S_1 = 6.5671 = S_2$ at $P = 1800$ kPa, we can determine if it is still a wet vapor. The saturated vapor entropy at 1800 kPa is 6.3751 kJ/(kg K); therefore, the steam is in the superheated region. Finding the enthalpy between 225 and 250 °C by interpolating

At 225 °C, $S = 6.4787$ kJ/(kg K), $H = 2845.5$ kJ/kg
At 250 °C, $S = 6.6071$ kJ/(kg K), $H = 2911.0$ kJ/kg

Then interpolating

$$H_2 = \frac{2911 - 2845.5}{6.6071 - 6.4787}(6.5671 - 6.6071) + 2911 = 2890.61\ \text{kJ/kg}$$

The isentropic work is

$$\left({}_1W_2\right)_S = H_2 - H_1 = 2890.61 - 2486.14 = 404.47\ \text{kJ/kg}$$

The actual work is

$$ {}_1W_2 = \left({}_1W_2\right)_S/\eta_C = 404.47/0.8 = 505.588\ \mathrm{kJ/kg} $$

The actual final enthalpy is

$$ H_2 = {}_1W_2 + H_1 = 505.6 + 2486.14 = 2991.72\ \mathrm{kJ/kg} $$

Using $P = 1800\ \mathrm{kPa}$ and $H_2 = 2991.72\ \mathrm{kJ/kg}$, we can find the final temperature between 275 and 300 °C

At 275 °C, $H = 2972.3\ \mathrm{kJ/kg}$

At 300 °C, $H = 3030.7\ \mathrm{kJ/kg}$

then

$$ T_2 = \frac{300 - 275}{3030.7 - 2972.3}(2991.72 - 3030.7) + 300 = 283.313\ {}^\circ\mathrm{C} $$

The mass flow rate is

$$ \dot{m} = \frac{\dot{W}}{{}_1W_2} = \frac{50\,000\ \mathrm{kg/s}}{505.58\ \mathrm{kJ/kg}} = 98.895\ \mathrm{kg/s} $$

Example 6.3

Calculate the work done by a pump and the final temperature at the exit if the inlet temperature and pressure are 70 °C and 200 kPa, respectively. The efficiency of the pump is 0.85 when the fluid is at 5000 kPa. Calculate the entropy change during the process. Assume that a first-order Taylor series expansion provides the change in pressure:

$$ V = V_0 + \left(\frac{\partial V}{\partial P}\right)_S\bigg|_{V_0}(P - P_0) $$

Data

$V = 1022.7\ \mathrm{cm^3/kg}$, $\beta = 5.838 \times 10^4\ \mathrm{K^{-1}}$, $C_P = 4.1899\ \mathrm{kJ/(kg\ K)}$, and $K_S = 4.2298 \times 10^{-7}\ \mathrm{kPa^{-1}}$

Solution

The isentropic work in a pump is

$$ \left({}_1W_2\right)_S = \Delta H = \int_{P_1}^{P_2} V dP $$

If we assume the change in pressure of the volume is a Taylor series expansion

$$ V = V_0 + \left(\frac{\partial V}{\partial P}\right)_S\bigg|_{V_0}(P - P_0) $$

In terms of the adiabatic compressibility, $K_S = -\frac{1}{V}\left(\frac{\partial V}{\partial P}\right)_S$

$$ V = V_1 - V_1 K_S (P - P_1) = V_1 \left[1 - K_S (P - P_1)\right] $$

then

$$ \left({}_1W_2\right)_S = \int_{P_1}^{P_2} V_1 \left[1 - K_S (P - P_1)\right] dP $$

$$= V_1 \left(P_2 - P_1\right) \left[1 + P_1 K_S\right] - V_1 K_S \left(P_2^2 - P_1^2\right)/2$$

$$\left({}_1W_2\right)_S = 1022.7\,(5000 - 200)\left[1 + (200)\,4.2298 \times 10^{-7}\right]$$
$$- (1022.7)\,4.2298 \times 10^{-7}\left(5000^2 - 200^2\right)/2$$

$$\left({}_1W_2\right)_S = 4.904 \text{ kJ/kg}$$

The actual work is

$${}_1W_2 = \left({}_1W_2\right)_S/\eta_C = 4.904/0.85 = 5.769 \text{ kJ/kg}$$

The temperature change comes from

$$dH = C_P dT + V\,(1 - \beta T)\,dP$$

Integrating considering heat capacity and the isothermal compressibility to be constants

$$\Delta H = C_P \Delta T + \left(1 - \beta T_{avg}\right)\int_{P_1}^{P_2} V_1\left[1 - K_S\left(P - P_1\right)\right] dP$$
$$\Delta H = C_P \Delta T + \left(1 - \beta T_{avg}\right) V_1 \left(P_2 - P_1\right)\left[1 + P_1 K_S\right] - V_1 K_S\left(P_2^2 - P_1^2\right)/2$$

or

$$5.769 = 4.1899\Delta T + 4.904\left(1 - \beta T_{avg}\right)$$

Solving for T_2

$$5.769 = 4.1899T_2 - 4.1899 \times 343.15 + 4.904 - 4.904 \times 2.919 \times 10^{-4}\left(T_2 + 343.15\right)$$
$$5.769 + 4.1899 \times 343.15 - 4.904 + 4.904 \times 2.919 \times 10^{-4} \times 343.15 = 4.1899T_2$$
$$- 4.904 \times 2.919 \times 10^{-4} T_2$$

$$T_2 = \frac{1439.12}{4.1885} = 343.591 \text{ K} = 70.44\,°\text{C}$$

Although the work of a compressor results from a change of enthalpy, sometimes, it is necessary to look in detail at the process of compression. The thermodynamic cycle consists of four steps:

- **Step 1→2 Isobaric Expansion:** The inlet valve is open, and the exhaust valve is closed to draw in gas at a constant pressure.
- **Step 2→3 Adiabatic Compression:** Gas is compressed to the outlet pressure.
- **Step 3→4 Isobaric Compression:** The inlet valve is closed, and the exhaust valve is open to push the gas at a constant outlet pressure
- **Step 4→1 Adiabatic Expansion:** The pressure drops to the inlet pressure. In this step, both valves are closed.

Figure 6.3 shows the process in a *PV* diagram.

Working with compressors, we define

$$\text{Intake volume}: V_I \equiv V_2 - V_1 \tag{6.75}$$

$$\text{Displacement volume}: V_D \equiv V_2 - V_4 \tag{6.76}$$

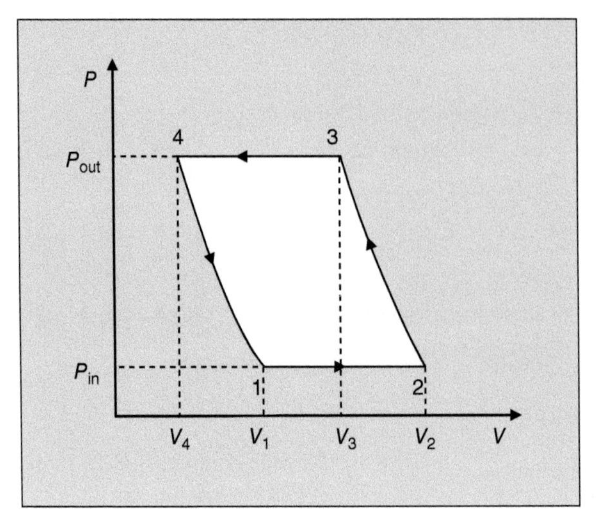

Figure 6.3 *PV* diagram for a compressor cycle.

$$\text{Clearance volume}: V_C \equiv V_4 \tag{6.77}$$

The clearance is $C = \frac{V_C}{V_D} = \frac{V_4}{V_2 - V_4}$. Looking at the intake volume

$$V_I \equiv V_D + V_4 - V_1 = V_D + V_4 - V_1 = V_D\left(1 + \frac{V_4}{V_D} - \frac{V_4}{V_4}\frac{V_1}{V_D}\right) \tag{6.78}$$

$$V_I - V_D = V_4 - V_1 \tag{6.79}$$

then

$$V_I = V_D\left(1 + C - C\frac{V_1}{V_4}\right) \tag{6.80}$$

If the fluid in the compressor is a perfect gas, then the above equation becomes

$$P_{in}V_1^{\gamma^{ig}} = P_{out}V_4^{\gamma^{ig}} \Rightarrow V_I = V_D\left(1 + C - C\left(\frac{P_{out}}{P_{in}}\right)^{1/\gamma^{ig}}\right) \tag{6.81}$$

The work for each step is

$${}_1W_2 = P_{in}\left(V_2 - V_1\right) = R\left(T_2 - T_1\right) \tag{6.82}$$

$${}_2W_3 = {}_2\Delta U_3 = C_V\left(T_3 - T_2\right) = \frac{R}{\gamma^{ig} - 1}\left(T_3 - T_2\right) \tag{6.83}$$

$${}_3W_4 = P_{out}\left(V_4 - V_3\right) = R\left(T_4 - T_3\right) \tag{6.84}$$

$${}_4W_1 = {}_4\Delta U_1 = C_V\left(T_1 - T_4\right) = \frac{R}{\gamma^{ig} - 1}\left(T_1 - T_4\right) \tag{6.85}$$

The total work of the isentropic step is

$$W_{isen} = \frac{R}{\gamma^{ig} - 1}\left\{T_3 - T_2 + T_1 - T_4\right\} \tag{6.86}$$

$$W_{isen} = \frac{RT_2}{\gamma^{ig} - 1}\left\{\frac{T_3}{T_2} - 1 + \frac{T_1}{T_2} - \frac{T_4}{T_2}\right\} \tag{6.87}$$

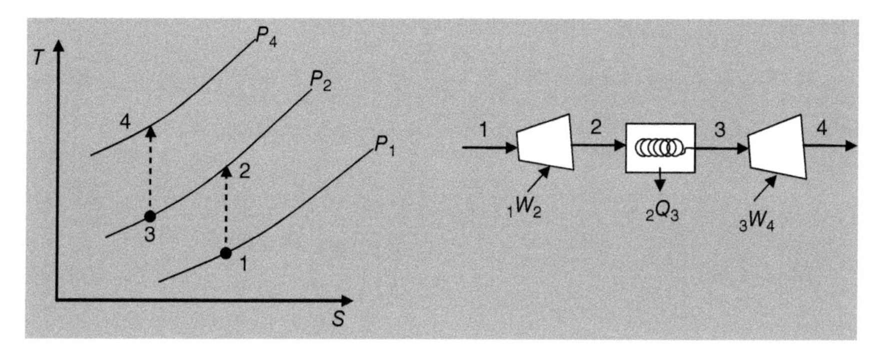

Figure 6.4 Compressor $T-S$ diagram.

with

$$W_{isen} = \frac{RT_2}{\gamma^{ig} - 1}\left\{\frac{T_3}{T_2} - 1 + \frac{T_1}{T_2} - \frac{T_4}{T_2}\right\} \tag{6.88}$$

$$\frac{T_3}{T_2} = \left(\frac{P_{out}}{P_{in}}\right)^{(\gamma^{ig}-1)/\gamma^{ig}} = r^{(\gamma^{ig}-1)/\gamma^{ig}} \tag{6.89}$$

$$\frac{T_1}{T_2} = \frac{V_1}{V_2} \tag{6.90}$$

$$\frac{T_4}{T_2} = \frac{P_4V_4}{P_2V_2} = \left(\frac{P_{out}}{P_{in}}\right)\frac{V_4}{V_2} = r\frac{V_4}{V_2} \tag{6.91}$$

Substituting

$$W_{isen} = \frac{P_{in}V_2}{\gamma^{ig} - 1}\left[r^{(\gamma^{ig}-1)/\gamma^{ig}} - 1 + \frac{V_1}{V_2} - r\frac{V_4}{V_2}\right] \tag{6.92}$$

or

$$W_{isen} = \frac{P_{in}}{\gamma^{ig} - 1}\left[V_2r^{(\gamma^{ig}-1)/\gamma^{ig}} - V_2 + V_1 - rV_4\right] \tag{6.93}$$

but $V_2 = V_D + V_C$

$$W_{isen} = \frac{P_{in}}{\gamma^{ig} - 1}\left[\left(V_C + V_D\right)r^{(\gamma^{ig}-1)/\gamma^{ig}} - V_I - rV_C\right] \tag{6.94}$$

When $V_C = 0$

$$W_{isen} = \frac{P_{in}}{\gamma^{ig} - 1}\left[V_Dr^{(\gamma^{ig}-1)/\gamma^{ig}} - V_I\right] \tag{6.95}$$

At high pressures, high temperatures occur in the cylinder of the compressor. A way to avoid this problem is to perform the compression process with multiple stages connected by intercoolers. Figure 6.4 illustrates a simple, single-stage compression system. The total work is

$$W_T = {}_1W_2 + {}_3W_4 \tag{6.96}$$

Considering an adiabatic process,

$$W_T = H_2 - H_1 + H_4 - H_3 \tag{6.97}$$

For a perfect gas (ideal gas with constant heat capacities),

$$W_T = C_P\left(T_2 - T_1\right) + C_P\left(T_4 - T_3\right) \tag{6.98}$$

with $S_1 = S_2$, $S_3 = S_4$, and $P_2 = P_3$, we can write the last equation as

$$W_T = C_P T_1\left(\frac{T_2}{T_1} - 1\right) + C_P T_3\left(\frac{T_4}{T_3} - 1\right) \tag{6.99}$$

$$W_T = C_P T_1\left[\left(\frac{P_2}{P_1}\right)^{(\gamma^{ig}-1)/\gamma^{ig}} - 1\right] + C_P T_3\left[\left(\frac{P_4}{P_3}\right)^{(\gamma^{ig}-1)/\gamma^{ig}} - 1\right] \tag{6.100}$$

The minimum total work results when $P_2/P_1 = P_4/P_3$.

6.2 Heat Engines and Refrigeration Units

6.2.1 Heat and Work

The Joule experiment converts energy transferred as work into energy transferred as heat. The conversion for this process is 100% because we can convert work into friction and then into energy transferred as heat at any temperature. However, the reverse process, converting energy transferred as heat entirely into energy transferred as work, is impossible because that violates the second law. We define energy transferred as work as energy of higher *quality* than energy transferred as heat. In Chapter 3, we defined heat engines and refrigerators. Also, we noted that a Carnot cycle engine delivers the maximum efficiency or coefficient of performance (*COP*) (pump or refrigerator) operating between two temperatures. In this chapter, we use pattern efficiencies to approximate the actual engine efficiencies. These isentropic efficiencies are useful for engines, nozzles, pumps, turbines, and compressors. In addition, they are parts of modern power systems that contain the above-mentioned equipment plus condensers, boilers, and throttling valves. In a thermodynamic analysis of these processes, it is useful to plot temperature vs entropy. Figure 6.5 demonstrates this plot for a Carnot engine. Were this plot a Carnot refrigerator, the arrows would point in the opposite directions.

We can illustrate the process in Figure 6.5 using a vapor power plant (this is an exercise to familiarize you with concepts, such as a cycle would always be impractical in an actual plant). This plant consists of a pump to increase the pressure, a reboiler to increase the temperature, a turbine to produce energy transferred as work, and a condenser to change the vapor into liquid. The vapor exiting the turbine and entering the pump is "wet," indicating that it contains liquid as a significant component. Figure 6.6 presents a schematic in which the cooling fluid passes through tubes. The idealized assumptions using the Carnot cycle are as follows:

1. Heat exchangers (i.e. boilers and condensers) operate at constant pressure ($dP = 0$).
2. Work devices (i.e. pumps, turbines, and compressors) operate adiabatically ($đQ = 0$).

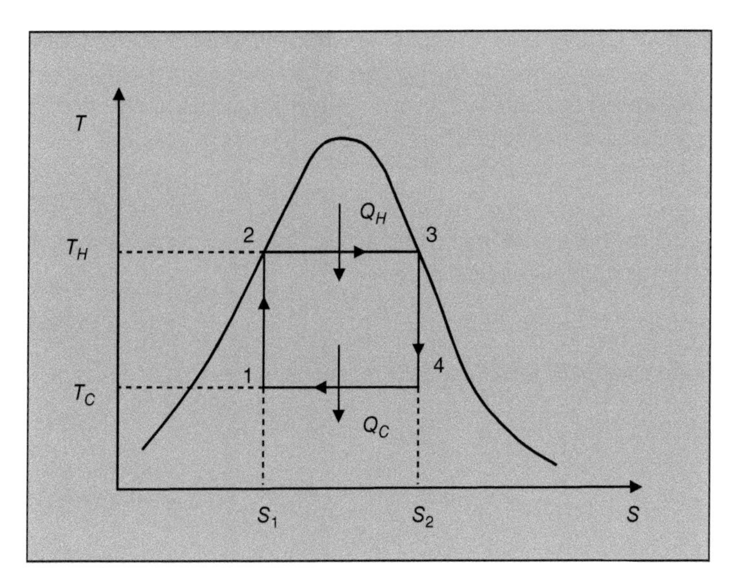

Figure 6.5 $T-S$ plot for the Carnot cycle.

Figure 6.6 Schematic diagram of a vapor power plant.

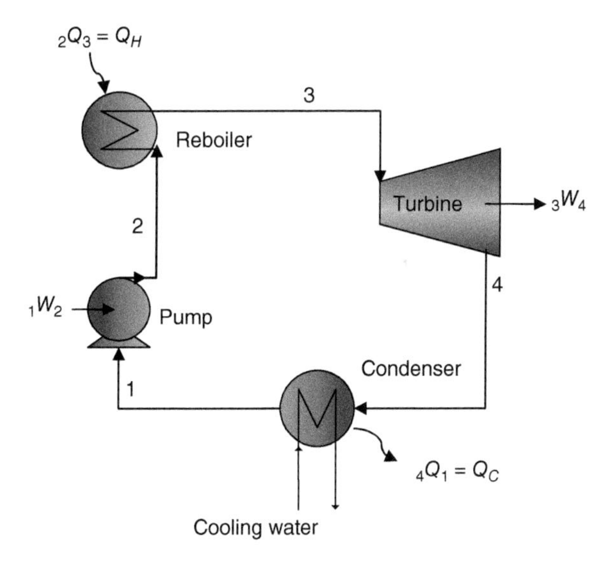

We can now apply the first law of thermodynamics to the complete system (vapor power plant) considering no nuclear reactions, nor any inlet and outlet steams. For the complete cycle $1 \rightarrow 2 \rightarrow 3 \rightarrow 4 \rightarrow 1$

Expanding the above equation

$$_2Q_3 + {_4Q_1} + {_1W_2} + {_3W_4} = 0 \tag{6.101}$$

Remember that $_1W_2$ and $_2Q_3$ must be positive because the energy transferred as work into the system occurs in the pump and the energy transferred as heat occurs in the reboiler. Also, ${}_3W_4$ and ${}_4Q_1$ are negative because the turbine produces energy transferred as work, and in the condenser, the energy transferred as heat leaves the system to reduce the temperature. For step 1 → 2, applying the first law (see application in Chapter 2 for the compressor) for an adiabatic process:

$$ {}_1W_2 = H_2 - H_1 \tag{6.102}$$

where H_2 and H_1 are the outlet and inlet enthalpies in the pump, respectively. For the condenser in a constant pressure flow process

$$ {}_2Q_3 = H_3 - H_2 \tag{6.103}$$

Now, applying the first law to the turbine and the reboiler provides similar equations:

$$ {}_3W_4 = H_4 - H_3 \tag{6.104}$$

and

$$ {}_4Q_1 = H_1 - H_4 \tag{6.105}$$

H_3 and H_4 are the inlet and outlet enthalpies of the turbine, respectively, and for the condenser, H_4 becomes the inlet steam enthalpy and H_1 is the outlet steam enthalpy. In summary,

$$\left.\begin{array}{l} \text{Pump work} \rightarrow {}_1W_2 = H_2 - H_1 \\ \text{Reboiler duty} \rightarrow {}_2Q_3 = H_3 - H_2 \\ \text{Turbine work} \rightarrow {}_3W_4 = H_4 - H_3 \\ \text{Condenser duty} \rightarrow {}_4Q_1 = H_1 - H_4 \end{array}\right\} {}_2Q_3 + {}_4Q_1 + {}_1W_2 + {}_3W_4 = 0 \tag{6.106}$$

The efficiency of this Carnot cycle is

$$\eta = \frac{|W|}{Q_{in}} = \frac{|{}_1W_2 + {}_3W_4|}{{}_2Q_3} = \frac{Q_{23} + Q_{41}}{Q_{23}} = 1 + \frac{Q_{41}}{Q_{23}} = \frac{Q_H - Q_C}{Q_H} = 1 - \frac{Q_C}{Q_H} = 1 - \frac{T_C}{T_H} \tag{6.107}$$

Remember that Q_C and Q_H represent numerical values with their signs and ${}_iQ_j$ are variables. In terms of enthalpies,

$$\eta = 1 + \frac{(H_1 - H_4)_S}{(H_3 - H_2)_S} = \left(\frac{H_3 - H_2 + H_1 - H_4}{H_3 - H_2}\right)_S = \left(\frac{H_3 + H_1 - H_2 - H_4}{H_3 - H_2}\right)_S \tag{6.108}$$

The enthalpies here apply to isentropic processes.

6.2.2 Rankine Cycle

In a Rankine cycle, a pump compresses a liquid to increase its pressure, which then feeds to a boiler to provide high temperature–high pressure vapor to drive a turbine. The low-pressure exhaust from the turbine is either superheated vapor or liquid in two phases (wet steam). A condenser liquefies the steam, which feeds back to the pump. The schematic diagram is identical to Figure 6.7. The only difference is that

Figure 6.7 Rankine cycle power plant.

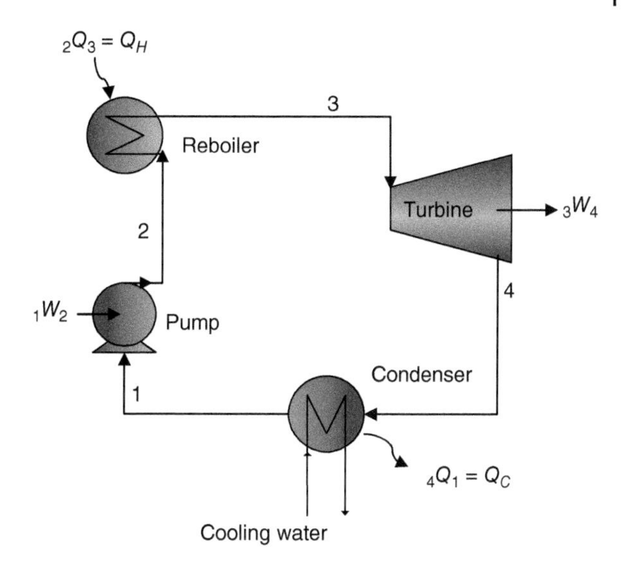

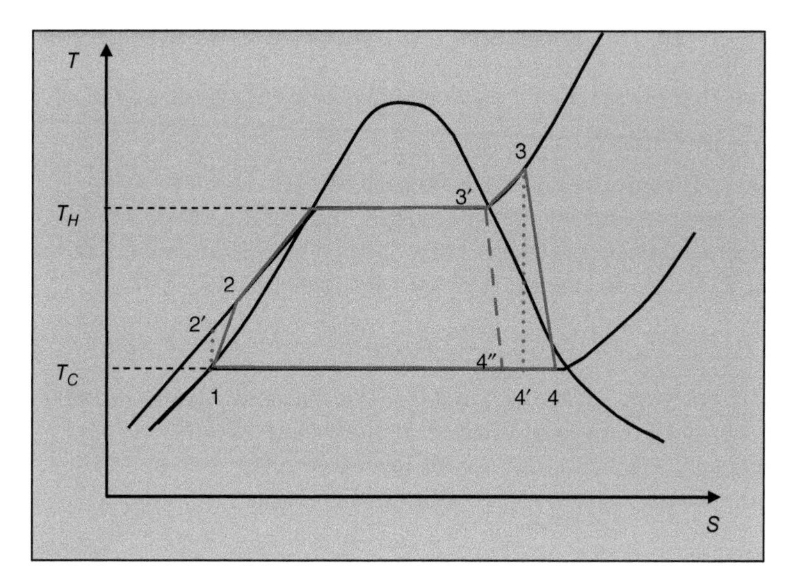

Figure 6.8 T–S plot for the Rankine cycle.

the pump does not operate with a two-phase fluid but with liquid. This avoids cavitation and corrosion in the tubes. Figure 6.8 is a T–S diagram of the Rankine cycle.

The cycle again consists of four processes:

- 1 to 2: Adiabatic compression (Pump). This step adiabatically compresses saturated liquid at the reboiler pressure to obtain a sub-cooled liquid. Step 1 to 2′ corresponds to an adiabatic, reversible compression (isentropic) process. When following step 1 to 2, use pattern efficiencies in the calculations because it is an irreversible path.

- 2 to 3: Isobaric heat supply (Reboiler). This constant pressure process absorbs heat to vaporize the liquid into superheated vapor or saturated vapor (point 3′) passing from saturated liquid to saturated vapor.
- 3 to 4: Adiabatic expansion (steam turbine). This is an adiabatic expansion of the vapor at the condenser pressure. This process can cross the saturation curve producing a wet vapor as in point 4 of Figure 6.8, but it can also produce saturated vapor. If the line goes from 3 to 4′, the process is adiabatic and reversible (isentropic).
- 4 to 1: Isobaric heat rejection (Condenser). This step liberates energy transferred as heat isothermally at T_C to produce a complete condensation. Sometimes, this process does not occur at constant temperature and instead produces compressed liquid.

In summary, if the turbine and pump are reversible, then the paths are $1 \rightarrow 2' \rightarrow 3 \rightarrow 4' \rightarrow 1$. The actual paths (irreversible) are $1 \rightarrow 2 \rightarrow 3' \rightarrow 4'' \rightarrow 1$, if saturated vapor enters the turbine and wet vapor exits the turbine. The irreversible path $1 \rightarrow 2 \rightarrow 3 \rightarrow 4 \rightarrow 1$ occurs when the superheated vapor enters the turbine and wet vapor exits it. Obviously, superheated or saturated vapor can exit the turbine.

For the thermodynamic calculations of the irreversible paths, we use pattern efficiencies under two scenarios:

1. All the thermodynamic conditions are available, and we can calculate the efficiency of the pump, the turbine, and the total efficiency of the cycle.
2. The most common situation is that we know the conditions at the entrance of the turbine, i.e. T_3 and P_3, and the outlet pressure P_4. Then, from Eq. (6.104)

$$ {}_3W_4 = H_4 - H_3 \quad (6.104)$$

We know H_3, but we do not know H_4 and W_S. Thus, we need another condition. If we consider the process to be adiabatic and reversible, $S_3 = S_4$, and we can use conditions (S_3, P_4) to calculate an enthalpy that corresponds to an isentropic process. Now, the isentropic work (as in the Carnot cycle) is

$$\left({}_3W_4\right)_S = H_4\left(P_4, S_3\right) - H_3\left(T_3, P_3\right) \tag{6.109}$$

This is the maximum work deliverable by the turbine. One should note whether the steam is saturated or wet. In the latter case, calculate the quality of the steam using P_4 and

$$\chi = \frac{S_3 - S^l}{S^v - S^l} \tag{6.110}$$

The actual work results from using the pattern efficiency

$$\eta_P = \frac{{}_3W_4}{\left({}_3W_4\right)_S} \Rightarrow {}_3W_4 = \eta_P\left({}_3W_4\right)_S \tag{6.111}$$

Now, we can calculate the actual enthalpy, H_4,

$$H_4 = {}_3W_4 + H_3 \tag{6.112}$$

We perform almost the same calculations for the pump, but we use the pattern efficiency for energy transferred as work consuming devices

$$\eta_C = \frac{W_S}{W}$$

Remember that for the pump, the differential enthalpy is

$$dH = VdP \ (S \text{ constant}) \tag{6.113}$$

and

$$(\Delta H)_S = \int_{P_1}^{P_2} VdP = V\left(P_2 - P_1\right) \tag{6.114}$$

Equation (6.114) considers an incompressible fluid and

$$(\Delta H)_S = \left({}_1W_2\right)_S \tag{6.115}$$

To compare the Rankine cycle with the Carnot cycle, one should consider only saturated states to have a more visual idea of the processes as shown in Figure 6.9.

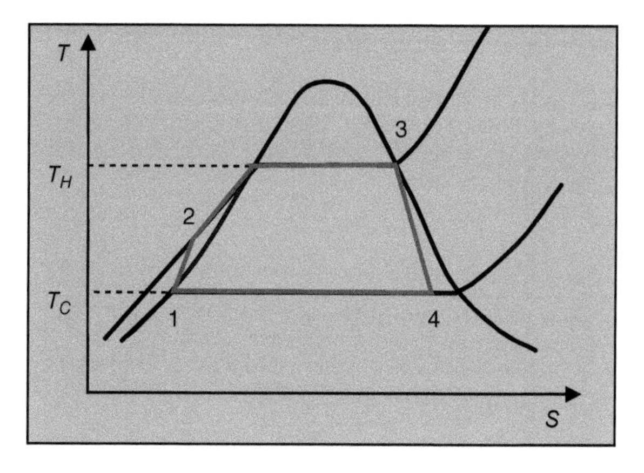

Figure 6.9 Comparison between Rankine and Carnot cycles.

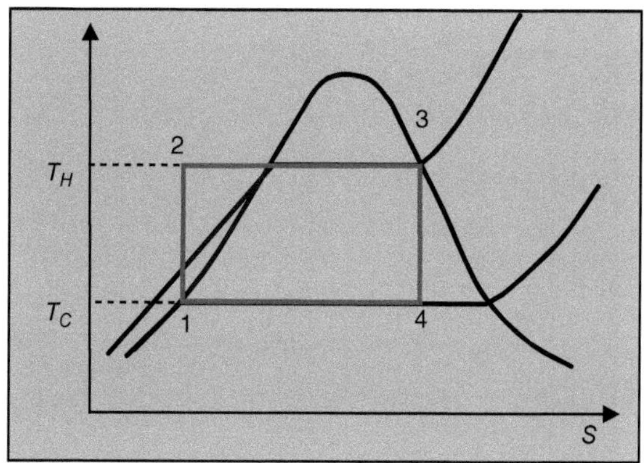

Then, when $T_3 = T_H$ and $T_1 = T_C$, it is obvious that the efficiency of the Carnot cycle $\eta = 1 - T_1/T_3$ is greater than that of the irreversible Rankine cycle or even of the reversible one. The reason is that the temperatures at which energy transfers as heat to and from the working fluid are more widely separated in the Carnot cycle.

Example 6.4

Steam enters a turbine at 7000 kPa and 550 °C. The exit steam from the turbine enters the condenser at 101 kPa, where it condenses to a saturated liquid. A pump transfers this liquid to the reboiler. Assume that the processes in the turbine and the pump are isentropic. What is the efficiency of the Rankine cycle? What is the temperature at the exit of the pump? What is the efficiency of the process if the efficiencies in the turbine and pump are 60% and 70%, respectively?

Solution

Data

$$P_3 = 7000\ \text{kPa}$$

$$T_3 = 550\ °\text{C}$$

$$P_4 = 101\ \text{kPa}$$

Following the procedure, the enthalpy and entropy at the entrance of the turbine come from the steam tables

$$H_3 = 3529.6\ \text{kJ/kg and } S_3 = 6.9485\ \text{kJ/(kg K)}$$

Now, we calculate the quality at the exit of the turbine because $S_3 = S_4$ and $P_4 = 101$ kPa using

$$\chi = \frac{S_3 - S^l}{S^v - S^l}$$

If the saturated conditions at $P_4 = 101$ kPa are

$$S^l = 1.3069\ \text{kJ/(kg K)}\ \text{and}\ S^v = 7.3554\ \text{kJ/(kg K)}$$

$$H^l = 419.064\ \text{kJ/kg and } H^v = 2676\ \text{kJ/kg}$$

then

$$\chi = \frac{6.9485 - 1.3069}{7.3554 - 1.3069} = 0.933$$

Now, we can calculate the enthalpy of steam exiting the turbine

$$H_4 = (H^v - H^l)\,\chi + H^l$$

$$H_4 = (2676 - 419.064)\,\chi + 419.064 = 2524.2\ \text{kJ/kg}$$

The energy transferred as work in the turbine is

$$({}_3W_4)_S = H_4 - H_3 = 2524.2 - 3529.6 = -1005.4\ \text{kJ/kg}$$

We must also calculate the isentropic work in the pump, where $P_4 = P_1 = 101$ kPa

$$({}_1W_2)_S = (\Delta H)_S = V\,(P_2 - P_1)$$

We need the saturated liquid volume at 101 kPa

$$V = 1044\ \text{cm}^3/\text{kg}$$

Then, when

$$P_2 = P_3\ \left(1\ \text{kJ} = 10^6\ \text{kPa cm}^3\right)$$

$$\left({}_1W_2\right)_S = (\Delta H)_S = 1044\,(7000 - 101) = 7.202556 \times 10^6\ \text{kPa cm}^3/\text{kg} = 7.20\ \text{kJ/kg}$$

and the net work is

$$W_{in} = \left({}_3W_4\right)_S + \left({}_1W_2\right)_S = -1005.4 + 7.20 = -998.2\ \text{kJ/kg}$$

The reboiler duty comes from

$${}_2Q_3 = H_3 - H_2$$

but the enthalpy, H_2, is not known; therefore, from $({}_1W_2)_S = (\Delta H)_S$,

$$H_2 = \left({}_1W_2\right)_S + H_1$$

in which H_1 is the enthalpy of the saturated liquid at 101 kPa, $H_1 = 419.064$ kJ/kg and

$$H_2 = 7.20 + 419.064 = 426.264\ \text{kJ/kg}$$

The reboiler duty is

$${}_2Q_3 = H_3 - H_2 = 3529.6 - 426.264 = 3103.336\ \text{kJ/kg}$$

and the efficiency for the reversible process is

$$\eta = \frac{|W_{net}|}{Q_{in}} = \frac{997.68}{3103.336} = 0.3215$$

We know that the pump is isentropic, so using

$$\Delta S = 0 = C_P \ln\left(T_2/T_1\right) - \beta V \Delta P$$

and

$$T_2 = T_1 \exp\left(\frac{\beta V \Delta P}{C_P}\right)$$

at $T_1 = 100\,°\text{C}$ for saturated liquid, we have $\beta = 7.5062 \times 10^{-4}\ \text{K}^{-1}$ and $C_P = 4.2157$ kJ/(kg K)

$$T_2 = 373.15 \exp\left[\frac{\left(7.5063 \times 10^{-4}\ \text{K}^{-1}\right)\left(1044\ \text{cm}^3/\text{kg}\right)(7000 - 101)\ \text{kPa}}{4.2157\ \text{kJ/}\left(\text{kg K}\right)} \frac{1\ \text{kJ}}{10^6\ \text{kPa cm}^3}\right] = 373.629\ \text{K}$$

To account for irreversibilities in the turbine and pump, we must calculate the real energy transferred as work in both devices using the pattern efficiencies,

$$\eta_P = \frac{{}_3W_4}{\left({}_3W_4\right)_S} \Rightarrow {}_3W_4 = \eta_P\left({}_3W_4\right)_S = (-1005.4) \times 0.6 = -603.24\ \text{kJ/kg}$$

and

$$\eta_C = \frac{({}_1W_2)_S}{{}_1W_2} \Rightarrow {}_1W_2 = \frac{({}_1W_2)_S}{\eta_C} = \frac{7.20}{0.7} = 10.29\ \text{kJ/kg}$$

Now, the net energy transferred as work is

$$W_{net} = {}_3W_4 + {}_1W_2 = -603.24 + 10.29 = -592.95\ \text{kJ/kg}$$

The enthalpy at the exit of the turbine is

$$H_4 = {}_3W_4 + H_3 = -592.95 + 3529.6 = 2936.6\ \text{kJ/kg}$$

and for the pump

$$H_2 = {}_1W_2 + H_1 = 10.29 + 419.064 = 429.354\ \text{kJ/kg}$$

The reboiler duty is

$${}_2Q_3 = H_3 - H_2 = 3529.6 - 429.354 = 3100.246\ \text{kJ/kg}$$

and the efficiency is

$$\eta_{real} = \frac{|W_{net}|}{Q_{in}} = \frac{592.95}{3100.246} = 0.1913$$

6.2.3 Modifications of the Rankine Cycle

One of the modifications of the Rankine cycle consists of adding energy as heat at high temperatures and rejecting energy as heat at low temperatures. The temperature at point 3 in Figure 6.9 should be as high as possible at pressure P_2. The sink temperature, essentially the boiling point of water, at P_4 may be reduced by lowering the cooling water temperature in the condenser. At least, a 10 °C difference between the cooling water and the condensing temperature at P_4 is necessary for economical rates of energy transfer as heat in the condenser.

Another modification that improves the efficiency of the Rankine cycle is reheating. Here, the steam expands in a turbine until it is nearly saturated, then it returns to a reboiler for reheating at constant P_2 to a temperature T_H. Finally, it expands in a second turbine to a pressure P_6. This arrangement prevents moisture formation in the turbines. Moisture is a major cause of corrosion. Figure 6.10 is a T–S diagram for this arrangement.

The efficiency of this Rankine cycle is

$$\eta = \frac{|W_{net}|}{Q_{in}} = \frac{|{}_3W_4 + {}_5W_6 + {}_1W_2|}{Q_{in}} = \frac{{}_6Q_1 + {}_2Q_5}{{}_2Q_5} \tag{6.116}$$

or from the first law at each device in terms of enthalpies

$$\eta = \frac{(H_4 - H_3) + (H_6 - H_5) + (H_2 - H_1)}{(H_3 - H_2) + (H_5 - H_4)} \tag{6.117}$$

Figure 6.11 is a diagram of the process.

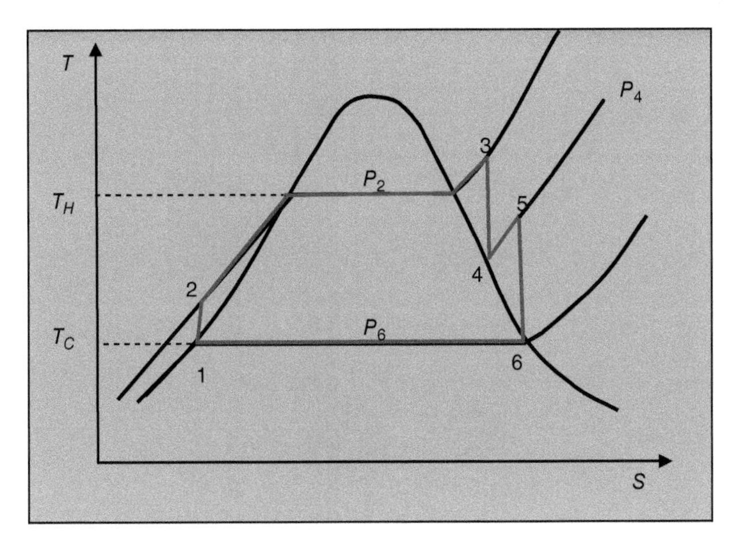

Figure 6.10 $T-S$ diagram of a reheat Rankine cycle.

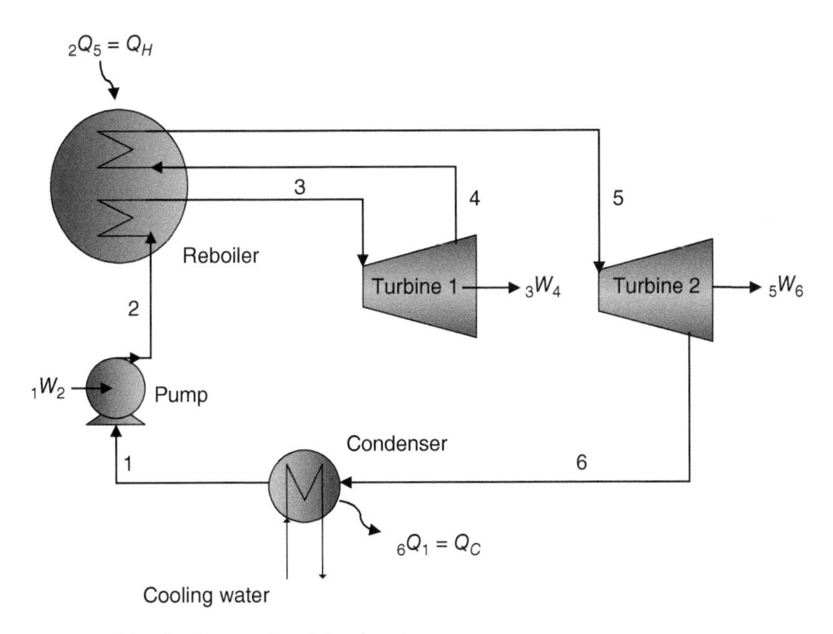

Figure 6.11 Rankine cycle with reheating.

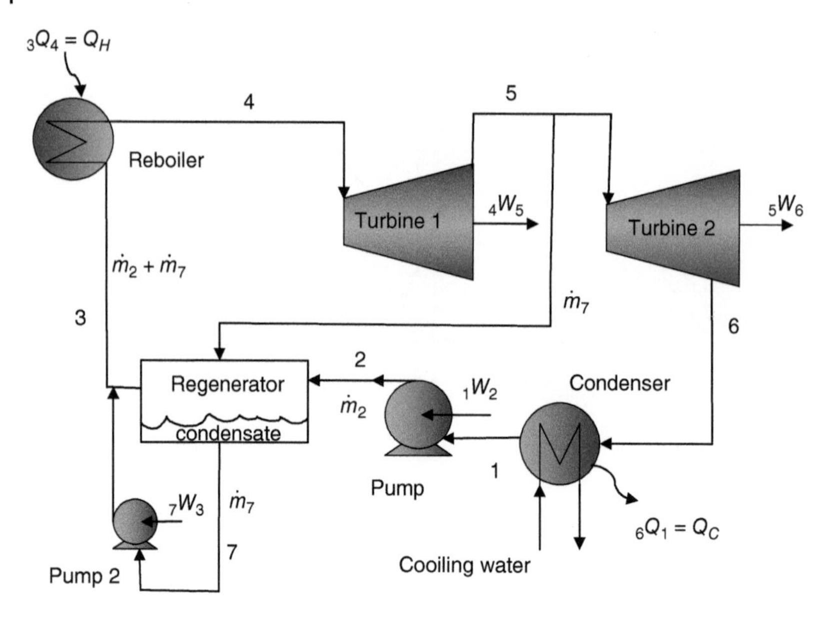

Figure 6.12 Rankine cycle with regeneration.

Another modification is to use regeneration, in which after the high-temperature steam expands in the turbine, it transfers some energy as heat to the liquid from the pump before it enters the reboiler. This allows the energy transferred as heat in the cycle to occur at an approximately constant temperature. Figure 6.12 illustrates this process. Figure 6.13 is a T–S diagram of this system. The efficiency in this regeneration process is

$$\eta = \frac{|W_{net}|}{Q_{in}} = \frac{|_4W_5 + {}_5W_6 + {}_1W_2 + {}_7W_8|}{Q_{in}} \tag{6.118}$$

Now, we must include the mass flow because we have recycle in the process. For each piece of equipment, the energy balance from the first law is

$$\text{Reboiler}: {}_3Q_4 = (\dot{m}_2 + \dot{m}_7)(H_4 - H_3) \tag{6.119}$$

$$\text{Turbine 1}: {}_4W_5 = (\dot{m}_2 + \dot{m}_7)(H_5 - H_4) \tag{6.120}$$

$$\text{Turbine 2}: {}_5W_6 = \dot{m}_2 (H_6 - H_5) \tag{6.121}$$

$$\text{Pump 1}: {}_1W_2 = \dot{m}_2 (H_2 - H_1) \tag{6.122}$$

$$\text{Pump 2}: {}_7W_3 = \dot{m}_7 (H_3 - H_7) \tag{6.123}$$

Then,

$$\eta = \frac{(H_5 - H_4) + \left(\frac{\dot{m}_2}{\dot{m}_2+\dot{m}_7}\right)[(H_6 - H_5) + (H_2 - H_1)] + \left(\frac{\dot{m}_7}{\dot{m}_2+\dot{m}_7}\right)(H_3 - H_7)}{(H_4 - H_3)} \tag{6.124}$$

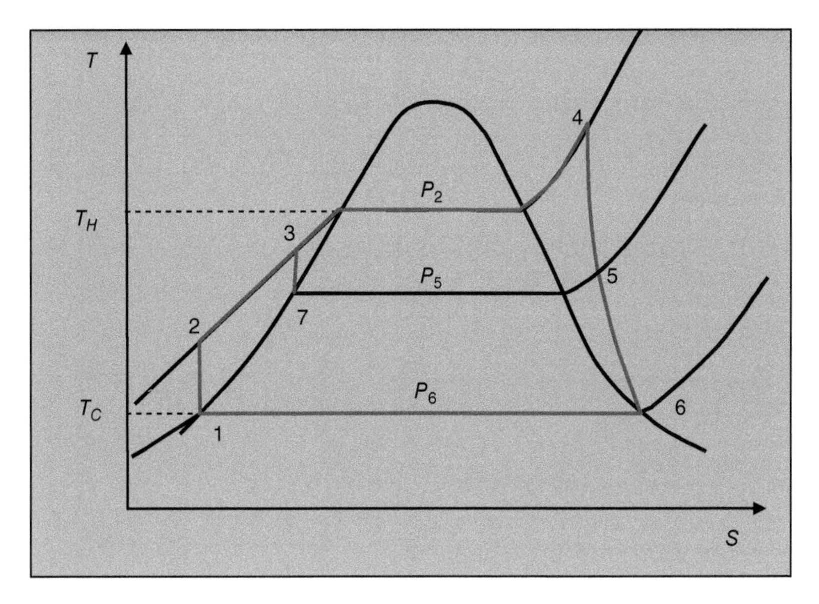

Figure 6.13 T–S diagram for the regeneration cycle.

A simple Rankine cycle operating $2 \rightarrow 4 \rightarrow 6 \rightarrow 1 \rightarrow 2$ has an efficiency of

$$\eta = \frac{(H_6 - H_4)}{(H_4 - H_2)} \tag{6.125}$$

Example 6.5

Calculate the efficiency of a power plant with regeneration as shown in Figure 6.11. The efficiencies of the turbines and pumps are 0.60 and 0.70, respectively. The conditions at the inlet of the turbine are 7000 kPa and 550 °C, and steam exits the turbine at 3000 kPa. Steam exits turbine 2 at 101 kPa and passes to a condenser in which it condenses to a saturated liquid. The regenerator temperature is 230 °C, and the steam temperature entering the reboiler is the exit temperature from pump 2.

Solution

Data for the turbine

$$P_4 = 7000\ \text{kPa}$$

$$T_4 = 550\ ^\circ\text{C}$$

$$P_5 = 3000\ \text{kPa}$$

$$P_6 = 101\ \text{kPa}$$

The same data as in example 6.4 are valid at the entrance of the turbine

$$H_4 = 3529.6\ \text{kJ/kg and } S_4 = 6.9485\ \text{kJ/(kg K)}$$

Now, at $P_4 = 3000$ kPa, the saturation conditions are

$$S^l = 2.6455\ \text{kJ}/(\text{kg K}) \text{ and } S^v = 6.1837\ \text{kJ}/(\text{kg K})$$

Then, the steam exiting turbine 1 is a superheated vapor. At S_4 and 3000 kPa, the steam tables provide

$$\text{At } T = 425\,°\text{C} \rightarrow S = 7.0067\ \text{kJ/(kg K)}, H = 3288.7\ \text{kJ/kg}$$
$$\text{At } T = 400\,°\text{C} \rightarrow S = 6.9246\ \text{kJ/(kg K)}, H = 3232.5\ \text{kJ/kg}$$

then from extrapolation,

$$H_5 = \frac{3288.7 - 3232.5}{7.0067 - 6.9246}(6.9485 - 7.0067) + 3288.7 = 3248.86\ \text{kJ/kg}$$

$$T_5 = \frac{425 - 400}{7.0067 - 6.9246}(6.9485 - 7.0067) + 400 = 407.28\,°\text{C}$$

The energy transferred as work in the turbine is

$$({}_4W_5)_S = H_5 - H_4 = 3248.86 - 3529.6 = -280.74\ \text{kJ/kg}$$

Considering the efficiency

$${}_4W_5 = ({}_4W_5)_S\eta = 0.6 \times (-280.74) = -168.44\ \text{kJ/kg}$$

the enthalpy of the steam discharged is

$$H_5 = {}_4W_5 + H_4 = -168.44 + 3529.6 = 3361.16\ \text{kJ/kg}$$

Using this value and $P = 3000$ kPa, we can find the entropy in the steam tables at superheated conditions

$$T = 450°\text{C} \rightarrow H = 3344.6\ \text{kJ/kg}, S = 7.0854\ \text{kJ}/(\text{kg K})$$

$$T = 475°\text{C} \rightarrow H = 3400.4\ \text{kJ/kg}, S = 7.1612\ \text{kJ}/(\text{kg K})$$

and from interpolation

$$S_5 = \frac{7.1612 - 7.0854}{3400.4 - 3344.6}(3361.16 - 3400.4) + 7.1612 = 7.1078\ \text{kJ}/(\text{kg K})$$

We must calculate the enthalpy values for steams that enter the regenerator. Saturated liquid at 230 °C condenses in the regenerator, and its properties are

$$H^l_{230} = 990.3\ \text{kJ/kg}, P = 2797.6\ \text{kPa}, V^l = 1209\ \text{cm}^3/\text{kg}, \beta = 1.6821 \times 10^{-3}\ \text{K}^{-1}$$

$$C_P = 4.6876\ \text{kJ}/(\text{kg K})$$

Pump 2 is adiabatic at a pressure of 7000 kPa. For the isentropic pump

$$({}_7W_3)_S = (\Delta H)_S = V\,(P_3 - P_7) = 1209\,(7000 - 2797.6) = 5.08\text{ kJ/kg}$$

and the actual energy transferred as work is

$${}_7W_3 = \frac{({}_7W_3)_S}{\eta_C} = \frac{5.08}{0.7} = 7.26\text{ kJ/kg}$$

The enthalpy at the exit of pump 2 is

$$H_3 = {}_7W_3 + H_7 = 7.26 + 990.3 = 997.56\text{ kJ/kg}$$

and the temperature in steam 3 exiting the pump 2 is

$$\Delta H = C_P \Delta T + V\,(1 - \beta T)\,\Delta P$$

In this equation, we consider water to be an incompressible liquid with a constant heat capacity, then

$$7.26\text{ kJ/kg} = 4.6876\text{ kJ/(kg K)}\,\Delta T\text{ (K)} - \frac{1209\text{ cm}^3\text{/kg}\ \ (1 - 1.6821 \times 10^{-3} \times 503.15)\,(7000 - 2797.6)\text{ kPa}}{1 \times 10^6 \frac{1\text{ kJ}}{\text{kPa cm}^3}}$$

$$= 4.6876\Delta T + 0.78066$$

and then

$$T_3 = 503.15 + \frac{(7.26 - 0.78066)}{4.6876} = 504.532\text{ K} = 231.382\,^\circ\text{C}$$

The saturated liquid conditions at $P_4 = 101$ kPa are

$$V^l = 1044\text{ cm}^3\text{/kg},\ S^l = 1.3069\text{ kJ/(kg K)},\text{ and } S^v = 7.3554\text{ kJ/(kg K)},$$
$$H^l = 419.064\text{ kJ/kg, and } H^v = 2676\text{ kJ/kg}$$

The isentropic energy transferred as work in the pump is

$$({}_1W_2)_S = (\Delta H)_S = V\,(P_2 - P_1)$$
$$= 1044\,(7000 - 101) = 7.202556 \times 10^6\text{ kPa cm}^3\text{/kg} = 7.20\text{ kJ/kg}$$

then

$${}_1W_2 = \frac{({}_1W_2)_S}{\eta_C} = \frac{7.20}{0.7} = 10.29\text{ kJ/kg}$$

and

$$H_2 = {}_1W_2 + H_1 = 10.29 + 419.064 = 429.354\text{ kJ/kg}$$

Applying the first law around the regenerator (see Figure below), assuming steady state, and considering 1 kg of water exiting the condenser per hour, $\dot{m}_2 = 1$ kg/hr,

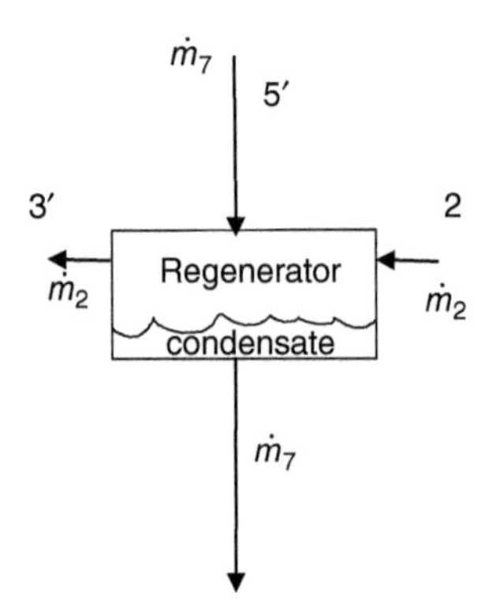

$$\sum \dot{m}_i H_i = 0 \Rightarrow \dot{m}_2 H_{3'} + \dot{m}_7 H_7 - \dot{m}_7 H_5 - \dot{m}_2 H_2 = 0$$

$$\dot{m}_2 \left(H_{3'} - H_2\right) + \dot{m}_7 \left(H_7 - H_5\right) = 0 \Rightarrow \dot{m}_7 = \dot{m}_2 \frac{\left(H_2 - H_{3'}\right)}{\left(H_7 - H_5\right)}$$

We must calculate the enthalpy at the exit of the regenerator. We have the enthalpy of a saturated liquid at 230 °C and $P = 2797.6$ kPa : $H^l_{230} = 990.3$ kJ/kg. We must calculate it at 230 °C and 7000 kPa. This process is isothermal, so for an incompressible liquid

$$\Delta H = V(1 - \beta T)\,\Delta P = 1209\left(1 - 503.15 \times 1.6821 \times 10^{-3}\right)$$
$$(7000 - 2797.6)/1 \times 10^6 = 0.781 \text{ kJ/kg}$$

and

$$H_{3'} = \Delta H + H^l_{230} = 990.3 + 0.781 = 991.081 \text{ kJ/kg}$$

The mass exiting as a condensate liquid is

$$\dot{m}_7 = \dot{m}_2 \frac{\left(H_{3'} - H_2\right)}{\left(H_7 - H_5\right)} = 1 \frac{429.354 - 991.081}{990.3 - 3361.16} = 0.2369 \text{ kg/hr}$$

Finally, we must calculate the enthalpy at the exit of turbine 2 considering the isentropic behavior

$$S_5 = S_6 = 7.1078 \text{ kJ/(kg K)}$$

Now, we calculate the quality at the exit of the turbine. Because $S_3 = S_4$ and $P_4 = 101$ kPa, using

$$\chi = \frac{S_6 - S^l}{S^v - S^l} = \frac{7.1078 - 1.3069}{7.3554 - 1.3069} = 0.959$$

the enthalpy in the isentropic process is

$$H_{6'} = H^l + \chi\left(H^v - H^l\right) = 419.064 + 0.959\,(2676 - 419.064) = 2583.6 \text{ kJ/kg}$$

The isentropic work is

$$\left({}_5W_{6'}\right)_S = H_{6'} - H_5 = (2583.6 - 3361.16) = -777.489 \text{ kJ/kg}$$

The negative signs indicate that the turbine transfers energy as work. The energy transferred as work and the enthalpy change are

$$_5W_6 = \eta\left(_5W_{6'}\right)_S = 0.6 \times (-777.489) = -466.494\ \text{kJ/kg}$$

and the actual enthalpy of steam 6 is

$$H_6 = {_5W_6} + H_5 = -466.494 + 3362.16 = 2895.666\ \text{kJ/kg}$$

The efficiency using Eq. (6.124) is

$$\eta = \frac{\left(H_5 - H_4\right) + \left(\frac{\dot{m}_2}{\dot{m}_2 + \dot{m}_7}\right)\left[\left(H_6 - H_5\right) + \left(H_2 - H_1\right)\right] + \left(\frac{\dot{m}_7}{\dot{m}_2 + \dot{m}_7}\right)\left(H_3 - H_7\right)}{\left(H_4 - H_3\right)}$$

$$= \frac{(-168.44) + \left(\frac{1}{1 + 0.2369}\right)(-466.494) + (10.29) + \left(\frac{0.2369}{1 + 0.2369}\right)(7.26)}{(3529.6 - 997.56)}$$

$$= 0.2109$$

The efficiency increases slightly for a single turbine step.

6.2.4 Internal Combustion Engines

Internal combustion engines are devices that produce work using combustion (generally of a mixture of air + hydrocarbons). The combustion occurs in a manner that produces high-pressure combustion products that expand through a turbine or piston to produce energy transfer as work. The products become the working fluid. Three major types of internal combustion engines exist:

The spark ignition engine used primarily in automobiles (i.e. Otto Engine).
The diesel engine used in large vehicles and industrial systems, in which the improvements in cycle efficiency give it advantages over the more compact and lightweight spark ignition engines.
Aircraft use the gas turbine because of its high power/weight ratio. Gas turbines also find use in stationary power generation.

6.2.4.1 Otto Cycle

Nicolaus Otto was born in Holzhausen, Germany, on 14 June 1832. As a traveling salesman, he developed his interest in building combustion engines based upon the work of Lenoir's two-stroke, gas-driven internal combustion engine. Otto and Eugen Langen started the first engine manufacturing company in 1864: N.A. Otto and Cie (now DEUTZ AG, Köln). In 1867, they received a Gold Medal at the Paris World Exhibition for their atmospheric gas engine.

In May 1876, Otto built a four-stroke piston cycle internal combustion engine that revolutionized the industry. He continued to develop his four-stroke engine after 1876, but he quit in 1884 after inventing the first magneto ignition system for low-voltage ignition. Otto's patent (see Figure 6.14) was overturned in 1886 in favor of the patent granted to Alphonse Beau de Roaches for his four-stroke engine. However, Otto built a working engine, while Roaches' design remained unproduced.

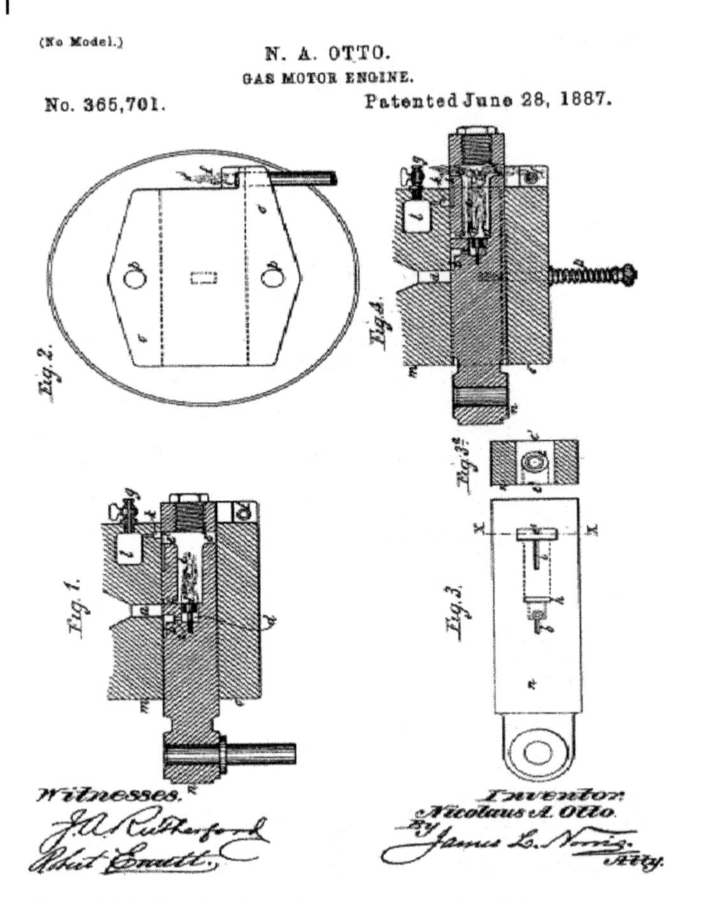

Figure 6.14 Drawing from the patent granted to N. Otto.

This landmark in engine design is the basis for many of the automobiles today. The four strokes of the cycle are intake, compression, combustion, and exhaust. Each corresponds to one full expansion/compression of the working fluid; therefore, the complete cycle requires two revolutions of the crankshaft to complete the cycle as shown in Figure 6.15.

Analyzing these strokes:

Intake Stroke (A → B): This is an almost constant pressure stroke in which the piston moves outward, drawing a fresh charge of vaporized fuel/air mixture. This step is A → B in Figure 6.16

Compression Stroke (B → C → D): With the valves closed, the flywheel momentum drives the piston upward compressing the fuel/air mixture from B → C. This path is almost adiabatic. At the top of the compression stroke, the sparkplug fires igniting the compressed mixture and increasing the pressure in a nearly constant volume process (step C → D).

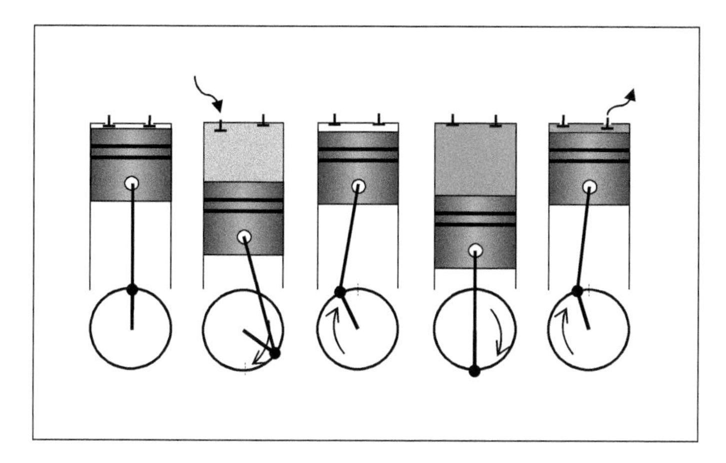

Figure 6.15 Schematic of the start and the four strokes in a piston.

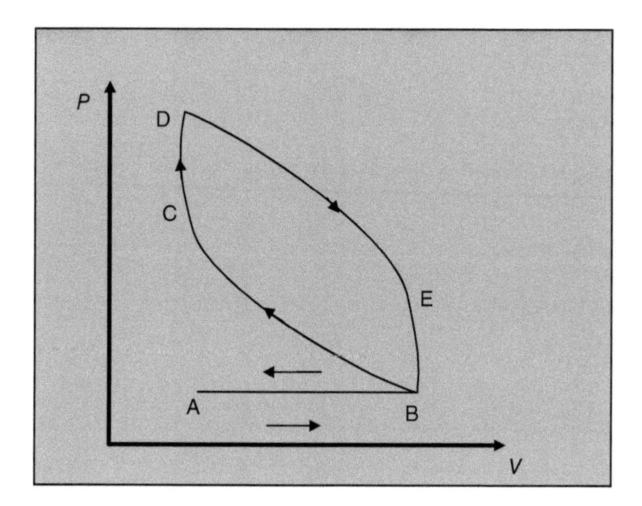

Figure 6.16 Schematic P–V diagram of the Otto engine cycle.

Combustion Stroke (D → E → B): As the fuel burns, it expands, driving the piston downward almost adiabatically following path D → E. The exhaust valve opens, and the pressure decreases following a constant volume process until point B.

Exhaust Stroke (B → A): In this stroke, the piston simply drives the exhausted fuel out of the chamber.

From the thermodynamic analysis, the general assumptions of the Otto engine cycle are as follows:

- The gasoline–air mixture is an ideal gas.
- All processes are quasi-static, so *mechanical equilibrium* exists (no unbalanced forces), *thermal equilibrium* exists (no temperature differences between parts of the system), and *chemical equilibrium* exists (no chemical reactions).
- The process is frictionless.

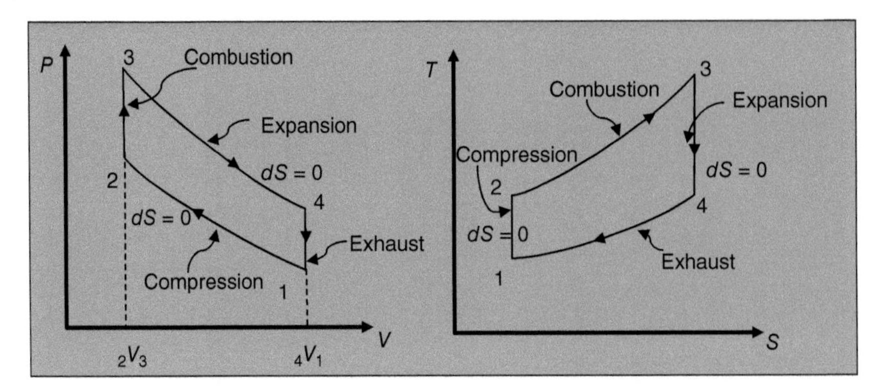

Figure 6.17 *P–V* and *T–S* diagrams for the air standard cycle (idealized Otto engine cycle).

A *PV* diagram or a *T–S* diagram assumes the cycle consists of four steps:

- 1 → 2 Isentropic compression
- 2 → 3 Constant volume compression
- 3 → 4 Isentropic expansion
- 4 → 1 Constant volume expansion.

Figure 6.17 represents all these steps on *P–V* and *T–S* diagrams.
From the first law,

$$0 = \sum W_i + \sum Q_i \Rightarrow \sum W_i = -\left({}_2Q_3 + {}_4Q_1\right) \tag{6.126}$$

The efficiency of the air standard cycle is

$$\eta = \frac{|W_{net}|}{Q_{in}} = \frac{{}_2Q_3 + {}_4Q_1}{{}_2Q_3} \tag{6.127}$$

and considering a perfect gas (constant heat capacities) in the constant volume processes

$${}_2Q_3 = C_V\left(T_3 - T_2\right) \text{ and } {}_4Q_1 = C_V\left(T_1 - T_4\right) \tag{6.128}$$

The efficiency is

$${}_4Q_1 = C_V\left(T_1 - T_4\right) \tag{6.129}$$

For the two isentropic steps using a perfect gas,

$$T_1 V_1^{\gamma^{ig}-1} = T_2 V_2^{\gamma^{ig}-1} \tag{6.130}$$

and

$$T_3 V_3^{\gamma^{ig}-1} = T_4 V_4^{\gamma^{ig}-1} \tag{6.131}$$

Because $V_1 = V_4 = {}_4V_1$ and $V_2 = V_3 = {}_2V_3$, the above equation becomes

$$T_4 V_1^{\gamma^{ig}-1} = T_3 V_2^{\gamma^{ig}-1} \tag{6.132}$$

Subtracting Eq. (6.132) from Eq. (6.130)

$$\left(T_1 - T_4\right) V_1^{\gamma^{ig}-1} = \left(T_2 - T_3\right) V_2^{\gamma^{ig}-1} \tag{6.133}$$

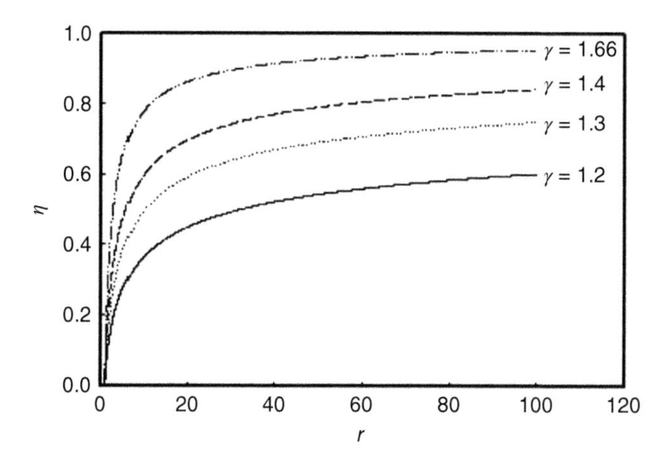

Figure 6.18 Effect of the compression ratio and γ^{ig} in the efficiency of the standard air engine.

Changing the order of the temperatures

$$\frac{T_1 - T_4}{T_3 - T_2} = -\left(\frac{V_2}{V_1}\right)^{\gamma^{ig}-1} = -\left(\frac{{}_2V_3}{{}_4V_1}\right)^{\gamma^{ig}-1} \tag{6.134}$$

The compression ratio, r, is ${}_4V_1/{}_2V_3$, which is the ratio of the volumes at the beginning and end of compression. Using this result and substituting Eqs. (6.128) and (6.129) into the efficiency, Eq. (6.127),

$$\eta = 1 - \left(\frac{1}{r}\right)^{\gamma^{ig}-1} = 1 - r^{1-\gamma^{ig}} \tag{6.135}$$

In this equation, γ^{ig} is always greater than 1; therefore, the thermal efficiency increases sharply at low values of the compression ratio, r, and slowly at higher values. To acquire a high efficiency at a given ratio of heat capacities requires very high compression ratios. Often, it is better to change the working fluid such that the ratio of heat capacities increases. A higher efficiency results are shown in Figure 6.18.

Example 6.6

An Otto engine compresses air isentropically at 1 bar and 25 °C. The compression ratio is 4 and adds 900 kJ/kg during combustion. Determine

The maximum temperature in the cycle
The air standard efficiency
The net energy transferred as work
The energy transferred as heat.

Use $C_P = 1.0063$ kJ/(kg K) and $\gamma^{ig} = 1.4$.

Solution

Data

$$P_1 = 1 \text{ bar}$$

$$r = 4 = \left(\frac{V_1}{V_2}\right)$$

$${}_2Q_3 = 900\ \text{kJ/kg}$$

$$T_1 = 25 + 273.15 = 298.15\ \text{K}$$

The temperature comes from

$${}_2Q_3 = C_V\left(T_3 - T_2\right) \Rightarrow T_3 = {}_2Q_3/C_V + T_2$$

however, we must know T_2. From the isentropic compression 1–2,

$$P_1V_1^{\gamma^{ig}} = P_2V_2^{\gamma^{ig}} \Rightarrow P_2 = P_1\left(\frac{V_1}{V_2}\right)^{\gamma^{ig}} = P_1r^{\gamma^{ig}} = 1 \times 4^{1.4} = 6.96\ \text{bar}$$

$$T_1V_1^{\gamma^{ig}-1} = T_2V_2^{\gamma^{ig}-1} \Rightarrow T_2 = T_1\left(\frac{V_1}{V_2}\right)^{\gamma^{ig}-1} = T_1r^{\gamma^{ig}-1} = 298.15 \times 4^{1.4-1} = 519.11\ \text{K}$$

The heat capacity at a constant volume is

$$C_V = C_P - R = 1.0063\ \text{kJ/(kg K)} - \frac{8.314\ \text{kJ/(kg mol K)}}{28.965\ \text{kg/(kg mol)}} = 0.7193\ \text{kJ/(kg K)}$$

so

$$T_3 = {}_2Q_3/C_V + T_2 = 900/0.7193 + 519.11 = 1770.389\ \text{K}$$

The thermal efficiency is

$$\eta = 1 - r^{1-\gamma^{ig}} = 1 - 4^{1-1.4} = 0.4256$$

Using the definition of efficiency,

$$\eta = \frac{|W_{net}|}{{}_2Q_3} \Rightarrow |W_{net}| = \eta_2Q_3 = 900 \times 0.426 = 383.086\ \text{kJ/kg}$$

The energy rejected as heat is available from the net work

$$|W_{net}| = {}_2Q_3 + {}_4Q_1 \Rightarrow {}_4Q_1 = |W_{net}| - {}_2Q_3 = 383.086 - 900 = -516.914\ \text{kJ/kg}$$

Of course, we can calculate T_4 and then the energy transferred as heat

$$T_4 = T_3\left(\frac{V_2}{V_1}\right)^{\gamma^{ig}-1} = T_3\left(\frac{1}{r}\right)^{\gamma^{ig}-1} = 1770.389\left(\frac{1}{4}\right)^{1.4-1} = 1016.821\ \text{K}$$

$${}_4Q_1 = C_V\left(T_1 - T_4\right) = 0.7193\,(298.15 - 1016.821) = -516.914\ \text{kJ/kg}$$

6.2.4.2 Diesel Engine Cycle

Rudolf Diesel was born on 18 March 1858, in Paris to German parents who sent him to Germany for his education. Diesel was graduated from the Royal Bavarian Polytechnic of Munich. After graduation, he worked with Carl von Linde designing and constructing refrigerators. In 1893, he published a paper about internal combustion within a cylinder. He developed his first successful engine in 1897, and he received patent #608,845 for his *internal combustion engine*, see Figure 6.19.

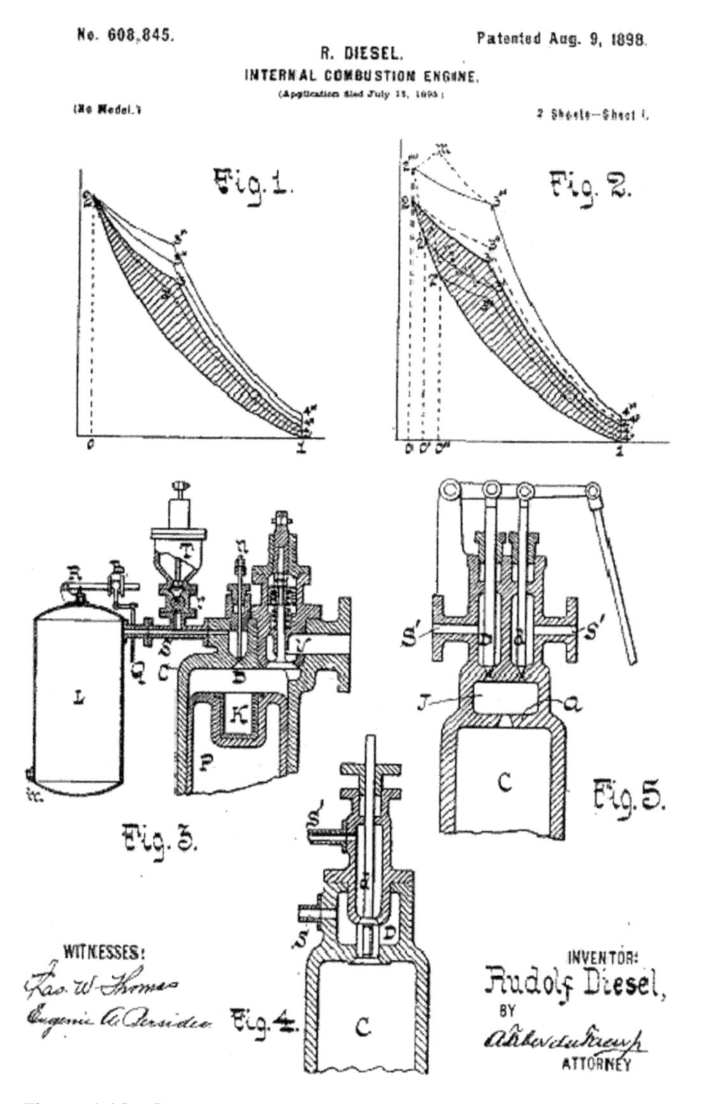

Figure 6.19 Patent #608,845 for the diesel engine.

The main difference between the Otto and the diesel engine is that in the latter, at the end of the compression step, the temperature is so high that combustion occurs spontaneously. The compression ratio is higher; therefore, the pressure and temperature are higher. In the diesel engine, fuel enters slowly at the end of the compression step, so combustion is isobaric. Figure 6.20 illustrates this cycle.

The equations for this engine are the same as for the Otto cycle except in the isobaric step for which

$$ {}_2Q_3 = C_P \left(T_3 - T_2\right) \tag{6.136} $$

Equation (6.137) provides the thermal efficiency

$$ \eta = 1 + \frac{{}_4Q_1}{{}_2Q_3} = 1 + \frac{C_V\left(T_1 - T_4\right)}{C_P\left(T_3 - T_2\right)} = \frac{1}{\gamma^{ig}} \frac{T_1 - T_4}{T_3 - T_2} \tag{6.137} $$

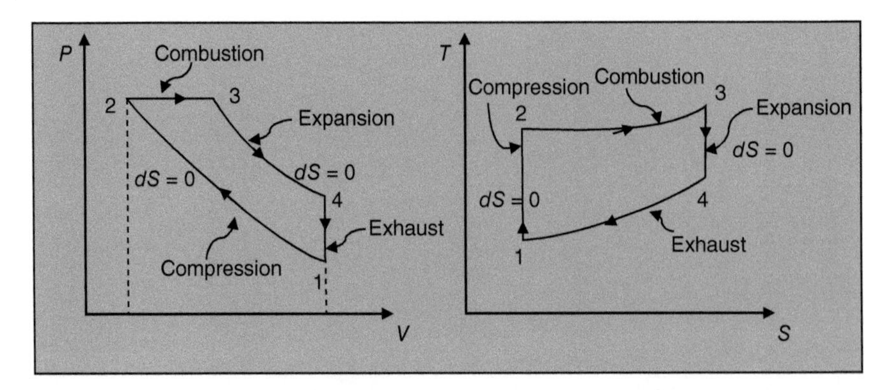

Figure 6.20 *P*–*V* and *T*–*S* diagram for the air standard cycle (idealized diesel engine cycle).

It is convenient to express the isentropic steps in terms of temperatures

$$T_1 V_1^{\gamma^{ig}-1} = T_2 V_2^{\gamma^{ig}-1} \tag{6.138}$$

and

$$T_3 V_3^{\gamma^{ig}-1} = T_4 V_4^{\gamma^{ig}-1} \tag{6.139}$$

Equation (6.137) in terms of the compression ratio is

$$T_1 = T_2 \left(\frac{V_2}{V_1} \right)^{\gamma^{ig}-1} = T_2 \left(\frac{1}{r} \right)^{\gamma^{ig}-1} \tag{6.140}$$

If we define an expansion ratio,

$$r_e = \frac{V_4}{V_3} \tag{6.141}$$

then Eq. (6.138) becomes

$$T_4 = T_3 \left(\frac{1}{r_e} \right)^{\gamma^{ig}-1} \tag{6.142}$$

Using these results, the efficiency is

$$\eta = 1 + \frac{1}{\gamma^{ig}} \frac{T_2 \left(\frac{1}{r}\right)^{\gamma^{ig}-1} - T_3 \left(\frac{1}{r_e}\right)^{\gamma^{ig}-1}}{T_3 - T_2} = 1 - \frac{1}{\gamma^{ig}} \left[\frac{\left(\frac{1}{r_e}\right)^{\gamma^{ig}-1} - \frac{T_2}{T_3}\left(\frac{1}{r}\right)^{\gamma^{ig}-1}}{1 - \frac{T_2}{T_3}} \right] \tag{6.143}$$

In the isobaric step using the ideal gas EOS,

$$\frac{V_2}{T_2} = \frac{V_3}{T_3} \Rightarrow \frac{T_2}{T_3} = \frac{V_2}{V_3} \tag{6.144}$$

and from the isochoric step $V_1 = V_4$, then dividing by V_1, the right hand side (RHS) becomes

$$\frac{T_2}{T_3} = \frac{V_2/V_1}{V_3/V_1} = \frac{V_2/V_1}{V_3/V_4} \tag{6.145}$$

and using the compression and expansion ratios,

$$\frac{T_2}{T_3} = \frac{r_e}{r} \tag{6.146}$$

Substituting this result into the efficiency

$$\eta = 1 - \frac{1}{\gamma^{ig}} \left[\frac{\left(\frac{1}{r_e}\right)^{\gamma^{ig}-1} - \left(\frac{r_e}{r}\right)\left(\frac{1}{r}\right)^{\gamma^{ig}-1}}{1 - \left(\frac{r_e}{r}\right)} \right] \tag{6.147}$$

and multiplying and dividing the term in brackets by $1/r_e$

$$\eta = 1 - \frac{1}{\gamma^{ig}} \left[\frac{\left(\frac{1}{r_e}\right)^{\gamma^{ig}} - \left(\frac{1}{r}\right)^{\gamma^{ig}}}{\frac{1}{r_e} - \frac{1}{r}} \right] \tag{6.148}$$

Example 6.7

Consider a diesel engine that uses air in which the highest temperature in the compression step is the same as that for an Otto engine with a compression ratio of 4. It absorbs 900 kJ/kg during combustion. Determine the expansion ratio and the thermal efficiency. If the compression ratio increases to 8, what is the thermal efficiency assuming the same amount of energy transferred as heat during the combustion?

Solution

Because we reach the same temperature and add the same amount of energy transferred as heat during combustion as in example 6.6

$$T_2 = T_1\left(\frac{V_1}{V_2}\right)^{\gamma^{ig}-1} = T_1 r^{\gamma^{ig}-1} = 298.15 \times 4^{1.4-1} = 519.11\ \text{K}$$

and

$$T_3 = {}_2Q_3/C_V + T_2 = 900/0.7193 + 519.11 = 1770.389\ \text{K}$$

Using the relationship between ratios,

$$\frac{T_2}{T_3} = \frac{r_e}{r} \Rightarrow r_e = r\left(\frac{T_2}{T_3}\right) = 4\left(\frac{519.11}{1770.389}\right) = 4 \times 0.2932 = 1.1729$$

so, the thermal efficiency is

$$\eta = 1 - \frac{1}{\gamma^{ig}} \left[\frac{\left(\frac{1}{r_e}\right)^{\gamma^{ig}} - \left(\frac{1}{r}\right)^{\gamma^{ig}}}{\frac{1}{r_e} - \frac{1}{r}} \right] = 1 - \frac{1}{1.4} \left[\frac{\left(\frac{1}{1.1729}\right)^{1.4} - \left(\frac{1}{4}\right)^{1.4}}{\frac{1}{1.1729} - \frac{1}{4}} \right] = 0.2220$$

For same compression ratio, the Otto engine has a higher efficiency. For a higher compression ratio,

$$T_2 = T_1\left(\frac{V_1}{V_2}\right)^{\gamma^{ig}-1} = T_1 r^{\gamma^{ig}-1} = 298.15 \times 8^{1.4-1} = 684.97\ \text{K}$$

and

$$T_3 = {}_2Q_3/C_V + T_2 = 900/0.7193 + 684.97 = 1936.186 \text{ K}$$

$$r_e = 8\left(\frac{684.97}{1936.186}\right) = 8 \times 0.3538 = 2.8302$$

then

$$\eta = 1 - \frac{1}{\gamma^{ig}}\left[\frac{\left(\frac{1}{r_e}\right)^{\gamma^{ig}} - \left(\frac{1}{r}\right)^{\gamma^{ig}}}{\frac{1}{r_e} - \frac{1}{r}}\right] = 1 - \frac{1}{1.4}\left[\frac{\left(\frac{1}{2.8302}\right)^{1.4} - \left(\frac{1}{8}\right)^{1.4}}{\frac{1}{2.8302} - \frac{1}{8}}\right] = 0.4412$$

The diesel engine performs better in this case because the compression ratio is higher.

6.2.4.3 Gas Turbine Cycle

The gas turbine is an internal combustion engine. Gas turbine engines are, theoretically, extremely simple having three parts:

- Compressor
- Combustion area
- Turbine

Figure 6.21 presents a schematic diagram of a gas turbine cycle. The turbine is an open cycle, but for thermodynamic simplicity, we consider the Brayton cycle, which is a closed cycle. Figure 6.22 presents this cycle in which:

Step 1 to 2: Ambient air enters the compressor, increasing its pressure and temperature. In some engines, the pressure of the air can rise by a factor of 30. This requires an energy transfer as work input by the compressor (${}_1W_2$). Ideally, this work should be as small as possible.

Step 2 to 3: Then, the high-pressure air passes into the combustion chamber, where it mixes with the fuel. This is a constant pressure process that requires addition of energy transfer as heat, (${}_2Q_3$). It is common to think pressure increases in a combustion process, but that is not the case.

Step 3 to 4: The high-temperature gases enter the turbine, in which they expand to a lower pressure. The turbine extracts energy as work from the fluid (${}_3W_4$), producing a torque on the shaft connected to the compressor. This energy transferred as work must be sufficient to drive the compressor and its auxiliary components.

Step 4 to 1: Finally, the exhaust gases leave the turbine. When modeling the whole process as a closed cycle, the product gases induce a constant pressure energy rejection as heat (${}_4Q_1$), as well as a throttling process. The objective in this step is to maximize the velocity of the exiting exhaust gases to create the highest pressure possible.

The fuel used in the combustion chamber is generally kerosene, jet fuel, propane, or natural gas. High-pressure air enters the combustion zone with velocities of hundreds of km/h. A flame-holder device keeps a flame burning continuously in the environment.

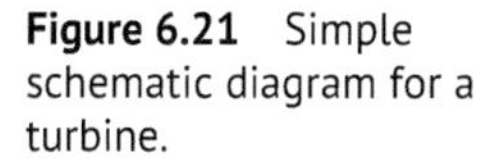

Figure 6.21 Simple schematic diagram for a turbine.

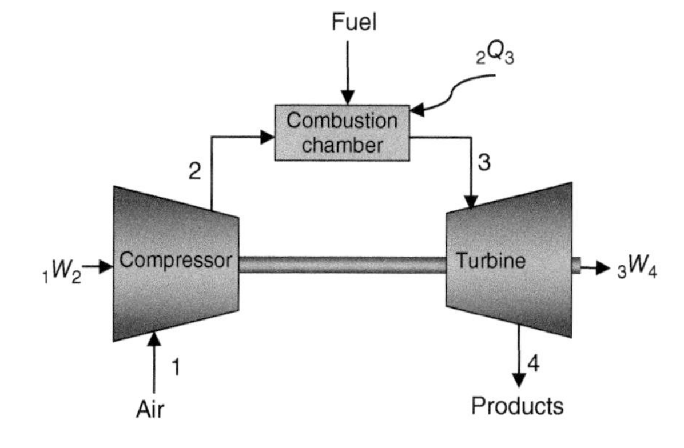

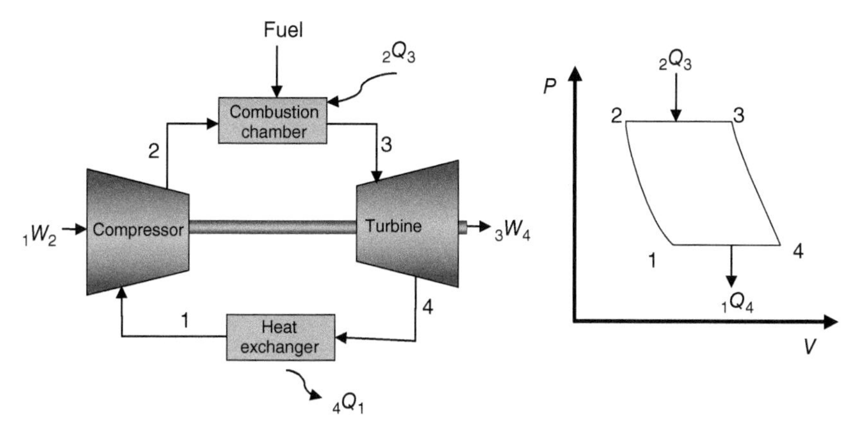

Figure 6.22 Simplified Brayton cycle.

The thermal efficiency is

$$\eta = \frac{|W_{net}|}{Q_{in}} = \frac{|{}_1W_2 + {}_3W_4|}{{}_2Q_3}$$

and using the first law for the cycle

$$\sum W_i = -\sum Q_i = -\left({}_2Q_3 + {}_4Q_1\right)$$

If we consider air to be an ideal gas with constant heat capacities (perfect gas),

$${}_2Q_3 = H_3 - H_2 = C_P\left(T_3 - T_2\right) \tag{6.149}$$

$${}_4Q_1 = H_4 - H_1 = C_P\left(T_1 - T_4\right) \tag{6.150}$$

then, the efficiency is

$$\eta = 1 + \frac{T_1 - T_4}{T_3 - T_2} \tag{6.151}$$

Using the isentropic paths $1 \rightarrow 2$ and $3 \rightarrow 4$

$$\left(\frac{T_2}{T_1}\right)^{\gamma^{ig}} = \left(\frac{P_2}{P_1}\right)^{\gamma^{ig}-1} \tag{6.152}$$

and

$$\left(\frac{T_4}{T_3}\right)^{\gamma^{ig}} = \left(\frac{P_4}{P_3}\right)^{\gamma^{ig}-1} \tag{6.153}$$

but $P_4 = P_1$ and $P_3 = P_2$, so

$$\left(\frac{T_4}{T_3}\right)^{\gamma^{ig}} = \left(\frac{P_1}{P_2}\right)^{\gamma^{ig}-1} \text{ or } \left(\frac{T_3}{T_4}\right)^{\gamma^{ig}} = \left(\frac{P_2}{P_1}\right)^{\gamma^{ig}-1} \tag{6.154}$$

and

$$T_2 = T_1\left(\frac{P_2}{P_1}\right)^{(\gamma^{ig}-1)/\gamma^{ig}} \tag{6.155}$$

with

$$T_3 = T_4\left(\frac{P_2}{P_1}\right)^{(\gamma^{ig}-1)/\gamma^{ig}}. \tag{6.156}$$

Combining these equations

$$T_3 - T_2 = (T_4 - T_1)\left(\frac{P_2}{P_1}\right)^{(\gamma^{ig}-1)/\gamma^{ig}} \text{ or } \frac{T_3 - T_2}{T_4 - T_1} = \left(\frac{P_2}{P_1}\right)^{(\gamma^{ig}-1)/\gamma^{ig}} \tag{6.157}$$

Rearranging the last equation

$$\frac{T_1 - T_4}{T_3 - T_2} = -\left(\frac{P_1}{P_2}\right)^{(\gamma^{ig}-1)/\gamma^{ig}} \tag{6.158}$$

then substituting into the efficiency equation

$$\eta = 1 - \left(\frac{P_1}{P_2}\right)^{(\gamma^{ig}-1)/\gamma^{ig}} \tag{6.159}$$

For an irreversible process, this means that we have efficiencies for the turbine and the compressor

$$_1W_2 = ({}_1W_2)_S/\eta_{com}$$

$$_3W_4 = ({}_3W_4)_S\eta_t$$

The exit temperatures of both devices are different from those of an isentropic process. In addition,

$$\sum W_i = -\sum Q_i = -({}_2Q_3 + {}_4Q_1)$$

A jet engine is similar to a gas turbine except the gases exit at high pressure through a nozzle. Also, the air passes through a diffuser to increase its pressure and to reduce the energy transferred as work. Here, the kinetic energy of the gases is not negligible.

Example 6.8

Air enters a compressor in a gas turbine at 101 kPa and 298.15 K. The air pressure rises to 800 kPa. After the compressor, the air enters the combustion chamber in which the temperature is 1500 K. Then, it passes into a turbine and expands to 400 kPa. Finally, a second burner raises the air temperature to 1400 K, and it enters a second turbine in which it expands to the inlet pressure of the compressor. The efficiency in the compressor is 80%, and the turbine efficiencies are 75%. The net power is 2 MW. Use $\gamma^{ig} = 1.4$. Calculate the mass flow rate of the air and the efficiency of the cycle. Plot a T–S diagram for the cycle.

Solution

A diagram of the cycle is

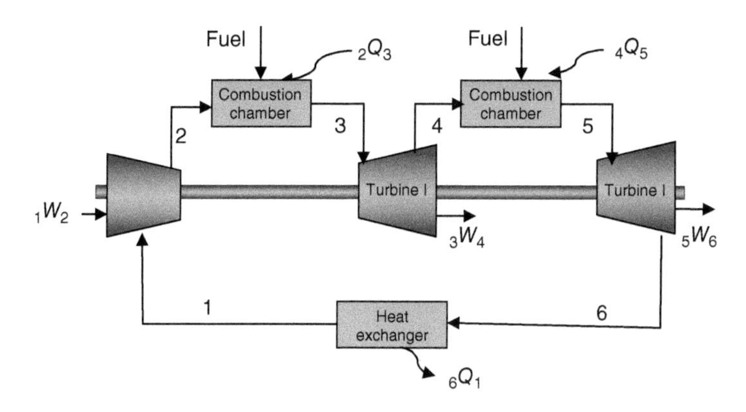

The T–S diagram is

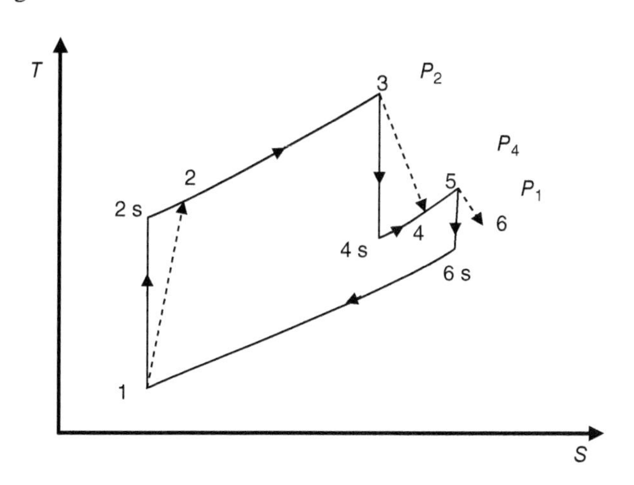

Data

$$P_1 = 101\ \text{kPa}$$

$$T_1 = 298.15\ \text{K}$$

$$P_2 = P_3 = 800\ \text{kPa}$$

$$T_3 = 1500\ \text{K}$$

$$P_3 = P_4 = 300\ \text{kPa}$$

$$T_5 = 1400\ \text{K}$$

$$P_6 = 101\ \text{kPa}$$

To calculate the exit temperature of the compressor, first calculate the temperature of the isentropic process

$$\frac{T_2}{T_1} = \left(\frac{P_2}{P_1}\right)^{(\gamma^{ig}-1)/\gamma^{ig}} \Rightarrow T_2 = T_1\left(\frac{P_2}{P_1}\right)^{(\gamma^{ig}-1)/\gamma^{ig}} = 298.15\left(\frac{800}{101}\right)^{(1.4-1)/1.4} = 538.549\ \text{K}$$

then,

$$({}_1W_2)_S = C_P\,(T_2 - T_1) = 1.005\,(538.549 - 298.15) = 241.6\ \text{kJ/kg}$$

and the actual work is

$${}_1W_2 = ({}_1W_2)_S/\eta_{comp} = 241.6/0.8 = 302\ \text{kJ/kg}$$

Now, the temperature is

$$T_2 = {}_1W_2/C_P + T_1 = 302/1.005 + 298.15 = 598.65\ \text{K}$$

and we can calculate the heat duty in the combustion chamber I,

$${}_2Q_3 = C_P\,(T_3 - T_2) = 1.005\,(1500 - 598.65) = 905.86\ \text{kJ/kg}$$

For turbine I, considering it an isentropic process,

$$\frac{T_4}{T_3} = \left(\frac{P_4}{P_3}\right)^{(\gamma^{ig}-1)/\gamma^{ig}} \Rightarrow T_4 = T_3\left(\frac{P_4}{P_3}\right)^{(\gamma^{ig}-1)/\gamma^{ig}} = 1500\left(\frac{300}{800}\right)^{(1.4-1)/1.4} = 1133.407\ \text{K}$$

and the isentropic work is

$$({}_3W_4)_S = C_P\,(T_4 - T_3) = 1.005\,(1133.407 - 1500) = -368.426\ \text{kJ/kg}$$

while the actual work is

$${}_3W_4 = ({}_3W_4)_S\eta_{t-I} = -368.426 \times 0.75 = -276.319\ \text{kJ/kg}$$

and the actual temperature is T_4

$$T_4 = {}_3W_4/C_P + T_3 = -276.319/1.005 + 1500 = 1225.055\ \text{K}$$

The heat duty in combustion chamber II is

$${}_4Q_5 = C_P\,(T_5 - T_4) = 1.005\,(1400 - 1225.055) = 175.819\ \text{kJ/kg}$$

Repeating the same procedure for turbine II, considering it an isentropic process,

$$\frac{T_6}{T_5} = \left(\frac{P_6}{P_5}\right)^{(\gamma^{ig}-1)/\gamma^{ig}} \Rightarrow T_6 = T_5\left(\frac{P_6}{P_5}\right)^{(\gamma^{ig}-1)/\gamma^{ig}} = 1400\left(\frac{101}{300}\right)^{(1.4-1)/1.4} = 1025.752\ \text{K}$$

the isentropic work is

$$({}_5W_6)_S = C_P\,(T_6 - T_5) = 1.005\,(1025.752 - 1400) = -376.119\ \text{kJ/kg}$$

and the actual work is

$${}_5W_6 = ({}_5W_6)_S\eta_{t-I} = -376.119 \times 0.75 = -282.089\ \text{kJ/kg}$$

The actual temperature, T_6, is

$$T_6 = {}_5W_6/C_P + T_5 = -282.089/1.005 + 1400 = 1119.314\ \text{K}$$

The duty in the heat exchanger is

$${}_6Q_1 = C_P\left(T_1 - T_6\right) = 1.005\,(298.15 - 1119.314) = -825.27\ \text{kJ/kg}$$

and the efficiency is

$$\eta = \frac{|W_{net}|}{Q_{in}} = \frac{|{}_1W_2 + {}_3W_4 + {}_5W_6|}{{}_2Q_3 + {}_4Q_5} = \frac{|302 - 276.319 - 282.089|}{905.857 + 175.819} = \frac{256.407}{1081.676} = 0.237$$

6.2.5 Refrigeration: The Carnot Cycle for a Refrigeration Unit

Refrigerators are cooling devices that consist of several parts. One of these parts, a heat pump, extracts energy transferred as heat. The extracted energy passes to the environment from the heat pump to maintain a lower temperature inside the device. As in engines, the Carnot cycle is the most efficient refrigeration cycle for operation between a low-temperature reservoir (refrigerator) and the high-temperature reservoir (ambient conditions). Figure 6.23 is a schematic of the Carnot cycle.

We have seen in Chapter 4 that instead of using the efficiency in refrigerators, we use the COP,

$$\eta_R = COP = \frac{T_C}{T_H - T_C}$$

Irreversibility in the adiabatic expansion and compression steps of a refrigeration cycle is a straightforward concept. If we look at the T–S diagram, Figure 6.24, the reversible path is $1 \to 2 \to 3 \to 4 \to 1$, and the irreversible path is $1 \to 2' \to 3' \to 4 \to 1$. If Q_C is the same for both paths, then we can express the COP for both cycles as

$$(COP)_{CARNOT} = \frac{Q_C}{W_S} \tag{6.160}$$

and

$$(COP)_{IRR} = \frac{Q_C}{W} \tag{6.161}$$

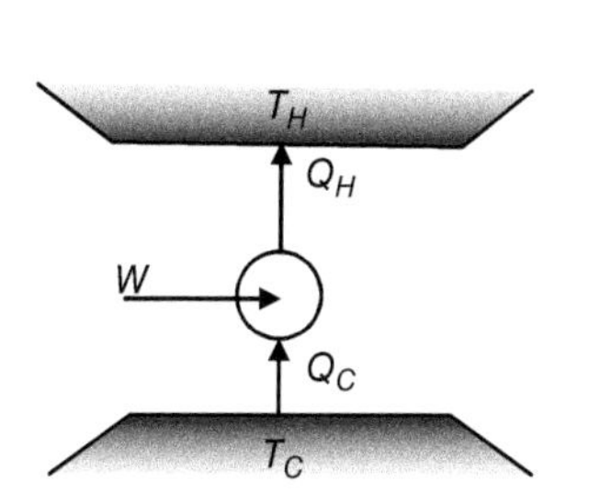

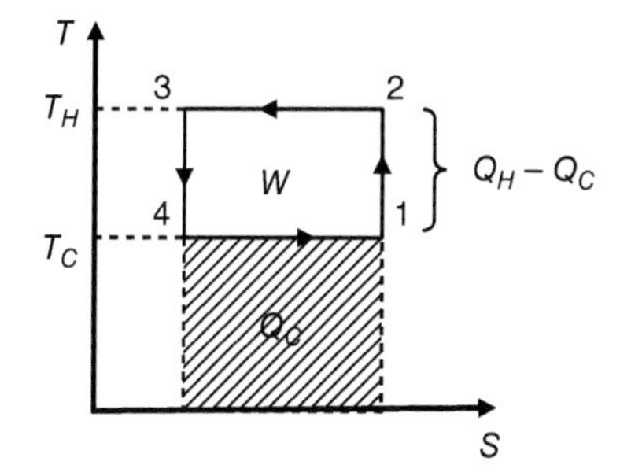

Figure 6.23 Carnot refrigerator cycle.

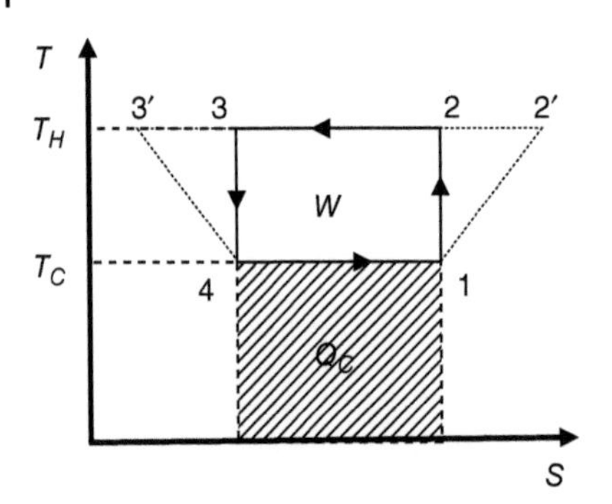

Figure 6.24 Carnot cycle for an irreversible refrigerator.

then

$$(COP)_{CARNOT}W_S = (COP)_{IRR}W \tag{6.162}$$

and

$$\frac{(COP)_{CARNOT}}{(COP)_{IRR}} = \frac{W}{W_S} \tag{6.163}$$

In this case, we can calculate the *COP* of the Carnot cycle and the reversible net energy transferred as work. Given the efficiency, we can calculate the irreversible work, which is similar to the efficiency in heat engines

$$\frac{\eta_{IRR}}{\eta_{IRR}\eta_{CARNOT}} = \frac{W}{W_S} \tag{6.164}$$

6.2.5.1 Vapor Compression Cycle

This is the most common cycle for refrigerators used in households and in commercial and industrial refrigeration systems. Figure 6.25 is a schematic of this cycle with typical devices.

A thermodynamic analysis performed on a T–S diagram appears in Figure 6.25. Step 1 → 2 compresses a refrigerant isentropically, which then exits the compressor as vapor at high temperature. Then, the vapor enters the condenser in which it cools until the refrigerant condenses to liquid removing energy as heat at a constant temperature and pressure (step 2 → 3). From 3 → 4, the vapor expands through a

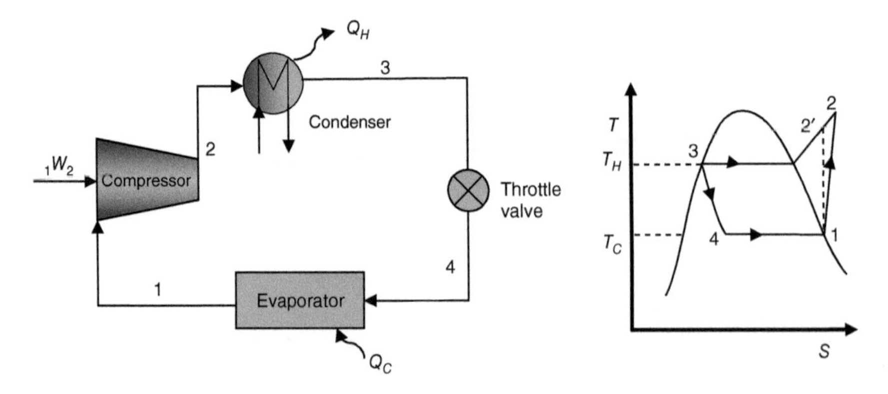

Figure 6.25 Vapor compression refrigeration.

throttle valve in which a sudden decrease in pressure causes a flash evaporation and auto-refrigeration. This process converts the single phase refrigerant into a vapor–liquid mixture at a lower temperature. The cold, two-phase refrigerant passes through the evaporator coils where it totally vaporizes from contact with warm air from the compartment of the refrigerator. A fan blows the air through the evaporator tube. Finally, in step 4 → 1, the vapor returns to the inlet of the compressor. The irreversibility occurs in the compressor because of a fractional pressure drop in the system step 1 → 2′.

From the first law of thermodynamics:

$$\text{Evaporator}: \ {}_4Q_1 = H_1 - H_4 \tag{6.165}$$

$$\text{Condenser}: \ {}_2Q_3 = H_3 - H_2 \tag{6.166}$$

$$\text{Compressor}: \ {}_1W_2 = H_2 - H_1 \tag{6.167}$$

$$\text{Throttle valve}: \ H_4 = H_3 \tag{6.168}$$

$$\text{Coefficient of performance}: \ COP = \frac{|Q_{in}|}{W_{net}} = \frac{H_1 - H_4}{H_2 - H_1} \tag{6.169}$$

Generally, $\ln P$ vs H diagrams are more common in refrigeration than the T–S diagrams because they present the values of the enthalpies. Figure 6.26 illustrates the refrigeration cycle in a $\ln P$–H diagram. The flow rate of the refrigerant results from the rate of the energy transferred as heat via absorption

$$\dot{m} = \frac{{}_4\dot{Q}_1}{H_1 - H_4} \tag{6.170}$$

An alternative for a gas compression refrigerator is to use a turbine instead of the throttle valve. The equations are the same for the condenser and evaporator, so the net energy transferred as work is

$$W_{net} = \left({}_1W_2\right)_{compressor} + \left({}_3W_4\right)_{turbine} = \left(H_2 - H_1\right) + \left(H_4 - H_3\right) \tag{6.171}$$

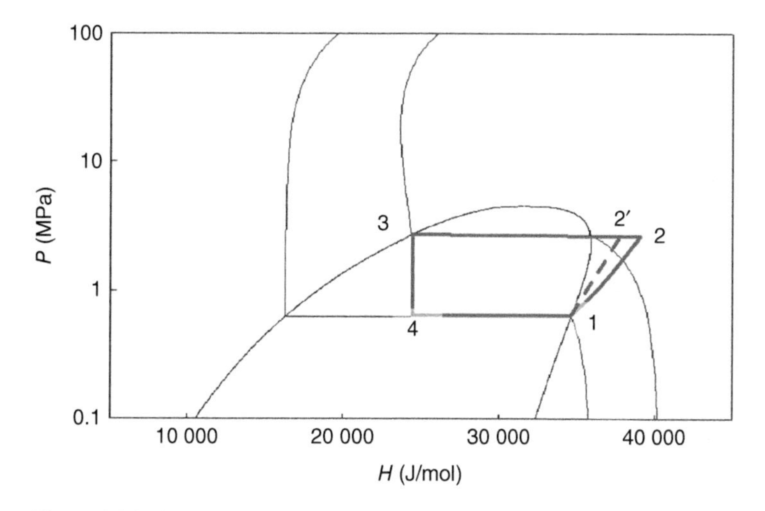

Figure 6.26 Vapor compression cycle on a ln P–H diagram using refrigerant R-152a.

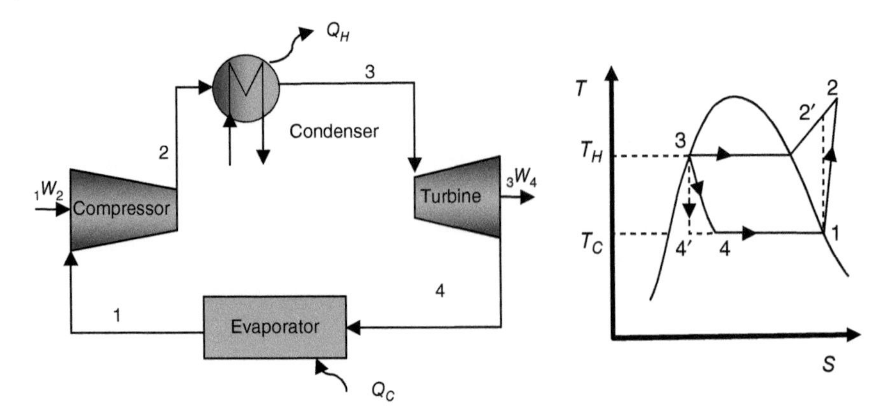

Figure 6.27 Vapor compression refrigeration with a turbine.

Remember that $W_{compressor} > 0$ and $W_{turbine} < 0$. Expansion engines are efficient, but they are much more difficult and expensive to construct. Expansion valves are common in household refrigerators and air conditioners. Figure 6.27 illustrates this arrangement. Dashed lines represent isentropic processes in the compressor and turbine.

Example 6.9

What temperature does a refrigerator maintain that uses R-134a refrigerant if the *COP* is 4 and the efficiency in the compressor is 0.7? The temperature at the exit of the condenser is 35 °C, and the temperature in the evaporator is 10° higher than the refrigerator temperature.

Solution

We have a

$$COP = \frac{H_1 - H_4}{H_2 - H_1}$$

but

$$_1W_2 = \left(_1W_2\right)_S / \eta_{comp} \Rightarrow H_2 - H_1 = \left(H_{2S} - H_1\right) / \eta_C \Rightarrow H_2 = \left(H_{2S} - H_1\right) / \eta_C + H_1$$

so,

$$COP = \frac{H_1 - H_4}{H_2 - H_1} = \frac{H_1 - H_4}{\left(H_{2S} - H_1\right) / \eta_C}$$

Now, $T_3 = 35\,°\text{C}$, and the refrigerant condenses as a saturated liquid at 8.8698 bar, so

$$H_3 = 249.01\ \text{kJ/kg}$$

Assuming that the temperature in the refrigerator is 0 °C, then $T_1 = 0\text{–}10\,°\text{C}$, and the liquid vaporizes at approximately 2 bar as a saturated vapor

$$H_1 = 392.66\ \text{kJ/kg and } S_1 = 1.7334\ \text{kJ/}\left(\text{kg K}\right)$$

If the compression step is isentropic, then $S_1 = S_{2S}$ and $P = 8.6898$,

$$H_{2S} = 423.04\ \text{kJ/kg}$$

In the throttle valve, $H_3 = H_4$, so

$$COP = \frac{H_1 - H_4}{(H_{2S} - H_1)/\eta_C} = \frac{392.66 - 249.01}{(423.04 - 392.66)/0.7} = 3.31$$

If T in the refrigerator is 10 °C, then, $T_1 = 10–10 = 0$ °C, then

$$H_1 = 398.6\ \text{kJ/kg and } S_1 = 1.7271\ \text{kJ/(kg K)} \text{ and } H_{2S} = 421.21\ \text{kJ/kg}$$

and

$$COP = \frac{H_1 - H_4}{(H_{2S} - H_1)/\eta_C} = \frac{398.6 - 249.01}{(421.21 - 398.6)/0.7} = 4.63$$

Upon extrapolation

$$T_{ref} = \frac{0 - 10}{3.31 - 4.63}(4 - 3.31) + 0 = 5.22$$

Assuming $T_{ref} = 5$ °C, then, $T_1 = 5–10 = -5$ °C

$$H_1 = 395.66\ \text{kJ/kg and } S_1 = 1.73\ \text{kJ/(kg K)}$$

Again, assuming the compression step is isentropic

$$H_{2S} = 422.11\ \text{kJ/kg}$$

and

$$COP = \frac{H_1 - H_4}{(H_{2S} - H_1)/\eta_C} = \frac{395.66 - 249.01}{(422.11 - 395.66)/0.7} = 3.88$$

The temperature is between 10 and 5 °C.

6.2.5.2 Air Refrigeration Cycle

In this refrigeration cycle, the working fluid is a compressed gas that expands, but it does not change phase. The refrigeration cycle is a *gas cycle*. Air is the most common working fluid; therefore, it is known as the air refrigeration cycle. In this system, the air does not condense, so the condenser and evaporator of a vapor compression cycle are heat exchangers. An schematic diagram of this cycle appears as Figure 6.28.

The air cycle is less efficient than the vapor compression cycle because the air cycle is the reverse Brayton cycle instead of the reverse Rankine cycle. As such, the working fluid does not receive and reject energy transferred as heat at a constant temperature. In the gas cycle, the refrigeration effect equals the product of the specific heat of the gas and the rise in temperature of the gas on the low-temperature side. Therefore, for the same cooling load, a gas refrigeration cycle requires a larger mass flow rate, which makes the unit bulky.

This cycle is very common on gas turbine-powered jet aircraft because compressed air is readily available from the engine compressor sections. These jet aircraft cooling

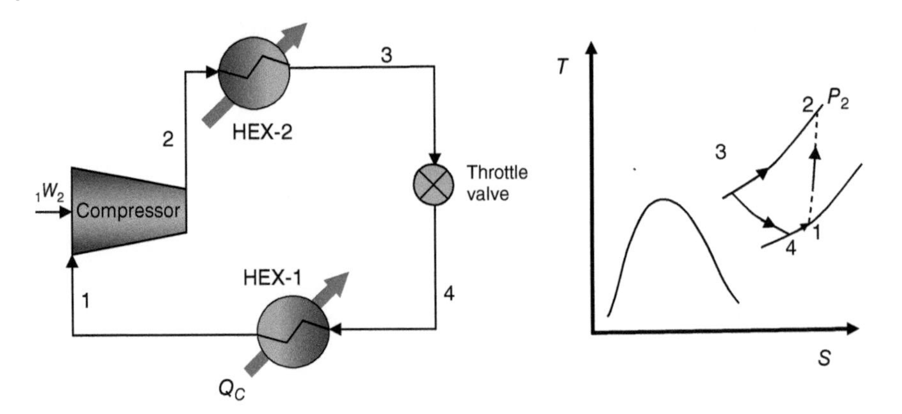

Figure 6.28 Schematic diagram of an air refrigeration cycle.

and ventilation units also serve the purpose of pressurizing the aircraft. Air cycle refrigerator units are not often used.

Using the first law and considering a perfect gas in the heat exchangers

$$\text{Heat exchanger 1: } COP = \frac{H_1 - H_4}{\left(H_{2S} - H_1\right)/\eta_C} = \frac{395.66 - 249.01}{(422.11 - 395.66)/0.7} = 3.88 \tag{6.172}$$

$$\text{Heat exchanger 2: } COP = \frac{H_1 - H_4}{\left(H_{2S} - H_1\right)/\eta_C} = \frac{395.66 - 249.01}{(422.11 - 395.66)/0.7} = 3.88 \tag{6.173}$$

$$\text{Compressor: } COP = \frac{H_1 - H_4}{\left(H_{2S} - H_1\right)/\eta_C} = \frac{395.66 - 249.01}{(422.11 - 395.66)/0.7} = 3.88 \tag{6.174}$$

The *COP* is (for constant heat capacities)

$$COP = \frac{|Q_{in}|}{W_{net}} = \frac{C_P\left(T_1 - T_4\right)}{C_P\left[\left(T_4 - T_1\right) + \left(T_2 - T_3\right)\right]} = \frac{\left(T_1 - T_4\right)}{\left(T_4 - T_1\right) + \left(T_2 - T_3\right)} \tag{6.175}$$

Now, considering the ideal gas in an isentropic process for the compressor and the expansion valve (or turbine),

$$\left(\frac{T_2}{T_1}\right)^{\gamma^{ig}} = \left(\frac{P_2}{P_1}\right)^{\gamma^{ig}-1} = \left(\frac{T_3}{T_4}\right)^{\gamma^{ig}} \Rightarrow \left(\frac{T_2}{T_1}\right) = \left(\frac{T_3}{T_4}\right) \tag{6.176}$$

and we can rearrange the *COP* expression

$$COP = \frac{T_1\left(1 - T_4/T_1\right)}{T_1\left(T_4/T_1 - 1\right) + T_2\left(1 - T_3/T_2\right)} \tag{6.177}$$

then using the above result,

$$COP = \frac{T_1\left(1 - T_4/T_1\right)}{T_1\left(T_4/T_1 - 1\right) + T_2\left(1 - T_4/T_1\right)} = \frac{T_1}{-T_1 + T_2} \tag{6.178}$$

therefore,

$$COP = \frac{T_1}{T_2 - T_1} \tag{6.179}$$

Example 6.10

An air refrigeration cycle compresses air adiabatically from 101 kPa and 540 K to 400 kPa. Air exits HEX-2 at 500 K and expands at 101 kPa. The isentropic efficiency of the compressor is 85% and $C_P^{ig}/R = 3.355 + 0.575 \times 10^{-3}T - 0.016 \times 10^5/T^2$.

Calculate the net energy transfer as work for the cycle, in J/mol
Determine the COP
Compare your COP result with the ideal COP

Solution

Data

$P_1 = P_4 = 101$ kPa, $P_2 = 400$ kPa, $T_1 = 540$ K, $T_3 = 500$ K, $H_3 = H_4$, and $\eta_{comp} = 0.85$.

For an ideal gas, the change in entropy is

$$\Delta S = R\int_{T_1}^{T_2} \frac{C_P^{ig}}{R}\frac{dT}{T} - R\ln\left(\frac{P_2}{P_1}\right)$$

and at the isentropic compressor,

$$0 = \int_{T_1}^{T_2} \frac{C_P^{*}}{R}\frac{dT}{T} - \ln\left(\frac{P_2}{P_1}\right)$$

$$0 = \left\{3.355\ln\frac{T_2}{T_1} + 0.575 \times 10^{-3}\left(T_2 - T_1\right) + \frac{0.016 \times 10^5}{2}\left(\frac{1}{T_2^{\,2}} - \frac{1}{T_1^{\,2}}\right)\right\} - \ln\frac{P_2}{P_1}$$

therefore,

$$0 = \left\{3.355\ln\frac{T_2}{540} + 0.575 \times 10^{-3}\left(T_2 - 540\right) + \frac{0.016 \times 10^5}{2}\left(\frac{1}{T_2^{\,2}} - \frac{1}{540^2}\right)\right\} - \ln\frac{400}{101}$$

Using Solver Excel, the temperature for the isentropic process is

$$T_2 = 781.2\text{ K}$$

Now, the irreversibilities become

$$\left({}_1W_2\right)_S = \Delta H = R\int_{T_1}^{T_2}\left(\frac{C_P^{ig}}{R}\right)dT$$

$$\Delta H = R\left\{3.355\left(T_2 - T_1\right) + 0.575 \times 10^{-3}\left(T_2^{\,2} - T_1^{\,2}\right) + 0.016 \times 10^5\left(\frac{1}{T_2} - \frac{1}{T_1}\right)\right\}$$

$$\Delta H = 8.314\left\{3.355\,(781.2 - 540) + 0.575 \times 10^{-3}\left(781.2^2 - 540^2\right) + 0.016 \times 10^5\left(\frac{1}{781.2} - \frac{1}{540}\right)\right\}$$

$$\left({}_1W_2\right)_S = \Delta H = 8245.3\ \text{J/mol}$$

$${}_1W_2 = \frac{\left({}_1W_2\right)_S}{\eta_{comp}} = \frac{6774.3}{0.85} = 9700.3\ \text{J/mol}$$

We can calculate the actual T_2 using the equation for enthalpy

$$9129.7 = R\left\{3.355\left(T_2 - 510\right) + 0.575 \times 10^{-3}\left({T_2}^2 - 510^2\right) + 0.016 \times 10^5\left(\frac{1}{T_2} - \frac{1}{510}\right)\right\}$$

$$T_2 = 822.2\ \text{K}$$

Then, in the throttling valve, $H_3 = H_4$, and with an ideal gas, $T_3 = T_4$. The heat duties are

$P_1 = P_4 = 101$ kPa, $P_2 = 400$ kPa, $T_1 = 540$ K, $T_3 = 500$ K, $H_3 = H_4$, and $\eta_{comp} = 0.85$.

$$\text{Heat exchanger 1:}\ {}_4Q_1 = \Delta H = \int_{T_4}^{T_1} C_P^{ig}\,dT$$

$$\text{Heat exchanger 2:}\ {}_2Q_3 = \Delta H = \int_{T_2}^{T_3} C_P^{ig}\,dT$$

$${}_4Q_1 = 8.314\left\{3.355\,(540 - 500) + 0.575 \times 10^{-3}\left(540^2 - 500^2\right) + 0.016 \times 10^5\left(\frac{1}{540} - \frac{1}{500}\right)\right\}$$

and

$${}_4Q_1 = 1312.6\ \text{J/mol}$$

This value is the same as for the irreversible or isentropic processes. For the isentropic process,

$$\left({}_2Q_3\right)_S = 8.314\left\{3.355\,(500 - 781.2) + 0.575 \times 10^{-3}\left(500^2 - 781.2^2\right) + 0.016 \times 10^5\left(\frac{1}{500} - \frac{1}{781.2}\right)\right\}$$

$$\left({}_2Q_3\right)_S = -9557.9\ \text{J/mol}$$

and

$$COP = \frac{|Q_{in}|}{W_{net}} = \frac{1312.6}{8245.3} = 0.159$$

For the irreversible process,

$${}_2Q_3 = 8.314\left\{3.355\,(500 - 822.2) + 0.575 \times 10^{-3}\left(500^2 - 822.2^2\right) + 0.016\right.$$

$$\times 10^5 \left(\frac{1}{500} - \frac{1}{822.2} \right) \Big\}$$
$$= -11013 \text{ J/mol}$$

and the *COP* is

$$COP = \frac{|Q_{in}|}{W_{net}} = \frac{1312.6}{8245.3} = 0.135$$

6.2.5.3 Absorption Refrigeration

An absorption refrigerator is almost identical to a vapor compression refrigerator with the difference being that a heat engine replaces the compressor. Absorption refrigerators are an alternative when electricity is unreliable, costly, or unavailable or when surplus energy transferred as heat is available (e.g. from turbine exhaust, or industrial processes, or from solar plants). Absorption and compressor refrigerators use a refrigerant with a very low boiling point. In both types, when this refrigerant evaporates (boils), it removes some energy transferred as heat with it providing a cooling effect. The main difference between the two types of refrigerators is the way the refrigerant changes from a gas back into a liquid so that the cycle can repeat. In vapor compressor cycles, refrigerators typically use an HCFC (hydrochlorofluorocarbon) or HFC (hydrofluorocarbon) as the refrigerant, while absorption refrigerators typically use ammonia.

Figure 6.29 is a schematic of an absorption refrigerator. The blue section (to the right) is the same as in a vapor compression cycle, but the left-hand side replaces the compressor using a nonvolatile liquid solvent to absorb the vapor refrigerant at the pressure of the evaporator and at low temperature. In this process, the energy transferred as heat passes to the surroundings at the liquid temperature (this is the lower temperature T_C section of the heat engine):

- The liquid solution exits the absorber as a refrigerant and passes to a heat exchanger at the pressure of the condenser.
- The heat exchanger increases the temperature of the compressed liquid.

At the regenerator, the energy transferred as heat enters the solution and evaporates the refrigerant from the solvent (this is the high-temperature T_H section of the heat engine). The vapor refrigerant separates from the solution allowing an almost pure vapor refrigerant to pass to the condenser, while the hot solvent returns to the absorber passing first through the heat exchanger to adjust the steam temperatures.

If we consider two Carnot cycles, an engine and a refrigerator, illustrated in Figure 6.30, then

$$\text{Carnot engine}: \frac{W}{Q_H} = \frac{T_H - T_{S-2}}{T_H} \tag{6.180}$$

$$\text{Carnot refrigerator}: \frac{Q_C}{W} = \frac{T_C}{T_{S-1} - T_C} \tag{6.181}$$

Combining these equations,

$$\frac{Q_C}{Q_H} = \left(\frac{T_C}{T_{S-1} - T_C} \right) \left(\frac{T_H - T_{S-2}}{T_H} \right) \tag{6.182}$$

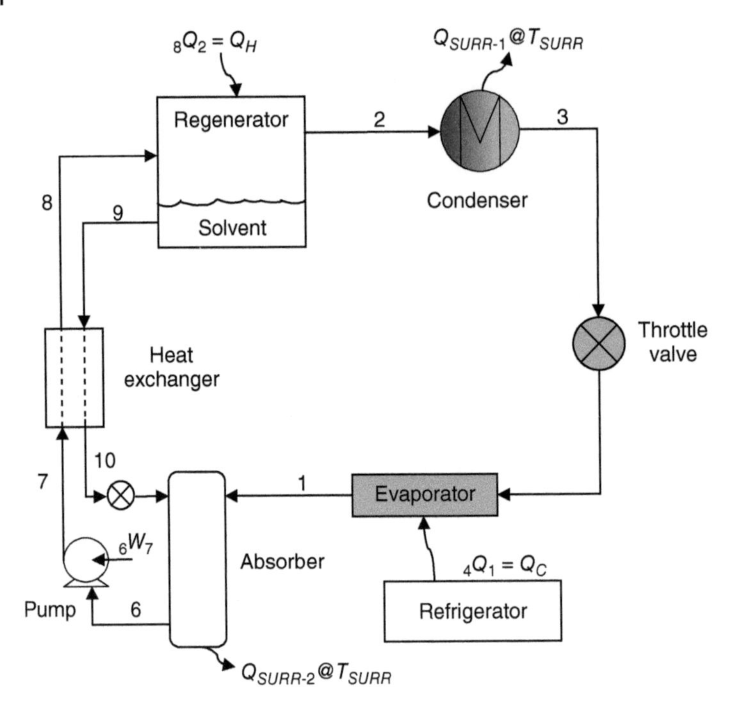

Figure 6.29 Schematic of an absorption refrigeration cycle.

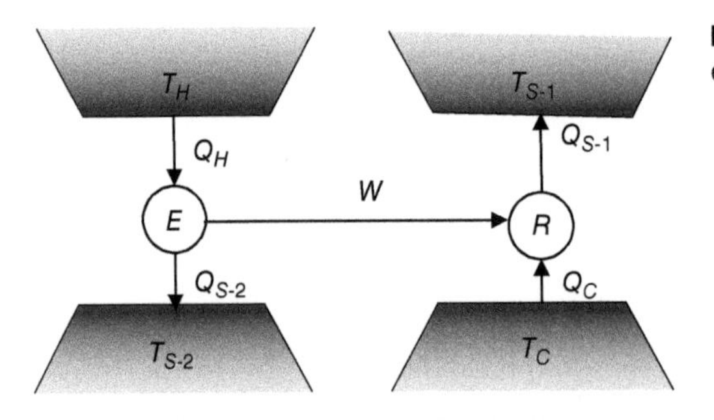

Figure 6.30 Diagram with an engine and a refrigerator.

An absorption refrigerator has three basic temperatures: the regenerator temperature, (T_H), the ambient temperature, $T_{S\text{-}1}$, and the temperature of the refrigerator, T_C. The condenser and the absorber temperatures of the refrigerant are usually about $T_{S\text{-}2} + 5\,°\text{C}$, whereas the refrigerant temperature is $T_C = 5\,°\text{C}$. The flow rates of the rich and lean refrigerant solutions, $\dot{m}_7$ and $\dot{m}_9$, come from material balances around either the regenerator or the absorber. For the regenerator,

$$\dot{m}_7 = \dot{m}_9 + \dot{m}_2 \tag{6.183}$$

and a component material balance provides

$$\dot{m}_7 x_7 = \dot{m}_9 x_9 + \dot{m}_2 \tag{6.184}$$

in which x_7 and x_9 are refrigerant compositions, and the refrigerant is "pure" in steam 2.

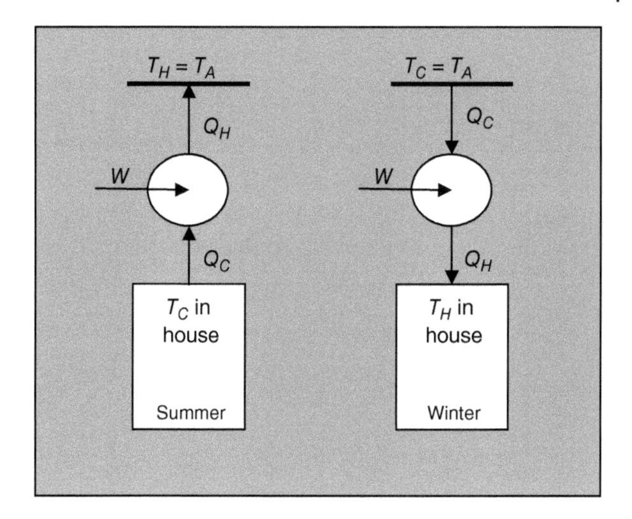

Figure 6.31 Heat pump diagram.

6.2.5.4 Heat Pump

The heat pump is a refrigerator that transfers energy as heat from a low temperature, T_C, to a high temperature, T_H (Figure 6.31). It can provide energy transferred as heat to a house in the winter and extract it in the summer. The energy transferred as heat passes from the house to the ambient temperature, T_A (usually the ground temperature) during the summer, while the opposite occurs during the winter. The desired home temperature is the same in both cases, but T_A is higher in the summer.

It is important to realize that a heat pump always acts as a refrigerator and never as a heat engine because of the difference in efficiencies. For example, if $T_A = 35\,°\mathrm{C}$ in the summer and $-5\,°\mathrm{C}$ in the winter and the desired house temperature is $21\,°\mathrm{C}$, then

$$(COP)_{\max,summer} = \left(\frac{Q_C}{W}\right)_{\max,summer} = \frac{T_C}{T_A - T_C} = \frac{294.15}{308.15 - 294.15} = 21$$

but

$$\left(\frac{Q_H}{W}\right)_{\max,winter} = \frac{T_H}{T_H - T_C} = \frac{294.15}{294.15 - 268.15} = 11.3$$

For the same heat pump, the maximum value of W would be the same for both the summer and the winter. Then, the above ratios show that the heat pump is more efficient as a refrigerator than a heater. A complete diagram of a heat pump is similar to the air refrigeration cycle, but it can run in either direction.

6.2.5.5 Liquefaction Process

Liquefaction in this chapter refers to the process of condensing a gas into a liquid. The process most often uses condensation from cooling. Liquefied gases are useful for scientific, industrial, and commercial purposes. The liquefaction process in this section is important for liquefaction of air, nitrogen, propane, oxygen, hydrogen, helium, etc. Liquid propane is stored and transported in cylinders before its use as a fuel in households. Liquid oxygen is valuable in hospitals, where it can assist

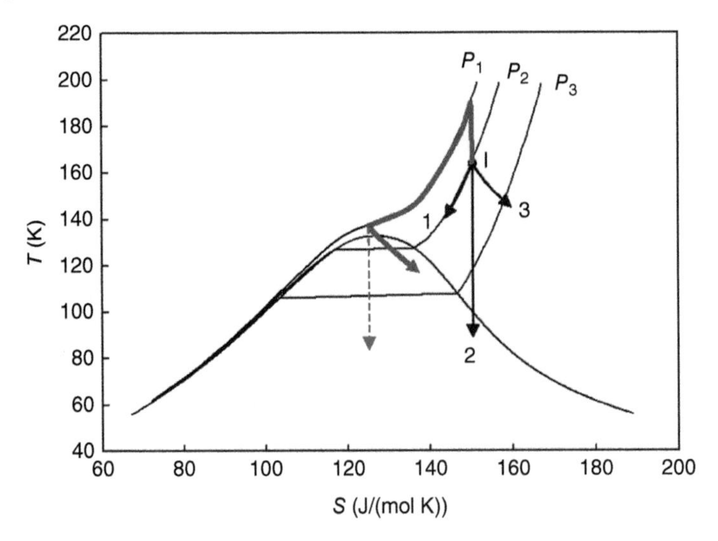

Figure 6.32 Cooling process for liquefaction.

patients with breathing problems. Liquid nitrogen is useful in microsurgery or as a cooling media in many research laboratories. Liquefied chlorine finds use in water purification and sanitation of industrial waste. Liquefaction of air provides nitrogen, oxygen, and argon and other atmospheric noble gases after separating the air components by fractional distillation in a cryogenic air separation unit.

The cooling process in liquefaction uses three steps:

1. Cooling at a constant pressure (heat exchanger).
2. Cooling at a constant entropy (expansion engine).
3. Cooling at a constant enthalpy (throttling valve).

Figure 6.32 presents these paths on a T–S diagram for air. The constant pressure process reaches the two-phase region producing liquid by reducing the temperature. Also, liquefaction could result using an expansion engine, but that would require much lower temperatures, as is apparent in the figure. Liquefaction at a constant enthalpy requires high enough temperature and pressure that the path crosses into the two-phase region. To produce liquid (see the red line in the figure), we must increase the initial pressure using a compressor (isentropic), followed by cooling at a constant pressure to cross into the two-phase region and introduce some liquid. Sometimes, using an expander can produce more liquid. The advantage of using an expansion valve is that it does not require temperatures as low as the expansion engine, and it is much less expensive.

The liquefaction process using an expansion valve is a Linde or Joule–Thomson process. Figure 6.33 shows this process in which the gas incurs

- compression at ambient pressure
- pre-cooling and then additional cooling in a heat exchanger
- expansion with a throttling valve (isenthalpic Joule–Thomson expansion) producing some liquid

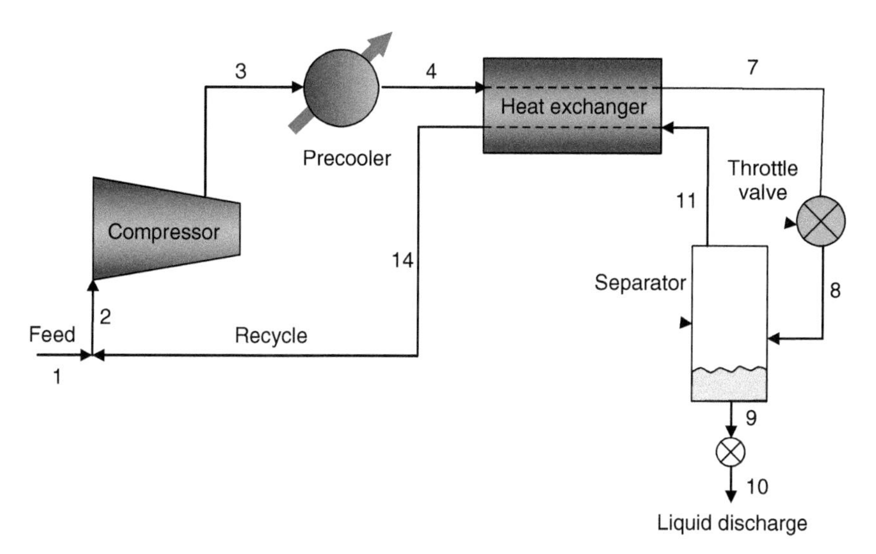

Figure 6.33 Diagram for the Linde liquefaction process.

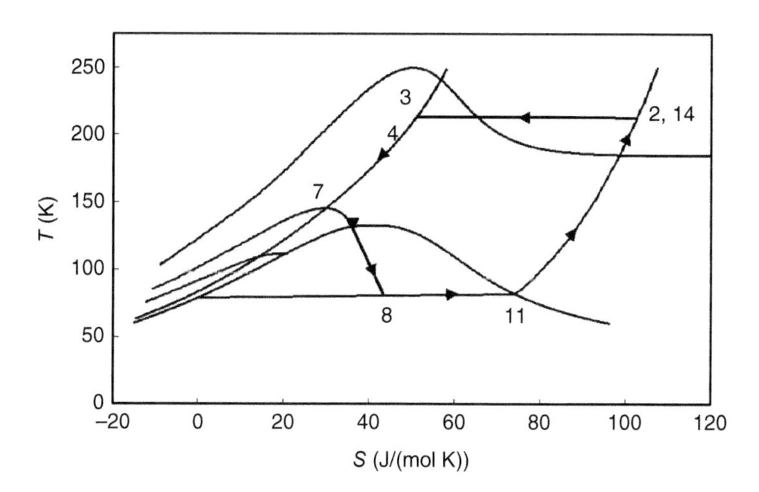

Figure 6.34 *T–S* diagram for the Linde liquefaction process.

- removal of liquid and return of the cool to the compressor via the heat exchanger.

Figure 6.34 is a *T–S* diagram for the Linde process. The throttle valve could be an expander if it does not operate in the two-phase region (that would be inefficient). The Claude method uses this idea. A portion of the gas from the heat exchanger passes to an expander and then mixes with the separator vapor. This gas is a saturated or superheated vapor that can recycle through the heat exchanger to the compressor inlet. The remaining gas goes through the normal process of liquefaction. Figure 6.35 illustrates the Claude process.

Applying the first law to the heat exchanger I and the separator,

$$_{15}\dot{W}_{16} = \dot{m}_{14}H_{14} + \dot{m}_{10}H_{10} - \dot{m}_4H_4 \tag{6.185}$$

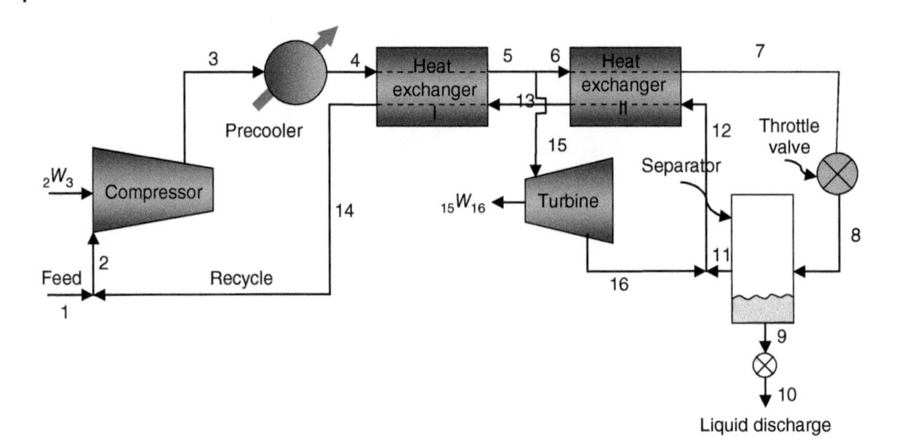

Figure 6.35 Diagram for the Claude liquefaction process.

Now, applying the first law to the turbine

$$ {}_{15}\dot{W}_{16} = \dot{m}_{15}\left(H_{16} - H_5\right) \tag{6.186} $$

Material balances around heat exchanger I and the separator provide

$$ \dot{m}_4 = \dot{m}_{14} + \dot{m}_{10} \tag{6.187} $$

Substituting Eqs. (6.186) and (6.187) into Eq. (6.185)

$$ \dot{m}_{15}\left(H_{16} - H_5\right) = \left(\dot{m}_4 - \dot{m}_{10}\right) H_{14} + \dot{m}_{10} H_{10} - \dot{m}_4 H_4 \tag{6.188} $$

and dividing by $\dot{m}_4$

$$ \left(\frac{\dot{m}_{15}}{\dot{m}_4}\right)\left(H_{16} - H_5\right) = \left(\frac{\dot{m}_4 - \dot{m}_{10}}{\dot{m}_4}\right) H_{14} + \left(\frac{\dot{m}_{10}}{\dot{m}_4}\right) H_{10} - H_4 \tag{6.189} $$

If we define $\ell = \left(\frac{\dot{m}_{10}}{\dot{m}_4}\right)$ as the liquefied fraction and $w = \left(\frac{\dot{m}_{15}}{\dot{m}_4}\right)$ as the withdrawn vapor fraction, then

$$ w\left(H_{16} - H_5\right) = (1 - \ell) H_{14} + \ell H_{10} - H_4 \tag{6.190} $$

The Linde process is a limiting case of the Claude process when $W = 0$. Figure 6.36 is a T–S diagram for the Claude process.

Example 6.11

A Claude process is useful for liquefying propane for transportation. Propane enters the compressor at 1 bar, leaves at 100 bar, and pre-cooling is at 300 K. Propane leaves the throttling valve at 1 bar. The expander exit is saturated vapor propane at 1 bar. The recycle steam leaves the exchanger at 298 K. The efficiency of the turbine is 0.75. Plot the amount of liquefaction vs the withdrawn quantity and specify the maximum withdrawn quantity.

Extra data

At 100 bar

$$ H = 157.6 - 2.922T + 1.401 \times 10^{-2} T^2 - 1.0065 \times 10^{-5} T^3 \tag{A} $$

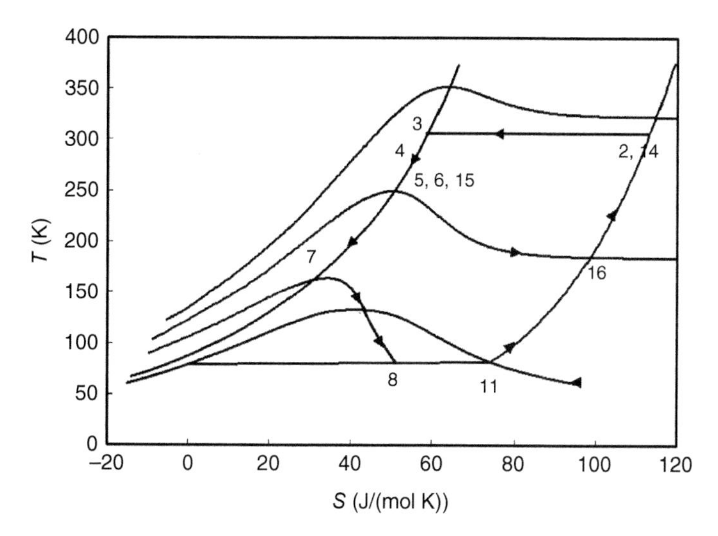

Figure 6.36 $T-S$ diagram for air and paths of the Claude liquefaction process.

$$S = -2.0991 + 1.28 \times 10^{-2}T - 5.848 \times 10^{-6}T^2 \tag{B}$$

At 1 bar

$$H = 561.91 - 2.477S + 9.54S^2 \tag{C}$$

and

$$H = 304.056 + 0.483T + 2.05 \times 10^{-3}T^2 \tag{D}$$

In Eqs. (A)–(D), the enthalpy is in kJ/kg; the entropy is kJ/(kg K); and T is in kelvin.

$$T^{sat} = 230.74\ \text{K}, H^v = 525.29\ \text{kJ/kg}, H^l = 99.686\ \text{kJ/kg, and } S^v = 2.45\ \text{kJ/(kg K)}$$

Solution

Following Figure 6.35, we have

$$P_2 = 1\ \text{bar}$$

$$P_3 = 100\ \text{bar}$$

$$P_4 = P_5 = P_6 = P_{15} = P_7 = 100\ \text{bar}$$

$$T_4 = 300\ \text{K}$$

$$P_{14} = 1\ \text{bar}$$

$$T_{14} = 298\ \text{K}$$

$$P_{16} = 1\ \text{bar}$$

$$P_8 = 1\ \text{bar}$$

From Eq. (A) and 300 K,

$$H_4 = 269.6\ \text{kJ/kg}$$

at Point 14 (using Eq. (A))

$$H_{14} = 630.04\ \text{kJ/kg}$$

At saturation (1 bar),

$$T = 230.74\ \text{K}, H^v = H_{16} = 525.29\ \text{kJ/kg, and } S^v = S_{16} = 2.45\ \text{kJ/(kg K)}$$

$$H_9 = H^l = 99.686\ \text{kJ/kg}$$

At the turbine,

$$\left({}_{15}W_{16}\right)_S = \left(H_{16S} - H_5\right)$$

and

$${}_{15}W_{16} = \left(H_{16S} - H_5\right) \times \eta_t = \left(H_{16S} - H_5\right) \times 0.75 = H_{16} - H_5$$

Now,

$${}_{15}W_{16} = \left(H_{16S} - H_5\right) \times \eta_t = \left(H_{16S} - H_5\right) \times 0.75 = H_{16} - H_5$$

and

$$H_{16} = \left(H_{16S} - H_5\right) \times 0.75 + H_5 = 525.29$$

For the isentropic process, $S_5 = S_{16}$. Now, we can solve for T_5 using the above equations with

$$H_5 = 157.6 - 2.922T_5 + 1.401 \times 10^{-2}{T_5}^2 - 1.0085 \times 10^{-5}{T_5}^3$$

and

$$H_{16S} = 561.91 - 2.477S_{16} + 9.54S_{16}^2$$

with

$$S_5 = -2.0991 - 1.28 \times 10^{-2}T_5 - 5.848 \times 10^{-6}{T_5}^2$$

The temperature comes from using SOLVER in Excel

$$T_5 = 331.548\ \text{K and } H_5 = 362.033\ \text{kJ/kg}$$

The following procedure can produce the temperature from property tables:

1. Assume T_5
2. Find with T_5 and $P = 100$ bar, H_5, S_5
3. With S_5 and 1 bar, find H_{16S}, then calculate H_{16}. If the above equation is satisfied, the temperature is correct, if not, assume another temperature.

The liquefied fraction comes from

$$w\left(H_{16} - H_5\right) = (1 - \ell)H_{14} + \ell H_{10} - H_4$$

because $H_9 = H_{10} = 99.686$ kJ/kg.

Now,

$$w\left(H_{16} - H_5\right) + H_4 - H_{14} = \ell\left(H_{10} - H_{14}\right)$$

with

$$\ell = \frac{w\left(H_{16} - H_5\right) + H_4 - H_{14}}{\left(H_{10} - H_{14}\right)}$$

The maximum amount of liquefaction occurs when everything that enters the separator leaves as a liquid. Using the material balance around the separator,

$$\dot{m}_{15} + \dot{m}_{11} = \dot{m}_{12} = \dot{m}_{14}$$

and dividing by $\dot{m}_4$,

$$\frac{\dot{m}_{15}}{\dot{m}_4} + \frac{\dot{m}_{11}}{\dot{m}_4} = \frac{\dot{m}_{14}}{\dot{m}_4} \Rightarrow w + \frac{\dot{m}_{11}}{\dot{m}_4} = 1 - \ell$$

The maximum withdrawal ratio occurs when $\dot{m}_{11}/\dot{m}_4 = 0$ and

$$w = 1 - \ell \Rightarrow w + \ell = 1$$

The final plot is

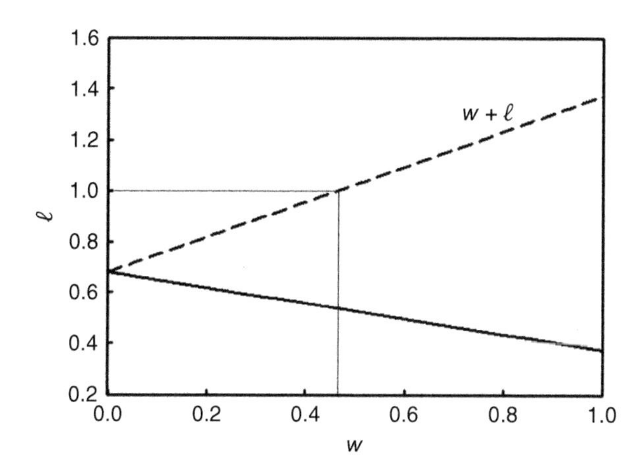

The maximum withdrawn vapor ratio is $w = 0.436$.

6.2.6 Process Simulators: Using Process Simulation for Fluid Flow Problems

The preceding plethora of equations and processes are all valid within their assumptions, but we now have a much more practical option to make the calculations – use of process simulators. You will learn much more about these tools in later classes, but we can "dip our toes" to get a taste of things to come. Let us return to compressor problems under various situations. Although several excellent process simulators are available to the industry (and academia), we shall use in this text the ProMax simulator offered by Bryan Research & Engineering because it is easy to apply and because the authors have ready access to it. We shall provide examples using an ideal gas as the process fluid, but we shall also examine real fluids. For process simulators, either fluid requires essentially equal efforts.

Process simulators, such as ProMax, apply to open systems. Unfortunately, it is extremely difficult to model closed systems such as the piston and cylinder compression example that your text covers initially. However, ProMax does model open system compressors. Process simulators apply to real-world industrial operations. That said, using the ideal gas law to demonstrate a compressor application is informative. Your text has investigated compression of an ideal gas in a closed piston/cylinder apparatus for isothermal and adiabatic compression that provides:

$$-W_{isothermal} = RT\ln\left(\frac{P_1}{P_2}\right) = RT\ln\left(\frac{V_2{}^{ig}}{V_1{}^{ig}}\right) \quad (3.53)$$

$$W_{adiabatic} = \Delta U^{ig} = \int C_V{}^{ig} dT \quad (3.55)$$

Although simulators normally deal with real gases in compressors, it is possible in ProMax to use an ideal gas as the working fluid. An open system compressor can also be adiabatic ($Q = 0$) or isothermal ($\Delta T = 0$). The following adiabatic compressor (Figure 6.37) compresses an ideal gas from 1 to 3 MPa, causing a temperature rise to satisfy the first law. Assuming the compressor is 100% efficient and adiabatic, the energy required as work to compress the ideal gas is 3109 J/mol. This example required about five minutes to construct, and it ran to completion in less than a second. The more complicated examples in this chapter might require more time to construct (although probably less than 10–15 minutes), but the run time would still be less than a minute.

In this example, the compressor is 100% efficient. A 100% efficient compressor would mean that it is a "perfect" piece of equipment with zero nonidealities, but the equipment is never 100% efficient. As the efficiency decreases, the amount of energy required as work increases, but the process remains adiabatic. The following example shown in Figure 6.38 uses a 75% efficient compressor. The inlet conditions are the same as in the previous case, and the final pressure is still 3 MPa. The energy required as work to compress the ideal gas in this compressor is 4310 J/mol. The ties required for construction and running remain essentially the same (for all examples in this section).

It is also possible to use ProMax to describe an isothermal compression. As compression occurs, the temperature increases, but it is possible to add a cooler to the simulation and return the temperature to the inlet temperature. Because internal

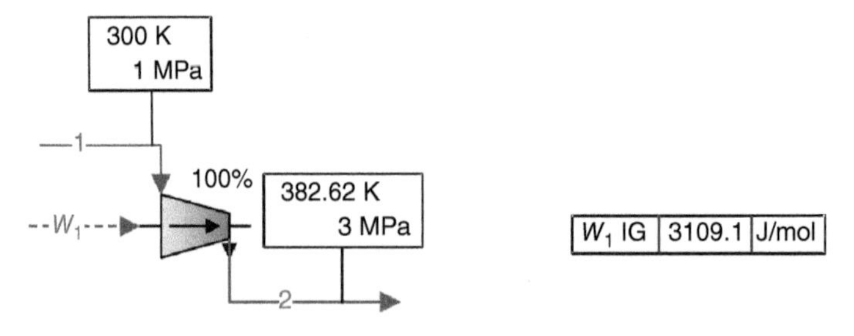

Figure 6.37 100% efficient, adiabatic compression of an ideal gas.

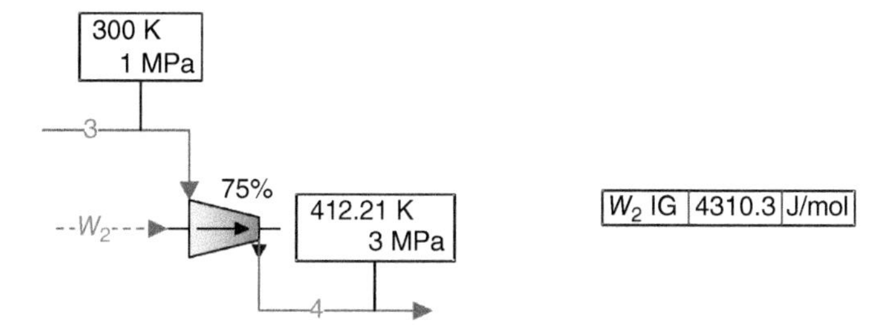

Figure 6.38 75% efficient, adiabatic compression of an ideal gas.

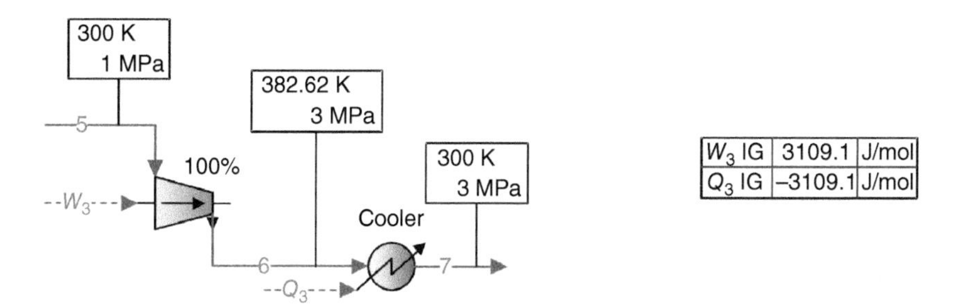

Figure 6.39 100% efficient, isothermal compression of an ideal gas.

energy and enthalpy are functions of only temperature for an ideal gas and the overall change in temperature is zero, the overall changes in internal energy and enthalpy are also zero for an isothermal process.

This first process in Figure 6.39 shows an overall isothermal process with a 100% efficient compression of the ideal gas. The energy required as work and the energy released as heat are equal and opposite.

The next example simulates isothermal compression of the ideal gas with a 75% efficient compressor as shown in Figure 6.40. In addition, there is a pressure drop of 0.1 MPa in the exchanger (this pressure drop does not affect the final results). The amounts of energy required as work and removed as heat again are equal, and the total change in energy is zero. This would be the case for all efficiencies and/or pressure drops for an ideal gas.

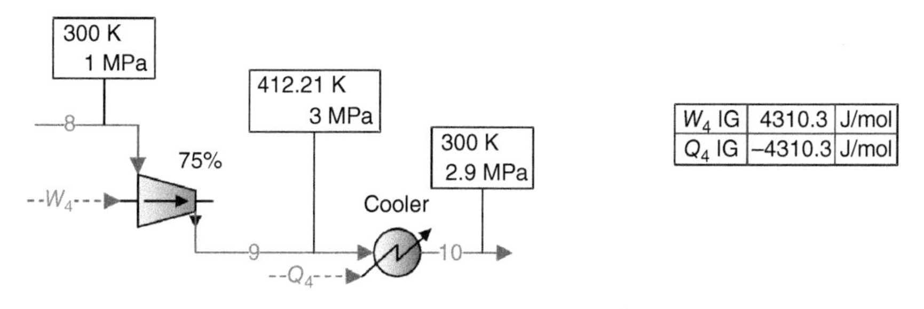

Figure 6.40 75% efficient, isothermal compression of an ideal gas.

Had the ideal gas compression been in a piston/cylinder apparatus, the isothermal and adiabatic work would have been:

$$-W_{isothermal} = RT\ln\left(\frac{P_1}{P_2}\right) = RT\ln\left(\frac{V_2{}^{ig}}{V_1{}^{ig}}\right) \qquad (3.53)$$

$$T = 300\ \text{K}$$

$$P_1 = 1\ \text{MPa}$$

$$P_2 = 3\ \text{MPa}$$

$$W_{isothermal} = -(8.314\ \text{J/(mol K)})(300\ \text{K})\ln\left(\frac{1\ \text{MPa}}{3\ \text{MPa}}\right)$$

$$W_{isothermal} = 2740.2\ \text{J/mol}$$

$$W_{adiabatic} = \Delta U^{ig} = \int C_V^{ig} dT \qquad (3.55)$$

$$T_2 = T_1\left(\frac{P_1}{P_2}\right)^{\frac{1-\gamma^{ig}}{\gamma^{ig}}} \qquad (3.64)$$

but $\gamma^{ig} = 1.3$ for simple polyatomic gases

$$T_2 = (300\ \text{K})\left(\frac{1\ \text{MPa}}{3\ \text{MPa}}\right)^{\frac{1-1.3}{1.3}} = 386.57\ \text{K}$$

Now using

$$C_V^{ig} = C_P^{ig} - R = \frac{C_P^{ig}}{R} - 1 \qquad (3.41)$$

then

$$C_V^{ig} = 4.217R - R = 3.217R \qquad (6.191)$$

and the adiabatic work

$$W_{adiabatic} = (26.746\ \text{J/(mol K)})(386.57 - 300\ \text{K}) = 2315.37\ \text{J/mol}$$

Previously, we use the ideal gas to obtain thermodynamic information, but it is always more accurate to use a real gas. This is not easy to do by hand, but in ProMax, it is easy to change the EOS to a real gas form. The tool has nine EOS available. For current purposes, it can be instructive to compare the Peng–Robinson (PR) and Groupe Européen de Recherches Gazières (GERG) equations of state for pure methane. Peng–Robinson is a well-known and relatively simple cubic EOS. The GERG equation, while complicated, is very accurate when dealing with natural gas and its components (methane being a major component of natural gas).

The first two configurations below (Figures 6.41 and 6.42) use the GERG property package. The first case demonstrates a 100% efficient adiabatic process. ProMax automatically accounts for the properties of the real gas and calculates the energy required as work or heat. A 100% efficient adiabatic compressor requires 3065.5 J/mol to compress methane.

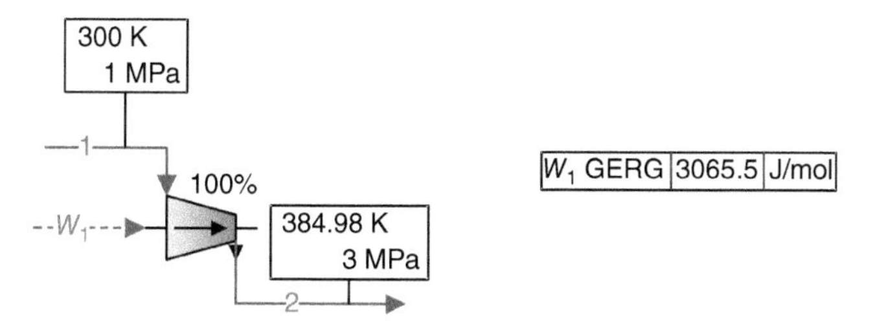

Figure 6.41 100% efficient, adiabatic compression of a real gas (GERG).

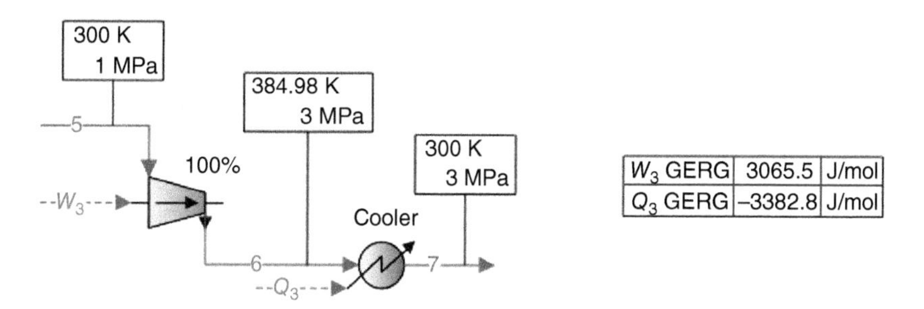

Figure 6.42 100% efficient, isothermal compression of a real gas (GERG).

Because the process fluid is not an ideal gas, the energy required as work and the energy released as heat are not equal in the isothermal arrangement. The process below is a 100% efficient compressor, and there is no pressure drop through the exchanger.

The next two configurations in Figures 6.43 and 6.44 show the same processes as above but use the Peng–Robinson EOS. The explanation for each configuration is identical to the GERG section. The first compressor is 100% efficient and adiabatic.

The next configuration contains a 100% efficient compressor and a cooler with a 0.1 MPa pressure drop. A schematic diagram is shown in Figure 6.44.

Table 6.1 compares the two configurations for the three equations of state (Ideal gas is from before). The results vary greatly from the ideal gas to the real gas equations of state, but the Peng–Robinson and GERG equations produce reasonably similar results. Both equations are accurate for methane at these conditions, but GERG is more accurate, while requiring more time to solve.

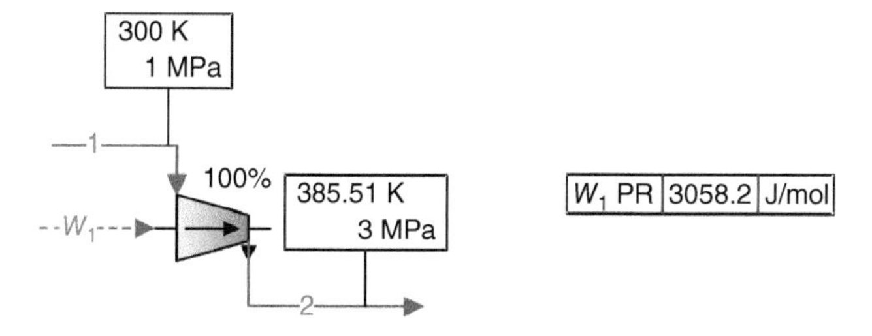

Figure 6.43 100% efficient, adiabatic compression of a real gas (PR).

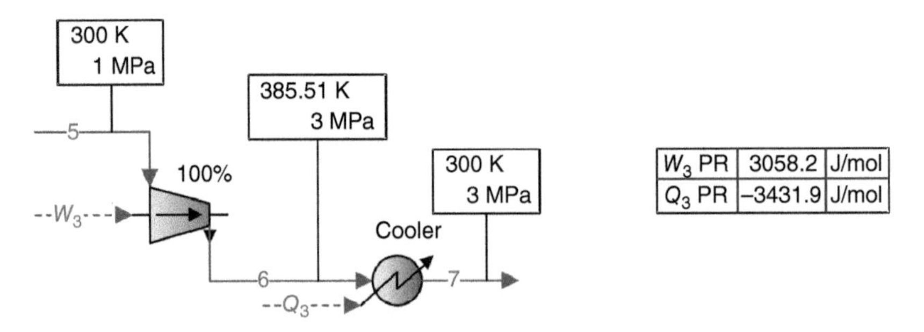

Figure 6.44 100% efficient, isothermal compression of a real gas (PR).

Table 6.1 Comparing 100% efficient compressors.

			Ideal gas	GERG	PR
	Efficiency	***W* or *Q***	**J/mol**	**J/mol**	**J/mol**
Adiabatic	100%	*W*	3109.1	3065.5	3058.2
Isothermal	100%	*W*	3109.1	3065.5	3058.2
		Q	−3109.1	−3382.8	−3431.9

Previously, compressors are 100% efficient, which render them adiabatic and isentropic. As the efficiency decreases, the amount of energy required as work increases, but the process is still adiabatic. The following examples use a 75% efficient compressor with ideal and real gas equations of state. The inlet conditions are the same as the previous cases and the final pressure is still 3 MPa.

For comparative purposes, the first two examples (Figures 6.45 and 6.46) use the ideal gas EOS. These processes simulate isothermal compression of an ideal gas with a 75% efficient compressor. In addition, there is a pressure drop of 0.1 MPa in the exchanger, but this pressure drop does not affect the final results. The amount of energy required as work and removed as heat are again equal and the total change in energy is zero. This would be the case for all efficiencies and/or pressure drops for an ideal gas.

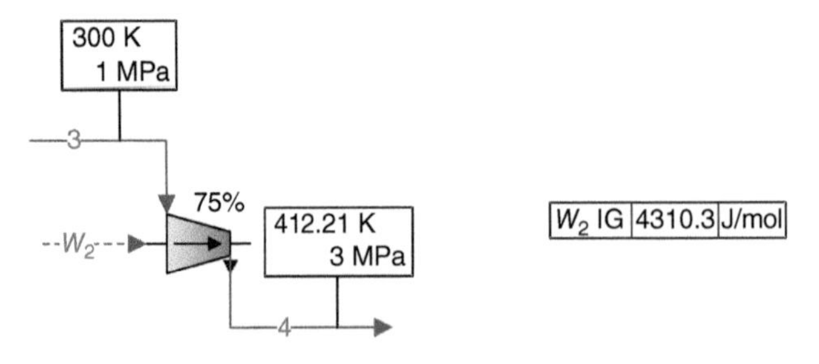

Figure 6.45 75% efficient, adiabatic compression of an ideal gas.

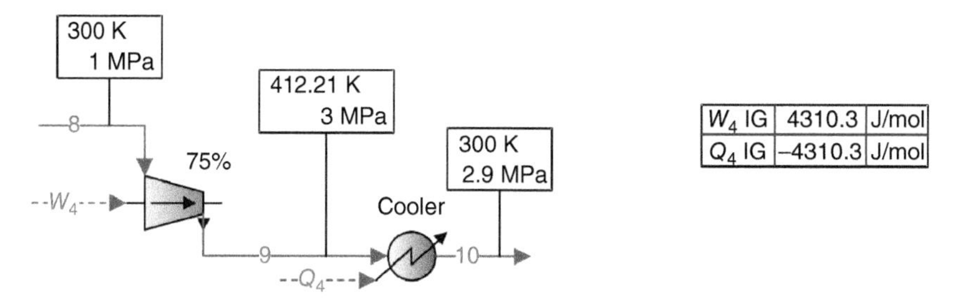

Figure 6.46 75% efficient, isothermal compression of an ideal gas.

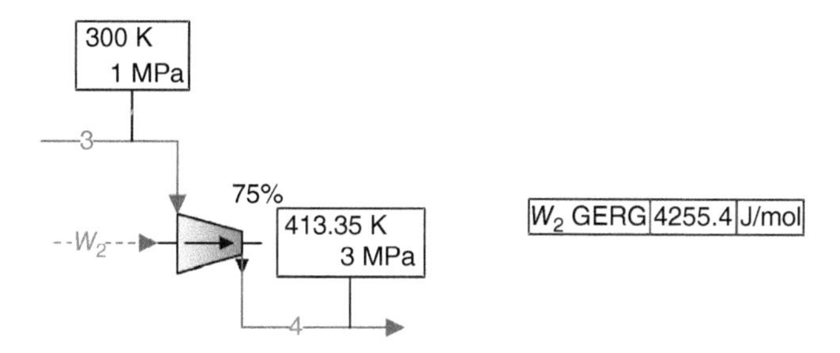

Figure 6.47 75% efficient, adiabatic compression of a real gas (GERG).

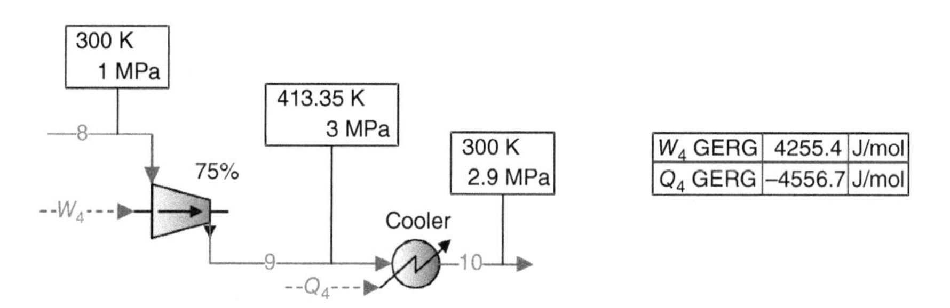

Figure 6.48 75% efficient, isothermal compression of a real gas (GERG).

The next two configurations, Figures 6.47 and 6.48, show the same processes as above, but use the GERG EOS. A 75% efficient adiabatic compressor requires 4245 J/mol to compress methane:

The next process utilizes a 75% efficient compressor with a pressure drop through the exchanger (Figure 6.48). The energy required as work and the energy released as heat are not equal because methane is a real gas. However, the energy required and released is greater because of the lower efficiency in the compressor (causing a greater temperature increase).

The next two configurations in Figures 6.49 and 6.50 are the same as those shown above, but use the Peng–Robinson EOS. The first compressor is 75% efficient and adiabatic.

The next configuration contains a 75% efficient compressor and a cooler with a 0.1 MPa pressure drop.

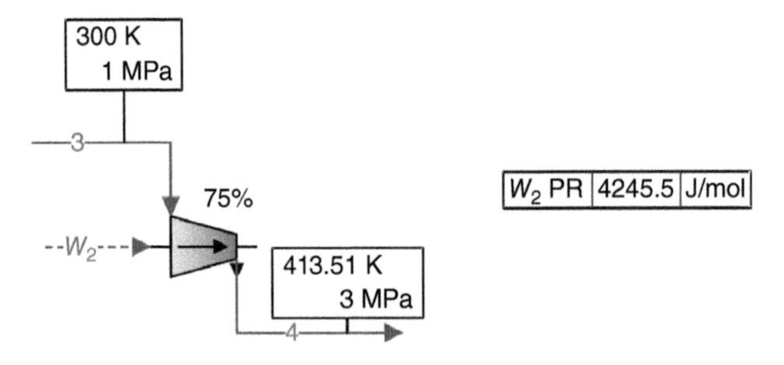

Figure 6.49 75% efficient, adiabatic compression of a real gas (PR).

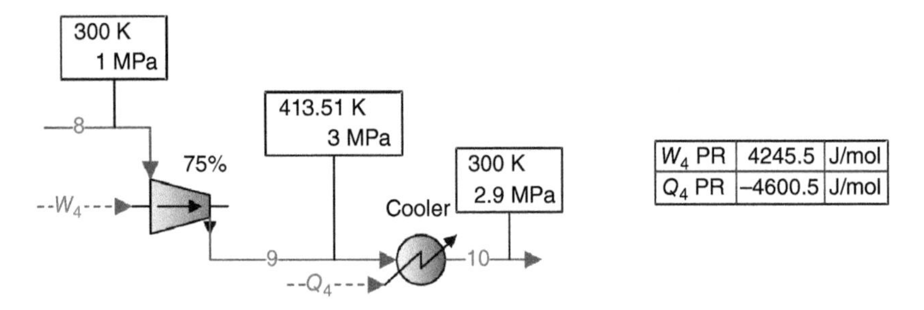

Figure 6.50 75% efficient, isothermal compression of a real gas (PR).

Table 6.2 Comparing 100% and 75% efficient compressors.

			Ideal gas	**GERG**	**PR**
	Efficiency	***W* or *Q***	**J/mol**	**J/mol**	**J/mol**
Adiabatic	100%	*W*	3109.1	3065.5	3058.2
	75%	*W*	4310.3	4255.4	4245.5
Isothermal	100%	*W*	3109.1	3065.5	3058.2
		Q	−3109.1	−3382.8	−3431.9
	75%	*W*	4310.3	4255.4	4245.5
		Q	−4310.3	−4556.7	−4600.5

Table 6.2 compares the four configurations for the three equations of state (including previous results). The results vary greatly from the ideal gas to the real gas equations of state, but the Peng–Robinson and GERG equations produce similar results. This is because both equations are accurate for methane at these conditions. Generally, GERG is more accurate (but it requires more time to solve).

Process simulators can handle any of the processes in this chapter, and do it much faster than the hand calculations we showed. You will use these tools extensively in courses later in the curriculum.

Problems for Chapter 6

6.1 An ideal gas passes through a horizontal pipe. Consider adiabatic flow to calculate the speed of sound of the ideal gas at T = 300 K and 1 bar with M = 40 g/gmol and $C_P^{ig}/R = 7/2$.

6.2 Steam passes through an orifice in an adiabatic, steady-state process. What is the exit velocity of the steam if the inlet conditions are T_1 = 300 °C and P_1 = 500 kPa and the exit conditions are T_2 = 400 °C and P_2 = 3000 kPa. $\eta = 0.5$.

6.3 A turbine receives steam at 5000 kPa and 600 °C. The steam exits the turbine at 200 kPa. Calculate the reversible work.

6.4 Calculate the pattern efficiency of a compressor that delivers air at 1000 kPa after receiving it at 20 °C and 100 kPa. The actual work done by the compressor is 12 kJ/mol and $C_P^{ig}/R = 3.5$.

6.5 Irreversibility is $i = T_0\Delta S - Q$, where T_0 is the ambient temperature. Calculate the irreversibility for an ideal gas that enters an adiabatic compressor at 101.325 kPa and 298.15 K and leaves at 350 kPa and 500 K. Consider $C_P^{ig}/R = 2.5$.

6.6 A venturi meter operates by measuring the difference in pressure between two locations with different cross-sectional areas for flow. Calculate the mass flow rate and the pressure difference between a point at which the pipe has a cross-sectional area of 40 cm^2 and a point inside the meter where the cross-sectional area is 20 cm^2, when the linear velocity of liquid water in the pipe is 1.5 m/s. Assume that no energy is transferred as heat and that the flow is isentropic. The temperature of the flowing water is 38 °C and the pressure is 3 bar.

6.7 An incompressible liquid with a density of 885 kg/m^3, a heat capacity, C_P, of 4.2 kJ/(kg K), and a thermal expansion, β, of 6.75×10^{-4} K^{-1} flows through a pipe as shown in the diagram. At location 1, the temperature is 35 °C, the pressure is 486 kPa, and the linear velocity is 10.6 m/s. The internal diameter is 5.0 cm at location 1 and 4.0 cm at location 2.

Calculate the difference between the pressures at locations 1 and 2, $\Delta P = P_2 - P_1$, and the difference in temperatures $\Delta T = T_2 - T_1$. Which of these two differences would be more useful as a measurement of flow? You may assume that the flow is isentropic.

6.8 Determine the amount of work required per mole and the outlet temperature for the following processes that compresses methane from 300 K and 2 to 20 bar. All compressors operate with pattern efficiencies of 0.85.

a. The compression occurs in a single stage in which the pressure increases from 2 to 20 bar.
b. A two-stage compression process where the methane is compressed from 2 to 7 bar in the first stage, then cooled back to 300 K at a constant pressure in an intercooler between the first and second stage, and finally compressed from 7 to 20 bar in the second stage.

Data

$$T_C = 190.564\text{ K},\quad P_C = 45.992\text{ bar},\quad \omega = 0.011$$

$$C_P^{ig}/R = 1.7025 + 9.0861 \times 10^{-3}T - 21.6528 \times 10^{-7}T^2$$

6.9 Consider a standard gas turbine cycle with a compression ratio of 4. Air enters the compressor at 1 atm and 300 K and exhausts from the turbine at 500 K and 1 atm.

a. Determine the thermodynamic efficiency of this cycle.
b. Determine the ratio of the work produced by the turbine to the work consumed by the compressor.
c. Determine the efficiency of a Carnot cycle that operates between the highest and the lowest temperature present in the turbine cycle.

Data: $C_P/R = 3.5$

6.10 Consider an air standard gas turbine cycle with a compression ratio of 4.75 in which both the compressor and expander (turbine) operate with 85% efficiency. The inlet pressure and temperature are 1.05 bar and 25.6 °C, respectively. The combustion process introduces 42.518 kJ/mol into the air steam. Determine

a. the thermal efficiency of this cycle;
b. the thermal efficiency that the cycle would have if both the compressor and the expander operated at 100% efficiency;
c. the amount of energy per mole of gas that could be recovered as heat if the exhaust gas was cooled to the temperature of the incoming air.

6.11 Consider a conventional steam power plant with a pump, a boiler/superheater combination, a turbine, and a condenser. The steam leaves the boiler/superheater at 480 °C and 2100 kPa, respectively, and the turbine as a saturated vapor. You may assume isentropic operation of the pump and the turbine.

a. Determine the pressure at the turbine exhaust.
b. Determine the boiler/superheater duty.
c. Determine the pump work required per kilogram of steam circulated.
d. Determine the thermodynamic efficiency of this power plant.

6.12 Consider the section of a conventional steam power cycle shown below. The feedwater steam enters the pump (1) at 30 °C and 300 kPa. After the boiler/superheater adds 3440 kJ/kg of energy, the steam passes to the turbine (3) at 3100 kPa and 550 °C. The turbine, which operates with a pattern efficiency of 0.82, exhausts the steam (4) at 100 kPa. Determine the thermal efficiency of this cycle.

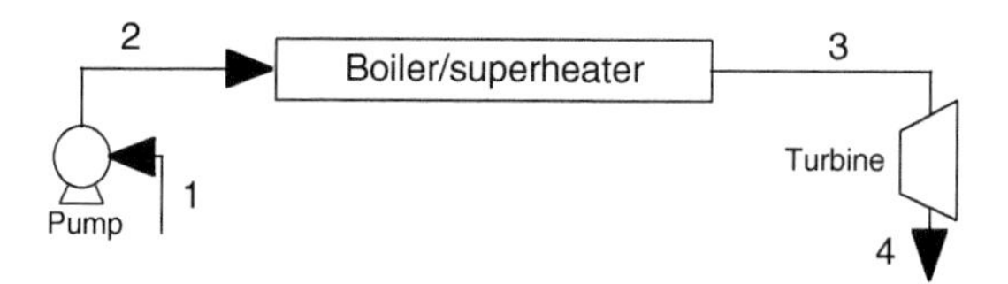

6.13 The following process is a method to increase the overall efficiency of a steam power plant by inserting a regenerator between two turbines. Compressed liquid exits a pump at 3400 kPa (state 2) and passes to a boiler/superheater. The steam leaves the boiler/superheater at 480 °C (state 3) and passes to the first turbine (I). The exhaust from turbine I, which is at 475 kPa (state 4), passes to the regenerator, which adds sufficient heat to increase the temperature back to 480 °C (state 5) before the steam passes to turbine II. The exhaust from turbine II, which is at 70 kPa (state 6), passes through a condenser, from which it exits as a saturated liquid (state 1). The efficiency of turbine I is 85% and that of turbine II is 80%.

a. Calculate the amount of energy transferred as work in turbines I and II.
b. Calculate the amount of energy transferred as heat in the regenerator.
c. Calculate the thermal efficiency of the entire process.

6.14 An air-standard Otto cycle (gasoline engine) absorbs 19 280 J/mol of air during the combustion process. The total volume swept by the pistons per revolution of the crankshaft (displacement) is 2.0 l and the compression ratio (by volume) is 6.2. Fresh air enters at atmospheric pressure and 300 K.

a. Determine the thermal efficiency of this engine.
b. Determine the temperature and pressure at the end of the compression step.
c. Noting that two revolutions of the crankshaft are required to complete each Otto cycle, determine the number of rpm (revolutions per minute) required to produce 50 hp from this engine. Treat air as an ideal gas with $C_P^{ig} = 3.5R$.

6.15 A Brayton cycle gas turbine has a compression ratio of 18 : 1. Air enters at 1 bar and 298 K. The temperature at the compressor exit is 776.2 K. Sufficient fuel is added and burned to raise the pressurized gas steam to give a turbine (expander) inlet temperature of 1533 K. After exiting the expander at 1 bar, the hot gases go to a heat exchanger to raise steam for a Rankine cycle turbine. The expander exit temperature is 843.6 K. This exhaust goes to a heat exchange to raise steam at 823 K and 80 bar (8.0 MPa). The steam turbine condenser temperature is 323 K, and the steam quality is 95%. The final temperature of the gas turbine exhaust at the exit of the heat exchanger is 343 K. Calculate the individual compressor and turbine efficiencies (both expander and steam turbine) and the efficiencies for each cycle. For the Brayton cycle, $C_P = 7R/2$ (ideal gas). Ignore changes in volume and C_P caused by combustion products and combustion chamber and heat exchanger pressure drops and heat exchange to surroundings from the equipment. Also ignore the pump.

6.16 Consider an air standard diesel cycle in which the intake air temperature is 310 K. You may assume that the compression and expansion steps are adiabatic and reversible and that the ideal gas heat capacity of air is 3.5*R*.

a. Determine the compression ratio required to obtain a temperature of 840 K at the end of the compression stroke.
b. Determine the amount of energy supplied as heat per mole of air during the combustion process if the temperature at the end of the combustion process is 1400 K.
c. Determine the work done per mole of air during the compression stroke, the combustion process, and the expansion stroke.
d. Determine the thermal efficiency of this cycle.

6.17 Consider a diesel engine described by the air standard diesel cycle. The intake air is at 20.5 °C and 0.9 bar. The temperature after the compression stroke is 650 °C. It rises to 975 °C during the fuel burn, and the exhaust temperature is 175 °C.

a. Calculate the compression ratio for this engine.
b. Calculate the energy transferred as heat per mole of air during the combustion step.
c. Calculate the energy transferred as work per mole of air during the entire cycle.
d. Calculate the thermal efficiency of the engine.

6.18 A conventional ammonia refrigeration (shown below) operates a system that condenses liquid xenon at −40 °C. The ammonia compressor operates reversibly and adiabatically. The ammonia condenser operates at 38 °C and the evaporator at −40 °C (there is no temperature gradient between the evaporating ammonia and the condensing xenon). The Xe vapor enters the condenser at 40 °C, and the liquid Xe is saturated at −40 °C when it leaves the condenser. The vapor pressure of Xe at −40 °C is 15.50 bar.

a. Determine the *COP* for the ammonia refrigeration cycle. Also calculate the *COP* of a Carnot cycle operating between the evaporator and condenser temperatures and compare the real *COP* to the Carnot *COP*.
b. Determine the work required and the total amount of ammonia required to condense 4 l of liquid xenon.

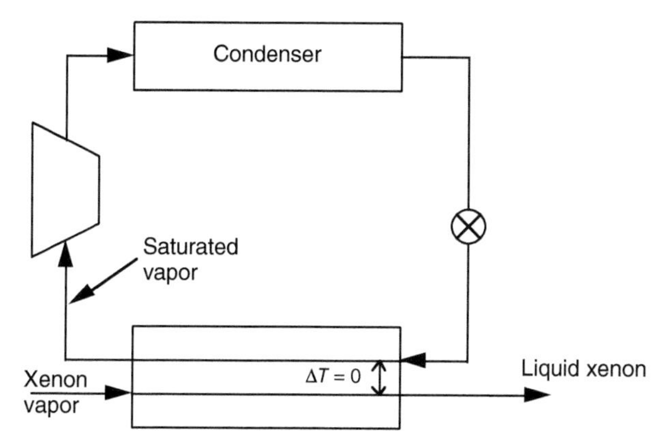

Data

Xenon properties:

$$T_C = 408.1\ \text{K},\ P_C = 36.48\ \text{bar},\quad \omega = 0.181,\ V_C = 262.7\ \text{cm}^3/\text{mol}$$

$$C_P^{ig} = 2.5R$$

$$B_2 = -842.543 + 3.90663T - 0.005063T^2 \quad 220\ \text{K} \leq T \leq 350\ \text{K}$$

Use Tables from National Institute of Standards and Technology (NIST) Webbook

6.19 Consider an ammonia refrigeration system that utilizes a two-stage compressor with intercooling between the stages. Each compressor stage has a pattern efficiency of 85% compared to reversible adiabatic operation. The intercooler operates at 5.5 bar and cools the ammonia steam to 26.85 °C. The evaporator operates at −35 °C and the condenser at 45 °C. The refrigeration capacity of this system is 4 tons (1 ton of refrigeration = 12661 kJ/h).

a. Determine the *COP* for this system.
b. Determine the power required to drive each stage of the compressor.
c. Estimate the density of the liquid discharged by the condenser.

6.20 A commercial freezer uses a standard vapor compression refrigeration cycle with ammonia as the working fluid.

a. Calculate the *COP* of this system if the evaporator operates at −29 °C, the condenser at 45 °C, and the compressor efficiency is 0.86.
b. Calculate the energy consumption rate of the compressor (***in horsepower***) required to achieve 4 tons of refrigeration capacity (1 ton of refrigeration = 12661 kJ/h).

6.21 HFC-134a is used as a working fluid both in household refrigerator/freezers and in automobile air conditioners. Both cases utilize a standard vapor compression refrigeration cycle with an expansion valve. Compare the working conditions for the two cases by determining the evaporator and condenser pressures, the quality of the two-phase steam entering the evaporator, the *COP*, and the heat release for each of the two cases below.

a. **Refrigerator/Freezer:** Evaporator temperature = −30 °C, condenser temperature = 30 °C
b. **Automobile:** Evaporator temperature = 4 °C, condenser temperature = 70 °C.

Use EOS for HFC-134a from the webbook of NIST.

6.22 Consider an air conditioning system that uses Freon-12 as the refrigerant in a standard vapor compression cycle with an expansion valve. The high side pressure is 14 bar, the low side pressure is 3 bar, and the compressor efficiency is 0.77. The cold air in the evaporator is at 10 °C and the outdoor air temperature is 40 °C. Use the EOS for R12 from the webbook of NIST.

a. Determine the temperatures of the fluids in the evaporator and the condenser.
b. Determine the efficiency that a Carnot cycle would provide when operating under these conditions.
c. Determine the actual *COP* for this cycle.
d. Determine the mass and volumetric flow rates required at the evaporator outlet to provide 3.5 tons of air conditioning capacity (1 ton of refrigeration = 12661 kJ/h).

6.23 A recent innovation to increase the efficiency of water distillation uses an arrangement as shown below, in which the water vapor leaving the still is compressed to a pressure sufficiently high to allow the vapor to condense at a temperature higher than the saturation temperature in the still. This allows the energy released upon condensation to be returned to the boiler to vaporize more water. The only input of energy to the apparatus during steady-state operation is the work supplied to the compressor. In a conventional distillation unit (also shown in the drawing below), energy is supplied directly to the boiler as heat.

Consider the case in which liquid water enters both units at 20 °C and 100 kPa. The state at point 2 is liquid water at 98 °C, while that at point B is liquid water at 90 °C. Both stills operate at a pressure of 100 kPa, and the vapor leaving the stills at points 3 and C is saturated at 100 kPa. In the heat pump unit, the vapor then is compressed reversibly and adiabatically to 150 kPa, so that points 4, 5, and 6 all are at 150 kPa. In the conventional unit, the pressure is 100 kPa throughout.

a. Calculate the amount of energy supplied as work to the compressor per kilogram of water distilled in the heat pump distillation unit.
b. Calculate the temperature of the water leaving the heat pump unit at 6, assuming no energy is lost as heat to the surroundings.
c. Calculate the amount of energy that must be supplied as heat to the still per kilogram of water distilled in the conventional unit.
d. Calculate the minimum amount of energy that must be removed as heat in the condenser to obtain a completely liquid steam at E.

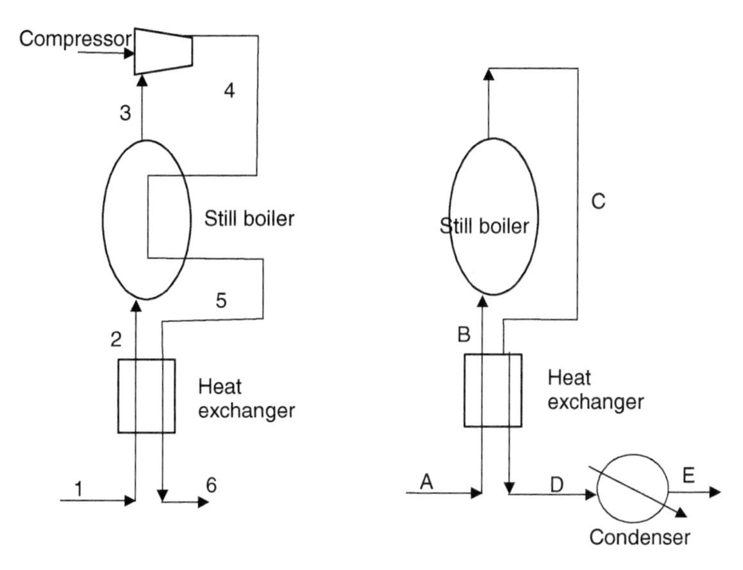

6.24 Consider an air liquefaction system that uses the Claude process as shown below. The following specifications apply: the air entering at the gas intake (1) is at 305 K and 1 atm; the air entering the cooler (3) is at 280 K and 15 MPa; the high pressure gas entering the expansion engine (4) is at −345 K; the separator operates at 1 atm; the expansion engine and the compressor operate isentropically; there are no significant pressure drops on the high-pressure side between the compressor and the expansion engine; the only significant pressure drop on the low-pressure side is in the cooler.

a. Calculate the fraction of air that condenses
b. Determine the compressor work required per kilogram of m_3

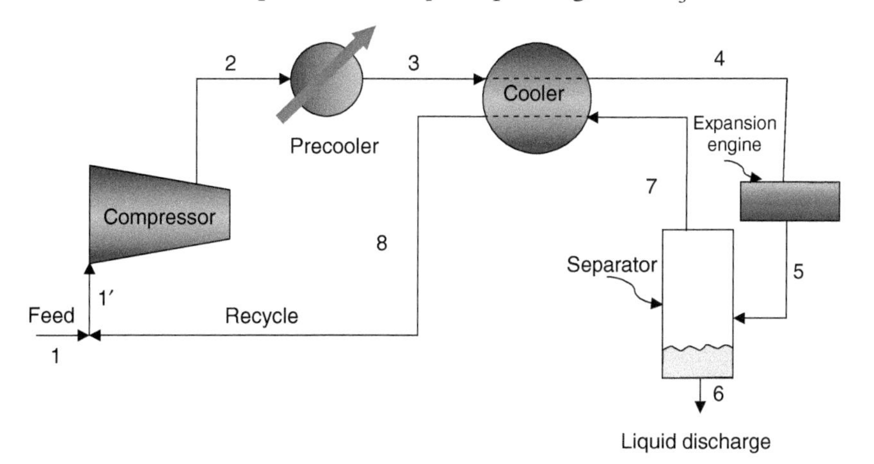

Data

Conditions of equilibrium at 1 atm

$P = 0.101325$ MPa

$H^L = 0.001$ kJ/kg, $H^V = 204.81$ kJ/kg, $T^L = 78.903$ K, $T^V = 81.72$ K

7

Solution Theory

7.1 Introduction

In this chapter, we continue with the mathematical development of the relationships among fundamental thermodynamic properties. We develop the thermodynamics for open systems, i.e. systems that can exchange matter with their surroundings or other systems. These systems can consist of two components (binary systems), three components (ternary systems), or multicomponent systems, in other words mixtures.

First, we focus upon definitions. Then, we apply those relationships to mixtures formed of ideal gases, ideal solutions, and real gases and liquids. The material that we cover here is indispensable for understanding the equilibrium between two or more phases in multicomponent mixtures.

7.2 Composition Variables

Before we start the thermodynamic treatment of open systems, we define a mixture as a substance produced by combining several pure components, pure components and a mixture, or several mixtures using physical forces *excluding* chemical reactions. It is usually possible to return a mixture to its original components. The mixtures can be homogeneous or heterogeneous.

Homogeneous mixtures are mixtures whose thermodynamic properties vary continuously throughout the mixture and that have the same composition throughout. The examples are air, water and sugar, water and alcohol, etc. When the mixtures contain particles with the size of atoms, small molecules, or small ions (less than 1 nm in all dimensions), we term them homogeneous mixtures or solutions. The common examples are salt dissolved in water, carbon dioxide in water, mercury in gold, etc. If the particle size increases from 1 to 1000 nm, we call the homogeneous mixture a colloid. Typical colloids are jelly, milk, blood, etc. In a solution or a colloid, the particle size is still invisible to the naked eye.

Heterogeneous mixtures are those whose thermodynamic properties vary discontinuously throughout. The examples are sand, water and oil, salad, granite, etc.

Thermodynamics for Chemical Engineers, First Edition. Kenneth R. Hall and Gustavo A. Iglesias-Silva.

When the particle size is greater than 1 μm, we call the heterogeneous mixtures suspensions.

We know that defining a state condition for a pure substance requires fixing two intensive variables, for example, temperature and pressure. For a mixture, these two properties are not sufficient, so we must also specify the composition of the mixture. In phase equilibria, the most practical composition variables are the *mole fractions* and the *weight fractions*. We name the mole fractions x_i (for liquid compositions of species i) and y_i (for vapor compositions). We use z_i when the equation can apply to either liquids or vapors. Similarly, the weight fraction is w_i for component i in the mixture. We can calculate these two variables using

$$x_i \text{ or } y_i = \frac{n_i}{\sum_{i=1}^{N} n_i} \tag{7.1}$$

and

$$w_i = \frac{m_i}{\sum_{i=1}^{N} m_i} \tag{7.2}$$

in which n_i is the number of moles of specie i in the liquid or vapor phase and m_i is the mass of specie i. Obviously, for both definitions

$$\sum_{i=1}^{N} x_i = 1 \text{ or } \sum_{i=1}^{N} y_i = 1 \tag{7.3}$$

$$\sum_{i=1}^{N} w_i = 1 \tag{7.4}$$

The molecular weight provides the relationship among them

$$w_i = \frac{z_i M_i}{\sum_{j=1}^{N} z_j M_j} \tag{7.5}$$

in which M_i and M_j are the molecular weights of species i and j, respectively. Other composition variables of interest to chemical engineers exist; one is the molality, which is the number of moles of solute in 1 kg of solvent. The solvent is the substance that it is in excess, and the solute is the other substance. For example, if we have water and ammonia in a proportion such that $n_{water} \gg n_{Ammonia}$, then the water is the solvent and the ammonia the solute. Thus,

$$\text{Molality}_i = \frac{n_i}{1 \text{ kg solvent}} \tag{7.6}$$

The molality is independent of temperature and pressure because its definition is in terms of mass. Another composition variable is the concentration denoted by C_i and defined as the number of moles divided by the volume of the solution,

$$C_i = n_i / V \tag{7.7}$$

Generally, the units of concentration are moles/liter of solution (SI does not recognize liter as a unit, it uses cubic decimeters instead, 1 L = 1 dm^3).

Example 7.1

What is the mole fraction of a binary mixture containing 100 g of water and 10 g of ethanol? Calculate the molality and the concentration of the solute. The densities of ethanol and water are 0.7839 and 0.9970 g/cm^3 at 25 °C, and the molar masses are 18.01528 and 46.068 g/mol, respectively.

Solution

Calculate the number of moles for each component:

$$n_i = \frac{m_i}{MW_i}$$

For ethanol:

$$n_{ethanol} = \frac{10}{46.068} = 0.2071$$

For water:

$$n_{water} = \frac{100}{18.01528} = 5.5508$$

Thus, the total moles are 0.2071 + 5.5508 = 5.7579 mol.
The mole fraction of ethanol is

$$x_{ethanol} = \frac{0.2071}{5.7579} = 0.036$$

and for water it is

$$x_{water} = \frac{5.5508}{5.7579} = 0.964 = 1 - x_{ethanol}$$

For the calculation of molality, the solvent is water and the solute is ethanol. Then, in 1 kg of water, we have 100 g of ethanol, so the mixture has the same mole fractions. The number of moles in 100 g of ethanol is 2.071; therefore, the molality of ethanol is using Eq. (7.6)

$$m_{ethanol} = \frac{2.071 \text{ mol}}{1 \text{ kg of water}} = 2.071 \text{ mol/kg}$$

The calculation of concentration is slightly more complex. First, calculate the total volume (denoted by superscript t) considering the two substances,

$$V^t = V^t_{ethanol} + V^t_{water}$$

We know that the volume and density are reciprocals of each other

$$\rho_i = \frac{m_i}{V^t_i}$$

The equation for the total volume becomes

$$V^t = \frac{m_{ethanol}}{\rho_{ethanol}} + \frac{m_{water}}{\rho_{water}} = \frac{10 \text{ g}}{0.783924 \text{ g/cm}^3} + \frac{100 \text{ g}}{0.997044 \text{ g/cm}^3} = 113.0528 \text{ cm}^3$$

So, we have 0.11305 l of solution, and the concentrations of each component are

$$C_{ethanol} = \frac{0.2071 \text{ mol}}{0.113053 \text{ l}} = 1.8319 \text{ mol/l}$$

and

$$C_{water} = \frac{5.5508 \text{ mol}}{0.113053 \text{ l}} = 49.1 \text{ mol/l}$$

7.3 Chemical Potential

In Chapter 5, we obtained the fundamental equations for closed systems. All of them are in differential form:

$$dF = \left[\frac{\partial F}{\partial \mathbb{X}}\right]_{\mathbb{Y}} d\mathbb{X} + \left[\frac{\partial F}{\partial \mathbb{Y}}\right]_{\mathbb{X}} d\mathbb{Y} \tag{7.8}$$

in which $\mathbb{X}$ and $\mathbb{Y}$ are two canonical or two intensive variables for each of the thermodynamic properties (*UHAGS*). Now, consider an open system with $n_1, n_2, \ldots, n_N$ moles of components 1, 2, ..., N. Any total thermodynamic property then depends on two variables and the number of moles:

$$nF = nF(\mathbb{X}, \mathbb{Y}, n_1, n_2, \ldots, n_N) \tag{7.9}$$

The total differential of nF is

$$d(nF) = \left[\frac{\partial(nF)}{\partial \mathbb{X}}\right]_{n,\mathbb{Y}} d\mathbb{X} + \left[\frac{\partial(nF)}{\partial \mathbb{Y}}\right]_{n,\mathbb{X}} d\mathbb{Y} + \sum_{i=1}^{N}\left[\frac{\partial(nF)}{\partial n_i}\right]_{\mathbb{X},\mathbb{Y},n_{j\neq i}} dn_i \tag{7.10}$$

The total number of moles is constant for the first two partial derivatives, so

$$d(nF) = n\left[\frac{\partial F}{\partial \mathbb{X}}\right]_{n,\mathbb{Y}} d\mathbb{X} + n\left[\frac{\partial F}{\partial \mathbb{Y}}\right]_{n,\mathbb{X}} d\mathbb{Y} + \sum_{i=1}^{N}\left[\frac{\partial(nF)}{\partial n_i}\right]_{\mathbb{X},\mathbb{Y},n_{j\neq i}} dn_i \tag{7.11}$$

Definitions for the derivatives in the above equation already have appeared in Chapter 5. When the two variables are the canonical variables, i.e. if $F = U$, then $\mathbb{X} = nS$ and $\mathbb{Y} = nV$. The term in the summation is the chemical potential of component i in the mixture

$$\mu_i \equiv \left[\frac{\partial(nF)}{\partial n_i}\right]_{\mathbb{X},\mathbb{Y},n_{j\neq i}} \tag{7.12}$$

The chemical potential has multiple definitions that depend upon the thermodynamic property and the canonical variables:

$$\mu_i \equiv \left[\frac{\partial(nU)}{\partial n_i}\right]_{nS,nV,n_{j\neq i}} = \left[\frac{\partial(nH)}{\partial n_i}\right]_{nS,P,n_{j\neq i}} = \left[\frac{\partial(nA)}{\partial n_i}\right]_{nV,T,n_{j\neq i}} = \left[\frac{\partial(nG)}{\partial n_i}\right]_{P,T,n_{j\neq i}} = \overline{G}_i \tag{7.13}$$

Historically, the last equality serves as the definition of the chemical potential because it is common to have the change of a thermodynamic property at a constant temperature and pressure for phase equilibrium. However, that is not appropriate for modern applications, as we shall see. Therefore, in this text, we use Eq. (7.13) to define the chemical potential.

Inserting the definition of chemical potential and using canonical variables from Eq. (7.10) or (7.11)

$$d(nU) = \left[\frac{\partial U}{\partial S}\right]_{n,nV} d(nS) + \left[\frac{\partial nU}{\partial nV}\right]_{n,nS} d(nV) + \sum_{i=1}^{N}\mu_i dn_i \tag{7.14}$$

$$d(nH) = \left[\frac{\partial H}{\partial S}\right]_{n,P} d(nS) + n\left[\frac{\partial H}{\partial P}\right]_{n,nS} dP + \sum_{i=1}^{N}\mu_i dn_i \tag{7.15}$$

$$d(nA) = n\left[\frac{\partial A}{\partial T}\right]_{n,nV} dT + \left[\frac{\partial A}{\partial V}\right]_{n,nS} d(nV) + \sum_{i=1}^{N} \mu_i dn_i \tag{7.16}$$

$$d(nG) = n\left[\frac{\partial G}{\partial T}\right]_{n,P} dT + n\left[\frac{\partial G}{\partial P}\right]_{n,T} dP + \sum_{i=1}^{N} \mu_i dn_i \tag{7.17}$$

If we substitute the definitions of the derivatives

$$d(nU) = Td(nS) - Pd(nV) + \sum_{i=1}^{N} \mu_i dn_i \tag{7.18}$$

$$d(nH) = Td(nS) + (nV)dP + \sum_{i=1}^{N} \mu_i dn_i \tag{7.19}$$

$$d(nA) = -(nS)dT - Pd(nV) + \sum_{i=1}^{N} \mu_i dn_i \tag{7.20}$$

$$d(nG) = -(nS)dT + (nV)dP + \sum_{i=1}^{N} \mu_i dn_i \tag{7.21}$$

7.3.1 More Maxwell Relations

Using the earlier procedure, the Maxwell relations resulted from applying the reciprocity criteria.

For the internal energy

$$\left[\frac{\partial \mu_i}{\partial (nV)}\right]_{nS,n} = -\left[\frac{\partial P}{\partial n_i}\right]_{nS,nV,n_{j\neq i}} \tag{7.22}$$

$$\left[\frac{\partial \mu_i}{\partial (nS)}\right]_{nV,n} = \left[\frac{\partial T}{\partial n_i}\right]_{nV,nS,n_{j\neq i}} \tag{7.23}$$

For the enthalpy

$$\left[\frac{\partial \mu_i}{\partial P}\right]_{nS,n} = \left[\frac{\partial (nV)}{\partial n_i}\right]_{nS,P,n_{j\neq i}} \tag{7.24}$$

$$\left[\frac{\partial \mu_i}{\partial (nS)}\right]_{P,n} = \left[\frac{\partial T}{\partial n_i}\right]_{P,nS,n_{j\neq i}} \tag{7.25}$$

For the Helmholtz energy

$$\left[\frac{\partial \mu_i}{\partial T}\right]_{nV,n} = -\left[\frac{\partial (nS)}{\partial n_i}\right]_{nV,T,n_{j\neq i}} \tag{7.26}$$

$$\left[\frac{\partial \mu_i}{\partial (nV)}\right]_{T,n} = -\left[\frac{\partial P}{\partial n_i}\right]_{T,nV,n_{j\neq i}} \tag{7.27}$$

For the Gibbs Energy

$$\left[\frac{\partial \mu_i}{\partial P}\right]_{T,n} = \left[\frac{\partial (nV)}{\partial n_i}\right]_{T,P,n_{j\neq i}} \tag{7.28}$$

$$\left[\frac{\partial \mu_i}{\partial T}\right]_{P,n} = -\left[\frac{\partial (nS)}{\partial n_i}\right]_{P,T,n_{j\neq i}} \tag{7.29}$$

Using the Euler theorem, we can write the thermodynamic properties as

$$F = F(X, Y, z_1, z_2, \dots, z_{N-1}) \tag{7.30}$$

in which X and Y are the canonical variables or any intensive property. The total differential is

$$dF = \left[\frac{\partial F}{\partial X}\right]_{z,Y} dX + \left[\frac{\partial F}{\partial Y}\right]_{z,X} dY + \sum_{i=1}^{N-1} \left[\frac{\partial F}{\partial z_i}\right]_{X,Y,z_{j\neq i}} dz_i \tag{7.31}$$

and the fundamental equations become

$$dU = \left[\frac{\partial U}{\partial S}\right]_{z,V} dS + \left[\frac{\partial U}{\partial V}\right]_{z,S} dV + \sum_{i=1}^{N-1} \left[\frac{\partial U}{\partial z_i}\right]_{V,S,z_{j\neq i}} dz_i \tag{7.32}$$

$$dH = \left[\frac{\partial H}{\partial S}\right]_{z,P} dS + \left[\frac{\partial H}{\partial P}\right]_{z,S} dP + \sum_{i=1}^{N-1} \left[\frac{\partial H}{\partial z_i}\right]_{V,S,z_{j\neq i}} dz_i \tag{7.33}$$

$$dA = \left[\frac{\partial A}{\partial T}\right]_{z,V} dT + \left[\frac{\partial A}{\partial V}\right]_{z,T} dV + \sum_{i=1}^{N-1} \left[\frac{\partial A}{\partial z_i}\right]_{T,V,z_{j\neq i}} dz_i \tag{7.34}$$

$$dG = \left[\frac{\partial G}{\partial T}\right]_{z,P} dT + \left[\frac{\partial G}{\partial P}\right]_{z,T} dP + \sum_{i=1}^{N-1} \left[\frac{\partial G}{\partial z_i}\right]_{T,P,x_{j\neq i}} dz_i \tag{7.35}$$

or

$$dU = TdS - PdV + \sum_{i=1}^{N-1} \left[\frac{\partial U}{\partial z_i}\right]_{V,S,x_{j\neq i}} dz_i \tag{7.36}$$

$$dH = TdS + VdP + \sum_{i=1}^{N-1} \left[\frac{\partial H}{\partial z_i}\right]_{V,S,x_{j\neq i}} dz_i \tag{7.37}$$

$$dA = -SdT - PdV + \sum_{i=1}^{N-1} \left[\frac{\partial A}{\partial z_i}\right]_{T,V,x_{j\neq i}} dz_i \tag{7.38}$$

$$dG = -SdT + VdP + \sum_{i=1}^{N-1} \left[\frac{\partial G}{\partial z_i}\right]_{T,P,z_{j\neq i}} dz_i \tag{7.39}$$

7.4 Partial Molar Properties

Equations (7.14)–(7.17) are fundamental equations for open systems. Partial derivatives with respect to n_i are important in the study of mixtures. These partial derivatives indicate how much an extensive thermodynamic property changes when adding or removing a chemical component from the system. We call these partial derivatives *partial molar quantities* when the intensive properties, temperature, pressure, and the moles of the other components, remain constant:

$$\overline{F}_i \equiv \left[\frac{\partial (nF)}{\partial n_i}\right]_{P,T,n_{j\neq i}} \tag{7.40}$$

We have partial molar properties from our fundamental equations and for other thermodynamic properties such as $\overline{U}_i$, $\overline{H}_i$, $\overline{A}_i$, $\overline{G}_i$, $\overline{S}_i$, $\overline{V}_i$, $\overline{C}_{P_i}$, etc. Now, because

$d(nF) = ndF + Fdn$ and $dn_i = z_i dn + ndz_i$, we can substitute these results into Eq. (7.11), considering X and Y as any two intensive properties

$$(ndF + Fdn) = n\left[\frac{\partial F}{\partial X}\right]_{n,Y} dX + n\left[\frac{\partial F}{\partial Y}\right]_{n,X} dY + \sum_{i=1}^{N}\left[\frac{\partial(nF)}{\partial n_i}\right]_{X,Y,n_{j\neq i}} (z_i dn + ndz_i) \tag{7.41}$$

Collecting terms in n and dn

$$\left(dF - \left[\frac{\partial F}{\partial X}\right]_{n,Y} dX - \left[\frac{\partial F}{\partial Y}\right]_{n,X} dY - \sum_{i=1}^{N}\left[\frac{\partial(nF)}{\partial n_i}\right]_{X,Y,n_{j\neq i}} dz_i\right) n = \left(\sum_{i=1}^{N} z_i \left[\frac{\partial(nF)}{\partial n_i}\right]_{X,Y,n_{j\neq i}} - F\right) dn \tag{7.42}$$

Because a system can be of any size, n and dn can be positive, negative, or zero. Under these circumstances, the only way this equation can be always valid is if the terms in parentheses are zero, which means

$$dF = \left[\frac{\partial F}{\partial X}\right]_{n,Y} dX + \left[\frac{\partial F}{\partial Y}\right]_{n,X} dY + \sum_{i=1}^{N}\left[\frac{\partial(nF)}{\partial n_i}\right]_{X,Y,n_{j\neq i}} dz_i \tag{7.43}$$

and

$$F = \sum_{i=1}^{N} z_i \left[\frac{\partial(nF)}{\partial n_i}\right]_{X,Y,z_{j\neq i}} \tag{7.44}$$

The above equations are not equivalent to considering $F = F(X, Y, z_1, z_2, \ldots, z_{N-1})$. The most convenient variables are either T and P or T and V. If we use T and P, the derivatives are the partial molar properties

$$F = \sum_{i=1}^{N} z_i \left[\frac{\partial(nF)}{\partial n_i}\right]_{T,P,n_{j\neq i}} = \sum_{i=1}^{N} z_i \overline{F}_i \tag{7.45}$$

The above equation demonstrates that any molar thermodynamic property equals the mole fraction average of the partial molar properties of each component in the mixture. It is not true that the molar property is a mole fraction average of the molar properties of the pure components except for ideal gases

$$F^{ig} = \sum_{i=1}^{N} z_i F_i^{ig} \tag{7.46}$$

With the definition of the partial molar properties, Eqs. (7.13), (7.28), and (7.29) become

$$\overline{G}_i = \mu_i \tag{7.13}$$

$$\left[\frac{\partial \overline{G}_i}{\partial P}\right]_{T,z} = \overline{V}_i \tag{7.28}$$

$$\left[\frac{\partial \overline{G}_i}{\partial T}\right]_{P,z} = -\overline{S}_i \tag{7.29}$$

These latter equations allow us to calculate the changes of the chemical potential with respect to temperature and pressure. Because all the definitions of the thermodynamic properties are linear functions, they have counterparts in terms of partial molar properties. For example,

$$H = U + PV$$

Multiplying by n and taking the derivative with respect to n_i at constant T and P,

$$\left[\frac{\partial(nH)}{\partial n_i}\right]_{T,P,n_{j\neq i}} = \left[\frac{\partial(nU)}{\partial n_i}\right]_{T,P,n_{j\neq i}} + P\left[\frac{\partial(nV)}{\partial n_i}\right]_{T,P,n_{j\neq i}} \tag{7.47}$$

Therefore,

$$\overline{H}_i = \overline{U}_i + P\overline{V}_i \tag{7.48}$$

The other definitions are

$$\overline{A}_i = \overline{U}_i - T\overline{S}_i \tag{7.49}$$

$$\overline{G}_i = \overline{H}_i - T\overline{S}_i \tag{7.50}$$

Also, for a constant composition or constant total number of moles

$$d\overline{U}_i = Td\overline{S}_i - Pd\overline{V}_i \tag{7.51}$$

$$d\overline{H}_i = Td\overline{S}_i + \overline{V}_i dP \tag{7.52}$$

$$d\overline{A}_i = -\overline{S}_i dT - Pd\overline{V}_i \tag{7.53}$$

$$d\overline{G}_i = -S_i dT + \overline{V}_i dP \tag{7.54}$$

Equation (7.45) presents the molar properties in terms of the partial molar properties. Thus, we can calculate the molar property using the definition of the partial molar property in terms of moles. Now, let us calculate the partial molar property using mole fractions. Starting with the definition of the partial molar property, Eq. (7.40),

$$\overline{F}_i \equiv \left[\frac{\partial(nF)}{\partial n_i}\right]_{T,P,n_{j\neq i}} = n\left[\frac{\partial F}{\partial n_i}\right]_{T,P,n_{j\neq i}} + F\left[\frac{\partial(n)}{\partial n_i}\right]_{T,P,n_{j\neq i}} \tag{7.55}$$

because $n = \sum_{i=1}^{N} n_i$, then

$$\left[\frac{\partial(n)}{\partial n_i}\right]_{T,P,n_{j\neq i}} = 1 \tag{7.56}$$

so

$$\overline{F}_i = F + n\left[\frac{\partial F}{\partial n_i}\right]_{T,P,n_{j\neq i}} \tag{7.57}$$

To calculate the derivative, use Eq. (7.31) at constant $X = T$ and $Y = P$

$$dF = \sum_{i=1}^{N-1}\left[\frac{\partial F}{\partial z_i}\right]_{T,P,z_{j\neq i}} dz_i \tag{7.58}$$

Taking the derivative with respect to n_i

$$\left(\frac{\partial F}{\partial n_i}\right)_{T,P,n_{j\neq i}} = \sum_{k=1}^{N-1}\left[\frac{\partial F}{\partial z_k}\right]_{T,P,z_{l\neq k}}\left(\frac{\partial z_k}{\partial n_i}\right)_{T,P,n_{j\neq i}} \tag{7.59}$$

and using $z_k = n_k/n$

$$\left(\frac{\partial z_k}{\partial n_i}\right)_{T,P,n_{j\neq i}} = \begin{cases} \dfrac{n-n_i}{n^2} & \text{for } k = i \\ -\dfrac{n_k}{n^2} & \text{for } k \neq i \end{cases} \tag{7.60}$$

and

$$n\left(\frac{\partial F}{\partial n_i}\right)_{T,P,n_{j\neq i}} = -\sum_{\substack{k=1\\k\neq i}}^{N-1} z_k\left[\frac{\partial F}{\partial z_k}\right]_{T,P,z_{l\neq k}} + (1-z_i)\left[\frac{\partial F}{\partial z_i}\right]_{T,P,z_{l\neq k}} \tag{7.61}$$

Substituting Eq. (7.61) into Eq. (7.57) provides an expression for the partial molar property in terms of molar properties

$$\overline{F}_i = F + \left[\frac{\partial F}{\partial z_i}\right]_{T,P,z_{l\neq i}} - \sum_{k=1}^{N-1} z_k\left[\frac{\partial F}{\partial z_k}\right]_{T,P,z_{l\neq k}} \tag{7.62}$$

Applying the above equation to binary systems

$$\overline{F}_1 = F + \left[\frac{dF}{dz_1}\right]_{T,P} - z_1\left[\frac{dF}{dz_1}\right]_{T,P} = F + (1-z_1)\left[\frac{dF}{dz_1}\right]_{T,P} = F + z_2\left[\frac{dF}{dz_1}\right]_{T,P} \tag{7.63}$$

For component 2 in Eq. (7.62), the first derivative is zero (this is always true for the Nth component) because the number of moles of component 2 is constant, and the summation added to the property F is

$$\overline{F}_2 = F - z_1\left[\frac{dF}{dz_1}\right]_{T,P} \tag{7.64}$$

The total derivatives are proper for a binary system because we have already $z_1 + z_2 = 1$ in the expressions. Because we start from Eq. (7.40), which is a linear equation for binary systems, Eqs. (7.63) and (7.64) are also straight lines on plots of F vs z_1, z_2 or $\overline{F}_i$ vs z_1, z_2. In the first case, we have

$$F = \overline{F}_1 - z_2\left[\frac{dF}{dz_1}\right]_{T,P} \quad \text{and} \quad F = \overline{F}_2 + z_1\left[\frac{dF}{dz_1}\right]_{T,P} \tag{7.65}$$

Looking at Eq. (7.65), we notice that the two intercepts are the partial molar properties. Figure 7.1 represents both equations, and equating the equations produces

$$\left[\frac{dF}{dz_1}\right]_{T,P} = \overline{F}_1 - \overline{F}_2 \tag{7.66}$$

We know that a mixture property must tend to the pure component value when $z_1 = 1$ or $z_2 = 1$

$$\lim_{\substack{z_1\to 1\\ z_2\to 0}} F = F_1 \tag{7.67}$$

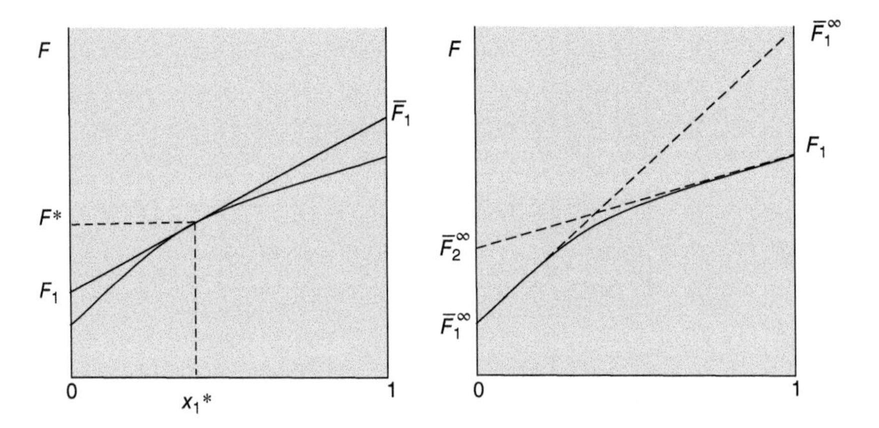

Figure 7.1 Graphical representation of the partial molar properties.

$$\lim_{\substack{z_1 \to 0 \\ z_2 \to 1}} F = F_2 \tag{7.68}$$

also

$$\lim_{z_1 \to 0} \overline{F}_1 = \overline{F}_1^{\infty} \tag{7.69}$$

$$\lim_{z_2 \to 0} \overline{F}_2 = \overline{F}_2^{\infty} \tag{7.70}$$

Example 7.2

Find the general relationship between $\overline{F}_i$ and $\left[\frac{\partial(nF)}{\partial n_i}\right]_{T,\mathbb{Y},n_{j\neq i}}$ and apply it when $nF = nV$.

Solution

Consider a general function $nF = nF(\mathbb{Y}, n_i)$ keeping constant T and $n_{j\neq i}$. The total differential is

$$d(nF) = \left[\frac{\partial(nF)}{\partial \mathbb{Y}}\right]_{T,n} d\mathbb{Y} + \left[\frac{\partial(nF)}{\partial n_i}\right]_{T,\mathbb{Y},n_{j\neq i}} dn_i$$

Taking the derivative with respect to n_i keeping constant T, P, and $n_{j\neq i}$,

$$\left[\frac{\partial(nF)}{\partial n_i}\right]_{T,P,n_{j\neq i}} = \left[\frac{\partial(nF)}{\partial \mathbb{Y}}\right]_{T,z} \left(\frac{\partial \mathbb{Y}}{\partial n_i}\right)_{T,P,n_{j\neq i}} + \left[\frac{\partial(nF)}{\partial n_i}\right]_{T,\mathbb{Y},n_{j\neq i}} \left(\frac{\partial n_i}{\partial n_i}\right)_{T,P,n_{j\neq i}}$$

and updating the terms

$$\overline{F}_i = \left[\frac{\partial(nF)}{\partial \mathbb{Y}}\right]_{T,n} \overline{\mathbb{Y}}_i + \left[\frac{\partial(nF)}{\partial n_i}\right]_{T,\mathbb{Y},n_{j\neq i}}$$

in which $\overline{\mathbb{Y}}_i$ is the partial molar property of the molar variable. Now, if $\overline{\mathbb{Y}} = nV$

$$\overline{F}_i = \left[\frac{\partial F}{\partial V}\right]_{T,n} \overline{V}_i + \left[\frac{\partial(nF)}{\partial n_i}\right]_{T,nV,n_{j\neq i}}$$

Example 7.3

The following table provides the molar volume of a mixture of nitrogen + carbon dioxide at 350 K and 10 MPa

z_{N_2}	Volume (m^3/kmol)
0.0	0.19235
0.1	0.21254
0.2	0.22881
0.3	0.24245
0.4	0.25409
0.5	0.26416
0.6	0.27294
0.7	0.28065
0.8	0.28746
0.9	0.29350
1.0	0.29884

An equation for the volume of the mixture is

$$V = 0.29884z_1 + 0.19235z_2 + z_1z_2(0.04864z_1 + 0.103137z_2)$$

Find an expression for $\overline{V}_1$ and $\overline{V}_2$. Also, find the values of $\overline{V}_1^\infty$ and $\overline{V}_2^\infty$.

Solution

Using Eq. (7.40)

$$\overline{V}_1 = \left[\frac{\partial nV}{\partial n_1}\right]_{T,P,n_2} \rightarrow nV = 0.29884n_1 + 0.19235n_2$$
$$+ \frac{n_1n_2}{n}\left(0.04864\frac{n_1}{n} + 0.103137\frac{n_2}{n}\right)$$

and

$$\overline{V}_1 = \left[\frac{\partial nV}{\partial n_1}\right]_{T,P,n_2} = 0.29884 + \left(\frac{n_2n - n_1n_2}{n^2}\right)\left(0.04864\frac{n_1}{n} + 0.103137\frac{n_2}{n}\right)$$
$$+ \left(\frac{n_1n_2}{n}\right)\left[0.04864\frac{n - n_1}{n^2} + 0.103137\left(-\frac{n_2}{n^2}\right)\right]$$

then

$$\begin{aligned}\overline{V}_1 &= 0.29884 + (z_2 - z_1z_2)(0.04864z_1 + 0.103137z_2) \\ &\quad + (z_1z_2)[0.04864(1 - z_1) + 0.103137(-z_2)] \\ &= 0.29884 + z_2^2(0.04864z_1 + 0.103137z_2) - \left(z_1z_2^2\right)0.054497 \\ &= 0.29884 + 0.103137z_2^3 - 0.005857z_1z_2^2\end{aligned}$$

$$\overline{V}_1^\infty = \lim_{\substack{x_1 \to 0 \\ x_2 \to 1}} \overline{V}_1 = 0.401977 \ \ m^3/kmol$$

Also,

$$\overline{V}_2 = \left[\frac{\partial nV}{\partial n_2}\right]_{T,P,n_1} = 0.19235 + 0.04864 z_1^3 + 0.157634 z_1^2 z_2$$

$$\overline{V}_2^{\infty} = 0.24099 \ \mathrm{m^3/kmol}$$

7.5 General Gibbs–Duhem Equation

If we differentiate Eq. (7.44)

$$dF = \sum_{i=1}^{N} z_i \, d\left[\frac{\partial (nF)}{\partial n_i}\right]_{X,Y,n_{j\neq i}} + \sum_{i=1}^{N} \left[\frac{\partial (nF)}{\partial n_i}\right]_{X,Y,n_{j\neq i}} dz_i \tag{7.71}$$

and substitute this result into Eq. (7.43)

$$\begin{aligned}\sum_{i=1}^{N} z_i \, d\left[\frac{\partial (nF)}{\partial n_i}\right]_{X,Y,n_{j\neq i}} + \sum_{i=1}^{N} \left[\frac{\partial (nM)}{\partial n_i}\right]_{X,Y,n_{j\neq i}} dz_i &= \left[\frac{\partial F}{\partial X}\right]_{z,Y} dX \\ + \left[\frac{\partial F}{\partial Y}\right]_{z,X} dY + \sum_{i=1}^{N} \left[\frac{\partial (nF)}{\partial n_i}\right]_{X,Y,n_{j\neq i}} dz_i \end{aligned} \tag{7.72}$$

Note that n can be replace by the mole fraction z when keeping constant the total number of mols. Deleting common terms produces

$$\left[\frac{\partial F}{\partial X}\right]_{z,Y} dX + \left[\frac{\partial F}{\partial Y}\right]_{z,X} dY - \sum_{i=1}^{N} z_i \, d\left[\frac{\partial (nF)}{\partial n_i}\right]_{X,Y,n_{j\neq i}} = 0 \tag{7.73}$$

Note that the mole fraction z can replace n when the number of moles is constant. Equation (7.73) is the most general fundamental consistency equation. If X and Y are constants, then

$$\sum_{i=1}^{N} z_i \, d\left[\frac{\partial (nF)}{\partial n_i}\right]_{X,Y,n_{j\neq i}} = 0 \tag{7.74}$$

This equation is a general consistency equation. Equations (7.73) and (7.74) involve only $X = T$ and $Y = P$ for convenience. Now, the summation term involves the definition of the partial molar property; therefore,

$$\left[\frac{\partial F}{\partial T}\right]_{z,P} dT + \left[\frac{\partial F}{\partial P}\right]_{z,T} dP - \sum_{i=1}^{N} z_i \, d\overline{F}_i = 0 \tag{7.75}$$

and

$$\sum_{i=1}^{N} z_i \, d\overline{F}_i = 0 \text{ at constant } T \text{ and } P \tag{7.76}$$

Equation (7.75) is the general Gibbs–Duhem equation, while Eq. (7.76) is an application of the Gibbs–Duhem equation for isobaric and isothermal systems. This equation is important because it allows thermodynamic consistency checks of an equation or experimental data. For binary systems, Eq. (7.76) is

$$z_1 d\overline{F}_1 + z_2 d\overline{F}_2 = 0 \tag{7.77}$$

which upon differentiation with respect to z_1 becomes

$$z_1 \frac{d\overline{F}_1}{dz_1} + z_2 \frac{d\overline{F}_2}{dz_1} = 0 \tag{7.78}$$

then

$$z_1 \frac{d\overline{F}_1}{dz_2} + z_2 \frac{d\overline{F}_2}{dz_2} = 0 \tag{7.79}$$

This indicates that if $\overline{F}_1$ increases as z_1 increases, then $\overline{F}_2$ should decrease because the slopes are of opposite sign.

Example 7.4
Check if the expressions for the following partial molar volumes are thermodynamically consistent according to the Gibbs–Duhem equation.

$$\overline{V}_1 = 0.29884 + 0.103137 z_2^3 - 0.005857 z_1 z_2^2$$
$$\overline{V}_2 = 0.19235 + 0.04864 z_1^3 + 0.157634 z_1^2 z_2$$

Solution
Differentiating both equations with respect to the composition

$$d\overline{V}_1 = 0.103137\left(3z_2^2 dz_2\right) - 0.005857 z_2^2 dz_1 - 0.005857 z_1 (2z_2 dz_2)$$

but $dz_2 = -dz_1$, so

$$\begin{aligned} d\overline{V}_1 &= -0.309411 z_2^2 dz_1 - 0.005857 z_2^2 dz_1 + 0.011714 z_1 z_2 dz_1 \\ &= -0.315268 z_2^2 dz_1 + 0.011714 z_1 z_2 dz_1 \end{aligned}$$

then

$$\begin{aligned} d\overline{V}_2 &= 0.04864 z_1^2 dz_1 + 0.157634\left(2z_1 z_2 dz_1 + z_1^2 dz_2\right) \\ &= 0.14592 z_1^2 dz_1 + 0.315268 z_1 z_2 dz_1 - 0.157634 z_1^2 dz_1 \\ &= -0.011714 z_1^2 dz_1 + 0.315268 z_1 z_2 dz_1 \end{aligned}$$

Now

$$z_1 d\overline{V}_1 = -0.315268 z_1 z_2^2 dz_1 + 0.011714 z_1^2 z_2 dz_1$$

and

$$z_2 d\overline{V}_2 = -0.011714 z_1^2 z_2 dz_1 + 0.315268 z_1 z_2^2 dz_1$$

so

$$\begin{aligned} z_1 d\overline{V}_1 + z_2 d\overline{V}_2 = {} & -0.315268 x_1 z_2^2 dz_1 + 0.011714 z_1^2 z_2 dz_1 - 0.011714 z_1^2 z_2 dz_1 \\ & + 0.315268 z_1 z_2^2 dz_1 \end{aligned}$$

or

$$z_1 d\overline{V}_1 + z_2 d\overline{V}_2 = 0$$

This means that the equation for the molar volume is thermodynamically consistent according to the Gibbs–Duhem criterion.

7.6 Differential Thermodynamic Properties in Open Systems in Terms of Measurables

Now, we have all the equations required to obtain the change of the thermodynamic properties for open systems using measurable parameters such as T, P, or nV by using Eq. (7.11), but for convenience, we use T and nV for the internal energy and Helmholtz free energy and T and P for the enthalpy and the Gibbs free energy, respectively.

7.6.1 Using *T* and *nV*

For the internal energy

$$d(nU) = n\left[\frac{\partial U}{\partial T}\right]_{n,nV} dT + \left[\frac{\partial U}{\partial V}\right]_{n,T} d(nV) + \sum_{i=1}^{N}\left[\frac{\partial(nU)}{\partial n_i}\right]_{T,nV,n_{j\neq i}} dn_i \tag{7.80}$$

We have already calculated the derivatives in Chapter 6, so

$$d(nU) = nC_V dT + \left[T\left(\frac{\partial P}{\partial T}\right)_{V,n} - P\right] d(nV) + \sum_{i=1}^{N}\left[\frac{\partial(nU)}{\partial n_i}\right]_{T,nV,n_{j\neq i}} dn_i \tag{7.81}$$

Using the same procedure as the example 7.2 in Section 7.4

$$\overline{U}_i = \left[\frac{\partial(nU)}{\partial n_i}\right]_{T,nV,n_{j\neq i}} + \left(\frac{\partial U}{\partial V}\right)_{T,n} \overline{V}_i \tag{7.82}$$

Substituting the derivative of U with respect to V in terms of measurable variables

$$\overline{U}_i = \left[\frac{\partial(nU)}{\partial n_i}\right]_{T,nV,n_{j\neq i}} + \left[T\left(\frac{\partial P}{\partial T}\right)_{V,n} - P\right] \overline{V}_i \tag{7.83}$$

then

$$\left[\frac{\partial(nU)}{\partial n_i}\right]_{T,nV,n_{j\neq i}} = \overline{U}_i - \left[T\left(\frac{\partial P}{\partial T}\right)_{V,n} - P\right] \overline{V}_i \tag{7.84}$$

Substituting this result into Eq. (7.81)

$$\begin{aligned} d(nU) = {} & nC_V dT + \left[T\left(\frac{\partial P}{\partial T}\right)_{nV,n} - P\right] d(nV) \\ & + \sum_{i=1}^{N}\left\{\overline{U}_i - \left[T\left(\frac{\partial P}{\partial T}\right)_{nV,n} - P\right]\overline{V}_i\right\} dn_i \end{aligned} \tag{7.85}$$

We can see in the above equation that we have changed V for nV in the derivatives of P for the variables kept constant because n is constant.

Similarly, we can derive expressions for the entropy and the Helmholtz free energy:

$$d(nS) = n\frac{C_V}{T}dT + \left[\frac{\partial P}{\partial T}\right]_{n,V} d(nV) + \sum_{i=1}^{N}\left\{\overline{S}_i - \left(\frac{\partial P}{\partial T}\right)_{V,n}\overline{V}_i\right\} dn_i \tag{7.86}$$

and

$$d(nA) = -(nS)dT - Pd(nV) + \sum_{i=1}^{N}\{\overline{A}_i + P\overline{V}_i\}dn_i \tag{7.87}$$

7.6.2 Using *T*, *P*

For the enthalpy

$$d(nH) = n\left[\frac{\partial H}{\partial T}\right]_{n,P} dT + n\left[\frac{\partial H}{\partial P}\right]_{n,T} dP + \sum_{i=1}^{N}\left[\frac{\partial(nH)}{\partial n_i}\right]_{T,P,n_{j\neq i}} dn_i \tag{7.88}$$

As in Section 5.5, using the Maxwell relations and the heat capacity definition

$$d(nH) = nC_P dT + n\left[V - T\left(\frac{\partial V}{\partial T}\right)_{x,P}\right] dP + \sum_{i=1}^{N}\overline{H}_i dn_i \tag{7.89}$$

For the changes in entropy and in the Gibbs free energy

$$d(nS) = nC_V\frac{dT}{T} - \left(\frac{\partial V}{\partial T}\right)_{n,P} dP + \sum_{i=1}^{N}\overline{S}_i dn_i \tag{7.90}$$

and

$$d(nG) = -(nS)dT + (nV)dP + \sum_{i=1}^{N}\overline{G}_i dn_i \tag{7.91}$$

7.7 Ideal Gas Mixtures

We have already defined an ideal gas as a hypothetical fluid for which the molecules interact with the walls of its container with no intermolecular forces. We have seen for a pure component

$$P = \frac{RT}{V_i^{ig}}$$

Then, for n moles of an ideal gas mixture confined in a container of a total volume nV, the pressure is

$$P = \frac{nRT}{nV^{ig}} = \frac{RT}{V^{ig}} \tag{7.92}$$

At the same temperature, n_i moles of specie i in the mixture with the same total volume would exert a pressure of

$$p_i = \frac{n_i RT}{nV^{ig}} \tag{7.93}$$

Dividing Eq. (7.93) by Eq. (7.92)

$$\frac{p_i}{P} = \frac{n_i}{n} = y_i \tag{7.94}$$

Thus, we can define the *partial pressure* of each specie as

$$p_i = Py_i \tag{7.95}$$

Therefore, $P = \sum_{i=1}^{N} p_i \rightarrow$ Dalton's law. Using Eq. (7.92), the molar volume of an ideal gas mixture is

$$V^{ig} = \frac{RT}{P}$$

Then, the partial molar property is

$$\overline{V}_i^{ig} = \left[\frac{\partial (nV^{ig})}{\partial n_i}\right]_{T,P,n_{j\neq i}} = \left[\frac{\partial (nRT/P)}{\partial n_i}\right]_{T,P,n_{j\neq i}} = \frac{RT}{P} = V_i^{ig} \tag{7.96}$$

and using Eq. (7.46), the molar volume of the mixture results from the ideal gas volume of the pure components

$$V^{ig} = \sum_{i=1}^{N} y_i \overline{V}_i^{ig} = \sum_{i=1}^{N} y_i V_i^{ig}$$

or

$$nV^{ig} = \sum_{i=1}^{N} n_i V_i^{ig} \rightarrow \text{Amagat's law}$$

Now, to calculate the remaining thermodynamic properties of a mixture, we use the Gibbs theorem.

Gibbs Theorem *A total thermodynamic property of an ideal gas mixture at a given T and P is the sum of the total properties of the individual species each evaluated at the mixture temperature and its own partial pressure.* Mathematically, this is

$$nF^{ig}(T,P) = \sum_{i=1}^{N} n_i F_i^{ig}(T,p_i) \tag{7.97}$$

if we divide by n,

$$F^{ig}(T,P) = \sum_{i=1}^{N} y_i F_i^{ig}(T,p_i) \tag{7.98}$$

Proof

Consider an ideal gas mixture confined to a vessel of total volume nV at temperature T. For this mixture, any extensive thermodynamic property equals the sum of the extensive properties for each of the pure ideal gas components of the mixture, that is

$$F^{ig}(T, nV^{ig}, n_1, n_2, \ldots, n_N) = \sum_{i=1}^{N} F_i^{ig}(T, nV^{ig}, n_i) \tag{7.99}$$

Using the Euler theorem for homogeneous functions

$$nF^{ig}(T, V^{ig}, y_1, y_2, \ldots, y_N) = \sum_{i=1}^{N} n_i F_i^{ig}(T, nV^{ig}/n_i)$$

but from Eq. (7.93)

$$\frac{nV^{ig}}{n_i} = \frac{RT}{p_i}$$

and

$$F^{ig}(T, V^{ig}, y_1, y_2, \ldots, y_N) = F^{ig}(T, P, y_1, y_2, \ldots, y_N)$$

Therefore,

$$nF^{ig}(T, P, y_1, y_2, \ldots, y_N) = \sum_{i=1}^{N} n_i F_i^{ig}(T, p_i)$$

$$F^{ig}(T, P) = \sum_{i=1}^{N} y_i F_i^{ig}(T, p_i)$$

Now, we can calculate the thermodynamic properties. The internal energy and the enthalpy of an ideal gas are independent of pressure (they depend only on temperature)

$$U^{ig} = \sum_{i=1}^{N} y_i U_i^{ig} \tag{7.100}$$

and

$$H^{ig} = \sum_{i=1}^{N} y_i H_i^{ig} \tag{7.101}$$

The entropy, the Gibbs free energy, and the Helmholtz free energy depend upon pressure, so we use the fundamental equations at a constant temperature. For the entropy

$$dS_i^{ig} = -R\frac{dP}{P} = -Rd\ln P \tag{7.102}$$

Integrating from p_i to P

$$S_i^{ig}(T, P) - S_i^{ig}(T, p_i) = -R\frac{dP}{P} = -R\{\ln P - \ln p_i\} = -R\ln\left(\frac{P}{p_i}\right) \tag{7.103}$$

Using Eq. (7.95)

$$S_i^{ig}(T,P) - S_i^{ig}(T,p_i) = -R\ln\left(\frac{P}{Py_i}\right) = R\ln(y_i) \tag{7.104}$$

Then

$$S_i^{ig}(T,p_i) = S_i^{ig}(T,P) - R\ln y_i \tag{7.105}$$

and the entropy of the ideal gas mixture is

$$S^{ig}(T,P) = \sum_{i=1}^{N} S_i^{ig} y_i - R\sum_{i=1}^{N} y_i \ln y_i \tag{7.106}$$

Again, for the Helmholtz energy, we have at constant temperature

$$dA_i^{ig} = -PdV_i^{ig} = -Pd(RT/P) = RT\frac{dP}{P} = RTd\ln P \tag{7.107}$$

Integrating from p_i to P

$$A_i^{ig}(T,P) - A_i^{ig}(T,p_i) = -RT\ln y_i \tag{7.108}$$

Now,

$$A_i^{ig}(T,p_i) = A_i^{ig}(T,P) + RT\ln y_i \tag{7.109}$$

The Helmholtz free energy of an ideal gas mixture is

$$A^{ig}(T,P) = \sum_{i=1}^{N} y_i A_i^{ig}(T,P) + RT\sum_{i=1}^{N} y_i \ln y_i \tag{7.110}$$

Following the same procedure, the Gibbs free energy of an ideal gas mixture is

$$G^{ig}(T,P) = \sum_{i=1}^{N} y_i G_i^{ig}(T,P) + RT\sum_{i=1}^{N} y_i \ln y_i \tag{7.111}$$

Comparing Eqs. (7.100), (7.101), (7.106), (7.110), and (7.111) to Eq. (7.45), the partial molar properties of an ideal gas in the mixture are

$$\overline{H}_i^{ig} = H_i^{ig} \tag{7.112}$$

$$\overline{U}_i^{ig} = U_i^{ig} \tag{7.113}$$

$$\overline{S}_i^{ig} = S_i^{ig} - R\ln y_i \tag{7.114}$$

$$\overline{A}_i^{ig} = A_i^{ig} + RT\ln y_i \tag{7.115}$$

and

$$\overline{G}_i^{ig} = \mu_i^{ig} = G_i^{ig} + RT\ln y_i \tag{7.116}$$

This equation is also the chemical potential of specie i in a mixture. Also, at a constant temperature

$$dG_i^{ig} = V_i^{ig}dP = RT\frac{dP}{P} = RTd\ln P \tag{7.117}$$

Differentiating Eq. (7.116)

$$d\overline{G}_i^{ig} = d\mu_i^{ig} = dG_i^{ig} + RTd\ln y_i \tag{7.118}$$

Substituting the value of the pure component

$$d\overline{G}_i^{ig} = d\mu_i^{ig} = RTd\ln P + RTd\ln y_i = RTd\ln(Py_i) = RTd\ln p_i \tag{7.119}$$

The equations for the Gibbs energy are important in the description of mixtures because their canonical variables are T and P, the rest of properties are easy to obtain from them.

7.8 Fugacity and Fugacity Coefficient for Pure Substances

We have introduced the thermodynamic properties for mixtures using the Gibbs energy because it is a fundamental function of temperature and pressure. Here, we use a concept introduced by Lewis in 1901 in his work: *The Law of Physico-chemical Change*. This thermodynamic variable is *fugacity*, and (according to Lewis) it is a measurement of the "escaping tendency" of a fluid. He makes an analogy with energy transferred as heat that tends to escape a system, and the temperature is a measurement of this escaping tendency. For example, water vapor and liquid at equilibrium will have the same tendency to escape. The question here is why do we not use the vapor pressure instead of the fugacity? The answer will become obvious later. Lewis created fugacity because he wanted to reduce the working domain of the free Gibbs energy, $(-\infty, +\infty)$ to a $(0, +\infty)$ domain. At low pressures, the Gibbs free energy tends to $-\infty$. We know that for a perfect gas, the Gibbs energy at a constant temperature is

$$dG_i^{ig} = RTd\ln P$$

Therefore, we can write an analogous equation for real fluids at a constant temperature

$$dG_i = RTd\ln f_i \text{ at constant } T \tag{7.120}$$

in which f_i is the fugacity of a pure substance, and it has units of pressure. The logarithm of the fugacity is dimensionless because $df_i/f_i = d\ln f_i$. Comparing Eq. (7.120) to the ideal gas equation,

$$\lim_{P\to 0} f_i = P = f_i^{ig} \text{ at constant } T \tag{7.121}$$

Thus, the fugacity equals the pressure when pure substances behave as ideal gases. If the vapor pressure of a substance is very low, it equals its fugacity. This is the answer to our previous question. Using the fundamental equation for the Gibbs energy, we find the pressure and temperature dependence of the fugacity. At a constant temperature

$$dG_i = V_i dP$$

Then, using Eq. (7.120)

$$RTd\ln f_i = V_i dP \tag{7.122}$$

and

$$\left(\frac{\partial \ln f_i}{\partial P}\right)_T = \frac{V_i}{RT} \tag{7.123}$$

The dependence of the fugacity on temperature is slightly more complicated. Integrating the definition of fugacity from the ideal gas to real gas at T and P,

$$G_i - G_i^{ig} = RT\ln f_i - RT\ln P \tag{7.124}$$

The derivative with respect to temperature at a constant pressure is

$$\left(\frac{\partial G_i}{\partial T}\right)_P - \left(\frac{\partial G_i^{ig}}{\partial T}\right)_P = R\ln f_i + RT\left(\frac{\partial \ln f_i}{\partial T}\right)_P - R\ln P - RT\left(\frac{\partial \ln P}{\partial T}\right)_P \tag{7.125}$$

The last term is zero because P is a constant, and using the fundamental equation $dG = -SdT$ and Eq. (7.124)

$$-S_i + S_i^{ig} = \frac{G_i - G_i^{ig}}{T} + R\ln P + RT\left(\frac{\partial \ln f_i}{\partial T}\right)_P - R\ln P \tag{7.126}$$

Eliminating the terms and using $H = G + TS$

$$RT\left(\frac{\partial \ln f_i}{\partial T}\right)_P = -\frac{G_i - G_i^{ig}}{T} - \left(S_i - S_i^{ig}\right) = -\frac{H_i - H_i^{ig}}{T} \tag{7.127}$$

or

$$\left(\frac{\partial \ln f_i}{\partial T}\right)_P = -\frac{H_i^R}{RT^2} \tag{7.128}$$

Here, we can define a new quantity, the fugacity coefficient

$$\varphi_i \equiv f_i/P \tag{7.129}$$

with

$$\lim_{P\to 0}\varphi_i = 1 \tag{7.130}$$

The fugacity coefficient and the residual Gibbs energy have a relationship using Eq. (7.124)

$$G_i^R = RT\ln\varphi_i \tag{7.131}$$

and from the definition of the fugacity coefficient,

$$\left(\frac{\partial \ln \varphi_i}{\partial P}\right)_T = \left(\frac{\partial \ln f_i}{\partial P}\right)_T - \left(\frac{\partial \ln P}{\partial P}\right)_T \tag{7.132}$$

Substituting Eq. (7.123)

$$\left(\frac{\partial \ln \varphi_i}{\partial P}\right)_T = \frac{V_i}{RT} - \frac{1}{P} = \frac{Z_i - 1}{P} = \frac{V_i - V_i^{ig}}{RT} = \frac{V_i^R}{RT} \tag{7.133}$$

Also, from Eq. (7.128)

$$\left(\frac{\partial \ln \varphi_i}{\partial T}\right)_P = -\frac{H_i^R}{RT^2} \tag{7.134}$$

All the equations derived in this section apply to mixtures. We can consider

$$\frac{G_i^R}{RT} = f(T, P) \tag{7.135}$$

The differential equation for residual Gibbs energy is

$$d\left(\frac{G_i^R}{RT}\right) = d(\ln \varphi_i) = \left[\frac{\partial\left(G_i^R/RT\right)}{\partial T}\right]_P dT + \left[\frac{\partial\left(G_i^R/RT\right)}{\partial P}\right]_T dP \tag{7.136}$$

Using Eqs. (7.133) and (7.134)

$$d\left(\frac{G_i^R}{RT}\right) = -\left(\frac{H_i^R}{RT^2}\right) dT + \left(\frac{Z_i - 1}{P}\right) dP \tag{7.137}$$

This equation is the general equation for the residual Gibbs energy. **All equations developed in this section apply to mixtures if the composition is constant (no chemical reactions)**. For a mixture, we drop the subindex *i*.

7.9 Equations for Calculating Fugacity

We can calculate the fugacity of a pure component at different conditions: in the vapor or liquid or solid phase, at saturation, etc. Figure 7.2 shows the different state conditions (point *A*, point *B*, …, etc.).

7.9.1 Fugacity of a Vapor (Point A)

We can calculate the fugacity from Eq. (7.131) using an equation of state (EOS)

$$\ln \varphi_i = \ln \frac{f_i}{P} = \frac{G_i^R}{RT} = \int_0^P (Z_i - 1)\frac{dP}{P} = \int_0^\rho (Z_i - 1)\frac{d\rho}{\rho} + Z_i - 1 - \ln Z_i \tag{7.138}$$

7.9.2 Fugacity of a Vapor or Saturated Liquid (Point B)

At equilibrium, the Gibbs energy of a fluid is equal in both phases then

$$dG^{vl} = 0 \tag{7.139}$$

Using the definition of fugacity

$$RT(d \ln f_i)^{vl} = 0 \tag{7.140}$$

and integrating,

$$\ln f_i^v = \ln f_i^l \Rightarrow f_i^v = f_i^l = f_i^{sat} \tag{7.141}$$

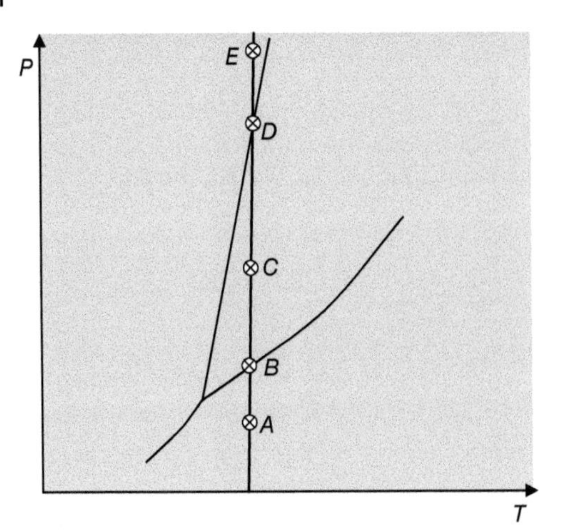

Figure 7.2 Selected state points to calculate the fugacity.

Also, because the pressure is the same for the liquid and the vapor

$$\varphi_i^v = \varphi_i^l = \varphi_i^{sat}$$

Then, to calculate the fugacity at saturation, we calculate the fugacity of the saturated vapor using Eq. (7.138)

$$\ln \varphi_i^v = \ln \frac{f_i^v}{P} = \int_0^{P^{sat}} (Z-1)\frac{dP}{P} = \int_0^{\rho^v} (Z-1)\frac{d\rho}{\rho} + Z^v - 1 - \ln Z^v \tag{7.142}$$

7.9.3 Fugacity of Liquid (Point C)

To calculate the fugacity in the liquid state, we can use Eq. (7.122) and integrate from saturation to point C,

$$\ln \frac{f_i}{f_i^{sat}} = \int_{P_i^{sat}}^{P} \frac{V_i^l}{RT} dP \tag{7.143}$$

then

$$f_i = f_i^{sat} \exp\left[\int_{P_i^{sat}}^{P} \frac{V_i^l}{RT} dP\right] \tag{7.144}$$

Using the definition of fugacity coefficient at saturation $f_i^{sat} = \varphi_i^{sat} P_i^{sat}$,

$$f_i = P_i^{sat} \varphi_i^{sat} \exp\left[\int_{P_i^{sat}}^{P} \frac{V_i^l}{RT} dP\right] \tag{7.145}$$

If the volume of the liquid is independent of pressure (often a good approximation for liquids), the equation becomes

$$f_i = P_i^{sat} \varphi_i^{sat} \exp\left[\frac{V_i^l \left(P - P_i^{sat}\right)}{RT}\right] \tag{7.146}$$

The exponential term is the Poynting Correction, and a reasonable approximation for the volume of the liquid is the saturated liquid volume.

7.9.4 Fugacity of Solid at the Melting Point (Point D)

At the melting point, the Gibbs energy is the same in the liquid and solid phases; therefore, an equivalent equation to Eq. (7.141) applies

$$f_i^{l,m} = f_i^{s,m} = f_i^{sat,m}$$

Because we know that the fugacity of the liquid equals the fugacity of the solid, we calculate the fugacity of the saturated liquid at the melting point using Eq. (7.146)

$$f_i^{sat,m} = P_i^{sat}\varphi_i^{sat} \exp\left[\frac{V_i^l\left(P_i^{sat,m} - P_i^{sat}\right)}{RT}\right] \tag{7.147}$$

7.9.5 Fugacity of a Solid (Point E)

For this result, we need to integrate Eq. (7.122) from the saturated solid at the melting point to Point E

$$\ln\frac{f_i}{f_i^{sat,m}} = \int_{P_{m,i}^{sat}}^{P} \frac{V_i^s}{RT} dP \tag{7.148}$$

therefore

$$f_i = f_i^{sat,m} \exp\left[\int_{P_{m,i}^{sat}}^{P} \frac{V_i^s}{RT} dP\right] \tag{7.149}$$

Substituting Eq. (7.147) into Eq. (7.149)

$$f_i = P_i^{sat}\varphi_i^{sat} \exp\left[\frac{V_i^l\left(P_i^{sat,m} - P_i^{sat}\right)}{RT}\right] \exp\left[\int_{P_{m,i}^{sat}}^{P} \frac{V_i^s}{RT} dP\right] \tag{7.150}$$

The volume of the solid is essentially independent of pressure, so Eq. (7.150) becomes

$$f_i = P_i^{sat}\varphi_i^{sat} \exp\left[\frac{V_i^l\left(P_i^{sat,m} - P_i^{sat}\right) + V_i^s\left(P - P_i^{sat,m}\right)}{RT}\right] \tag{7.151}$$

Example 7.5

A quadratic function of pressure can represent the compressibility factor of oxygen at 300 K

$$Z = 1 - 6.6378 \times 10^{-3}P + 2.4475 \times 10^{-4}P^2$$

in which P has units of MPa. Calculate the fugacity and fugacity coefficient at 0.1, 1, 2, 5, 10, 20, 30, and 40 MPa.

Solution

Using Eq. (7.138) (because the conditions do not cross the saturation curve)

$$\ln\varphi_i = \int_0^P (-6.6378 \times 10^{-3}P + 2.4475 \times 10^{-4}P^2)\frac{dP}{P} = -6.6378 \times 10^{-3}P + \frac{2.4475 \times 10^{-4}}{2}P^2$$

or

$$\ln \varphi_i = -6.6378 \times 10^{-3}P + 1.22375 \times 10^{-4}P^2$$

The following table contains the results:

P (MPa)	$\ln\varphi_i$	φ_i	$f = P\varphi_i$ (MPa)
0.1	−0.00065	0.99935	0.0999
1.	−0.00652	0.99351	0.9935
2.	−0.01303	0.98705	1.9741
5.	−0.03258	0.96795	4.8397
10.	−0.06515	0.93692	9.3692
20.	−0.13031	0.87783	17.5565
30.	−0.19546	0.82245	24.6736
40.	−0.26062	0.77058	30.8230

7.10 Application of Fugacity Equation to Gases and Liquids

We have used Eq. (7.138) to find an expression for the fugacity of a pure component. Now, we need an EOS applicable to gases. We can use the second virial coefficient EOS truncated after the second virial in either the Berlin or the Leiden form. The easiest (but least accurate) application is the Berlin form because it is a function of pressure

$$Z = 1 + \frac{B_2 P}{RT}$$

Because the second virial coefficient is a function of temperature, substituting the EOS into Eq. (7.138) produces

$$\ln \varphi_i = \ln \frac{f_i}{P} = \int_0^P \frac{B_2 P}{RT} \frac{dP}{P} = \frac{B_2 P}{RT} \tag{7.152}$$

which equals

$$\ln \varphi_i = \ln \frac{f_i}{P} = Z - 1 \tag{7.153}$$

Using the Leiden form

$$Z = 1 + B_2 \rho$$

we have

$$\ln \varphi_i = \ln \frac{f_i}{P} = 2B_2\rho - \ln\left(1 + \frac{B_2 \rho}{RT}\right) \tag{7.154}$$

or

$$\ln \varphi_i = 2(Z - 1) - \ln Z \tag{7.155}$$

It is important to notice that this equation usually is more accurate than Eq. (7.153). Now, we can calculate the fugacity in the liquid phase using the fugacity coefficient at saturation from (7.152)

$$f_i = P_i^{sat} \exp\left[\frac{B_2 P_i^{sat} + V_i^l\left(P - P_i^{sat}\right)}{RT}\right] \tag{7.156}$$

If we use Eq. (7.154)

$$f = \frac{P_i^{sat}}{1 + B_2\rho^v} \exp\left[2B_2\rho^v + \frac{V_i^l\left(P - P_i^{sat}\right)}{RT}\right] \tag{7.157}$$

Equation (7.138) is valid to obtain the fugacity of a pure substance at any state condition, gas, liquid, or fluid, using an EOS if the equation is valid at the given state condition. The Lee–Kesler corresponding state correlation of the compressibility factor can provide a fugacity coefficient correlation:

$$Z_i = Z_i^{(0)} + \omega Z_i^{(1)}$$

Here, Z is a function of T and P, so we can use Eq. (7.138) to obtain

$$\ln \varphi_i = \ln \varphi_i^{(0)} + \omega \ln \varphi_i^{(1)} \text{ or } \phi_i = \left[\phi_i^{(0)}\right]\left[\phi_i^{(1)}\right]^{\omega} \tag{7.158}$$

in which

$$\ln \varphi_i^{(0)} = \int_0^P \left(Z_i^{(0)} - 1\right)\frac{dP}{P} \tag{7.159}$$

and

$$\ln \varphi_i^{(1)} = \int_0^P Z_i^{(1)}\frac{dP}{P} \tag{7.160}$$

We have included an add-In for Excel® to calculate these quantities with the Lee–Kesler EOS at any state condition.

Unfortunately, equations of state that are functions of pressure and temperature usually apply over very restricted ranges of pressure. Generally, EOS, such as those mentioned in Chapter 3, are functions of temperature and volume/density. To calculate the fugacity coefficient or the fugacity using an EOS, we start with the definition of fugacity:

$$dG_i = RTd\ln f_i$$

The Helmholtz function, A_i, has natural variables of T and V, but the Gibbs function is a natural function of T and P. Practical equations of state (e.g. Soave–Redlich–Kwong [SRK], Peng–Robinson [PR]) are also natural functions of T and V (or ρ), so the Helmholtz function is a better source of equations for fugacity. Using the Legendre transform of G, H, and A

$$G = H - TS$$
$$H = U + PV$$
$$A = U - TS$$

we can substitute H into G to obtain

$$G = U + PV - TS$$

so

$$G = A + PV \tag{7.161}$$

Then, for a pure component

$$RTd\ln f_i = dG_i = d(A_i + PV_i) \tag{7.162}$$

Dividing through by RT

$$d\ln f_i = d\left(\frac{A_i}{RT} + \frac{PV_i}{RT}\right) = d\left(\frac{A_i}{RT}\right) + dZ_i \tag{7.163}$$

For this equation, the variables are P and T (because of the definition of fugacity in terms of G). So, integrating from an ideal gas to real conditions

$$\frac{A_i(T,P)}{RT} - \frac{A_i^{ig}(T,P)}{RT} + Z_i - 1 = \ln f_i - \ln P \tag{7.164}$$

Changing the variables to T and V (or ρ) involves using Eq. (5.100) to arrive at

$$\frac{A_i^r}{RT} - \ln Z_i + Z_i - 1 = \ln\frac{f_i}{P} = \ln\varphi_i$$

The above equation is valid for any EOS in any phase, and this is equivalent to using the definition of G^R/RT in terms of A^r/RT.

Example 7.6

Calculate the fugacity and fugacity coefficient of nitrogen at 400 K and $P = 60$ MPa using the PR EOS.

Solution

The PR EOS is

$$P = \frac{RT}{V-b} - \frac{a}{[V + (1-\sqrt{2})b][V + (1+\sqrt{2})b]}$$

In terms of the compressibility factor,

$$Z = \frac{V}{V-b} - \frac{a}{RT}\frac{V}{[V + (1-\sqrt{2})b][V + (1+\sqrt{2})b]}$$

with

$$a = 0.45724\alpha\frac{R^2T_C^2}{P_C}, \quad b = 0.07780\frac{RT_C}{P_C}$$

and

$$\alpha = [1 + (0.37464 + 1.54226\omega - 0.26992\omega^2)(1 - \sqrt{T/T_C})]^2$$

The EOS must be explicit in molar density to obtain an expression for the fugacity coefficient

$$Z = \frac{1}{1-b\rho} - \frac{a}{RT}\frac{\rho}{[1 + (1-\sqrt{2})b\rho][1 + (1+\sqrt{2})b\rho]}$$

and we must integrate the expression $(Z-1)/\rho$,

$$\frac{Z-1}{\rho} = \frac{b}{1-\rho b} - \frac{a}{RT}\frac{1}{[1+(1-\sqrt{2})b\rho][1+(1+\sqrt{2})b\rho]}$$

Substituting this expression into Eq. (7.138)

$$\ln \varphi_i = \int_0^\rho \frac{b}{1-\rho b} d\rho - \frac{a}{RT}\int_0^\rho \frac{1}{[1+(1-\sqrt{2})b\rho][1+(1+\sqrt{2})b\rho]} d\rho + Z - 1 - \ln Z$$

Using the method of partial fractions, the second integral becomes

$$\text{second integral} = \frac{a}{RTb}\frac{1}{2\sqrt{2}}\int_0^\rho \frac{(1-\sqrt{2})b}{[1+(1-\sqrt{2})b\rho]} d\rho - \frac{a}{RTb}\int_0^\rho \frac{(1+\sqrt{2})b}{[1+(1+\sqrt{2})b\rho]} d\rho$$

Then, the logarithm of the fugacity coefficient becomes

$$\ln \varphi_i = -\ln(1-\rho b) + \frac{a}{RTb}\frac{1}{2\sqrt{2}}\int_0^\rho \frac{(1-\sqrt{2})b}{[1+(1-\sqrt{2})b\rho]} d\rho - \frac{a}{RTb}\frac{1}{2\sqrt{2}}\int_0^\rho \frac{(1+\sqrt{2})b}{[1+(1+\sqrt{2})b\rho]} d\rho + Z - 1 - \ln Z$$

Solving the integrals, we obtain

$$\ln \varphi_i = -\ln(1-\rho b) + \frac{a}{RTb}\frac{1}{2\sqrt{2}}\ln[1+(1-\sqrt{2})b\rho] - \frac{a}{RTb}\frac{1}{2\sqrt{2}}\ln[1+(1+\sqrt{2})b\rho] + Z - 1 - \ln Z$$

and grouping terms,

$$\ln \varphi_i = -\ln(1-\rho b) + \frac{a}{RTb}\frac{1}{2\sqrt{2}}\ln\left(\frac{1+(1-\sqrt{2})b\rho}{1+(1+\sqrt{2})b\rho}\right) + Z - 1 - \ln Z$$

The parameters for the pure component are

Specie	**T_C (K)**	**P_C (bar)**	ω_i
N_2	126.2	34.0	0.038

Specie	**$T_{R,i} = T/T_C$**	α_i	**b_i (cm^3/mol)**	**a_i (bar cm^6/mol^2)**
N_2	3.169	0.43855	24.0088	649 264.1

The fugacity coefficient is a function of density; therefore, we must find the density at the given temperature and pressure. Using Excel, the density is

$$\rho = 0.013696 \text{ mol/cm}^3$$

In the equation

$$Z = 1.31734, \quad -\ln(1-\rho b) = 0.39871, \quad \frac{a}{RTb} = 0.81317, \quad \text{and}$$

$$\ln\left(\frac{1+(1-\sqrt{2})b\rho}{1+(1+\sqrt{2})b\rho}\right) = -0.73077$$

and the fugacity coefficient is

$$\ln\varphi_i = 0.23034, \varphi_i = 1.259, \quad \text{and } f = \varphi P = 755.42\,\text{bar}$$

7.11 Fugacity and Fugacity Coefficient in a Solution

Earlier, we defined the fugacity of a pure substance by observing the functionality of the Gibbs energy of the ideal gas. In a mixture, the chemical potential is the Gibbs property of component i in the mixture. For an ideal gas, Eq. (7.119) relates the chemical potential to the partial pressure

$$d\overline{G}_i^{ig} = d\mu_i^{ig} = RTd\ln p_i \tag{7.119}$$

For real fluids, we can define the fugacity of component i in the mixture as

$$d\overline{G}_i \equiv d\mu_i \equiv RTd\ln\hat{f}_i \quad \text{for } T = \text{constant} \tag{7.165}$$

This equation must satisfy

$$\lim_{P\to 0}\hat{f}_i = p_i = \hat{f}_i^{ig}$$

At a constant temperature and composition, Eq. (7.54) becomes

$$d\overline{G}_i = d\mu_i = RTd\ln\hat{f}_i = \overline{V}_i dP \tag{7.166}$$

then

$$\left[\frac{\partial\ln\hat{f}_i}{\partial P}\right]_{T,x} = \frac{\overline{V}_i}{RT} \tag{7.167}$$

Also, following the same procedure used to obtain Eq. (7.128), we have

$$\left(\frac{\partial\ln\hat{f}_i}{\partial T}\right)_{P,x} = -\frac{\overline{H}_i^R}{RT^2} \tag{7.168}$$

If we integrate Eq. (7.165) from ideal gas to real fluid behavior

$$\overline{G}_i - \overline{G}_i^{ig} = RT\ln\hat{f}_i - RT\ln p_i \tag{7.169}$$

and define the fugacity coefficient of component i in the mixture as

$$\hat{\varphi}_i \equiv \frac{\hat{f}_i}{\hat{f}_i^{ig}} = \frac{\hat{f}_i}{p_i} \tag{7.170}$$

then

$$\overline{G}_i^R = RT\ln\hat{\varphi}_i \tag{7.171}$$

Equation (7.171) indicates that the logarithm of the fugacity coefficient of component i in the mixture is a partial molar quantity. From Eqs. (7.167) and (7.168), the derivatives with respect to pressure and temperature are

$$\left[\frac{\partial \ln \hat{\varphi}_i}{\partial P}\right]_{T,x} = \frac{\overline{V}_i}{RT} - \frac{1}{P} = \frac{\overline{Z}_i - 1}{P} = \frac{\overline{V}_i - V_i^{ig}}{RT} = \frac{\overline{V}_i - \overline{V}_i^{ig}}{RT} = \frac{\overline{V}_i^R}{RT} \tag{7.172}$$

using the same procedure from Eq. (7.128)

$$\left(\frac{\partial \ln \hat{\varphi}_i}{\partial T}\right)_{P,x} = -\frac{\overline{H}_i^R}{RT^2} \tag{7.173}$$

We can write a differential equation for the dimensionless residual Gibbs energy and the partial molar residual Gibbs energy. Consider

$$\frac{nG^R}{RT} = f(T, P, n_1, n_2, \ldots, n_c) \tag{7.174}$$

The total differential is

$$d\left(\frac{nG^R}{RT}\right) = \left[\frac{\partial(nG^R/RT)}{\partial T}\right]_{P,x} dT + \left[\frac{\partial(nG^R/RT)}{\partial P}\right]_{T,x} dP + \sum_{i=1}^{N}\left[\frac{\partial(nG^R/RT)}{\partial n_i}\right]_{T,P,n_{j\neq i}} dn_i$$

or

$$d\left(\frac{nG^R}{RT}\right) = n\left[\frac{\partial(G^R/RT)}{\partial T}\right]_{P,x} dT + n\left[\frac{\partial(G^R/RT)}{\partial P}\right]_{T,x} dP + \sum_{i=1}^{N}\left[\frac{\partial(nG^R/RT)}{\partial n_i}\right]_{T,P,n_{j\neq i}} dn_i \tag{7.175}$$

Using Eqs. (7.133) and (7.134), but for mixtures

$$d\left(\frac{nG^R}{RT}\right) = -\frac{nH^R}{RT^2}dT + \frac{nV^R}{RT}dP + \sum_{i=1}^{N}\left[\frac{\partial(G^R/RT)}{\partial n_i}\right]_{T,P,n_{j\neq i}} dn_i \tag{7.176}$$

Using the definition of partial molar property for the last term

$$d\left(\frac{nG^R}{RT}\right) = -\frac{nH^R}{RT^2}dT + \frac{nV^R}{RT}dP + \sum_{i=1}^{N}\frac{\overline{G}_i^R}{RT}dn_i \tag{7.177}$$

or

$$d\left(\frac{nG^R}{RT}\right) = -\frac{nH^R}{RT^2}dT + \frac{nV^R}{RT}dP + \sum_{i=1}^{N}\ln \hat{\varphi}_i dn_i \tag{7.178}$$

A similar equation can be obtained for the partial molar reduced Gibbs energy,

$$d\left(\frac{\overline{G}_i^R}{RT}\right) = -\frac{\overline{H}_i^R}{RT^2}dT + \frac{\overline{V}_i^R}{RT}dP + \sum_{i=1}^{N}\left[\frac{\partial(\ln \hat{\varphi}_i)}{\partial n_i}\right]_{T,P,n_{j\neq i}} dn_i \tag{7.179}$$

As shown for the pure components using EOS to calculate the fugacity coefficient, the fugacity in a solution using an EOS starts with Eq. (7.165)

$$d\overline{G}_i \equiv d\mu_i \equiv RTd\ln \hat{f}_i$$

However, this traditional definition is incomplete. As we have seen in Eq. (7.13)

$$\mu_i \equiv \left[\frac{\partial(nU)}{\partial n_i}\right]_{nS,nV,n_{j\neq i}} = \left[\frac{\partial(nH)}{\partial n_i}\right]_{nS,P,n_{j\neq i}} = \left[\frac{\partial(nA)}{\partial n_i}\right]_{nS,P,n_{j\neq i}} = \left[\frac{\partial(nG)}{\partial n_i}\right]_{P,T,n_{j\neq i}} = \overline{G}_i$$

Therefore, we should use the Helmholtz energy

$$d\ln\hat{f}_i \equiv d\left(\frac{\overline{G}_i}{RT}\right) = d\left(\frac{\overline{A}_i + P\overline{V}_i}{RT}\right) = d\left(\frac{\overline{A}_i}{RT}\right) + d\overline{Z}_i \tag{7.180}$$

Integrating from ideal gas to real fluid conditions

$$\ln\hat{f}_i - \ln p_i = \frac{\overline{A}_i}{RT} - \frac{\overline{A}_i^{ig}}{RT} + \overline{Z}_i - \overline{Z}_i^{ig} \tag{7.181}$$

Because the conditions are at a given temperature and pressure and $\overline{Z}_i^{ig} = 1$

$$\ln\hat{\varphi}_i = \frac{\overline{A}_i^R}{RT} + \overline{Z}_i - 1 \tag{7.182}$$

We must calculate the partial molar residual dimensionless Helmholtz energy as a function of T and V (or r). Using Eq. (5.100) for a mixture and multiplying by the number of moles

$$\frac{nA^R}{RT} = \frac{nA^r}{RT} - n\ln Z \tag{7.183}$$

Then, obtaining the partial molar property using its definition, Eq. (7.40)

$$\frac{\overline{A}_i^R}{RT} = \frac{\overline{A}_i^r}{RT} - \left[\frac{\partial\left(n\ln\frac{nZ}{n}\right)}{\partial n_i}\right]_{T,P,n_{j\neq i}} = \frac{\overline{A}_i^r}{RT} - \ln Z - \frac{\overline{Z}_i}{Z} + 1 \tag{7.184}$$

and

$$\ln\hat{\varphi}_i = \frac{\overline{A}_i^r}{RT} - \ln Z - \frac{\overline{Z}_i}{Z} + \overline{Z}_i \tag{7.185}$$

However, $\overline{A}_i^r/RT$ is still a derivative at constant T and P, and we need to change it to a derivative that involve the volume. Using the same procedure as in example in Section 7.4

$$\frac{\overline{A}_i^r}{RT} = \left[\frac{\partial\left(\frac{nA^r}{RT}\right)}{\partial n_i}\right]_{T,nV,n_{j\neq i}} + \left[\frac{\partial\left(\frac{A^r}{RT}\right)}{\partial V}\right]_{T,n} \overline{V}_i \tag{7.186}$$

but using the fundamental equation for the Helmholtz energy at a constant temperature for real and ideal conditions

$$\left[\frac{\partial A}{\partial V}\right]_{T,n} = -P$$

for an ideal gas

$$\left[\frac{\partial A^{ig}}{\partial V}\right]_{T,n} = -\frac{RT}{V}$$

Subtracting these equations

$$\left[\frac{\partial\left(\frac{A^r}{RT}\right)}{\partial V}\right]_{T,n} = -\left(\frac{P}{RT} - \frac{1}{V}\right) = -\frac{(Z-1)}{V} \tag{7.187}$$

and substituting this result into Eq. (7.186)

$$\frac{\overline{A}_i^r}{RT} = \left[\frac{\partial\left(\frac{nA^r}{RT}\right)}{\partial n_i}\right]_{T,nV,n_{j\neq i}} - (Z-1)\frac{\overline{V}_i}{V} \tag{7.188}$$

but the ratio of volumes is

$$\frac{\overline{V}_i}{V} = \frac{\overline{Z}_i RT/P}{ZRT/P} = \frac{\overline{Z}_i}{Z} \tag{7.189}$$

Then, replacing this result in Eq. (7.186)

$$\frac{\overline{A}_i^r}{RT} = \left[\frac{\partial\left(\frac{nA^r}{RT}\right)}{\partial n_i}\right]_{T,nV,n_{j\neq i}} - (Z-1)\frac{\overline{Z}_i}{Z} = \left[\frac{\partial\left(\frac{nA^r}{RT}\right)}{\partial n_i}\right]_{T,nV,n_{j\neq i}} - \overline{Z}_i + \frac{\overline{Z}_i}{Z} \tag{7.190}$$

Using the above equation, the fugacity coefficient becomes

$$\ln\hat{\varphi}_i = \left[\frac{\partial\left(\frac{nA^r}{RT}\right)}{\partial n_i}\right]_{T,nV,n_{j\neq i}} - \ln Z \tag{7.191}$$

The derivative in terms of the compressibility factor using Eq. (5.91) is

$$\left[\frac{\partial\left(\frac{nA^r}{RT}\right)}{\partial n_i}\right]_{T,nV,n_{j\neq i}} = \left[\frac{\partial}{\partial n_i}\int_0^\rho n(Z-1)\frac{d\rho}{\rho}\right]_{T,nV,n_{j\neq i}} = \int_0^\rho \left\{\left(\frac{\partial nZ}{\partial n_i}\right)_{T,nV,n_{j\neq i}} - 1\right\}\frac{d\rho}{\rho} \tag{7.192}$$

providing

$$\ln\hat{\varphi}_i = \int_0^\rho \left\{\left(\frac{\partial nZ}{\partial n_i}\right)_{T,nV,n_{j\neq i}} - 1\right\}\frac{d\rho}{\rho} - \ln Z \tag{7.193}$$

This equation can calculate the fugacity coefficient using any EOS in any phase.

7.12 Calculation of the Fugacity and Fugacity Coefficient in Solution

Until now, we have defined the fugacity and fugacity coefficient for component i in a mixture. Now, we must calculate them using an EOS. Because the fugacity coefficient is a constant temperature property, we can use Eq. (7.172) in an integration from ideal gas to real conditions

$$\ln \hat{\varphi}_i = \int_0^P \frac{\overline{Z}_i - 1}{P} dP \tag{7.194}$$

in which $\overline{Z}_i$ is the partial molar compressibility factor defined by

$$\overline{Z}_i = \left[\frac{\partial nZ}{\partial n_i}\right]_{T,P,n_{j\neq i}} \tag{7.195}$$

with Z representing the EOS of a mixture. Equation (7.194) provides the fugacity coefficient if the EOS is a function of pressure. Generally, this is not the case. EOS are most often functions of density. Therefore, we must use an equation similar to Eq. (7.138). From Eq. (7.178), the derivative with respect to n_i at constant T, nV, and $n_{j\neq i}$ is

$$\left[\frac{\partial(nG^R/RT)}{\partial n_i}\right]_{T,nV,n_{j\neq i}} = -\frac{nH^R}{RT^2}\left[\frac{\partial T}{\partial n_i}\right]_{T,nV,n_{j\neq i}} + \frac{nV^R}{RT}\left[\frac{\partial P}{\partial n_i}\right]_{T,nV,n_{j\neq i}} + \sum_{i=1}^{N} \ln \hat{\varphi}_i \left[\frac{\partial n_i}{\partial n_i}\right]_{T,nV,n_{j\neq i}} \tag{7.196}$$

Replacing definitions and eliminating terms

$$\left[\frac{\partial(n\ln\varphi)}{\partial n_i}\right]_{T,nV,n_{j\neq i}} = \frac{n(Z-1)}{P}\left[\frac{\partial P}{\partial n_i}\right]_{T,nV,n_{j\neq i}} + \ln \hat{\varphi}_i \tag{7.197}$$

The derivative of P is

$$\left[\frac{\partial P}{\partial n_i}\right]_{T,nV,n_{j\neq i}} = \left[\frac{\partial}{\partial n_i}\left(\frac{nZRT}{nV}\right)\right]_{T,nV,n_{j\neq i}} = \frac{RT}{nV}\left[\frac{\partial(nZ)}{\partial n_i}\right]_{T,nV,n_{j\neq i}} = \frac{P}{nZ}\left[\frac{\partial(nZ)}{\partial n_i}\right]_{T,nV,n_{j\neq i}} \tag{7.198}$$

Using this result in Eq. (7.197)

$$\ln \hat{\varphi}_i = \left[\frac{\partial(n\ln\varphi)}{\partial n_i}\right]_{T,nV,n_{j\neq i}} - \frac{(Z-1)}{Z}\left[\frac{\partial(nZ)}{\partial n_i}\right]_{T,nV,n_{j\neq i}} \tag{7.199}$$

We can calculate the first derivative from Eq. (7.138). Multiplying by n and taking the derivative with respect to n_i at constant T, nV, and $n_{j\neq i}$,

$$\left[\frac{\partial(n\ln\varphi)}{\partial n_i}\right]_{T,nV,n_{j\neq i}} = \left[\frac{\partial}{\partial n_i}\left(\int_0^\rho (nZ-n)\frac{d\rho}{\rho} + nZ - n - n\ln Z\right)\right]_{T,nV,n_{j\neq i}} \tag{7.200}$$

or

$$\left[\frac{\partial(n\ln\varphi)}{\partial n_i}\right]_{T,nV,n_{j\neq i}} = \int_0^\rho \left(\left[\frac{\partial nZ}{\partial n_i}\right]_{T,nV,n_{j\neq i}} - 1\right)\frac{d\rho}{\rho} + \left(\frac{Z-1}{Z}\right)\left[\frac{\partial nZ}{\partial n_i}\right]_{T,nV,n_{j\neq i}} - \ln Z \tag{7.201}$$

Substituting this equation into Eq. (7.199)

$$\ln \hat{\varphi}_i = \int_0^\rho \left(\left[\frac{\partial nZ}{\partial n_i}\right]_{T,nV,n_{j\neq i}} - 1\right)\frac{d\rho}{\rho} - \ln Z \tag{7.202}$$

We can calculate the fugacity of component i in the mixture using the virial EOS as in Section 7.10. The equation for the mixture is

$$Z = 1 + \frac{B_2 P}{RT} \quad \text{or} \quad Z = 1 + B_2/V$$

We have for the second virial coefficient in the mixture

$$B_2 = \sum_{i=1}^{N} \sum_{i=1}^{N} y_i y_j B_{2,ij} \tag{7.203}$$

in which y_j is the mole fraction of component j in the mixture and $B_{2,\,ij}$ is the interaction second virial coefficient, which represents the interaction between molecules i and j. In Eq. (7.203),

$$B_{2,ij} = B_{2,ji} \tag{7.204}$$

For a binary mixture,

$$\begin{aligned} B_2 &= \sum_{i=1}^{2} \sum_{i=1}^{2} y_i y_j B_{2,ij} = \sum_{i=1}^{2} y_i \{y_1 B_{2,i1} + y_2 B_{2,i2}\} \\ &= y_1 \{y_1 B_{2,11} + y_2 B_{2,12}\} + y_2 \{y_1 B_{2,21} + y_2 B_{2,22}\} \end{aligned} \tag{7.205}$$

Collecting terms and using Eq. (7.204)

$$B_2 = y_1^2 B_{2,11} + 2 y_1 y_2 B_{2,12} + y_2^2 B_{2,22} \tag{7.206}$$

If we use the Berlin form, we must calculate the partial molar second virial coefficient because

$$\overline{Z}_i = 1 + \frac{\overline{B}_{2,i} P}{RT} \tag{7.207}$$

and from Eq. (7.194), the fugacity coefficient of component i in the mixture is

$$\ln \hat{\varphi}_i = \int_0^P \frac{\overline{B}_{2,i}}{RT} dP = \frac{\overline{B}_{2,i} P}{RT} \tag{7.208}$$

The partial molar second virial coefficient is

$$\overline{B}_{2,i} = \left[\frac{\partial (n B_2)}{\partial n_i} \right]_{T,P,n_{j \neq i}} \tag{7.209}$$

and for a binary mixture from Eq. (7.206) is

$$n B_2 = \frac{n_1^2}{n} B_{2,11} + \frac{2 n_1 n_2}{n} B_{2,12} + \frac{n_2^2}{n} B_{2,22} \tag{7.210}$$

therefore

$$\overline{B}_{2,1} = \left[\frac{\partial (n B_2)}{\partial n_1} \right]_{T,P,n_2} = \frac{2 n_1 n - n_1^2}{n^2} B_{2,11} + 2 \frac{n_2 n - n_1 n_2}{n^2} B_{2,12} - \frac{n_2^2}{n^2} B_{2,22} \tag{7.211}$$

or using $y_i = n_i/n$

$$\begin{aligned} \overline{B}_{2,1} &= \left(2 y_1 - y_1^2\right) B_{2,11} + 2(y_2 - y_1 y_2) B_{2,12} - y_2^2 B_{2,22} \\ &= y_1 (2 - y_1) B_{2,11} + 2 y_2 (1 - y_1) B_{2,12} - y_2^2 B_{2,22} \end{aligned} \tag{7.212}$$

Now, replacing $y_1 = 1 - y_2$ and $y_2 = 1 - y_1$ to make everything in terms of y_2

$$\overline{B}_{2,1} = (1 - y_2)(1 + y_2)B_{2,11} + 2y_2^2 B_{2,12} - y_2^2 B_{2,22}$$

and finally

$$\overline{B}_{2,1} = \left(1 - y_2^2\right) B_{2,11} + 2y_2^2 B_{2,12} - y_2^2 B_{2,22} = B_{2,11} + y_2^2 \delta_{12} \tag{7.213}$$

with

$$\delta_{12} = 2B_{2,12} - B_{2,11} - B_{2,22} \tag{7.214}$$

Similarly,

$$\overline{B}_{2,2} = B_{2,22} + y_1^2 \delta_{12} \tag{7.215}$$

therefore

$$\ln \hat{\varphi}_1 = \frac{P}{RT}\left[B_{2,11} + y_2^2 \delta_{12}\right] \tag{7.216}$$

$$\ln \hat{\varphi}_2 = \frac{P}{RT}\left[B_{2,22} + y_1^2 \delta_{12}\right] \tag{7.217}$$

For multicomponent mixtures

$$\ln \hat{\varphi}_k = \frac{P}{RT}\left[B_{2,kk} + \frac{1}{2}\sum_i \sum_j y_i y_j (2\delta_{ik} - \delta_{ij})\right] \tag{7.218}$$

in which $\delta_{ii} = 0$ and $\delta_{ij} = \delta_{ji}$. For a ternary mixture,

$$\ln \hat{\varphi}_1 = \frac{P}{RT}\left[B_{2,11} + y_2^2 \delta_{12} + y_3^2 \delta_{13} + y_2 y_3 \delta_{12} + y_2 y_3 \delta_{13} - y_2 y_3 \delta_{23}\right] \tag{7.219}$$

$$\ln \hat{\varphi}_2 = \frac{P}{RT}\left[B_{2,22} + y_1^2 \delta_{21} + y_3^2 \delta_{23} + y_1 y_3 \delta_{21} + y_1 y_3 \delta_{23} - y_1 y_3 \delta_{13}\right] \tag{7.220}$$

$$\ln \hat{\varphi}_3 = \frac{P}{RT}\left[B_{2,33} + y_1^2 \delta_{31} + y_2^2 \delta_{32} + y_1 y_2 \delta_{31} + y_1 y_2 \delta_{32} - y_1 y_2 \delta_{12}\right] \tag{7.221}$$

If we use the Leiden form of the virial EOS, we should use Eq. (7.202). Then, we must obtain the derivative of nZ with respect to T, nV, and $n_{j \neq i}$. So, multiplying the EOS by n,

$$nZ = n + \frac{nB_2 \cdot n}{nV} \tag{7.222}$$

The derivative is

$$\left[\frac{\partial nZ}{\partial n_i}\right]_{T,nV,n_{j\neq i}} = 1 + \frac{1}{nV}\left\{nB_2 + n\left[\frac{\partial (nB_2)}{\partial n_i}\right]_{T,nV,n_{j\neq i}}\right\} \tag{7.223}$$

Because the virial coefficients do not depend on nV, the derivative of nB_2 equals the partial molar property,

$$\left[\frac{\partial nZ}{\partial n_i}\right]_{T,nV,n_{j\neq i}} = 1 + \frac{1}{V}\{B_2 + \overline{B}_{2,i}\} \tag{7.224}$$

Substituting this result into Eq. (7.202)

$$\ln \hat{\varphi}_i = \int_0^{\rho} (B_2 + \overline{B}_{2,i}) d\rho - \ln Z = (B_2 + \overline{B}_{2,i})\rho - \ln Z \tag{7.225}$$

If we use Eq. (7.212), then

$$B_2 + \overline{B}_{2,1} = 2y_1B_{2,11} + 2y_2B_{2,12} - y_1^2B_{2,11} - 2y_1y_2B_{2,12} - y_2^2B_{2,22} + B_2$$
$$= 2y_1B_{2,11} + 2y_2B_{2,12} \quad (7.226)$$

Similarly,

$$B_2 + \overline{B}_{2,2} = 2y_2B_{2,22} + 2y_1B_{2,12} \quad (7.227)$$

Hall and Iglesias-Silva [1] have provided a general formulation for multicomponent mixtures using any number of virial coefficients.

7.12.1 Using Cubic EOS

As seen before, Eq. (3.24), for a pure substance, the EOS is

$$f(P, V, T) = 0 \Rightarrow P = P(V, T) \quad (3.24)$$

However, for mixtures, this function is related to the composition of $N - 1$ components in the mixture,

$$f(P, V, T, z_1, z_3, z_2, \dots, z_N) = 0 \quad (7.228)$$

with and $\sum_{j=1}^{N} z_j = 1$, or

$$f(P, V, T, z_1, z_3, z_2, \dots, z_{N-1}) = 0 \quad (7.229)$$

One of the biggest problems is that the principle of corresponding states applies to mixtures; however, this is practically impossible because even for simple substances, the number of combinations of mixtures of the same components with different compositions is practically infinite.

For a fixed composition, equations of state for mixtures are applicable as in the case of pure substances:

$$\left.\begin{array}{l}\text{BWR}\\ \text{Peng}-\text{Robinson}\\ \text{Redlich}-\text{Kwong}\\ \text{Virial EOS}\\ \text{etc.}\end{array}\right\}\text{composition-dependent parameters}$$

For example, the RK EOS for mixtures is

$$P = \frac{RT}{V_m - b_m} - \frac{a_m}{T^{1/2}V_m(V_m + b_m)} \quad (7.230)$$

where a_m and b_m are functions of the composition, z_i. Therefore, it is important to define two concepts: mixing rules and combination rules.

7.12.1.1 Mixing Rules

These are methods to combine the parameters of pure components to estimate the parameters of the mixture in the EOS. For example, using cubic equations,

$$a_m = \sum_{i=1}^{N}\sum_{i=1}^{N} z_i z_j a_{ij} \tag{7.231}$$

$$b_m = \sum_{i=1}^{N} z_i b_i \tag{7.232}$$

in which the parameters a_{ij} represent the interaction between component i and component j of the mixture, which requires that we define the term *combination rules.* These are methods to calculate the parameter a_{ij} from the pure component parameters,

$$a_{ij} = \sqrt{a_{ii}a_{jj}}(1 - k_{ij}) \tag{7.233}$$

$$b_{ij} = \frac{b_{ii} + b_{jj}}{2} \tag{7.234}$$

in this case, k_{ij} is the interaction parameter that results from curve fitting the experimental data using optimization procedures to represent the thermodynamic behavior of mixtures. The problem with the interaction parameter is that it is purely empirical and, with a single value, it is impossible to represent all the thermodynamic properties (equilibrium, gaseous behavior, liquid behavior, etc.).

We already have used a mixing rule for the virial EOS truncated at the second virial coefficient. Equation (7.203) is the mixing rule for this EOS. For the virial equation, Eq. (3.76), for mixtures, becomes

$$Z = 1 + \sum_{k=1}^{\infty} B_{k+1}\rho^k = 1 + \sum_{k=1}^{\infty} B'_{k+1}P^k \tag{3.76}$$

The virial EOS is a function of temperature, density (volume), pressure, and composition expressed by the virial coefficients. For mixtures, the virial coefficients follow a mixing rule established by quantum mechanics,

$$B_{z+1} = \sum_{i=1}^{N}\sum_{j=1}^{N}\cdots\sum_{z=1}^{N} x_i x_j \cdots x_z B_{z+1,ij\cdots z} \tag{7.235}$$

in which B is the virial coefficient (2 (second), 3 (third), 4 (fourth), etc.), and the subscripts $ij...z$ reflect molecular interactions. For example, for interactions such as i–i, j–j, ... z–z, correspond to interactions for molecules of the same kind, that is pure components. As you can see, the mixing rules for virial coefficients above the fourth become complicated. Hall et al. [2] have proposed an assumption under Eq. (7.235) can become a quadratic mixing rule. For the third virial coefficient in a binary mixture

$$B_3 = x_1^3 B_{3,111} + x_2^3 B_{3,222} + 3x_1^2 x_2 B_{3,112} + 3x_1 x_2^2 B_{3,122} \tag{7.236}$$

but to be quadratic,

$$B_3 = x_1^3 B_{3,111} + x_2^3 B_{3,222} + 2x_1 x_2[(3/2)x_1 B_{3,112} + (3/2)x_2 B_{3,122}] \tag{7.237}$$

$$= x_1^2 B_{3,111} + x_2^2 B_{3,222} + 2x_1x_2\left[x_1\{(3/2)B_{3,112} - (1/2)B_{3,111}\}\right.$$
$$\left. + x_2\{(3/2)B_{3,122} - (1/2)B_{3,222}\}\right] \tag{7.238}$$

If we assume,

$$B^*_{3,12} = (3B_{3,112} - B_{3,111})/2 = (3B_{3,122} - B_{3,222})/2 \tag{7.239}$$

then the third virial coefficient reduces to a quadratic form,

$$B_3 = x_1^2 B_{3,111} + x_2^2 B_{3,222} + 2x_1x_2 B^*_{3,12} \tag{7.240}$$

The combining rules for the critical parameters used in the calculation of virial coefficients are

$$T_{B,12} = \frac{2(T_{B,11}T_{B,22})}{T_{B,11} + T_{B,22}} \tag{7.241}$$

$$T_{C,12} = \sqrt{T_{C,11}T_{C,22}}(1 - k_{12}) \tag{7.242}$$

$$V_{C,12} = \left[\frac{(V_{C,11})^{1/3} + (V_{C,22})^{1/3}}{2}\right]^3 \tag{7.243}$$

$$\omega_{12} = \frac{\omega_{11} + \omega_{22}}{2} \tag{7.244}$$

$$Z_{C,12} = \frac{Z_{C,11} + Z_{C,22}}{2} \tag{7.245}$$

$$P_{C,12} = Z_{C,12}RT_{C,12}/V_{C,12} \tag{7.246}$$

$$\mu_{R,12} = \frac{\mu_{R,11} + \mu_{R,22}}{2} \tag{7.247}$$

Example 7.7

Calculate the fugacity coefficient of each component of a binary mixture of 25% carbon monoxide and 75% of ethane at 373.15 K and 5 bar.

Solution

The fugacity coefficient of each component in the mixture can come from the virial equation because the pressure is low and the temperature is above the critical temperature for both pure substances. Using the second virial correlation developed by Hall–Iglesias-Silva previously used in Chapter 3,

$$B_2/b_0 = \left(\frac{T_B}{T}\right)^{0.2}\left[1 - \left(\frac{T_B}{T}\right)^{0.8}\right]\left[B_C\left(b_0\left[(T_B/T_C)^{0.2} - (T_B/T_C)\right]\right)\right]^{(T_C/T)^e}$$

$$B_C/V_C = -1.1747 - 0.3668\omega - 0.00061\mu_R$$

$$e = 1.4187 + 1.2058\omega$$

$$b_0/V_C = 0.1368 - 0.4791\omega + 13.81(T_B/T_C)^2\exp[-1.95(T_B/T_C)]$$

$$\mu_R = (10^5)\mu^2 P_C/T_C^2$$

The combining rules for the cross-second virial coefficient are

$$T_{B,12} = \frac{T_{B,11} + T_{B,22}}{2(T_{B,11} T_{B,22})}, \quad T_{C,12} = \sqrt{T_{C,11} T_{C,22}}$$

$$V_{C,12} = \left[\frac{(V_{C,11})^{1/3} + (V_{C,22})^{1/3}}{2} \right]^3, \quad \omega_{12} = \frac{\omega_{11} + \omega_{22}}{2}, \quad \mu_{R,12} = \frac{\mu_{R,11} + \mu_{R,22}}{2}$$

The following table contains pure component parameters,

Substance	T_C (K)	ω	V_C (cm^3/mol)	μ (D)	T_B (K)	P_C (bar)	μ_R
CO	132.85	0.0663	92.17	0.1	345.5	34.9	19.77
C_2H_6	305.33	0.0990	147.06	0.0	757.1	48.72	0.0

Calculated values for the interaction variables are

$T_{B,12} = 474.475$ K, $T_{C,12} = 201.403$ K, $V_{C,12} = 117.487$ K, $\omega_{12} = 0.0827$, and $\mu_{R,12} = 9.887$

The following figure is the prediction of the interaction second virial coefficient of $CO + C_2H_6$ at different temperatures. The experimental data come from McElroy and Ababio [3].

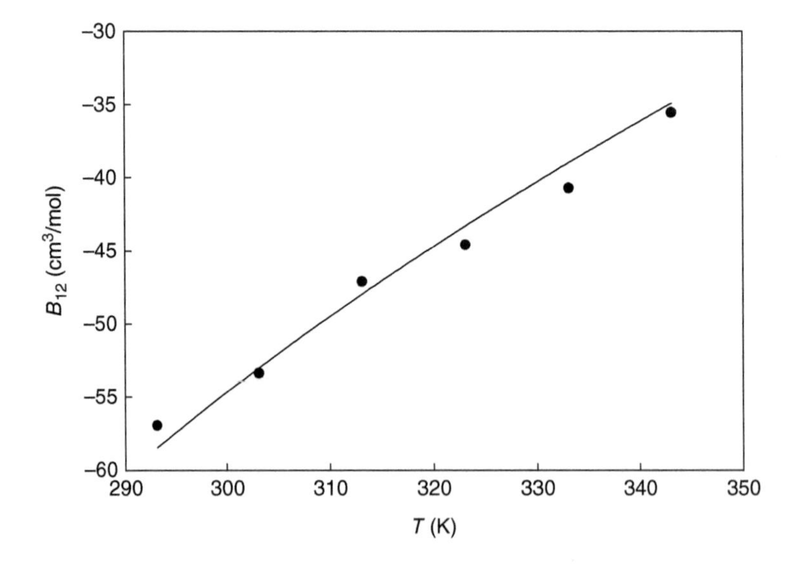

The terms for the calculation of the second virial coefficient for the pure components and the interaction coefficient are

Terms	CO	C_2H_6	CO/C_2H_6
T_R	2.80881	1.22212	1.85276
B_C/V_C	−1.21108	−1.21101	−1.21105
n	1.49865	1.53807	1.51836
T_B/T	0.92590	2.02894	1.27154
T_B/T_C	2.60068	2.47961	2.35585
b_0/V_C	0.69107	0.76396	0.87234
T_C/T	0.35602	0.81825	0.53974
b_0	63.69553	112.3482	102.4884
$(T_C/T)^n$	0.21273	0.73454	0.39206
B/b_0	0.06179	−1.02582	−0.23783
B (cm^3/mol)	3.936	−115.249	−24.375

The fugacity coefficient calculated using the pressure-dependent virial coefficient and Eqs. (7.213)–(7.217) is

$$\delta_{12} = 2B_{2,12} - B_{2,11} - B_{2,22} = 2(-24.375) - (3.936) - (-115.249) = 62.565$$

and

Quantity	CO	C_2H_6
$\overline{B}_{2,i}$ (cm^3/mol)	39.12785	−111.339
$\ln \hat{\varphi}_i$	0.006306	−0.01794
$\hat{\varphi}_i$	1.006326	0.982216

If we wish to use the more accurate density-dependent virial coefficient, we must calculate the density from the virial EOS and calculate the second virial coefficient of the mixture

$$\begin{aligned} B_2 &= y_1^2 B_{2,11} + 2y_1 y_2 B_{2,12} + y_2^2 B_{2,22} \\ &= (0.25)^2(3.936) + 2(0.25)(0.75)(-24.345) + (0.75)^2(-115.249) \\ &= -73.722 \ \text{cm}^3/\text{mol} \end{aligned}$$

From the virial EOS

$$P = \frac{RT}{V} + \frac{RTB_2}{V^2}$$

This is a quadratic equation in volume and the roots are

$$V = \frac{(RT/P)}{2} \pm \frac{(RT/P)\sqrt{1 + (4B_2P)/(RT)}}{2}$$

The higher root is the correct one, so $V = 6130.119\,\text{cm}^3/\text{mol}$. The compressibility factor is $Z = 0.98797$. From Eqs. (7.225) to (7.227), we obtain

Quantity	CO	C_2H_6
$B_2 + \overline{B}_{2,i}$	−34.594	−185.061
$\ln \hat{\varphi}_i$	0.00646	−0.01809
$\hat{\varphi}_i$	1.00648	0.98207

Example 7.8

Find the expression for the fugacity coefficient of component i in the mixture from the SRK EOS.

Solution

We need to use Eq. (7.193)

$$\ln \hat{\varphi}_i = \int_0^{\rho} \left(\left[\frac{\partial nZ}{\partial n_i} \right]_{T,nV,n_{j\neq i}} - 1 \right) \frac{d\rho}{\rho} - \ln Z$$

The compressibility factor is for mixture, so the SRK EOS,

$$P = \frac{RT}{V-b} - \frac{a}{V(V+b)} \Rightarrow Z = \frac{V}{V-b} - \frac{1}{RT}\frac{a}{V+b}$$

where V is the molar volume of the mixture at a given T and P, and the characteristic parameters are

$$b = \sum x_i b_i \text{ and } a = \sum_i \sum_j x_i x_j a_{ij}$$

Now, if we multiply the EOS by n

$$nZ = \frac{nV \cdot n}{nV - nb} - \frac{1}{RT}\frac{na \cdot n}{nV + nb}$$

Then, the derivative with respect to n_i keeping constant $n_{j\neq i}$ and nV is

$$\left[\frac{\partial nZ}{\partial n_i} \right]_{T,nV,n_{j\neq i}} = \frac{nV}{nV - nb} - \frac{nV \cdot n\left(\frac{\partial nb}{\partial n_i}\right)_{T,nV,n_{j\neq i}}}{(nV-nb)^2} - \frac{1}{RT}\frac{na + n\left(\frac{\partial na}{\partial n_i}\right)_{T,nV,n_{j\neq i}}}{nV+nb}$$
$$+ \frac{1}{RT}\frac{na \cdot n \cdot \left(\frac{\partial nb}{\partial n_i}\right)_{T,nV,n_{j\neq i}i}}{(nV+nb)^2}$$

Here, a and b do not depend on the molar volume; therefore, the derivatives of the characteristic parameters are also the partial molar characteristic parameters, $\overline{a}_i$ and $\overline{b}$,

$$\left[\frac{\partial nZ}{\partial n_i} \right]_{T,nV,n_{j\neq i}} = \frac{V}{V-b} + \frac{V\widetilde{b}_i}{(V-b)^2} - \frac{1}{RT}\frac{a + \widetilde{a}_i}{V+b} + \frac{1}{RT}\frac{a\widetilde{b}_i}{(V+b)^2}$$

In terms of density,

$$\left[\frac{\partial nZ}{\partial n_i} \right]_{T,nV,n_{j\neq i}} = \frac{1}{1-\rho b} + \frac{\rho\widetilde{b}_i}{(1-\rho b)^2} - \frac{\rho}{RT}\frac{a + \widetilde{a}_i}{1+\rho b} + \frac{1}{RT}\frac{\rho^2 a\widetilde{b}_i}{(1+\rho b)^2}$$

and

$$\left[\frac{\partial nZ}{\partial n_i}\right]_{T,nV,n_{j\neq i}} - 1 = \frac{\rho b}{1-\rho b} + \frac{\rho \widetilde{b}_i}{(1-\rho b)^2} - \frac{\rho}{RT}\frac{a+\widetilde{a}_i}{1+\rho b} + \frac{1}{RT}\frac{\rho^2 a\widetilde{b}_i}{(1+\rho b)^2}$$

The integral in Eq. (7.193) is formed of several terms

$$\int_0^\rho \frac{\rho b}{1-\rho b}\frac{d\rho}{\rho} = -\ln(1-\rho b)$$

$$\int_0^\rho \frac{\rho \widetilde{b}_i}{(1-\rho b)^2}\frac{d\rho}{\rho} = \frac{\widetilde{b}_i}{b}\left(\frac{1}{1-\rho b} - 1\right) = \frac{\widetilde{b}_i}{b}\frac{\rho b}{1-\rho b}$$

$$\int_0^\rho \frac{\rho}{RT}\frac{a+\widetilde{a}_i}{1+\rho b}\frac{d\rho}{\rho} = \frac{1}{RT}\frac{a+\widetilde{a}_i}{b}\ln(1+\rho b)$$

and

$$\frac{1}{RT}\int_0^\rho \frac{\rho^2 a\widetilde{b}_i}{(1+\rho b)^2}\frac{d\rho}{\rho} = \frac{a\widetilde{b}_i}{RT}\int_0^\rho \frac{\rho}{(1+\rho b)^2}\frac{d\rho}{\rho} = \frac{a\widetilde{b}_i}{RT}\frac{1}{b^2(1+\rho b)} - \frac{a\widetilde{b}_i}{RTb^2} + \frac{a\widetilde{b}_i}{RT}\frac{1}{b^2}\ln(1+b\rho)$$

$$= -\frac{a\widetilde{b}_i}{RT}\frac{\rho}{b(1+\rho b)} + \frac{a\widetilde{b}_i}{RT}\frac{1}{b^2}\ln(1+b\rho)$$

Therefore

$$\ln\widehat{\varphi}_i = \int_0^\rho \left(\left[\frac{\partial nZ}{\partial n_i}\right]_{T,nV,n_{j\neq i}} - 1\right)\frac{d\rho}{\rho} - \ln Z$$

$$\ln\widehat{\varphi}_i = -\ln(1-\rho b) + \frac{\widetilde{b}_i}{b}\frac{\rho b}{1-\rho b} - \frac{1}{RT}\frac{a+\widetilde{a}_i}{b}\ln(1+\rho b) - \frac{a\widetilde{b}_i}{RT}\frac{\rho}{b(1+\rho b)} + \frac{a\widetilde{b}_i}{RT}\frac{1}{b^2}\ln(1+b\rho) - \ln Z$$

or,

$$\ln\widehat{\varphi}_i = -\ln(1-\rho b) + \frac{\widetilde{b}_i}{b}\left[\frac{\rho b}{1-\rho b} - \frac{a}{RT}\frac{\rho}{(1+\rho b)}\right] + \left(-\frac{1}{RT}\frac{a+\widetilde{a}_i}{b} + \frac{a\widetilde{b}_i}{RT}\frac{1}{b^2}\right)\ln(1+\rho b) - \ln Z$$

Now, we need to calculate

$$\widetilde{a}_i = \left[\frac{\partial na}{\partial n_i}\right]_{T,nV,n_{j\neq i}}, \quad \widetilde{b}_i = \left[\frac{\partial nb}{\partial n_i}\right]_{T,nV,n_{j\neq i}}$$

Since $b = \sum x_i b_i$,

$$\widetilde{b}_i = \left[\frac{\partial nb}{\partial n_i}\right]_{T,nV,n_{j\neq i}} = b_i$$

and $a = \sum_i\sum_j x_i x_j a_{ij}$,

$$\widetilde{a}_i = \left[\frac{\partial na}{\partial n_i}\right]_{T,nV,n_{j\neq i}} = -a + 2\sum_{k=1}^{n} x_k a_{ki}$$

In this case, $\widetilde{a}_i = \overline{a}_i$ and $\widetilde{b}_i = \overline{b}_i$ since they do not depend upon the molar volume. For a binary mixture,

$$\overline{a}_i = \left[\frac{\partial na}{\partial n_i}\right]_{T,nV,n_{j\neq i}} = -a + 2\sum_{k=1}^{n} x_k a_{ki} = -a + 2x_1 a_{1i} + 2x_2 a_{2i}$$

In the equation for the fugacity coefficient of component i in the mixture, the molar volume results if given T and P are given. Sometimes, one prefers to have it in terms of the compressibility factor, if

$$\frac{\overline{b}_i}{b}\left[\frac{\rho b}{1-\rho b} - \frac{a}{RT}\frac{\rho}{(1+\rho b)}\right] = \frac{\overline{b}_i}{b}(Z-1)$$

$$\widetilde{D}_i = \left(\frac{1}{RT}\frac{a+\overline{a}_i}{b} - \frac{a\overline{b}_i}{RT}\frac{1}{b^2}\right) = \frac{a}{bRT}\left(\frac{a+\overline{a}_i}{a} - \frac{\overline{b}_i}{b}\right) = \frac{A}{B}\left(\frac{a+\overline{a}_i}{a} - \frac{\overline{b}_i}{b}\right)$$

where A and B are given by Eqs. (3.151) and (3.152) but a and b are mixture parameters

$$A = \frac{aP}{(RT)^2} \text{ and } B = \frac{Pb}{RT}$$

$$(1+\rho b) = \frac{Z+B}{Z}$$

and

$$\ln(Z-B) = \ln\left(\frac{Z-B}{Z}\right) + \ln Z = \ln(1-\rho b) + \ln Z$$

Then

$$\ln \hat{\varphi}_i = \frac{b_i}{b}(Z-1) - \ln(Z-B) - \overline{D}_i \ln\left(\frac{Z+B}{Z}\right)$$

7.13 Ideal Solutions

In many engineering processes, we deal with liquid mixtures and solutions. We can calculate the thermodynamic properties using an EOS, but this requires the equation to be accurate and to know the mixing rules. One way to avoiding using an EOS is to treat liquid mixtures as a separate entity. This requires formulating new property changes, specific concepts, and even reference states.

In gas mixtures, we can use the ideal gas behavior as a limiting value. Similarly, for liquid mixtures a reference state can be an ideal solution. This new reference state need not be at a given pressure, temperature, or volume as in the ideal gas (real gases behave as ideal gases at low pressures). Rather, it must incorporate chemical structures and intermolecular forces. In solutions, molecules interact with one another through these forces. Then, the definition of an ideal solution should be a solution in which the intermolecular forces of the components are equal. For example, if a solution contains components A and B, the intermolecular forces between A–A, B–B, and A–B configurations are the same. For example, mixtures of 1-propanol + 2-propanol, n-pentane + n-hexane, and R11 + R12 behave as ideal solutions. In an ideal solution, molecules are distinguishable in contrast to the ideal gas mixture. The ideal solution and ideal gas behavior are hypothetical, so they only approximate real mixtures.

In an ideal gas mixture

$$\overline{G}_i^{ig} = \mu_i^{ig} = G_i^{ig} + RT\ln y_i \tag{7.116}$$

Guided by this equation, we can define the partial molar property of an ideal solution as

$$\overline{G}_i^{is} = \mu_i^{is} \equiv G_i + RT\ln x_i \tag{7.248}$$

in which G_i is the Gibbs energy of the pure component at T and P of the mixture and x_i is the mole fraction. It is customary to use x because this equation applies to liquid mixtures. Taking the differential at a constant temperature

$$d\overline{G}_i^{is} = d\mu_i^{is} \equiv dG_i + RTd\ln x_i \text{ at } T = \text{constant} \tag{7.249}$$

Then, using Eq. (7.120), the definition of fugacity

$$d\overline{G}_i^{is} = d\mu_i^{is} \equiv RTd\ln f_i + RTd\ln x_i = RTd\ln(f_i x_i) \tag{7.250}$$

Applying the definition for the fugacity coefficient of component i in the mixture, Eq. (7.165), to an ideal solution

$$d\overline{G}_i^{is} \equiv d\mu_i^{is} \equiv RTd\ln \hat{f}_i^{is} \tag{7.251}$$

Comparing Eqs. (7.250) and (7.251)

$$\hat{f}_i^{is} = f_i x_i \tag{7.252}$$

This equation is the **Lewis–Randall rule**. It posits that the fugacity coefficient of component i in the mixture is a straight line with composition that passes through zero with a slope equal to the fugacity of the pure component. Note the resemblance to the equation for component i in a mixture of ideal gases, $\hat{f}_i^{ig} = Py_i$. Because gases are also subject to intermolecular forces, they often can be treated as ideal solutions. The definition of fugacity coefficient for component i in an ideal solution is

$$\hat{\varphi}_i^{is} = \frac{\hat{f}_i^{is}}{p_i} = \frac{f_i x_i}{P x_i} = \frac{f_i}{P} = \varphi_i \tag{7.253}$$

In an ideal solution, the fugacity coefficient of the species in the mixture equals the fugacity of the pure component at the T and P of the mixture independent of composition. Because Eq. (7.249) is similar to Eq. (7.116), the thermodynamic relationship should be similar. Equation (7.28) for an ideal solution is

$$\left[\frac{\partial \overline{G}_i^{is}}{\partial P}\right]_{T,x} = \overline{V}_i^{is} \tag{7.254}$$

Taking the derivative with respect to P at constant T and n using Eq. (7.248),

$$\left[\frac{\partial \overline{G}_i^{is}}{\partial P}\right]_{T,x} = \left[\frac{\partial G_i}{\partial P}\right]_{T,x} \tag{7.255}$$

Comparing Eq. (7.255) and (7.254)

$$\overline{V}_i^{is} = V_i \tag{7.256}$$

Also, Eq. (7.29) applied to an ideal solution is

$$\left[\frac{\partial \overline{G}_i^{is}}{\partial T}\right]_{P,x} = -\overline{S}_i^{is} \tag{7.257}$$

If we obtain the same derivative from Eq. (7.248), we have

$$\left[\frac{\partial \overline{G}_i^{is}}{\partial T}\right]_{P,x} = \left[\frac{\partial G_i}{\partial T}\right]_{P,x} - R\ln x_i = -S_i - R\ln x_i \tag{7.258}$$

Comparing the last two equations

$$\overline{S}_i^{is} = S_i + R\ln x_i \tag{7.259}$$

Then, we can have similar equations to those of an ideal gas mixture for ideal solutions

$$\overline{H}_i^{is} = H_i \tag{7.260}$$

$$\overline{U}_i^{is} = U_i \tag{7.261}$$

$$\overline{A}_i^{is} = A_i + RT\ln x_i \tag{7.262}$$

and using the summation relationship, Eq. (7.45), for an ideal solution mixture

$$V^{is} = \sum_{i=1}^{N} \overline{V}_i^{is} x_i = \sum_{i=1}^{N} V_i x_i \tag{7.263}$$

$$S^{is} = \sum_{i=1}^{N} \overline{S}_i^{is} x_i = \sum_{i=1}^{N} S_i x_i + R\sum_{i=1}^{N} x_i \ln x_i \tag{7.264}$$

$$H^{is} = \sum_{i=1}^{N} \overline{H}_i^{is} x_i = \sum_{i=1}^{N} H_i x_i \tag{7.265}$$

$$U^{is} = \sum_{i=1}^{N} \overline{U}_i^{is} x_i = \sum_{i=1}^{N} U_i x_i \tag{7.266}$$

$$A^{is} = \sum_{i=1}^{N} \overline{A}_i^{is} x_i = \sum_{i=1}^{N} A_i x_i + RT\sum_{i=1}^{N} x_i \ln x_i \tag{7.267}$$

and

$$G^{is} = \sum_{i=1}^{N} \overline{G}_i^{is} x_i = \sum_{i=1}^{N} \mu_i^{is} x_i \equiv \sum_{i=1}^{N} G_i x_i + RT\sum_{i=1}^{N} x_i \ln x_i \tag{7.268}$$

7.14 Excess Properties. Activity Coefficients

We define an excess property as

$$F^E \equiv F - F^{is} \tag{7.269}$$

which is the difference between the molar property of the mixture and the value of the same property as an ideal solution. This definition is similar to the definition of a residual property at T, P but it uses as a reference the ideal solution. For example, we can have $G^E, U^E, H^E, A^E, V^E, S^E, C_P^E$, etc.

The excess properties can be related to the residual property because $F^R = F - F^{ig}$; therefore,

$$F^E \equiv F^R - (F^{is} - F^{ig}) \tag{7.270}$$

but $F^{ig} = \sum_{i=1}^{N} \overline{F}_i^{ig} x_i$ and $F^{is} = \sum_{i=1}^{N} \overline{F}_i^{is} x_i$, and replacing the partial properties for each component causes the logarithmic terms to cancel leaving

$$F^E = F^R - \sum_{i=1}^{N} \left(F_i - F_i^{ig}\right) x_i = F^R - \sum_{i=1}^{N} F_i^R x_i \tag{7.271}$$

All the above equations are linear; therefore, the concept of partial molar properties applies to all of them. For example,

$$\overline{F}_i^E \equiv \overline{F}_i - \overline{F}_i^{is} \tag{7.272}$$

and

$$\overline{F}_i^E = \overline{F}_i^R - F_i^R \tag{7.273}$$

For the excess Gibbs energy

$$\overline{G}_i^E = \overline{G}_i - \overline{G}_i^{is} \tag{7.274}$$

Integrating Eq. (7.165) from the ideal solution to the real state,

$$\overline{G}_i = \overline{G}_i^{is} + RT \ln \hat{f}_i - RT \ln \hat{f}_i^{is} = \overline{G}_i^{is} + RT \ln \frac{\hat{f}_i}{f_i x_i} \tag{7.275}$$

We define the activity coefficient as

$$\gamma_i \equiv \frac{\hat{f}_i}{\hat{f}_i^{is}} = \frac{\hat{f}_i}{f_i x_i} \tag{7.276}$$

with $\gamma_i^{is} \equiv 1$. Substituting Eq. (7.275) into Eq. (7.274)

$$\overline{G}_i^E = RT \ln \gamma_i \tag{7.277}$$

and differentiating at a constant temperature

$$d\overline{G}_i^E = RTd \ln \gamma_i \tag{7.278}$$

This equation is similar to the equation developed for the fugacity coefficient, Eq. (7.171). Because the definition of an excess property is a linear relation, all the previous equations developed for the fugacity coefficient of component i in the mixture apply to the activity coefficient after replacing the residual by excess. That is, Eqs. (7.172), (7.173), (7.178), and (7.179) become

$$\left(\frac{\partial \ln \gamma_i}{\partial P}\right)_{T,x} = \frac{\overline{V}_i^E}{RT} \tag{7.279}$$

$$\left(\frac{\partial \ln \gamma_i}{\partial T}\right)_{P,x} = -\frac{\overline{H}_i^E}{RT^2} \tag{7.280}$$

$$d\left(\frac{nG^E}{RT}\right) = -\frac{nH^E}{RT^2}dT + \frac{nV^E}{RT}dP + \sum_{i=1}^{N} \ln \gamma_i dn_i \tag{7.281}$$

and

$$d\left(\frac{\overline{G}_i^E}{RT}\right) = -\frac{\overline{H}_i^E}{RT^2}dT + \frac{\overline{V}_i^E}{RT}dP + \sum_{i=1}^{N}\left[\frac{\partial(\ln \gamma_i)}{\partial n_i}\right]_{T,P,n_{j\neq i}} dn_i \tag{7.282}$$

Finally, we can calculate the activity coefficient if we know the functionality of the excess Gibbs energy with respect to composition.

$$\frac{\overline{G}_i^E}{RT} = \ln \gamma_i = \left[\frac{\partial(nG^E/RT)}{\partial n_i}\right]_{T,P,n_{j\neq i}} \tag{7.283}$$

From the definition of fugacity coefficient in solution (Eq. (7.170))

$$\hat{f}_i = \hat{\varphi}_i P x_i$$

Equating this result to the fugacity in solution from activity coefficients

$$\gamma_i f_i x_i = \hat{\varphi}_i P x_i \tag{7.284}$$

then

$$\gamma_i = \frac{\hat{\varphi}_i P}{f_i} = \frac{\hat{\varphi}_i}{f_i/P} = \frac{\hat{\varphi}_i}{\varphi_i} \tag{7.285}$$

The activity coefficient is the ratio of the fugacity coefficient of component i in solution and the fugacity coefficient of the pure component. We can calculate the activity coefficient from a pressure-dependent EOS by subtracting Eq. (7.138) from Eq. (7.194)

$$\ln \hat{\varphi}_i - \ln \varphi_i = \int_0^P \frac{\overline{Z}_i - 1}{P}dP - \int_0^P \frac{Z_i - 1}{P}dP$$

$$\ln \gamma_i = \ln \frac{\hat{\varphi}_i}{\varphi_i} = \int_0^P \frac{\overline{Z}_i - Z_i}{P}dP \tag{7.286}$$

Also, for density-dependent EOS, subtracting Eq. (7.138) from Eq. (7.202),

$$\ln \gamma_i = \int_0^{\rho}\left(\left[\frac{\partial nZ}{\partial n_i}\right]_{T,nV,n_{j\neq i}} - Z_i\right)\frac{d\rho}{\rho} - \ln \frac{Z}{Z_i} - (Z_i - 1) \tag{7.287}$$

Using the Legendre transform of the Helmholtz energy as a partial molar property, the partial molar excess Gibbs energy is

$$\overline{G}_i^E = \overline{A}_i^E + P\overline{V}_i^E \tag{7.288}$$

Then, the activity coefficient becomes

$$\ln \gamma_i = \left(\frac{\overline{A}_i^E}{RT}\right) + \left(\overline{Z}_i^E\right) \tag{7.289}$$

Expressing this equation in terms of residual properties using Eq. (7.273)

$$\ln \gamma_i == \left(\frac{\overline{A}_i^R}{RT}\right) - \left(\frac{A_i^R}{RT}\right) + \overline{Z}_i^R - Z_i^R \tag{7.290}$$

in which A_i^R and Z_i^R are the residual Helmholtz energy and the residual compressibility factor of the pure component i. Then, substituting the definitions of the partial properties in terms of compressibility factor, Eq. (7.184)

$$\ln \gamma_i = \left(\frac{\overline{A}_i^r}{RT} - \ln Z - \frac{\overline{Z}_i}{Z} + 1\right) - \left\{\frac{A_i^r}{RT} - \ln Z_i\right\} + \{\overline{Z}_i - 1\} - \{Z_i - 1\} \tag{7.291}$$

or using Eq. (7.190)

$$\ln \gamma_i = \left(\left[\frac{\partial\left(\frac{nA^r}{RT}\right)}{\partial n_i}\right]_{T,nV,n_{j\neq i}} - \overline{Z}_i - \ln Z + 1\right) - \left\{\frac{A_i^r}{RT} - \ln Z_i\right\} + \overline{Z}_i - Z_i \tag{7.292}$$

Now, using the integral equations or the Helmholtz energy,

$$\begin{aligned} \ln \gamma_i = {} & \left(\int_0^\rho \left\{\left(\frac{\partial nZ}{\partial n_i}\right)_{T,nV,n_{j\neq i}} - 1\right\}\frac{d\rho}{\rho} - \overline{Z}_i - \ln Z + 1\right) \\ & - \left\{\int_0^\rho \{Z_i - 1\}\frac{d\rho}{\rho} - \ln Z_i\right\} + \overline{Z}_i - Z_i \end{aligned} \tag{7.293}$$

Rearranging the terms

$$\ln \gamma_i = \int_0^\rho \left\{\left(\frac{\partial nZ}{\partial n_i}\right)_{T,nV,n_{j\neq i}} - Z_i\right\}\frac{d\rho}{\rho} - \ln\frac{Z}{Z_i} - (Z_i - 1) \tag{7.294}$$

The above equation is the formal derivation for Eq. (7.287).

7.15 Activity Coefficients with Different Standard States

In Section 7.10, we calculated the fugacity coefficient using an EOS valid for the mixture. We did not require a standard state (a particular state of a species at a given pressure, composition, and physical state gas, liquid, or solid) because implicitly we used the ideal gas at T and P of the mixture. Also, in our definition of the activity coefficient, we have assumed the standard state as the pure liquid at T and P of the mixture. In general, we can define the activity coefficient as

$$\gamma_i^{\dagger} \equiv \frac{\hat{f}_i}{\hat{f}_i^{is,\dagger}} = \frac{\hat{f}_i}{f_i^{\dagger} x_i} \tag{7.295}$$

in which $\gamma_i^{\dagger}$ is the activity coefficient with respect to the standard state and $\hat{f}_i^{is,\dagger}$ is the ideal solution fugacity of component i in the mixture with respect to the standard state $f_i^{\dagger}$.

If we use Lewis and Randall rule, $f_i^{\dagger} = f_i$. The quantity $\hat{f}_i/x_i$ does not depend upon the standard state, so obtaining the limit when the mole fraction composition tends to unity

$$\lim_{x_i \to 1} \left(\frac{\hat{f}_i}{x_i} \right) = \frac{f_i}{1} = f_i \tag{7.296}$$

therefore

$$\lim_{x_i \to 1} \gamma_i^{LR} = 1 \tag{7.297}$$

Notice that the Lewis–Randall rule is valid near mole fractions equaling unity. There are occasions when the mixture does not exist over the entire composition range. In such cases, the Lewis–Randall rule is not an adequate standard state because it is valid at concentrations for which the mixture does not exist, and the pure components are not at equilibrium. In this situation, another standard state is necessary, and the choice must be valid at low concentrations. We call this standard state *Henry's law*, and it must satisfy

$$\lim_{x_i \to 0} \left(\frac{\hat{f}_i}{x_i} \right) = H_i = f_i^{\dagger} \tag{7.298}$$

in which H_i is Henry's law constant. Also

$$\lim_{x_i \to 0} \gamma_i^{HL} = 1 \tag{7.299}$$

and

$$\hat{f}_i^{is,HL} = H_i x_i \tag{7.300}$$

Figure 7.3 is a representation of Lewis–Randall and Henry laws.

Sometimes, it is necessary to change the activity coefficient in a different standard state to the one that follows the Lewis–Randall rule, if one uses Eqs. (7.295) and (7.298)

$$\gamma_i^{\dagger} \equiv \frac{\hat{f}_i}{f_i^{\dagger} x_i} = \frac{\gamma_i f_i}{H_i} \tag{7.301}$$

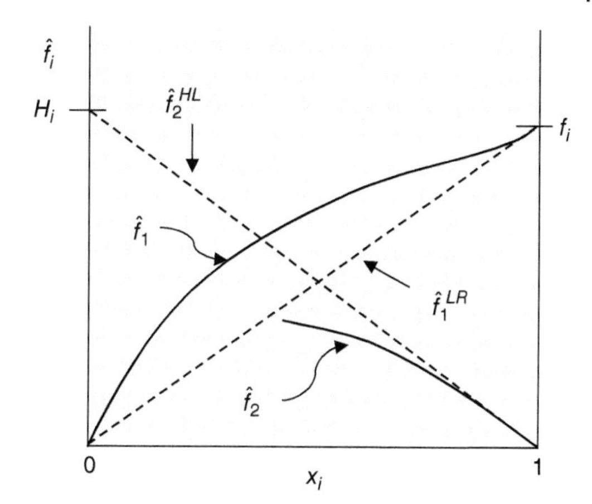

Figure 7.3 Representation of the Lewis–Randall Rule and Henry Law.

Now, we know that

$$\gamma_i \equiv \frac{\hat{f}_i}{x_i f_i}$$

From the definition of Henry's constant, Eq. (7.298)

$$H_i = \lim_{x_i \to 0} \left(\frac{\hat{f}_i}{x_i} \right) = \lim_{x_i \to 0} (\gamma_i f_i) = \gamma_i^{\infty} f_i \tag{7.302}$$

Substituting the above result in Eq. (7.301)

$$\gamma_i^{\dagger} = \frac{\gamma_i f_i}{\gamma_i^{\infty} f_i} = \frac{\gamma_i}{\gamma_i^{\infty}} \tag{7.303}$$

This is the relation of the activity coefficient in a different standard state with respect to the one that follows the definition of the activity coefficient (Lewis–Randall rule).

7.16 Effect of Pressure on the Fugacity in Solution and Activity Coefficients Using the Lewis–Randall Rule

As mentioned in Section 7.11, we can calculate the fugacity of a component in a solution using an EOS (in fact, this is the practical path). Also, we can calculate the fugacity coefficient from the definition of the activity coefficient

$$\hat{f}_i(T, P) = \gamma_i(T, P) f_i(T, P) x_i \tag{7.304}$$

Here, we have used the pressure because the fugacity of a pure component depends upon the pressure of the standard state. Let us assume that the fugacity of the pure component is a different pressure from that of the mixture, that is, $f_i(T, P^0)$, then we should calculate the fugacity at the conditions of the mixture using Eq. (7.123)

$$f_i(T, P) = f_i(T, P^0) \exp \left[\frac{1}{RT} \int_{P^0}^{P} V_i dP \right] \tag{7.305}$$

and Eq. (7.304) becomes

$$\hat{f}_i(T,P) = \gamma_i(T,P)x_i f_i(T,P^0)\exp\left[\frac{1}{RT}\int_{P^0}^{P} V_i dP\right] \tag{7.306}$$

Also, we can encounter the situation in which a value of the activity coefficient has its basis in a different standard state with $f_i(T, P^0)$, then

$$\gamma_i^0(T,P) = \frac{\hat{\varphi}_i(T,P)P}{f_i(T,P^0)} \tag{7.307}$$

If we multiply and divide by $f_i(T, P)$,

$$\gamma_i^0(T,P) = \frac{\hat{\varphi}_i(T,P)P}{f_i(T,P)}\frac{f_i(T,P)}{f_i(T,P^0)} = \frac{\hat{\varphi}_i(T,P)}{\varphi_i(T,P)}\frac{f_i(T,P)}{f_i(T,P^0)} \tag{7.308}$$

then using Eqs. (7.305) and (7.285)

$$\gamma_i^0(T,P) = \gamma_i(T,P)\exp\left[\frac{1}{RT}\int_{P^0}^{P} V_i dP\right] \tag{7.309}$$

and

$$\gamma_i(T,P) = \gamma_i^0(T,P)\exp\left[-\frac{1}{RT}\int_{P^0}^{P} V_i dP\right] \tag{7.310}$$

This equation provides the activity coefficient at the temperature and pressure of the mixture. We can calculate the activity coefficient at a different pressure of the mixture, $\gamma_i(T, P^0)$, but the standard state is the fugacity of the pure component at the conditions of the mixture. Simply calculate the activity at any pressure using Eq. (7.279)

$$\gamma_i(T,P) = \gamma_i(T,P^0)\exp\left[\frac{1}{RT}\int_{P^0}^{P} \overline{V}_i^E dP\right] = \gamma_i(T,P^0)\exp\left[\frac{1}{RT}\int_{P^0}^{P} (\overline{V}_i^E - V_i)dP\right] \tag{7.311}$$

then calculate the value of the fugacity of the solution at P^0 from

$$\hat{f}_i(T,P^0) = \gamma_i(T,P^0)f_i(T,P^0)x_i \tag{7.312}$$

If we want to change this value to a different pressure, we can use Eq. (7.167)

$$\hat{f}_i(T,P) = \hat{f}_i(T,P^0)\exp\left[\frac{1}{RT}\int_{P^0}^{P} \overline{V}_i dP\right] \tag{7.313}$$

Example 7.9

Calculate the percentage change in the activity coefficient of nitrogen in an equimolar mixture of nitrogen + carbon dioxide at 350 K, if there is a pressure change of 20 MPa. Assume that the partial molar excess volume of nitrogen is not a function of pressure.

Solution

Using the results from Example 7.3 in Section 7.4

$$\overline{V}_1 = 0.29884 + 0.103137x_2^3 - 0.005857x_1x_2^2$$

From the molar volume of the mixture, we can obtain $V_1 = 0.29884$. Assuming that the partial molar excess volume $(\overline{V}_1 - V_1)$ is independent of pressure, then

$$\overline{V}_1 - V_1 = 0.103137x_2^3 - 0.005857x_1x_2^2$$

and

$$\frac{\gamma_1(T,P)}{\gamma_1(T,P^0)} = \exp\left[\frac{(\overline{V}_1 - V_1)}{RT}\Delta P\right]$$

or

$$\frac{\gamma_1(T,P)}{\gamma_1(T,P^0)} = \exp\left[\frac{\{0.103137(0.5)^3 - 0.005857(0.5)^3\}(1000)}{(83.14)(350)}100\right] = 0.0836$$

and the percentage change is 8.36%.

7.17 Property Change on Mixing

Let us consider a mixing process in which a membrane separates two pure components at constant T and P. Then, by removing the membrane, the two components mix at the same temperature and pressure. Figure 7.4 is a diagram of this process.

A contraction or expansion of the system occurs, enabling the pressure to remain constant. Also, energy must flow into or out of the system as heat to maintain a constant temperature. The total volume change in the system after mixing is

$$\Delta V^{total} = (n_1 + n_2)V - n_1V_1 - n_2V_2 \tag{7.314}$$

Dividing the equation by $n_1 + n_2$

$$\Delta V = \frac{\Delta V^{total}}{n_1 + n_2} = V - x_1V_1 - x_2V_2 \tag{7.315}$$

Applying the first law to a closed system at a constant pressure and temperature

$$q = \Delta H^{total} = (n_1 + n_2)H - n_1H_1 - n_2H_2 \tag{7.316}$$

then

$$Q = \Delta H = \frac{\Delta H^{total}}{n_1 + n_2} = H - x_1H_1 - x_2H_2 \tag{7.317}$$

Now, we can generalize our results and define any property change of mixing as

$$\Delta_m F(T,P) \equiv F(T,P) - \sum_{i=1}^{N} z_i F_i(T,P) \tag{7.318}$$

or

$$\Delta_m F(T,P) \equiv \sum_{i=1}^{N} z_i \overline{F}_i(T,P) - \sum_{i=1}^{N} z_i F_i(T,P) = \sum_{i=1}^{N} z_i \{\overline{F}_i(T,P) - F_i(T,P)\} \tag{7.319}$$

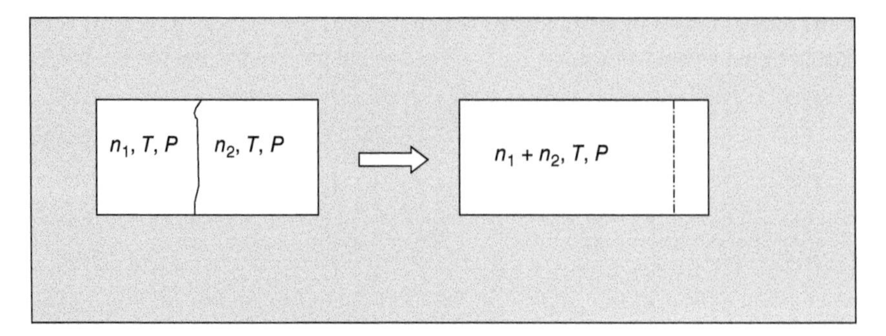

Figure 7.4 Schematic diagram of a mixing process.

in which z_i is the mole fraction and F and F_i are the property of the mixture and the pure component i, respectively. All the properties are at the same temperature and pressure. Then, property changes on mixing are

$$\Delta_m U(T,P) \equiv U(T,P) - \sum_{i=1}^{N} z_i U_i(T,P) \tag{7.320}$$

$$\Delta_m H(T,P) \equiv H(T,P) - \sum_{i=1}^{N} z_i H_i(T,P) \tag{7.321}$$

$$\Delta_m A(T,P) \equiv A(T,P) - \sum_{i=1}^{N} z_i A_i(T,P) \tag{7.322}$$

$$\Delta_m G(T,P) \equiv G(T,P) - \sum_{i=1}^{N} z_i G_i(T,P) \tag{7.323}$$

$$\Delta_m V(T,P) \equiv V(T,P) - \sum_{i=1}^{N} z_i V_i(T,P) \tag{7.324}$$

Now, we can find the general relationship with the excess property using Eq. (7.269)

$$\Delta_m F \equiv F^E + F^{is} - \sum_{i=1}^{N} z_i F_i(T,P) \tag{7.325}$$

The last two terms are the ideal solution property change upon mixing and

$$\Delta_m F \equiv F^E + \Delta_m F^{is} \tag{7.326}$$

In some cases, the summation term equals the ideal solution term, and

$$\Delta_m F \equiv F^E + \Delta_m F^{is} \text{ with } \Delta_m F^{is} = \begin{cases} 0 & \text{for } V, U, H \\ RT\sum_{i=1}^{N} z_i \ln z_i & \text{for } G, A, -ST \end{cases} \tag{7.327}$$

If the pure components are at different temperatures and pressures, we must know at what temperature and pressure the mixing occurs, and we must use the pure components at the temperature and pressure of mixing to apply the equations. The same occurs for gases,

$$\Delta_m F \equiv F^R + F^{ig} - \sum_{i=1}^{N} y_i F_i(T,P)$$

and

$$\Delta_m F \equiv F^R - \sum_{i=1}^{N} y_i F_i^R + \Delta_x \text{ with } \Delta_x = \begin{cases} 0 & \text{for } V, U, H \\ RT\sum_{i=1}^{N} y_i \ln y_i & \text{for } G, A, -ST \end{cases} \quad (7.328)$$

Example 7.10

Methane at 1 bar and 298.15 K and ethylene at 273.15 and 2 bar mix at 373.15 K and 5 bar. Find the total entropy change of the process to form an equimolar mixture of methane and ethylene. Use

$$\frac{C_{P,CH_4}^{ig}}{R} = 1.7 + 9.08 \times 10^{-3}T - 2.16 \times 10^{-6}T^2$$

$$\frac{C_{P,C_2H_4}^{ig}}{R} = 1.42 + 1.44 \times 10^{-2}T - 4.39 \times 10^{-6}T^2$$

Solution

Because the pressure is low, we can consider this to be a system of ideal gases and ideal gas mixtures. From Eq. (7.328)

$$\frac{\Delta_m S}{R} \equiv -\sum_{i=1}^{N} y_i \ln y_i$$

First, we raise the temperature of the pure components to the mixing temperature, and then we can mix at a constant temperature. Schematically, the process is

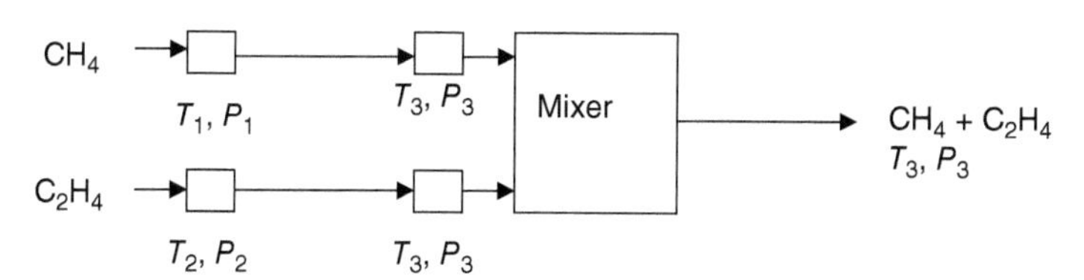

Then, we have

$$\frac{\Delta S^{ig}}{R} = n_{CH_4}\frac{\Delta S_{CH_4}^{ig}}{R} + n_{C_2H_4}\frac{\Delta S_{C_2H_4}^{ig}}{R} + (n_{CH_4} + n_{C_2H_4})\frac{\Delta_m S}{R}$$

From Eqs. (4.71) and (7.328)

$$\frac{\Delta S_{CH_4}^{ig}}{R} = 1.7\ln\left(\frac{373.15}{298.15}\right) + 9.08 \times 10^{-3}(373.15 - 298.15) - \frac{2.16 \times 10^{-6}}{2}(373.15^2 - 298.15^2) - \ln\frac{5}{1}$$

$$\frac{\Delta S_{CH_4}^{ig}}{R} = -0.60139$$

and

$$\frac{\Delta S^{ig}_{C_2H_4}}{R} = 1.42 \ln\left(\frac{373.15}{273.15}\right) + 1.44 \times 10^{-2}(373.15 - 273.15)$$
$$- \frac{4.39 \times 10^{-6}}{2}(373.15^2 - 273.15^2) - \ln\frac{5}{2}$$

$$\frac{\Delta S^{ig}_{C_2H_4}}{R} = 0.82416$$

Also, the entropy change upon mixing is,

$$\frac{\Delta_m S}{R} \equiv -(y_{CH_4} \ln y_{CH_4} + y_{C_2H_4} \ln y_{C_2H_4}) = -(0.5 \ln 0.5 + 0.5 \ln 0.5) = 0.69315$$

If the basis is 1 mol of mixture,

$$\frac{\Delta S^{ig}}{R} = 0.5 \cdot \frac{\Delta S^{ig}_{CH_4}}{R} + 0.5 \cdot \frac{\Delta S^{ig}_{C_2H_4}}{R} + 1 \cdot \frac{\Delta_m S}{R}$$

Therefore, the dimensionless entropy change of the process is

$$\frac{\Delta S^{ig}}{R} = 0.5(-0.60139) + 0.5(0.82416) + 1(0.69315) = 0.80454$$

or

$$\Delta S^{ig} = 0.80454 \times 8.314 = 6.69 \text{ J/K}$$

7.18 Excess Gibbs Energy Models

As we have seen, the logarithm of the activity coefficient is the partial molar excess Gibbs energy. Also, we know that the excess Gibbs energy is a function of temperature, pressure, and composition. Usually, these models are solution models. Several expressions for the Gibbs energy of binary mixtures have appeared in the literature. The simplest is

$$\frac{G^E}{RT} = 0 \Rightarrow \gamma_i = 1 \Rightarrow \text{ideal solution} \tag{7.329}$$

The next simplest form introduces a constant

$$\frac{G^E}{RTz_1z_2} = \text{constant} = B \tag{7.330}$$

If Eq. (7.330) represents the excess Gibbs energy behavior, then it is a "Regular Solution." The shape of the Gibbs energy is symmetrical with respect to the composition, and the activity coefficient results from Eq. (7.283)

$$\ln \gamma_i = \frac{\overline{G}_i^E}{RT} = \left[\frac{\partial}{\partial n_i}\left(\frac{nG^E}{RT}\right)\right]_{T,P,n_{j\neq i}} = \frac{\partial}{\partial n_i}\left[B\frac{n_i n_j}{n}\right] = B\left[\frac{n_j n - n_i n_j \frac{\partial n}{\partial n_i}}{n^2}\right] \tag{7.331}$$

but $(\partial n/\partial n_i) = 1$, so

$$\ln \gamma_i = B(z_j - z_i z_j) = B z_j(1 - z_i) \tag{7.332}$$

Because for a binary mixture $(1 - z_i) = z_j$,

$$\ln \gamma_i = B z_j^2 \tag{7.333}$$

The next composition function would be a straight line

$$\frac{G^E}{RTz_1 z_2} = A_{12} + (A_{21} - A_{12})z_1 \tag{7.334}$$

in which A_{12} is the intercept at $x_1 = 0$ and A_{21} is the intercept at $x_1 = 1$. Also, the above equation could be

$$\frac{G^E}{RTz_1 z_2} = A_{12}(z_1 + z_2) + (A_{21} - A_{12})z_1 \tag{7.335}$$

because $(z_1 + z_2) = 1$. After rearrangement, Eq. (7.335) becomes

$$\frac{G^E}{RTz_1 z_2} = A_{21} z_1 + A_{12} z_2 \tag{7.336}$$

This equation is the "Margules equation [4]." To obtain the activity coefficient as in Eq. (7.331), we execute the following steps:

Express the Gibbs energy as a function of composition,

$$\frac{G^E}{RT} = A_{21} z_1^2 z_2 + A_{12} z_1 z_2^2 \tag{7.337}$$

Multiply the excess Gibbs energy by n,

$$\frac{nG^E}{RT} = A_{21}\frac{n_1^2 n_2}{n^2} + A_{12}\frac{n_1 n_2^2}{n^2} \tag{7.338}$$

Obtain the derivative with respect to n_1 or n_2 at a constant T, P, and n_2 or n_1,

$$\ln \gamma_1 = \frac{\overline{G}_1^E}{RT} = \left[\frac{\partial}{\partial n_1}\left(\frac{nG^E}{RT}\right)\right]_{T,P,n_2} = A_{21}\frac{2n_1 n_2 n^2 - n_1^2 n_2 2n}{n^4} + A_{12}\frac{n_2^2 n^2 - n_1 n_2^2 2n}{n^3} \tag{7.339}$$

$$= 2A_{21}\frac{n_1 n_2 n - n_1^2 n_2}{n^3} + A_{12}\frac{n_2^2 n - 2n_1 n_2^2}{n^3} \tag{7.340}$$

$$= 2A_{21}\left(z_1 z_2 - z_1^2 z_2\right) + A_{12}\left(z_2^2 - 2z_1 z_2^2\right) \tag{7.341}$$

$$= 2A_{21} z_1 z_2 (1 - z_1) + A_{12} z_2^2 (1 - 2z_1) = 2A_{21} z_1 z_2^2 + A_{12} z_2^2 (1 - 2z_1) \tag{7.342}$$

$$\ln \gamma_1 = z_2^2[2A_{21} z_1 + A_{12}(1 - 2z_1)] \tag{7.343}$$

$$\ln \gamma_1 = z_2^2[A_{12} + 2(A_{21} - A_{12})z_1] \tag{7.344}$$

Similarly,

$$\ln \gamma_2 = z_1^2[A_{21} + 2(A_{12} - A_{21})z_2] \tag{7.345}$$

The next equation for the excess Gibbs energy could be a polynomial function

$$\frac{G^E}{RTz_1z_2} = a + bz_1 + cz_1^2 + \cdots \tag{7.346}$$

Redlich and Kister [5] suggested writing this equation as

$$\frac{G^E}{RTz_1z_2} = B + C(z_1 - z_2) + D(z_1 - z_2)^2 + \cdots \tag{7.347}$$

This equation is a generalization of the equations already mentioned. For example, if $B = C = D = \cdots = 0$, it becomes an ideal solution. If $B \neq 0$ but $C = D = \cdots = 0$, it becomes a regular solution. Sometimes, we call Eq. (7.330) the one-suffix Margules equation. Finally, if $B \neq 0$, $C \neq 0$ but $D = \cdots = 0$, it becomes the equation known as the two-suffix Margules equation. Many more complex equations have appeared in the literature. Here, we present those most important for or those most commonly used in the industry. An alternate representation of the excess Gibbs energy, proposed by Johannes Jacobus van Laar [6], describes the reciprocal of the excess Gibbs energy as a Redlich–Kister-type equation

$$\left[\frac{G^E}{RTz_1z_2}\right]^{-1} = B' + C'(z_1 - z_2) = \frac{z_1}{A'_{21}} + \frac{z_2}{A'_{12}} \tag{7.348}$$

in which $\left[A'_{21}\right]^{-1} = B' + C'$ and $\left[A'_{12}\right]^{-1} = B' - C'$. The excess Gibbs function is

$$\frac{G^E}{RTz_1z_2} = \frac{A'_{12}A'_{21}}{A'_{12}z_1 + A'_{21}z_2} \tag{7.349}$$

and the activity coefficients are

$$\ln \gamma_1 = \frac{A'_{12}\left(A'_{21}\right)^2 z_2^2}{\left(A'_{12}z_1 + A'_{21}z_2\right)^2} = A'_{12}\left[1 + \frac{A'_{12}z_1}{A'_{21}z_2}\right]^{-2} \tag{7.350}$$

$$\ln \gamma_2 = \frac{\left(A'_{12}\right)^2 A'_{21}z_1^2}{\left(A'_{12}z_1 + A'_{21}z_2\right)^2} = A'_{21}\left[1 + \frac{A'_{21}z_2}{A'_{12}z_1}\right]^{-2} \tag{7.351}$$

Later, we will learn how to obtain the parameters A'_{21} and A'_{12}, but a simple way is to use a value of the activity coefficient at an infinite dilution

$$\ln \gamma_1^{\infty} = A'_{12} \text{ and } \ln \gamma_2^{\infty} = A'_{21} \tag{7.352}$$

The Redlich–Kister-type equations and the van Laar equation are isothermal equations, so the parameters are temperature independent. They could be a function of temperature if vapor–liquid equilibrium data are available at different temperatures to obtain the temperature functionality.

In 1964, G.M. Wilson [7] proposed the following representation for the excess Gibbs energy considering local compositions around molecules of two different types

$$\frac{G^E}{RT} = -z_1 \ln(z_1 + z_2\Lambda_{12}) - z_2 \ln(z_2 + z_1\Lambda_{21}) \tag{7.353}$$

The activity coefficients calculated from this equation are

$$\ln \gamma_1 = -\ln(z_1 + z_2\Lambda_{12}) + z_2 \left[\frac{\Lambda_{12}}{z_1 + z_2\Lambda_{12}} - \frac{\Lambda_{21}}{z_2 + z_1\Lambda_{21}}\right] \tag{7.354}$$

$$\ln \gamma_2 = -\ln(z_2 + z_1\Lambda_{21}) - z_1 \left[\frac{\Lambda_{12}}{z_1 + z_2\Lambda_{12}} - \frac{\Lambda_{21}}{z_2 + z_1\Lambda_{21}}\right] \tag{7.355}$$

with

$$\Lambda_{ij} = \frac{V_j^L}{V_i^L} \exp\left[-\frac{a_{ij}}{RT}\right] \quad \text{for } i \neq j \tag{7.356}$$

in which V_i^L is the liquid molar volume of the pure component and a_{ij} is a characteristic parameter that in the local composition theory represents molecular interactions of the i–j molecules. The Wilson equation (Eq. (7.353)) can represent vapor liquid behavior as well as the van Laar equation and the Margules equation. The problem with the Wilson equation is that it cannot represent liquid–liquid immiscibility, nor reproduce results when the activity coefficient has an extremum (maximum or minimum) in the γ_i–x_i diagram. At infinite dilution, the activity coefficients are

$$\ln \gamma_1^\infty = -\ln(\Lambda_{12}) + 1 - \Lambda_{21} \tag{7.357}$$

$$\ln \gamma_2^\infty = -\ln(\Lambda_{21}) + 1 - \Lambda_{12} \tag{7.358}$$

Solving this system of equations involves using a trial-and-error procedure, and it results in multiple solutions. Applying the Wilson equation to multicomponent mixtures results in the Gibbs energy as

$$\frac{G^E}{RT} = -\sum_{i=1}^{N} z_i \ln\left(\sum_{j=1}^{N} z_j \Lambda_{ij}\right) \tag{7.359}$$

The general expression for the activity coefficients of component i is

$$\ln \gamma_i = 1 - \ln\left[\sum_{j=1}^{N} z_j \Lambda_{ij}\right] - \sum_{k=1}^{N} \frac{z_k \Lambda_{ki}}{\sum_{j=1}^{N} z_j \Lambda_{kj}} \tag{7.360}$$

This equation describes multicomponent mixtures using the parameters obtained from binary mixtures. The binary parameters of mixtures are in the vapor–liquid compilation data from DECHEMA [8] (1977).

The next model is the NRTL (nonrandom two liquid) equation. Renon et al. [9] proposed it in1968 with the idea to overcome the problems of the Wilson equation using local compositions. They represented the Gibbs energy as

$$\frac{G^E}{z_1 z_2 RT} = \frac{G_{21}\tau_{21}}{z_1 + z_2 G_{21}} + \frac{G_{12}\tau_{12}}{z_2 + z_1 G_{12}} \tag{7.361}$$

in which

$$G_{12} = \exp[-\alpha\tau_{12}] \text{ and } G_{21} = \exp[-\alpha\tau_{21}]$$

with

$$\tau_{12} = b_{12}/RT \text{ and } \tau_{21} = b_{21}/RT$$

The activity coefficients obtained from Eq. (7.361) are

$$\ln \gamma_1 = z_2^2 \left[\tau_{21} \left(\frac{G_{21}}{z_1 + z_2 G_{21}} \right)^2 + \frac{G_{12}\tau_{12}}{(z_2 + z_1 G_{21})^2} \right] \tag{7.362}$$

$$\ln \gamma_2 = z_1^2 \left[\tau_{12} \left(\frac{G_{12}}{z_2 + z_1 G_{12}} \right)^2 + \frac{G_{21}\tau_{21}}{(z_1 + z_2 G_{21})^2} \right] \tag{7.363}$$

The NRTL model is one of the best models to represent the nonideal behavior of liquids including partially miscible liquids. Therefore, it is the first choice for representing liquid behavior. This model has three characteristic parameters: b_{12}, b_{21}, and a obtained from experimental measurements. Adequate values for a range from −1 to 0.5, but generally, we use an arbitrary value of 0.3. Table 7.1 contains the values of α for different systems. Infinite dilution activity coefficients from Eqs. (7.362) and (7.363) are

$$\ln \gamma_1^\infty = \tau_{21} + \tau_{12} \exp[-\alpha \tau_{12}] \tag{7.364}$$

and

$$\ln \gamma_2^\infty = \tau_{12} + \tau_{21} \exp[-\alpha \tau_{21}] \tag{7.365}$$

For multicomponent mixtures, the NRTL equation and the activity coefficients are

$$\frac{G^E}{RT} = \sum_{i=1}^{N} z_i \frac{\sum_{j=1}^{N} \tau_{ji} G_{ji} z_j}{\sum_{k=1}^{n} G_{ki} z_k} \tag{7.366}$$

and

$$\ln \gamma_i = \frac{\sum_{j=1}^{N} \tau_{ji} G_{ji} z_j}{\sum_{k=1}^{n} G_{ki} z_k} + \sum_{j=1}^{N} \frac{z_j G_{ij}}{\sum_{k=1}^{n} G_{ki} z_k} \left(\tau_{ij} - \frac{\sum_{l=1}^{n} z_l \tau_{lj} G_{lj}}{\sum_{k=1}^{n} G_{kj} z_k} \right) \tag{7.367}$$

Finally, a different type of equation for the excess Gibbs energy is the *UNIQUAC* (*UNI*versal *QUA*si-*C*hemical) equation. Abrams and Prausnitz [10] proposed this equation in 1968. They separate the excess Gibbs energy into two contributions (a combinatorial and a residual):

$$\frac{G^E}{RT} = \left[\frac{G^E}{RT} \right]^{com} + \left[\frac{G^E}{RT} \right]^{res} \tag{7.368}$$

in which the contribution caused by differences in the sizes and shapes of the molecules is the combinatorial contribution, denoted by $[G^E/(RT)]^{com}$, and the contribution caused by interactions among molecules is the residual contribution denoted by $[G^E/(RT)]^{res}$. Then,

$$\left[\frac{G^E}{RT} \right]^{com} = \sum_{i=1}^{N} z_i \ln \left(\frac{\Phi_i}{z_i} \right) + \frac{Z}{2} \sum_{i=1}^{N} q_i z_i \ln \left(\frac{\theta_i}{\Phi_i} \right) \tag{7.369}$$

and

$$\left[\frac{G^E}{RT} \right]^{res} = -\sum_{i=1}^{N} q_i z_i \ln \left[\sum_{j=1}^{N} \theta_i \tau_{ij} \right] \tag{7.370}$$

Table 7.1 Values for the parameter α in the NRTL model.

System type	Mixtures	Example	α
Ia	Nonpolar + nonpolar	Hydrocarbons + tetrachloride carbon (excluding fluorocarbons)	0.3
Ib	Nonpolar + polar nonassociated	*n*-Heptane-methyl ethyl-ketone, benzene-acetone, and carbon tetrachloride-nitroethane	0.3
Ic	Polar + polar	Acetone-chloroform, chloroform-dioxane (some with negative excess Gibbs energy); acetone-methyl acetate; and ethanol-water (some with small positive excess Gibbs energy)	0.3
II	Saturated hydrocarbons with polar nonassociated liquids	*n*-Hexane-acetone or isooctane-nitroethane	0.2
III	Saturated hydrocarbons and the homolog perfluorocarbons	*n*-Hexane-perfluoro-*n*-hexane	0.4
IV	Strongly self-associated substance with a nonpolar substance	Alcohol-hydrocarbon and alcohol-carbon tetrachloride	0.47
V	Two polar substances	Acetonitrile-carbon tetrachloride and nitromethane-carbon tetrachloride	0.47
VI	Water plus a polar, nonassociated substance	*n*-Hexane-acetone and isooctane-nitroethane	0.3
VII	Water plus a polar self-associated substance	Water-butyl glycol and water-pyridine	0.47

The coordination number Z equals 10, $\tau_{ij} = \exp\left[-\frac{(u_{ji}-u_{ij})}{RT}\right]$, and the segment and area fractions are

$$\Phi_i = \frac{z_i r_i}{\sum_{j=1}^{N} z_j r_j} \tag{7.371}$$

and

$$\theta_i = \frac{z_i q_i}{\sum_{j=1}^{N} z_j q_j} \tag{7.372}$$

The size and surface parameters r_i and q_i are pure component parameters when evaluated using a group contribution method, and the UNIQUAC model becomes the UNIFAC (*UNI*versal quasichemical *F*unctional group *A*ctivity *C*oefficients) [11] model. Table 7.2 shows the values of r and q for various pure components. The

Table 7.2 Values of size and surface parameters.

Substance	r	q
Water	0.92	1.4
Carbon dioxide	1.3	1.12
Acetaldehyde	1.9	1.8
Ethane	1.8	1.7
Dimethylamine	2.33	2.09
Methyl acetate	2.8	2.58
Furfural	2.8	2.58
Benzene	3.19	2.4
Toluene	3.87	2.93
Aniline	3.72	2.83
Triethylamine	5.01	4.26
n-Octane	5.84	4.93
n-Decane	7.2	6.02
n-Hexadecane	11.24	9.26
Acetone	2.57	2.34
Chloroform	2.87	2.41

Source: Data from Abrams and Prausnitz [10].

characteristic parameters of the UNIQUAC model are the energy parameters u_{ij} and u_{ji} obtained from vapor–liquid experimental measurements. Abrams and Prausnitz provide the conditions when the UNIQUAC excess Gibbs energy model reduces to the Wilson, van Laar, and NRTL models for the excess Gibbs energy.

Example 7.11

The value of the activity coefficients at low pressures results from the *Pxy* data using $\gamma_i = (Py_i)/\left(P_i^{sat}x_i\right)$. The *P–x–y* data [12] for an ethanol (1) + n-octane (2) mixture at 328.15 K are as follows:

P (kPa)	x_1	y_1	P (kPa)	x_1	y_1	P (kPa)	x_1	y_1
8.43	0	0	38.72	0.3078	0.7981	40.61	0.8544	0.843
10.31	0.0014	0.1683	39.25	0.3938	0.8026	40.49	0.8866	0.856
18.91	0.0185	0.5611	39.52	0.4614	0.8057	40.17	0.9268	0.8823
19.85	0.0225	0.5791	39.96	0.5790	0.8116	39.69	0.9511	0.9051
31.32	0.0666	0.7416	40.17	0.6519	0.8150	39.08	0.9718	0.932
35.13	0.1233	0.7692	40.42	0.7371	0.8239	38.53	0.9833	0.956
37.46	0.206	0.7866	40.52	0.7896	0.8281	37.28	1	1

Find the parameters of the Margules equation and compare the activity coefficients to experimental values.

Solution

First, calculate the activity coefficient from the experimental values using

$$\gamma_1 = (Py_1)/\left(P_1^{sat} z_1\right)$$

The vapor pressures are in the table as are the pressures of the pure components at equilibrium: $P_1^{sat} = 37.48$ and $P_2^{sat} = 8.43$ kPa. The activity coefficients are

P (kPa)	x_1	γ_1	γ_2
8.43	0		1.0000
10.31	0.0014	33.246	1.0186
18.91	0.0185	15.385	1.0031
19.85	0.0225	13.704	1.0139
31.32	0.0666	9.3549	1.0285
35.13	0.1233	5.8787	1.0971
37.46	0.2060	3.8369	1.1943
38.72	0.3078	2.6931	1.3397
39.25	0.3938	2.1458	1.5162
39.52	0.4614	1.8511	1.6912
39.96	0.5790	1.5025	2.1213
40.17	0.6519	1.3471	2.5325
40.42	0.7371	1.2119	3.2117
40.52	0.7896	1.1399	3.9271
40.61	0.8544	1.0748	5.1945
40.49	0.8866	1.0486	6.0992
40.17	0.9268	1.0258	7.6620
39.69	0.9511	1.0132	9.1372
39.08	0.9718	1.0054	11.179
38.53	0.9833	1.0048	12.042
37.28	1.0000	1.0000	

from Eqs. (7.343) and (7.344), at infinite dilution

$$\ln \gamma_1^{\infty} = A_{12}$$

Similarly,

$$\ln \gamma_2^{\infty} = A_{21}$$

From the table, a linear interpolation between the last two points gives

$$\gamma_1^{\infty} = \frac{33.246 - 15.385}{0.0014 - 0.0185}(0 - 0.0014) + 33.246 = 34.708$$

$$\gamma_2^\infty = \frac{12.042 - 11.179}{(1 - 0.9833) - (1 - 0.9718)}(0 - (1 - 0.9833)) + 12.042 = 13.295$$

Therefore,

$$\ln \gamma_1^\infty = A_{12} = \ln(34.708) = 3.54697$$

$$\ln \gamma_1^\infty = A_{21} = \ln(13.295) = 2.58739$$

The Margules equation for the activity coefficients is

$$\ln \gamma_1 = x_2^2[3.54697 - 1.91916x_1]$$

$$\ln \gamma_2 = x_1^2[2.58739 + 1.91916x_2]$$

Plotting the equations

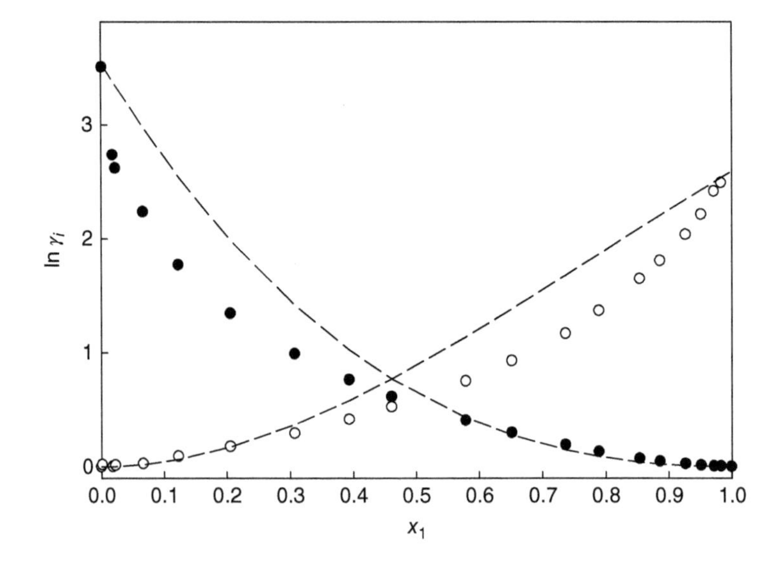

Problems for Chapter 7

7.1 Consider a mixture for which the molar volume is a function of composition

$$V_{mix}\ (\text{cm}^3/\text{mol}) = 57.8z_1 + 98.3z_2 - 31.9z_1z_2$$

a. Determine the molar volumes of species 1 and 2 as pure materials.
b. Determine the mixture volume and the partial molar volumes for each species at z_1 0.374.

7.2 Derive expressions for the partial molar third virial coefficients, $\overline{C}_1$ and $\overline{C}_2$ for a binary mixture (two chemical species present). The third virial coefficient of the mixture is

$$C_m = \sum_i \sum_j \sum_k x_i x_j x_k C_{ijk}$$

in which the parameters C_{ijk} depend only on temperature.

7.3 Find $\overline{H}_1$, $\overline{H}_2$, and $\overline{H}_3$ from the following expression for the enthalpy of a mixture

$$H = H_1x_1 + H_2x_2 + H_3x_3 + Cx_1x_2 + Dx_1x_3 + Ex_2x_3$$

where H_1, H_2, and H_3 are values of the enthalpies of the pure components and C, D, and E are constants. Check if the expression is thermodynamically consistent.

7.4 The vapor pressure of propane at 310 K is 12.75 bar.

a. Estimate the fugacity of saturated propane vapor at 310 K.
b. Assuming that liquid propane at 310 K is incompressible, at what pressure is the fugacity of compressed liquid propane 5% larger than that of the saturated vapor?

$$T_C = 369.825\ \text{K},\quad P_C = 42.48\ \text{bar},\quad V_C = 200\ \text{cm}^3/\text{mol},\quad \text{and}\ \omega = 0.152$$

7.5 Determine the fugacity of methane at −35 °C and 28 bar using generalized correlations.
Data

$$T_C = 190.564\ \text{K},\quad P_C = 45.992\ \text{bar},\ \omega = 0.011$$

7.6 Calculate the fugacity of saturated liquid water at 365.70 °C using

a. generalized correlations;
b. the steam tables. (You may assume that steam is an ideal gas at 1 kPa.)

Data

$$T_C = 647.096\ \text{K},\quad P_C = 220.64\ \text{bar},\quad \text{and}\ \omega = 0.344$$

7.7 Do not assume ideal gas behavior for acetone in the vapor phase for the following:

a. Calculate the fugacity of saturated liquid acetone at 76 °C.
b. Calculate the ratio of the fugacity of liquid acetone at 76 °C and 10 bar to the fugacity of saturated liquid acetone at 76 °C.

Data

$$T_C = 508.1\ \text{K},\quad P_C = 47\ \text{bar},\quad V_C = 212.766\ \text{cm}^3/\text{mol},\quad \text{and}\ \omega = 0.3071$$

$$\ln P^{sat}\ (\text{bar}) = 10.03 - \frac{-2940.47}{T\ (\text{K}) - 35.92}$$

7.8 Calculate the fugacity coefficient of chloroform at its normal boiling temperature and 60 MPa. Neglect the change of the liquid volume with pressure.
Data

$$\ln P^{sat}\ (\text{kPa}) = 13.7324 - \frac{2548.74}{t\ (°\text{C}) + 218.552}$$

$$T_C = 526.4,\ \ V_C = 536.4,\ \ P_C = 54.72,\ \omega = 0.222$$

7.9 Use the van der Waals equation to calculate the fugacity coefficient of propane at 400 K and 60 bar.

$$T_C = 369.8\ \text{K},\ \ P_C = 42.48\ \text{bar},\ \ \text{and}\ R = 83.14\ \text{bar}\ \ \text{cm}^3/(\text{mol K})$$

7.10 A ternary mixture consisting of 10 mol of component 1, 50 mol of component 2, 30 mol of component 3, and 40 mol of component 4 has the following fugacities: $\hat{f}_1 = 3$ bar, $\hat{f}_2 = 5$ bar, and $\hat{f}_4 = 8$ bar. What is the value of $\hat{f}_3$ if the fugacity of the mixture is 15 bar at a pressure of 3.5 bar?

7.11 Consider a gas mixture at 250 K that contains 60 mol% nitrogen and 40 mol% propane. Use 0.12 as the numerical value of the binary interaction parameter, k_{ij}.
a. Estimate the second virial coefficient of this mixture.
b. Calculate the fugacity of propane in this mixture at a pressure of 2.62 bar.
For nitrogen,

$$T_{C,11} = 126.192\ \text{K},\ \ P_{C,11} = 33.958\ \text{bar},\ \ V_{C,11} = 89.414\ \ \text{cm}^3/\text{mol},\ \ \text{and}$$
$$\omega_{11} = 0.037$$

For propane,

$$T_{C,22} = 369.825\ \text{K}, P_{C,22} = 42.48\ \text{bar}, V_{C,22} = 200\ \ \text{cm}^3/\text{mol}, \text{and}\ \omega_{22} = 0.152$$

7.12 Calculate the fugacities of nitrogen and hydrogen of the mixture in a stream that is 25 mol% N_2 and 75 mol% H_2 (stoichiometric feed for ammonia production), if the stream is at 450 K and 325 bar. You may assume that the mixture behaves as an ideal solution but not as an ideal gas.
Nitrogen

$$T_{C,11} = 126.192\ \text{K},\ \ P_{C,11} = 33.958\ \text{bar},\ \ V_{C,11} = 89.414\ \ \text{cm}^3/\text{mol},\ \ \text{and}$$
$$\omega_{11} = 0.037$$

Hydrogen

$$T_{C,22} = 33.145\ \text{K}, P_{C,22} = 12.964\ \text{bar}, V_{C,22} = 64.48\ \ \text{cm}^3/\text{mol}, \text{and}$$
$$\omega_{22} = -0.219$$

7.13 In a vapor mixture described by $Z = 1 + BP/RT$, the quantity $\delta_{12} = 2B_{2,12} - (B_{2,11} + B_{2,22})$ provides a measure of the departure of the mixture from ideal solution behavior. Use the Hall–Iglesias-Silva correlation for the virial coefficients to estimate δ_{12} for carbon dioxide + ethane mixtures at 315.2 K. The binary interaction parameter, k_{ij}, has a numerical value of 0.08 for the carbon dioxide + ethane interaction.

Data

Substance	T_C (K)	ω	V_C (cm^3/mol)	μ (D)	T_B (K)	P_C (bar)	μ_R
CO_2	304.128	0.224	94.119	0.0	702.1	73.773	0.0
C_2H_6	305.33	0.0990	147.06	0.0	757.1	48.72	0.0

7.14 Calculate the fugacity coefficient of methane in a binary mixture of 40% methane (1) and 60% propane (2) at 473.15 K and 5 bar.
Data

$$T_{C,11} = 190.564\ \text{K},\ P_{C,11} = 45.992\ \text{bar},\ V_{C,11} = 94.119\ \text{cm}^3/\text{mol},\ \text{and}\ \omega_{11} = 0.011$$

$$T_{C,22} = 369.825\ \text{K},\ P_{C,22} = 42.48\ \text{bar},\ V_{C,22} = 200\ \text{cm}^3/\text{mol},\ \text{and}\ \omega_{22} = 0.152$$

7.15 Calculate the fugacity coefficient for component A in an equimolar binary mixture. The mixture is at 300 atm and 200 K. The pure components and the mixture obey the following EOS:

$$P(V-b) = RT\frac{1}{b} = \frac{y_A}{b_A} + \frac{y_B}{b_B} b_A = 21.6, \quad b_B = 12.5\ \text{cm}^3/\text{mol}$$

7.16 The following EOS is valid for an equimolar binary mixture

$$P = \frac{RT}{V-b}$$

in which $b = y_1 b_1 + y_2 b_2$ with $b_1 = 40\,\text{cm}^3/\text{mol}$ and $b_1 = 60\,\text{cm}^3/\text{mol}$. If $T = 298.15$ K and the pressure is 40 bar, find the value of $\hat{\varphi}_1$.

7.17 Consider a mixture of methane (1) and a heavy hydrocarbon (2). The heavy hydrocarbon as a pure material has a vapor pressure given by

$$\ln P_{HC}^{sat}\ (\text{bar}) = 10.949 - \frac{3652}{T\ (\text{K}) - 37}$$

Both liquid and vapor phases are present at 305 K and 276 bar, and the mole fraction of methane in the liquid phase is 0.0837. You may assume that the vapor phase is an ideal solution but not an ideal gas.

a. Determine Henry's constant for methane in this oil at 305 K.
b. Calculate the fugacity of the heavy oil in the vapor phase. (Provide a numerical answer with units of bars.)

Consider that the total pressure at equilibrium is

$$P = \frac{x_1 H_1}{\hat{\varphi}_1} + x_2 P_2^{sat}$$

$$T_C = 190.564\,\text{K},\ \ P_C = 45.992\,\text{bar},\ \ \text{and}\ \omega = 0.011$$

7.18 Streams of methane and carbon dioxide mix to provide a mixture that is 35 mol% methane. All streams are at 350 K and 25 bar.

a. Assuming the mixture is an ideal solution (but not an ideal gas), calculate the amount of energy (if any) that must be transferred as heat to maintain the outlet temperature at 350 K. (No energy is transferred as work in this process.)
b. Calculate the total change in entropy for the universe for this process per mole of mixture produced.

7.19 Two pure liquids, A and B, are combined in a constant pressure, steady-state flow process with the following specifications: 1.6 mol/s of pure A at 60 °C mixes with 0.4 mol/s of pure B at 25 °C and the mixture passes through a heat exchanger that brings the mixture temperature to 40 °C. The mixture is an ideal solution, and the specific heats at a constant pressure of the pure liquids are $C_{P,\text{A}} = 45\,\text{J/(mol K)}$ and $C_{P,\text{B}} = 57\,\text{J/(mol K)}$. (Both specific heats are independent of temperature and pressure.)

a. Determine the heat exchanger duty (rate of energy transfer as heat).
b. Determine the rate at which this process increases the entropy of the universe.

7.20 A stream of ethylene (A) is throttled to atmospheric pressure and mixed with a stream of argon (B) that has also been throttled to atmospheric pressure. At (C), the mixture has 2 mol of ethylene per mole of argon, the temperature is 35 °C, and the pressure is 1.05 bar. The temperature and pressure of ethylene at (A) are 30 °C and 150 bar, respectively, and the pressure of the argon at (B) is 2.30 bar. If no energy is transferred as heat or work in this steady-state flow process, what is the temperature of argon at (B), and what is the total entropy change of the universe per mole of mixture exiting at (C)?

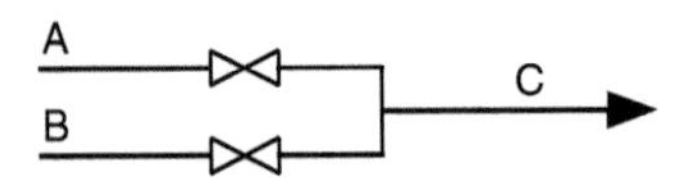

Data

$$T_{C,\text{C}_2\text{H}_4} = 282.35\,\text{K},\ \ P_{C,\text{C}_2\text{H}_4} = 50.418\,\text{bar},\ \ \text{and}\ \ \omega_{\text{C}_2\text{H}_4} = 0.087$$

$$C_{P,\text{C}_2\text{H}_4}/R = 1.4241 + 14.3925 \times 10^{-3}T - 43.9107 \times 10^{-7}T^2$$

$$T_{C,\text{Ar}} = 150.687\,\text{K},\ \ P_{C,\text{Ar}} = 48.630\,\text{bar},\ \ \text{and}\ \ \omega_{\text{Ar}} = 0$$

$$C_{P,\text{Ar}}/R = 2.5$$

7.21 Consider that the change of volume on mixing of a ternary mixture is

$$\Delta_m V \text{ (cm}^3\text{/mol)} = 10z_1z_2 + 40z_1z_3$$

in which the volumes of the pure components 1, 2, and 3 are 70, 90, and 100 cm^3/mol, respectively. Determine the expressions for the partial molar volume of each species and $\overline{V}_1^E$, $\overline{V}_2^E$.

7.22 Consider a mixture for which the excess Gibbs function is a function of composition given by

$$\frac{G^E}{RT} = z_1z_2[1.326 + 0.721(z_1 - z_2)]$$

a. Calculate numerical values for γ_1^∞ and γ_2^∞.
b. Calculate numerical values for g_1 and g_2 for $x_1 = 0.462$.

7.23 Use the Wilson equation to calculate the numerical values of (G^E/RT), g_A, and g_B at $x_A = 0.38$ if the activity coefficients at infinite dilution are $\gamma_A^\infty = 0.414$ and $\gamma_B^\infty = 0.306$.

7.24 The following vapor–liquid equilibrium data apply to a system at 42.5 °C.

P (kPa)	x_1	y_1	$\frac{G^E}{RTx_1x_2}$
24.56	0.000	0.000	—
73.89	0.197	0.716	1.57
87.93	0.408	0.785	1.57
92.89	0.615	0.816	1.57
96.44	0.793	0.859	1.57
97.62	1.000	1.000	—

a. Using the data given in the table, calculate g_A and g_B at $x_A = 0.408$.
b. Determine γ_A^∞ and γ_B^∞ at 42.5 °C for this system

7.25 Calculate the excess Gibbs energy of a mixture formed with 10 mol of component 1 and 30 mol of component 2. Use the NRTL model with $\ln\gamma_1^\infty = 3$, $\ln\gamma_2^\infty = 5$, $\alpha = 0.2$.

References

1 Hall, K.R. and Iglesias-Silva, G.A. (1995). Generalized derivations for mixtures using the virial equation: application to fugacity. *Chem. Eng. Comm.* 137: 211–214.

2 Hall, K.R., Iglesias-Silva, G.A., and Mansoori, G.A. (1993). Quadratic mixing rules for equations of state. Origins and relationships to the virial expansion. *Fluid Phase Equilib.* 91: 67–76.
3 McElroy, P.J. and Ababio, B.D. (1994). Compression factors and virial equation of state coefficients for the system carbon monoxide + ethane. *J. Chem. Eng. Data* 39: 327–329.
4 Margules, M. (1985). Über die Zusammensetzung der gesättigten Dämpfe von Misschungen. *Sitzungsber. Akad. Wiss. Wien Math. Naturw. Klasse II* 104: 1243–1278.
5 Redlich, O. and Kister, A.T. (1948). Algebraic representation of thermodynamic properties and the classification of solutions. *Ind. Eng. Chem.* 40: 345–348.
6 Van Laar, J.J. (1906). *Sechs Vorträgen über das thermodynamische Poten-tial. (Six Lectures on the Thermodynamic Potential).* Braunschweig: Friedr. Vieweg & Sohn.
7 Wilson, G.M. (1964). Vapor–liquid equilibrium. XI. A new expression for the excess free energy of mixing. *J. Am. Chem. Soc.* 86: 127–130.
8 Gmehling, J. and Onken, U. (1977). *Vapor–liquid Equilibrium Data Collection.* Franckfurt/Main: DECHEMA.
9 Renon, H. and Prausnitz, J.M. (1968). Local compositions in thermodynamic excess functions for liquid mixture. *AIChE J.* 14: 135–144.
10 Abrams, D.S. and Prausnitz, J.M. (1975). Statistical thermodynamics of liquid mixtures: a new expression for the excess Gibbs energy of partly or completely miscible systems. *AIChE J.* 21: 116–128.
11 Aage, F., Jones, R.L., and Prausnitz, J.M. (1975). Group-contribution estimation of activity coefficients in nonideal liquid mixtures. *AIChE J.* 21: 1086–1099.
12 Boublikova and Lu (1969). Isothermal vapour–liquid equilibria for the ethanol-*n*-octane system. *J. Appl. Chem.* 19: 89–92.

8

Phase Equilibrium

8.1 Introduction

Phase equilibrium is possibly the most important phenomena in chemical engineering. We encounter it in many unit operations such as distillation, absorption, crystallization, etc. For example, distillation is the most commonly used separation process in the chemical and petrochemical industries, and it utilizes equilibrium between two or more phases. Calculations associated with phase equilibrium use thermodynamic expressions from Chapter 7. Because these calculations are computer intensive, simulation programs provide a variety of possibilities to calculate equilibrium in process synthesis, design, and optimization. The results of these computer programs depend upon the quality of the models and model parameters they use. Some of these models are equations of state (EOS) or excess Gibbs energy models, which we explore later in this chapter. This chapter introduces equilibrium criteria, excess Gibbs models, and conditions for equilibrium.

8.2 Equilibrium

The second law has many practical applications. The concept of equilibrium is enormously important in thermodynamics, and the criteria that denote equilibrium derive from the second law. The definition of equilibrium is

> *An isolated system is in a state of equilibrium with respect to possible variations if no change in state of the system occurs over time*

J. Willard Gibbs first postulated the criteria for equilibrium. He stated the concept as two inequalities:

$$(dS)_{U,V} \leq 0 \text{ and } (dU)_{S,V} \geq 0 \tag{8.1}$$

Consider a closed system that it is not in equilibrium at T and P. Any change in this system must be irreversible. However, we can place the system in surroundings such that it achieves thermal equilibrium (constant temperature) and mechanical

Thermodynamics for Chemical Engineers, First Edition. Kenneth R. Hall and Gustavo A. Iglesias-Silva.

equilibrium (constant pressure). Energy exchange as heat and work is irreversible. According to the second law, the change of entropy of the surroundings of a system is

$$dS_{surr} = \frac{đQ_{surr}}{T_{surr}} = -\frac{đQ_{sys}}{T_{sys}} \tag{8.2}$$

in which $đQ$ represents the energy transfer as heat, and its value for the system equals that of the surroundings but with an opposite sign. Because thermal equilibrium exists, the temperature of the system T equals T_{surr}. The second law requires that the total entropy change of the universe increases:

$$dS_{univ}^{total} = dS_{sys}^{total} + dS_{surr}^{total} \geq 0 \tag{8.3}$$

where S^{total} is the total entropy of the system. Substituting Eq. (8.2) into Eq. (8.3)

$$dS_{sys}^{total} - \frac{đQ_{sys}}{T} \geq 0 \tag{8.4}$$

Rearranging,

$$dS_{sys}^{total} \geq \frac{đQ_{sys}}{T} \Rightarrow đQ_{sys} \leq TdS_{sys}^{total} \tag{8.5}$$

From the first law

$$d(nU) = đ(nQ) + đ(nW) \tag{8.6}$$

in which

$$-đ(nW) = Pd(nV) \tag{8.7}$$

Then,

$$đ(nQ) = d(nU) + Pd(nV) \tag{8.8}$$

Substituting this into Eq. (8.5)

$$d(nU) + Pd(nV) - TdS_{sys}^{total} \leq 0 \tag{8.9}$$

Also, because this equation involves properties of the system, it must apply to changes in a closed system, and we can use $dS_{sys}^{total} = d(nS)$ for a closed system. By inspection

$$d(nU)_{nV,nS} \leq 0 \tag{8.10}$$

and

$$d(nS)_{nV,nU} \geq 0 \tag{8.11}$$

Then, if we desire an equilibrium condition such that changes cannot occur, the contrary equations must be true at equilibrium

$$d(nU)_{nV,nS} \geq 0 \tag{8.12}$$

$$d(nS)_{nV,nU} \leq 0 \tag{8.13}$$

These two expressions are equivalent. This inequality is a necessary and sufficient condition for equilibrium to exist. It basically states that the entropy of a system is

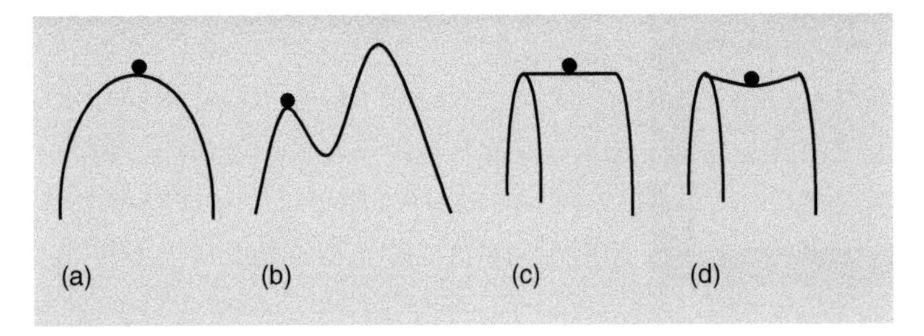

Figure 8.1 Different types of equilibrium states in the entropy surface: (a) Stable, (b) Metastable, (c) Neutral, (d) Unstable.

maximum at equilibrium and that the energy (in any form) of a system is minimum at equilibrium. We can say $(dS)_U < 0$ is sufficient because the second law forbids any change that decreases the entropy at constant energy. Thus, if all changes that could occur would cause a decrease in entropy, then none can occur. This indicates that a system in equilibrium does not change.

$(dS)_U = 0$ is sufficient for equilibrium as evidenced by the surfaces depicting various types of equilibrium, as shown in Figure 8.1. The ball in each case is at an extremum of the surface, but the equilibrium is stable, metastable, neutral, and unstable for the situations depicted.

For example, in case (a), we have stable equilibrium because any infinitesimal force applied to the ball causes the entropy to decrease, thus violating the second law. Case (b) is the same as case (a), but there exists a higher entropy equilibrium state (more stable). The point where the ball rests is a local maximum. In case (c), it is possible to apply a force to the ball along a line that would not cause the entropy to drop, but applying the force in any other direction would cause a decrease in entropy. Finally, if we apply a force on the ball along the curve in (d), it can move back and forth at entropies greater than the equilibrium point until it settles at the minimum.

We can now deduce the consequences of the criteria by examining a Massieu function

$$d(nS) = \frac{d(nU)}{T} + \frac{P}{T}d(nV) - \sum_{i=1}^{N} \frac{\mu_i}{T} dn_i \tag{8.14}$$

and applying it to a system of constant energy, volume, and moles that has two subdivisions separated by a transmitting, flexible, semipermeable wall as shown in Figure 8.2. In this system, the applicable expressions are

$$nS = n_1 S_1 + n_2 S_2 \tag{8.15}$$

$$nU = n_1 U_1 + n_2 U_2 = constant \tag{8.16}$$

$$nV = n_1 V_1 + n_2 V_2 = constant \tag{8.17}$$

$$n_i = n_i^1 + n_i^2 = constant \tag{8.18}$$

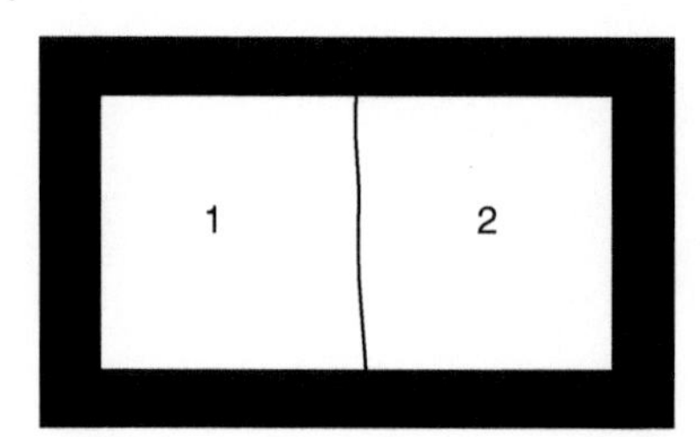

Figure 8.2 System of constant energy, volume, and moles.

The Massieu equation applies to each section of the system as well as to the whole system.

$$d(n_1S_1) = \frac{d(n_1U_1)}{T_1} + \frac{P_1}{T_1}d(n_1V_1) - \sum_{i=1}^{N}\frac{\mu_i^1}{T_1}dn_i^1 \tag{8.19}$$

$$d(n_2S_2) = \frac{d(n_2U_2)}{T_2} + \frac{P_2}{T_2}d(n_2V_2) - \sum_{i=1}^{N}\frac{\mu_i^2}{T_2}dn_i^2 \tag{8.20}$$

The differentials of Eqs. (8.15)–(8.18) are

$$d(nS) = d(n_1S_1) + d(n_2S_2) \tag{8.21}$$

$$d(nU) = d(n_1U_1) + d(n_2U_2) = 0 \Rightarrow d(n_2U_2) = -d(n_1U_1) \tag{8.22}$$

$$d(nV) = d(n_1V_1) + d(n_2V_2) = 0 \Rightarrow d(n_2V_2) = -d(n_1V_1) \tag{8.23}$$

$$dn_i = dn_i^1 + dn_i^2 = 0 \Rightarrow dn_i^2 = -dn_i^1 \tag{8.24}$$

If we add Eqs. (8.19) and (8.20) and use Eqs. (8.21)–(8.24)

$$d(nS)_{nV,nU} = \left[\frac{1}{T_1} - \frac{1}{T_2}\right]d(n_1U_1) + \left[\frac{P_1}{T_1} - \frac{P_2}{T_2}\right]d(n_1V_1) - \sum_{i=1}^{N}\left[\frac{\mu_i^1}{T_1} - \frac{\mu_i^2}{T_1}\right]dn_i^1 \leq 0 \tag{8.25}$$

Because $d(n_1U_1)$, $d(n_1V_1)$, and dn_i^1 can be (+, 0, −) arbitrarily, the only way $d(nS)_{nV,\,nU}$ can always be less than or equal to zero is for the zero to prevail. That is, the terms in brackets must be zero

$$\left[\frac{1}{T_1} - \frac{1}{T_2}\right] = 0 \Rightarrow T_1 = T_2 \tag{8.26}$$

$$\left[\frac{P_1}{T_1} - \frac{P_2}{T_2}\right] = 0 \Rightarrow P_1 = P_2 \tag{8.27}$$

$$\left[\frac{\mu_i^1}{T_1} - \frac{\mu_i^1}{T_1}\right] = 0 \Rightarrow \mu_i^1 = \mu_i^2 \quad \text{for } i = 1,2,\ldots,C \tag{8.28}$$

Thus, at equilibrium temperature, the pressure and chemical potential of each component are uniform throughout the system.

Equations (8.26) and (8.27) apply to pure components, but Eq. (8.28) becomes equality of the Gibbs energy in the vapor and liquid phases. Then, for pure components at uniform pressure and temperature, we have

$$G^v = G^l \tag{8.29}$$

This result results in the Clausius/Clapeyron equation.

8.3 Gibbs Phase Rule

The phase rule restricts the number of independent variables required to fix the state of one or more phases in equilibrium. Let F equal this number of independent, intensive properties or degrees of freedom. F equals the total number of unknown variables in the equilibrium state minus the total number of equations

$$F = \text{total number of intensive variables} - \text{number of equations} \tag{8.30}$$

In the system, the unknown variables are the temperature and pressure and the $N-1$ compositions in each phase. The number of phases is P, so

$$\text{Total number of intensive variable} = 2 + (N-1)P \tag{8.31}$$

We can now establish the total number of equations that describe equilibrium at a given T and P. The equality of the chemical potentials for each component in each phase is

$$\mu_i^\alpha = \mu_i^\beta = \mu_i^\gamma = \cdots \tag{8.32}$$

therefore

$$\text{Total number of equations} = (P-1)N \tag{8.33}$$

Finally, the number of degrees of freedom is

$$F = 2 + N - P \tag{8.34}$$

The above equation is the Gibbs phase rule. The most important consideration when applying the phase rule is that all phases are in equilibrium.

Examples:

- **Pure Water at Equilibrium:** For vapor and liquid water at equilibrium, $N=1$, $P=2$, so $F=1$. This indicates that if we fix one variable, e.g. temperature, then we have fixed the state of the system. Thus, we have fixed the vapor pressure, the saturated entropy of the steam or the liquid, the saturated liquid density or the saturated vapor density, etc. Also, $G^v_{pure} = G^l_{pure}$.
- **A Pure Component at the Triple Point:** Here, three phases, liquid, solid, and gas, exist. Then, $N=1$, $P=3$, and $F=0$. This indicates that no degrees of freedom exist, and it is an invariant condition, in this case, an invariant point. This point is a unique characteristic of every pure substance.
- **Binary Distillation:** Consider a binary mixture at vapor and liquid equilibrium. Here, $N=2$, $P=2$, so $F=2$. We must specify two variables from among T, P, x_i, and y_i. If we fix the temperature and pressure, then we have fixed the state of the system. Thus, the compositions of the liquid and the vapor cannot change.

Example 8.1
Use the phase rule to find how many variables are invariant for the following cases:

(a) A liquid–liquid extraction using a soluble component.
(b) A humidification process for a gas phase mixture of air and water.

Solution
For the first case, we have two immiscible phases with a mutually soluble third component, so

$$N = 3, P = 2, \text{ and } F = 2 + 3 - 2 = 3$$

In this case, we must set the temperature, pressure, and the amount of the third component in one phase to fix the amount of the third component in the second phase.

For case (b), the liquid phase is pure water, and the gas phase is air and water vapor. If we consider air as a single component, we have

$$N = 2, P = 2, \text{so } F = 2 + 2 - 2 = 2$$

Fixing the temperature and pressure fixes the amount of water vapor (humidity) in the gas phase.

8.4 Pure Components and Phase Equilibria

According to the Gibbs phase rule, a pure component in phase equilibrium has one degree of freedom. Thus, setting the temperature or pressure fixes the state of the system. The most common phase diagrams are pressure vs volume or pressure vs temperature, as shown in Section 3.2. Thermodynamically, we can start with Eq. (8.29) written in an integral form

$$\int_{G^l}^{G^v} dG = 0 \tag{8.35}$$

Using the fundamental equation for the Gibbs energy at a constant temperature and considering that $d(PV) = PdV + VdP$

$$dG = VdP = d(PV) - PdV \tag{8.36}$$

Integrating from the saturated liquid to the saturated vapor,

$$\int_{G^l}^{G^v} dG = \int_{P^lV^l}^{P^vV^v} d(PV) - \int_{V^l}^{V^v} PdV = 0 \tag{8.37}$$

or

$$P^vV^v - P^lV^l - \int_{V^l}^{V^v} PdV = 0 \tag{8.38}$$

Because $P^v = P^l = P^{sat}$

$$P^{sat}(V^v - V^l) - \int_{V^l}^{V^v} PdV = 0 \tag{8.39}$$

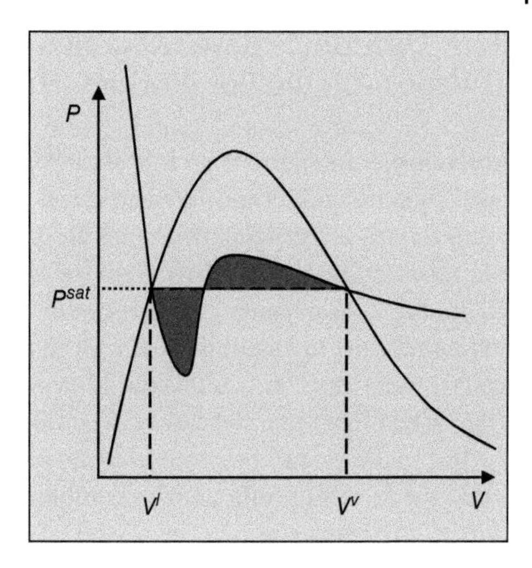

Figure 8.3 Maxwell equal area rule.

This equation is the *Maxwell Equal Area Rule* (MEAR). The first term is a rectangle that we subtract from the area under the curve. Thus, the area formed by the van der Waals loops must equal zero. Figure 8.3 shows the MEAR. Equation (8.39) is a necessary and sufficient condition for vapor–liquid equilibrium in pure components. Another way of representing the condition of equilibrium for a pure component is via residual properties. If we subtract the ideal gas Gibbs energy from each side of Eq. (8.29) and dividing by RT

$$\left(\frac{G^R}{RT}\right)^v = \left(\frac{G^R}{RT}\right)^l \tag{8.40}$$

Also, we must satisfy

$$\frac{P^v}{RT} = \frac{P^l}{RT} \text{ or } Z^v\rho^v = Z^l\rho^l \tag{8.41}$$

to have established equilibrium. Equations (8.40) and (8.41) is a system of equations in which the unknowns are the saturated vapor and liquid densities. Solving this system of equations can use any root-finding numerical technique. However, a simple procedure results from rearranging Eq. (8.39) into a recursive formula [1]

$$P^{sat}_{j+1} = \left[\frac{\int_{V^l}^{V^v} PdV}{(V^v - V^l)}\right]_j \tag{8.42}$$

The algorithm to find the vapor pressure and the saturated volumes is

1. Calculate the integral of Eq. (8.42) using an EOS.
2. Determine the maximum and minimum of the isotherm and calculate

$$P^* = (P_{\max} + P_{\min})/2$$

3. Calculate from the EOS three volume roots: The smallest is the volume of the liquid, and the greatest is the volume of the vapor.
4. Calculate a new pressure using Eq. (8.42)
5. Return to 3 and repeat the procedure until $\left|P^{sat}_{j+1} - P^{sat}_j\right| \leq \varepsilon$. This is the convergence criteria and generally $\varepsilon \approx 1 \times 10^{-3}$.

8.5 Different Phase Diagrams for Binary Mixtures at Vapor–Liquid Equilibrium (VLE)

According to the Gibbs phase rule, we must fix two intensive variables when a binary mixture is in equilibrium. We can select from among temperature, pressure, and compositions in the liquid and vapor for the variables. From them, we can construct a number of diagrams such as pressure vs x, y at a given temperature, as shown in Figure 8.4. This diagram contains two curves that limit the region in which vapor and liquid exist in equilibrium at a given temperature. The solid line is the boundary between vapor in equilibrium (dew point curve) and vapor in a single phase. The dashed line separates liquid at equilibrium from the liquid phase. Looking at the line of the global composition of the mixture, z_i, at a given pressure, P^*, we have a mixture in equilibrium with an equilibrium liquid composition x and an equilibrium vapor composition y. The mixture does not separate into two phases at points above the dew point curve or below the bubble point curve. The points at compositions 0 and 1 correspond to the vapor pressures of the pure components at the given temperature.

Figure 8.4 contains phases in equilibrium at a single temperature, while Figure 8.5 contains equilibria at several temperatures. Each of these temperatures contains a region in which vapor–liquid equilibrium exists. At some temperatures, like T_3, a vapor pressure does not exist for one of the components in the mixture. In this situation, the curve does reach a composition of 1. This curve touches the critical locus.

Another diagram is T vs x, y. These diagrams are upside down with respect to the P–x, y ones. The dew point curve lies above the bubble point curve as shown in Figure 8.6. Binary phase diagrams can have a point at which the composition of the

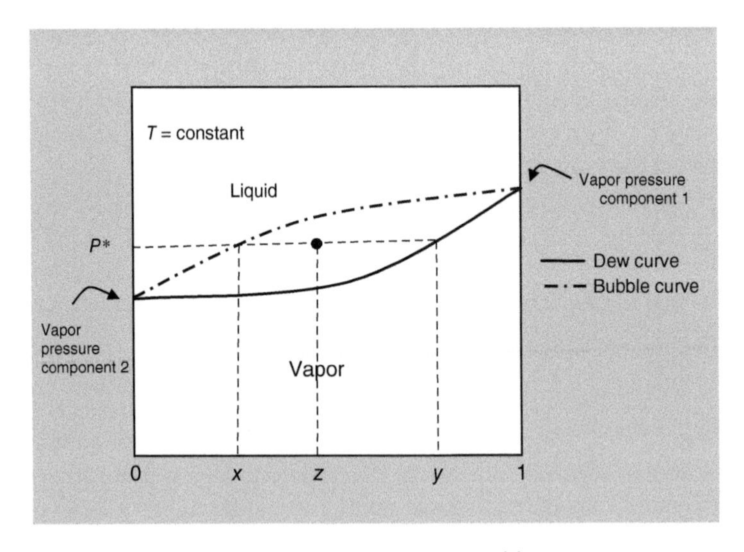

Figure 8.4 Phase diagram for pressure vs composition.

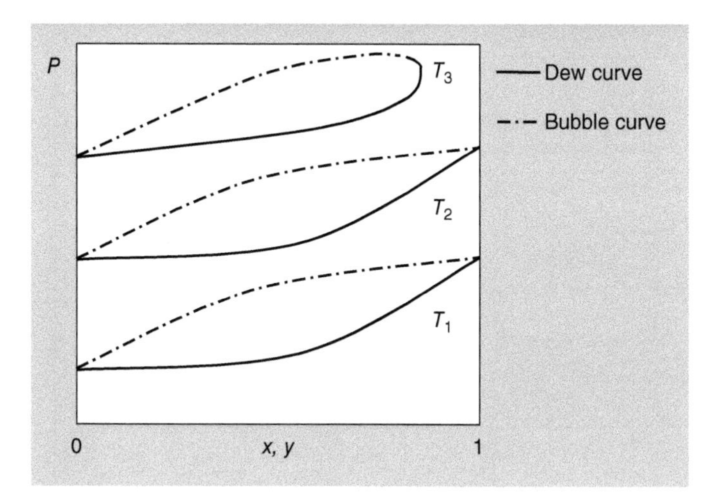

Figure 8.5 Phase equilibrium P–x, y diagram for a binary mixture at different temperatures.

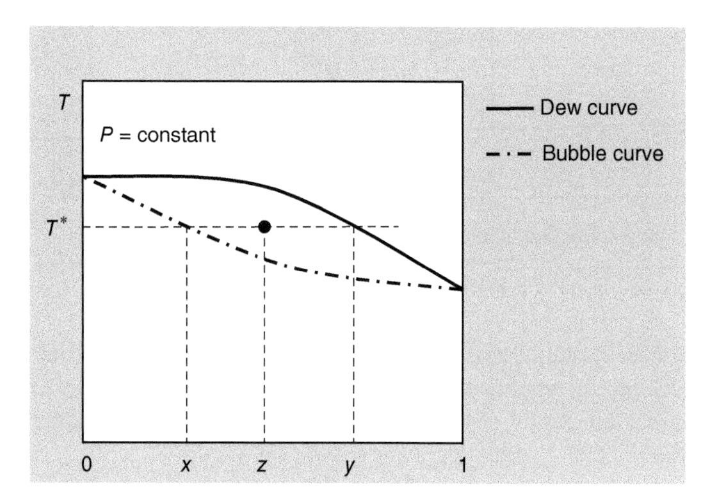

Figure 8.6 Phase diagram of temperature vs composition.

vapor equals the composition of the liquid. These points are azeotropes. Their boiling point temperature can also serve to name azeotropes. Minimum and maximum boiling point azeotropes appear in Figure 8.7.

As with pure components, we can have a P–T diagram for the binary mixture. Figure 8.8 shows this behavior. This diagram contains the vapor pressure curves for the pure components. It also contains curves at constant global composition of the mixture that shows equilibrium points of the dew and bubble curves. These curves, called phase envelopes, touch the critical locus. Point E is a saturated liquid of one composition and a saturated vapor of another composition in equilibrium at T and P.

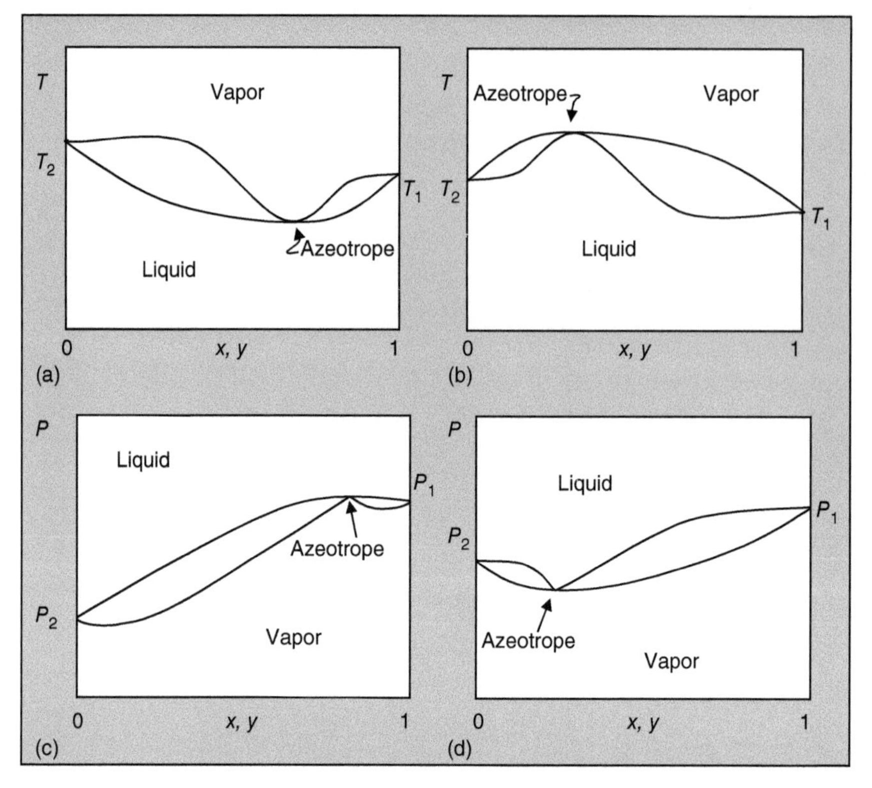

Figure 8.7 Phase diagrams for binary mixtures. ***T*–*x*, *y*:** (a) minimum boiling point azeotrope, (b) maximum boiling point azeotrope. ***P*–*x*, *y*:** (c) minimum boiling point azeotrope, (d) maximum boiling point azeotrope.

This point corresponds to the equilibrium point (P^*) shown in Figure 8.4. The composition of each envelope corresponds to the equilibrium composition x and y of Figure 8.4. The point that touches the critical locus is the critical point of the mixture with a global composition z.

Figure 8.9 zooms in on the phase envelope. Point C is the critical point of the mixture. Point A is the cricondenbar or point of maximum pressure, while point B is the cricondentherm or point of maximum temperature.

The dash-point curve is the saturated liquid with equilibrium compositions equal to the global composition of the mixture, and the curve of points is the saturated vapor with equilibrium compositions equal to the global composition of the mixture. Inside these boundaries is the two-phase region in which the global composition of the mixture is different from the one on the phase envelope. Small-point lines are isochoric lines that bend inside the two-phase region. The isochore that passes through the cricondentherm is collinear to the line in the two-phase region. The plot demonstrates different types of condensation: passing from the liquid at point D and going to point E, a vapor phase appears in the two-phase region until all the liquid disappears, and then we have a vapor mixture at point E, as expected. If we follow

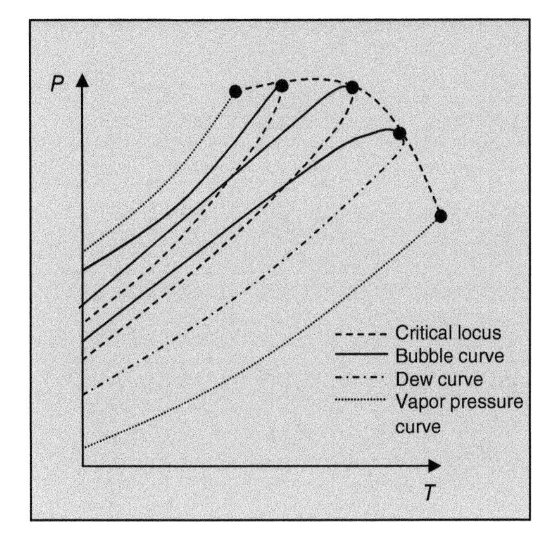

Figure 8.8 *P–T* diagram showing the phase envelopes of a binary mixture.

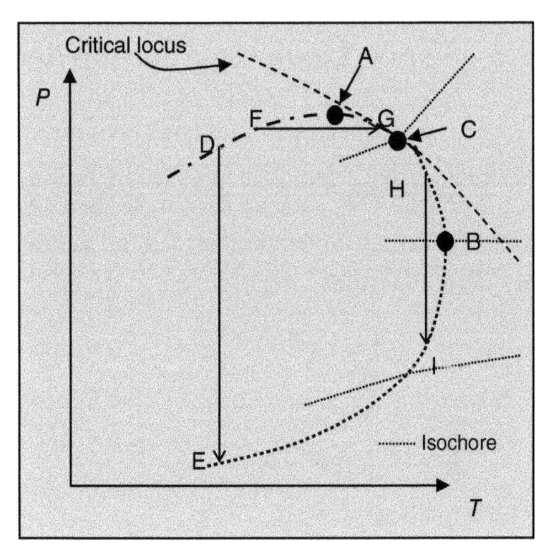

Figure 8.9 Phase envelope with retrograde condensation.

the path from F to G, we have liquid at point F that then condenses into two phases, but at point G, it becomes liquid again. This process is "retrograde vaporization." Following path H–I demonstrates "retrograde condensation" because we have vapor at point H and liquid appears until reaching a vapor phase at point I.

Another type of diagram is the composition diagram y vs x. Figure 8.10 shows the different behaviors at a given pressure. Points crossing the dashed line indicates an azeotrope because $x = y$. This type of diagram is important to check the thermodynamic consistency of data.

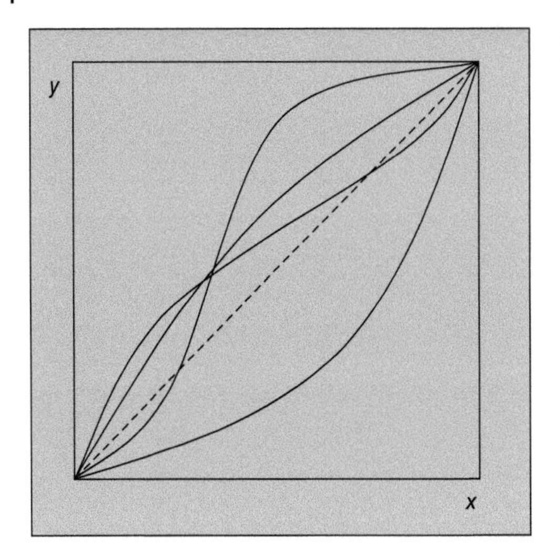

Figure 8.10 Different x–y diagrams at atmospheric pressure.

8.6 Vapor/Liquid Equilibrium Relationship

We have seen the vapor–liquid behavior on a diagram, and we know that equality of chemical potentials for each component in the mixture is a necessary condition for equilibrium. Also, we have seen in Section 7.11 that the chemical potential is related to the fugacities for component i in the mixture. Then, for vapor–liquid equilibrium in differential form,

$$d\mu_i^v = d\mu_i^l \quad \text{for} \quad i = 1, 2, \ldots, N \tag{8.43}$$

Now, using Eq. (7.165), $d\overline{G}_i \equiv d\mu_i \equiv RTd\ln\widehat{f}_i$,

$$RTd\ln\widehat{f}_i^v = RTd\ln\widehat{f}_i^l \tag{8.44}$$

which implies

$$\widehat{f}_i^v = \widehat{f}_i^l \quad \text{for} \quad i = 1, 2, \ldots, N \tag{8.45}$$

Using the definitions of the fugacity of component i in the mixture, for the vapor

$$\widehat{f}_i^v = y_i\widehat{\varphi}_i^v P \tag{8.46}$$

For the liquid, we have options:

(a) To use activity coefficients.
(b) To use the same definition.

Condition (a) is applicable at moderate and low pressures because many expressions for the activity coefficient are not functions of pressure. Condition (b) applies at low or high pressures. First, we use the definition of activity coefficient, Eq. (7.276)

$$\widehat{f}_i^l = x_i\gamma_i f_i^l \tag{8.47}$$

Next, substituting the fugacity of the liquid, Eq. (7.146)

$$\hat{f}_i^l = x_i \gamma_i P_i^{sat} \varphi_i^{sat} \exp\left[\frac{V_i^l \left(P - P_i^{sat}\right)}{RT}\right] \tag{8.48}$$

and at equilibrium

$$y_i \hat{\varphi}_i^v P = x_i \gamma_i P_i^{sat} \varphi_i^{sat} \exp\left[\frac{V_i^l \left(P - P_i^{sat}\right)}{RT}\right] \quad \text{for } i = 1, 2, \ldots, N \tag{8.49}$$

Equation (8.49) represents a system of nonlinear equations equating the fugacity of component i of the mixture in both phases (equality of chemical potential). The unknowns of this system can be the liquid and vapor compositions, temperature, or pressure depending on the variables selected. This equation is the gamma–phi ($\gamma - \hat{\varphi}$) formulation.

Next, if we use the fugacity coefficient for the liquid side, the fugacity of component i in the mixture is

$$\hat{f}_i^l = x_i \hat{\varphi}_i^l P \tag{8.50}$$

and the equilibrium relationship becomes

$$y_i \hat{\varphi}_i^v = x_i \hat{\varphi}_i^l \quad \text{for } i = 1, 2, \ldots, N \tag{8.51}$$

Equation (8.51) is applicable at any value of the pressure, and it is useful at high pressures with EOS for calculating the fugacity coefficient. This equation is called the *phi–phi* ($\hat{\varphi}^v - \hat{\varphi}^l$) formulation.

Equation (8.49) becomes simpler depending upon the condition of the vapor or liquid, that is, if the mixture is an ideal gas or an ideal solution or both. For most equilibrium calculations, we define the ratio of the vapor composition and liquid composition as

$$K_i \equiv \frac{y_i}{x_i} \tag{8.52}$$

Equation (8.52) defines the "equilibrium constant," but obviously, it is not a true constant because it depends on the composition, temperature, and pressure. Using Eq. (8.49), we find that

$$K_i = \frac{\gamma_i P_i^{sat} \varphi_i^{sat} \exp\left[\frac{V_i^l \left(P - P_i^{sat}\right)}{RT}\right]}{\hat{\varphi}_i^v P} \tag{8.53}$$

Several assumptions are possible:

1. If the Poynting correction is negligible, then the exponential term of Eq. (8.49) equals unity, and Eq. (8.49) becomes

   $$y_i \hat{\varphi}_i^v P = x_i \gamma_i P_i^{sat} \varphi_i^{sat} \quad \text{for } i = 1, 2, \ldots, N \tag{8.54}$$

   This equation generally applies to equilibrium calculations because the error involved is of the order of 0.1%.

2. For an ideal gas mixture with a negligible Poynting correction, the fugacity coefficients equal unity, $\hat{\varphi}_i^v = 1$ and $\varphi_i^{sat} = 1$, and the equilibrium relationship becomes

$$y_i P = x_i \gamma_i P_i^{sat} \quad \text{for } i = 1, 2, \ldots, N \tag{8.55}$$

This equation is the "modified" Raoult's law, and it only considers nonideality in the liquid phase.

3. For an ideal gas mixture with a negligible Poynting correction and an ideal solution, Eq. (8.55) applies with $\gamma_i = 1$

$$y_i P = x_i P_i^{sat} \quad \text{for } i = 1, 2, \ldots, N \tag{8.56}$$

Equation (8.56) is Raoult's law, and it only applies at low and moderate pressures. It may be valid when the species in the mixture are chemically similar. A limitation of Raoult's law is that it requires the vapor pressure of the pure components; therefore, it is applicable only if the mixture has equilibrium behavior at temperatures below the lowest critical temperature.

8.7 Phase Calculations Using the Gamma–Phi Formulation

Next, we investigate the different types of phase equilibria calculations. According to the phase rule, given two fixed variables, we can find the rest of the variables. This leads to the following cases:

Phase equilibria calculation	Known variables	Unknown variables
Dew point (Point a)	$T,\ y_i = z_i$	$P,\ x_i$
	$P,\ y_i = z_i$	$T,\ x_i$
Bubble point (Point b)	$T,\ x_i = z_i$	$P,\ y_i$
	$P,\ x_i = z_i$	$T,\ y_i$
Flash (Point c)	T, P	x_i, y_i

Figure 8.11 shows the known variables, and the arrows indicate the variables that we find in the phase equilibrium calculations. Figure 8.11 is a P–x, y diagram and a similar plot applies to a T–x, y diagram. When the known variable is pressure, the unknown variable is temperature. As seen in the diagram, the overall composition is the equilibrium composition at point (a) for the vapor and at point (b) for the liquid. Next, we establish the algorithm for calculating the unknown variables. Depending upon the condition of the mixture, the algorithm complexity can decrease. The simplest case considers that the mixture behavior follows Raoult's law.

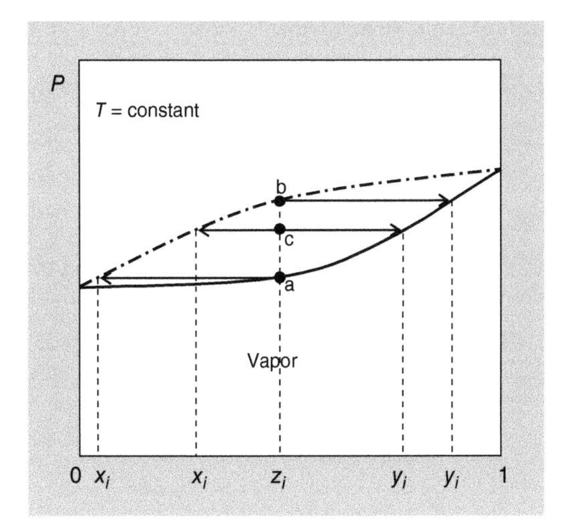

Figure 8.11 Phase equilibria calculations in a P–x, y diagram.

In the algorithms, we must include the material balance as the summation of the equilibrium compositions. Vapor and liquid compositions could come from the equilibrium relationships

$$y_i = K_i x_i = \frac{x_i \gamma_i P_i^{sat} \varphi_i^{sat} \exp\left[\frac{V_i^l \left(P - P_i^{sat}\right)}{RT}\right]}{\hat{\varphi}_i^v P} \tag{8.57}$$

$$x_i = \frac{y_i}{K_i} = \frac{y_i \hat{\varphi}_i^v P}{\gamma_i P_i^{sat} \varphi_i^{sat} \exp\left[\frac{V_i^l \left(P - P_i^{sat}\right)}{RT}\right]} \tag{8.58}$$

but if we consider the functionalities of the terms in the equilibrium relationships

$$\Phi_i = \frac{\hat{\varphi}_i^v}{\varphi_i^{sat}} = f(T, P, \{y_i\}) \tag{8.59}$$

$$\gamma_i = f(T, \{x_i\}) \tag{8.60}$$

and

$$P^{sat} = f(T) \tag{8.61}$$

we can conclude by observing the functionalities of Eqs. (8.59)–(8.61) that the phase equilibria calculations are iterative procedures. To simplify the procedure, we do not consider the Poynting correction in the following procedures.

8.7.1 Bubble Pressure

Specify: T,$\{x_i\}$ and determine: P,$\{y_i\}$.

Calculation Procedure:

1. The working equation is $\sum_{i=1}^{N} y_i = 1$. If we apply this equation to Eq. (8.57) neglecting the Poynting correction

$$\sum_{i=1}^{N} \frac{x_i \gamma_i P_i^{sat}}{\Phi_i P} = 1 \Rightarrow P = \sum_{i=1}^{N} \frac{x_i \gamma_i P_i^{sat}}{\Phi_i} \tag{A1}$$

2. Calculate each vapor pressure $\{P_i^{sat}\}$ using the specified T.
3. Calculate each activity coefficient $\{\gamma_i\}$ using the specified T and liquid composition $\{x_i\}$.
4. Find the root by solving a system of equations using any root-finding technique:

$$f_i(P, \{y_i\}) = y_i - \frac{x_i \gamma_i P_i^{sat}}{\Phi_i P} \quad \text{for } i = 1, 2, \ldots, N \tag{A2}$$

$$f_{N+1}(P, \{y_i\}) = 1 - \sum_{i=1}^{N} \frac{x_i \gamma_i P_i^{sat}}{\Phi_i P} = 0 \tag{A3}$$

If Raoult's law applies, we can calculate the pressure directly using $P = \sum_{i=1}^{N} x_i P_i^{sat}$ and $y_i = \frac{x_i P_i^{sat}}{P}$. If the modified Raoult's law applies, we can calculate the pressure from $P = \sum_{i=1}^{N} x_i \gamma_i P_i^{sat}$ and $y_i = \frac{x_i \gamma_i P_i^{sat}}{P}$ if the activity coefficient is independent of pressure.

Initial values are necessary to solve the system of equations. The initial pressure and compositions are those obtained when the mixture follows Raoult's or modified Raoult's laws.

8.7.2 Bubble Temperature

Specify: $\{x_i\}$, P and to determine: $\{y_i\}$, T.

Calculation Procedure:

1. Working equation: $\sum_{i=1}^{N} y_i = 1$. If we apply this equation to Eq. (8.57) neglecting the Poynting correction

$$\sum_{i=1}^{N} \frac{x_i \gamma_i P_i^{sat}}{\Phi_i P} = 1 \Rightarrow P = \sum_{i=1}^{N} \frac{x_i \gamma_i P_i^{sat}}{\Phi_i} \tag{B1}$$

2. Initial estimate of temperature: $T = \sum_{i=1}^{N} x_i T_i^{sat}$ in which the saturated temperature can come from the Antoine vapor pressure equation: $T_i^{sat} = \frac{B_i}{A_i - \ln P} - C_i$.
3. Calculate new vapor pressures $\left\{P_i^{sat}\right\}$.
4. Calculate activity coefficient $\{\gamma_i\}$ for each substance.
5. Find the unknowns using any root-finding technique by solving the system:

$$f_i(T, \{y_i\}) = y_i - \frac{x_i \gamma_i P_i^{sat}}{\Phi_i P} \quad \text{for } i = 1, 2, \ldots, N \tag{B2}$$

$$f_{N+1}(T,\{y_i\}) = 1 - \sum_{i=1}^{N} \frac{x_i \gamma_i P_i^{sat}}{\Phi_i P} = 0 \tag{B3}$$

If Raoult's law applies, we need to solve only one equation for the temperature

$$f_{N+1}(T) = P - \sum_{i=1}^{N} x_i P_i^{sat} = 0 \tag{B4}$$

and the vapor compositions are

$$y_i = \frac{x_i P_i^{sat}}{P}$$

When the mixture follows the modified Raoult's law, we can solve for the temperature using

$$f_{N+1}(T) = P - \sum_{i=1}^{N} x_i \gamma_i P_i^{sat} = 0 \tag{B5}$$

and the vapor compositions are

$$y_i = \frac{x_i \gamma_i P_i^{sat}}{P}$$

8.7.3 Dew Pressure

Specify: T, $\{y_i\}$, determine: P, $\{x_i\}$.

Calculation Procedure:

1. Working equation: $\sum_{i=1}^{N} x_i = 1$. If we apply this equation to Eq. (8.58) neglecting the Poynting correction

$$1 = \sum_{i=1}^{N} \frac{y_i \Phi_i P}{\gamma_i P_i^{sat}} \Rightarrow \frac{1}{P} = \sum_{i=1}^{N} \frac{y_i \Phi_i}{\gamma_i P_i^{sat}} \Rightarrow P = \left[\sum_{i=1}^{N} \frac{y_i \Phi_i}{P_i^{sat} \gamma_i} \right]^{-1} \tag{C1}$$

2. Calculate $\{P_i^{sat}\}$ (once only because T is available).
3. Find the unknowns solving a system of equations:

$$f_i(P,\{x_i\}) = x_i - \frac{y_i \Phi_i P}{\gamma_i P_i^{sat}} = 0 \quad \text{for } i = 1, 2, \ldots, N \tag{C2}$$

$$f_{N+1}(T,\{x_i\}) = 1 - \sum_{i=1}^{N} \frac{y_i \Phi_i P}{\gamma_i P_i^{sat}} = 0 \tag{C3}$$

If the mixture follows Raoult's law $\{\Phi_i = 1\}$ and $\{\gamma_i = 1\}$. We can solve for the pressure directly using

$$P = \frac{1}{\sum_{i=1}^{N} \frac{y_i}{P_i^{sat}}} \tag{C4}$$

and the liquid compositions are

$$x_i = \frac{y_i P}{P_i^{sat}} = 0$$

If the activity coefficients are independent of pressure, and the mixture follows the modified Raoult's law, we can calculate the pressure and the liquid compositions solving the system with $\{\Phi_i = 1\}$.

8.7.4 Dew Temperature

Specified: $\{y_i\}$, P, determined: $\{x_i\}$, T.

Calculation Procedure:

1. Working equation: $\sum_{i=1}^{N} x_i = 1$. If we apply this equation to Eq. (8.58) neglecting the Poynting correction

$$1 = \sum_{i=1}^{N} \frac{y_i \Phi_i P}{\gamma_i P_i^{sat}} \Rightarrow \frac{1}{P} = \sum_{i=1}^{N} \frac{y_i \Phi_i}{\gamma_i P_i^{sat}} \tag{D1}$$

2. Initial estimate for temperature $T = \sum_{i=1}^{N} y_i T_i^{sat}$ in which the saturated temperature comes from the Antoine vapor pressure equation $T_i^{sat} = \frac{B_i}{A_i - \ln P} - C_i$.
3. Calculate new vapor pressures $\{P_i^{sat}\}$.
4. Find the unknowns using any root-finding technique by solving the system:

$$f_i(P, \{x_i\}) = x_i - \frac{y_i \Phi_i P}{\gamma_i P_i^{sat}} = 0 \quad \text{for } i = 1, 2, \ldots, N \tag{D2}$$

$$f_{N+1}(T, \{x_i\}) = 1 - \sum_{i=1}^{N} \frac{y_i \Phi_i P}{\gamma_i P_i^{sat}} = 0 \tag{D3}$$

If Raoult's law applies, then we need to solve only one equation for the temperature

$$f_{N+1}(T) = 1 - \sum_{i=1}^{N} \frac{y_i P}{P_i^{sat}} = 0 \tag{D4}$$

and the liquid compositions are

$$x_i = y_i P / P_i^{sat}$$

When the mixture follows the modified Raoult's law, we can solve for the temperature using

$$f_{N+1}(T) = \frac{1}{P} - \sum_{i=1}^{N} \frac{y_i}{\gamma_i P_i^{sat}} \tag{D5}$$

and the liquid compositions are

$$x_i = \frac{y_i P}{\gamma_i P_i^{sat}} \tag{D6}$$

8.7.5 Flash

Assume that a mixture is in a container where it separates into two phases, vapor and liquid, as shown in Figure 8.12. F number of moles per time enter the separator

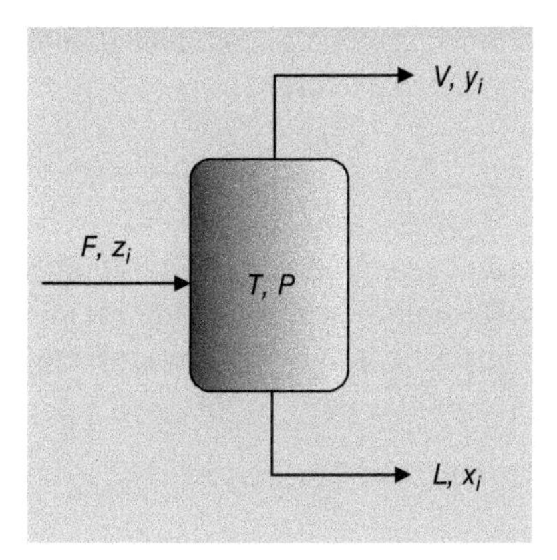

Figure 8.12 Flash separator.

and produce V moles of vapor and L moles of liquid. The equilibrium compositions for the vapor and the liquid are $\{y_i\}$, $\{x_i\}$ and the mixture enters with an overall composition $\{z_i\}$. If we perform a general material balance:

$$F = L + V \Rightarrow 1 = \frac{L}{F} + \frac{V}{F} \equiv L + V \tag{8.62}$$

in which L and V are the overall liquid and vapor mole fractions. A component material balance provides

$$z_i = x_i L + y_i V \tag{8.63}$$

Substituting for the liquid fraction

$$z_i = x_i(1 - V) + y_i V \tag{8.64}$$

Equation (8.64) can provide the liquid or the vapor composition. If we use the equilibrium relationship, Eq. (8.52), for the liquid composition, then

$$z_i = \frac{y_i}{K_i}(1 - V) + y_i V \tag{8.65}$$

and the vapor composition becomes

$$y_i = z_i K_i / V(K_i - 1) \tag{8.66}$$

Similarly, the liquid composition becomes (using $x_i = K_i y_i$)

$$x_i = z_i / [1 + V(K_i - 1)] \tag{8.67}$$

Using our working equations provides

$$\sum_{i=1}^{N} y_i = 1 \Rightarrow \sum_{i=1}^{N} z_i K_i / [1 + V(K_i - 1)] = 1 \tag{8.68}$$

$$\sum_{i=1}^{N} x_i = 1 \Rightarrow \sum_{i=1}^{N} z_i / [1 + V(K_i - 1)] = 1 \tag{8.69}$$

Equations (8.68) and (8.69) can provide the vaporization fraction, V, if the equilibrium constant is independent of the equilibrium compositions. From a numerical solution standpoint, a combination of these equations is more robust. Therefore, it is preferable to use

$$\sum_{i=1}^{N} y_i - \sum_{i=1}^{N} x_i = 0 \Rightarrow \sum_{i=1}^{N} z_i(K_i - 1)/[1 + V(K_i - 1)] = 0 \tag{8.70}$$

to find the equilibrium compositions and the vaporization fraction. As before, we can solve directly for V, if the equilibrium constants are independent of the equilibrium compositions; otherwise, we need to solve a system of equations or use an iterative procedure.

Calculation Procedure:

Specify: P, T, $\{z_i\}$ and determine: $\{x_i\}$, $\{y_i\}$, V.

1. Calculate P_{DP}, $\{x_i\}$ at T
2. Calculate P_{BP}, $\{y_i\}$ at T
3. Proceed with flash calculation only if $P_{DP} < P < P_{BP}$
4. Solve for $\{x_i\}$, $\{y_i\}$ and P using

$$K_i[\{y_i\}, \{x_i\}, V] = y_i - \frac{x_i \gamma_i P_i^{sat}}{\Phi_i P} = 0 \quad \text{for } i = 1, 2, \ldots, N \tag{F1}$$

$$f_{i+1}[\{y_i\}, \{x_i\}, V] = z_i - x_i(1 - V) - y_i V = 0 \quad \text{for } i = 1, 2, \ldots, N \tag{F2}$$

with

$$\sum_{i=1}^{N} y_i = 1 \text{ and } \sum_{i=1}^{N} x_i = 1$$

The initial guesses could be the values obtained from a similar mixture that behaves according to Raoult's law. Then, the equilibrium constant is independent of composition

$$K_i = \frac{P_i^{sat}}{P} \tag{8.71}$$

and we solve for V using Eq. (8.70). If the mixture follows the modified Raoult's law, we can solve the system of equations using $\Phi_i = 1$. Appendix A.7 presents iterative procedures that do not require use of root finding techniques.

Example 8.2

Consider a binary mixture of components A and B. Perform the following phase equilibrium calculations if the overall composition of the mixture is 45% of component A and 55% of component B:

(a) Bubble pressure at 350 K
(b) Dew pressure at 350 K
(c) Bubble temperature at 101.325 kPa
(d) Dew temperature at 101.325 kPa
(e) Flash calculation at 350 K and $P = (P_{DP} + P_{BP})/2$

Data

$$\ln P_A^{sat}\,(\text{kPa}) = 16.1 - \frac{2900}{T\,(\text{K}) - 40}$$

$$\ln P_B^{sat}\,(\text{kPa}) = 15.9 - \frac{3000}{T\,(\text{K}) - 60}$$

Virial coefficients:

$$B_{2,AA} = -1400 + 280\,000/T$$

$$B_{2,BB} = -2800 + 560\,000/T$$

$$B_{2,AB} = -2700 + 560\,000/T$$

Nonideal liquid phase:

$$\frac{G^E}{RT} = -0.529x_Ax_B$$

Solution

First, calculate the vapor pressure of the pure components at 350 K

$$P_A^{sat} = 849.936\,\text{kPa}, P_B^{sat} = 258.572\,\text{kPa}$$

Next, calculate the activity coefficient from the excess Gibbs energy

$$\frac{G^E}{RT} = -0.529x_Ax_B \Rightarrow \ln\gamma_i = \left[\partial\left(\frac{nG^E}{RT}\right)\Big/\partial n_i\right]_{T,P,n_{j\neq i}}$$

$$\times\begin{cases}\ln\gamma_A = -0.529x_B^2 = 0.85212\\ \ln\gamma_B = -0.529x_A^2 = 0.89841\end{cases}$$

and the virial coefficients at $T = 350$ K

$$B_{2,AA} = -600\,\text{cm}^3/\text{mol}$$

$$B_{2,BB} = -1200\,\text{cm}^3/\text{mol}$$

$$B_{2,AB} = -1100\,\text{cm}^3/\text{mol}$$

Calculate $\{\Phi_i\}$ using Eq. (7.152) at saturation and Eqs. (7.216) and (7.217)

$$\ln\varphi_i^{sat} = \frac{B_{2,ii}P^{sat}}{RT}$$

$$\ln\widehat{\varphi}_i^v = \frac{P}{RT}\left[B_{2,ii} + y_j^2\delta_{ij}\right]$$

If, $\Phi_i = \widehat{\varphi}_i^v/\varphi_i^{sat}$, then

$$\ln\Phi_i = \ln\left(\widehat{\varphi}_i^v/\varphi_i^{sat}\right) = \frac{\left[B_{2,ii}\left(P - P_i^{sat}\right) + y_j^2\delta_{ij}\right]}{RT}$$

or

$$\Phi_A = \exp\left\{\frac{\left[B_{2,AA}\left(P - P_A^{sat}\right) + y_B^2 P\delta_{AB}\right]}{RT}\right\}$$

$$\Phi_B = \exp\left\{\frac{\left[B_{2,BB}\left(P - P_B^{sat}\right) + y_A^2 P\delta_{AB}\right]}{RT}\right\}$$

with

$$\delta_{AB} = 2B_{2,AB} - B_{2,AA} - B_{2,AA} = -400\ \text{cm}^3/\text{mol}$$

(a) *Bubble pressure:* $x_A = z_A = 0.45, x_B = z_B = 0.55$
Calculate the pressure and the vapor compositions using

$$f_1 = y_A - \frac{x_A\gamma_A P_A^{sat}}{\Phi_A P} = 0$$

$$f_2 = y_B - \frac{x_B\gamma_B P_B^{sat}}{\Phi_B P} = 0$$

$$f_3 = 1 - \sum_{i=1}^{N}\frac{x_i\gamma_i P_i^{sat}}{\Phi_i P} = 0$$

Use as initial guesses the results obtained from the modified Raoult's law

$$P = x_A\gamma_A P_A^{sat} + x_B\gamma_B P_B^{sat} = (0.45)(0.8521)(849.936)$$
$$+ (0.55)(0.8984)(258.571) = 453.681\ \text{kPa}$$

$$y_A = \frac{(0.45)(0.8521)(849.936)}{453.681} = 0.7184,$$

$$y_B = \frac{(0.55)(0.8984)(258.571)}{453.681} = 0.2816$$

To solve the system of equations using an optimization technique that minimizes an objective function, we form the objective function following the Euclidean norm

$$F_{OBJ} = \sqrt{\sum_{i=1}^{nf} f_i^2}$$

Using Solver in Excel®, we obtain

$$P = 443.414, y_A = 0.68014, \text{ and } y_B = 0.31986$$

(b) *Dew pressure:* $y_A = z_A = 0.45, y_B = z_B = 0.55$
Now, the system becomes

$$f_1 = x_A - \frac{y_A\Phi_A P}{\gamma_A P_A^{sat}} = 0$$

$$f_2 = x_B - \frac{y_B\Phi_B P}{\gamma_B P_B^{sat}} = 0$$

$$f_3 = 1 - \sum_{i=1}^{N} \frac{y_i \Phi_i P}{\gamma_i P_i^{sat}} = 0$$

An initial guess can be the results obtained using Raoult's law

$$P = \frac{1}{\frac{y_A}{P_A^{sat}} + \frac{y_B}{P_B^{sat}}} = \frac{1}{\frac{0.45}{849.936} + \frac{0.55}{258.571}} = 376.432\ \text{kPa}$$

$$x_A = \frac{y_A P}{P_A^{sat}} = \frac{(0.45)(376.432)}{849.936} = 0.1993$$

$$x_B = \frac{y_B P}{P_B^{sat}} = \frac{(0.55)(376.432)}{258.571} = 0.8007$$

Using Solver in Excel provides

$$P = 339.043\ \text{kPa},\ x_A = 0.23078,\ y_B = 0.76922$$

(c) *Bubble temperature:* $x_A = z_A = 0.45, x_B = z_B = 0.55$
Calculate the temperature and vapor compositions using

$$f_1 = y_A - \frac{x_A \gamma_A P_A^{sat}}{\Phi_A P} = 0$$

$$f_2 = y_B - \frac{x_B \gamma_B P_B^{sat}}{\Phi_B P} = 0$$

$$f_3 = 1 - \sum_{i=1}^{N} \frac{x_i \gamma_i P_i^{sat}}{\Phi_i P} = 0$$

$$T_A^{sat} = \frac{2900}{16.1 - \ln(101.325)} - 40 = 292.577\ \text{K}$$

$$T_B^{sat} = \frac{3000}{15.9 - \ln(101.325)} - 60 = 325.918\ \text{K}$$

and the initial temperature is

$$T = x_A T_A^{sat} + x_B T_B^{sat} = (0.45)(292.577) + (0.55)(325.918) = 310.914\ \text{K}$$

With this temperature, $P_A^{sat} = 220.415$ kPa and $P_B^{sat} = 51.611$ kPa, and the compositions using the expression from the modified Raoult's law become

$$y_A = \frac{(0.45)(0.8521)(220.415)}{101.325} = 0.8341,$$

$$y_B = \frac{(0.55)(0.8984)(51.611)}{101.325} = 0.2517$$

Now, use Solver in Excel and minimize the function to obtain

$$f_3 = 1 - \sum_{i=1}^{N} \frac{x_i \gamma_i P_i^{sat}}{\Phi_i P} = 0$$

(d) *Dew temperature*: $y_A = z_A = 0.45, y_B = z_B = 0.55$

Now, we can calculate the temperature and liquid compositions solving

$$f_1 = x_A - \frac{y_A \Phi_A P}{\gamma_A P_A^{sat}} = 0$$

$$f_2 = x_B - \frac{y_B \Phi_B P}{\gamma_B P_B^{sat}} = 0$$

$$f_3 = 1 - \sum_{i=1}^{N} \frac{y_i \Phi_i P}{\gamma_i P_i^{sat}} = 0$$

The initial temperature and vapor pressures are the same as in the bubble temperature calculation $T = 310.914$ K, $P_A^{sat} = 220.415$ kPa, and $P_B^{sat} = 51.611$ kPa, and we obtain the liquid compositions using the expression from Raoult's law

$$x_A = \frac{y_A P}{P_A^{sat}} = \frac{(0.45)(101.325)}{220.415} = 0.20686,$$

$$x_B = \frac{y_B P}{P_B^{sat}} = \frac{(0.55)(101.325)}{51.611} = 1.07978$$

Normalize the compositions, so they are not greater than 1

$$x_A = \frac{0.20686}{0.20686 + 1.07978} = 0.16078, \; x_B = \frac{1.07978}{0.20686 + 1.07978} = 0.83922$$

Use Solver in Excel and minimize the function to obtain

$$T = 318.264 \text{ K}, \quad x_A = 0.22213, \quad x_B = 0.77787$$

(e) *Flash*: $T = 350$ K, $P = (453.681 + 339.043)/2$ kPa

Now, calculate the vapor and liquid compositions. We need only three equations. Because the unknowns are y_A, x_A, and V. Select the three equations

$$f_1 = y_A - \frac{x_A \gamma_A P_A^{sat}}{\Phi_A P} = 0$$

$$f_2 = y_B - \frac{x_B \gamma_B P_B^{sat}}{\Phi_B P} = 0$$

$$f_3 = z_A - x_A(1 - V) - y_A V = 0$$

with

$$y_B = 1 - y_A \text{ and } x_B = 1 - x_A$$

Using as initial guesses

$$y_A = 0.1, \; x_A = 0.1, \; V = 0.5$$

use Solver in Excel and minimize the function to obtain

$$y_A = 0.58532, \; x_A = 0.36702, \quad V = 0.38$$

8.8 Phase Calculations Using the Phi–Phi Formulation

We have learned how to calculate the bubble and dew points and flash calculations using the gamma–phi formulation. In this section, we use the phi–phi formulation. This formulation requires using an EOS, so it is valid only for conditions at which the EOS is valid. The system of equations is

$$y_i\hat{\varphi}_i^v = x_i\hat{\varphi}_i^l \quad \text{for } i = 1, 2, \dots, N \tag{8.51}$$

and

$$\sum_{i=1}^{N} y_i = 1 \text{ and } \sum_{i=1}^{N} x_i = 1$$

The fugacity coefficients in the vapor and the liquid depend upon temperature, vapor and liquid densities, and vapor and liquid compositions. Therefore, at any given temperature and pressure, we must solve the EOS for the densities of the vapor and the liquid. One simplification of Eq. (8.51) is when the mixture is an ideal solution, the fugacity coefficient of component i equals the fugacity coefficient of the pure component, and Eq. (8.51) becomes

$$\varphi_i^v y_i = \varphi_i^l x_i \quad \text{for } i = 1, 2, \dots, N \tag{8.72}$$

and the equilibrium constant is independent of the composition

$$K_i = \frac{y_i}{x_i} = \frac{\varphi_i^l}{\varphi_i^v} \tag{8.73}$$

Equilibrium constants calculated in this manner appear for hydrocarbons in nomograms called: DePriester Nomograms or DePriester charts. Figures 8.13 and 8.14 present these nomograms. To use them, we draw a straight line between the temperature and pressure, and the K_i value is the intersection with the curves.

Now, we can apply the phi–phi formulation to phase equilibrium calculations.

Procedure for Bubble Point Calculations:

We have specified: T, $\{x_i\}$ and we must determine: P, $\{y_i\}$ by

1. Providing an estimate of the pressure (or temperature) and vapor composition
2. Calculating densities from the EOS for the liquid and vapor
3. Solving the system of equations for a new P or T and $\{y_i\}$ using Eq. (8.51)

$$y_i - \frac{\hat{\varphi}_i^l}{\hat{\varphi}_i^v} x_i = y_i - K_i x_i = 0 \quad \text{for } i = 1, 2, \dots, N$$

   and

$$\sum_{i=1}^{N} y_i = 1 \text{ and } \sum_{i=1}^{N} x_i = 1$$

4. Returning to step 2 until values of the pressure or temperature and compositions do not change.

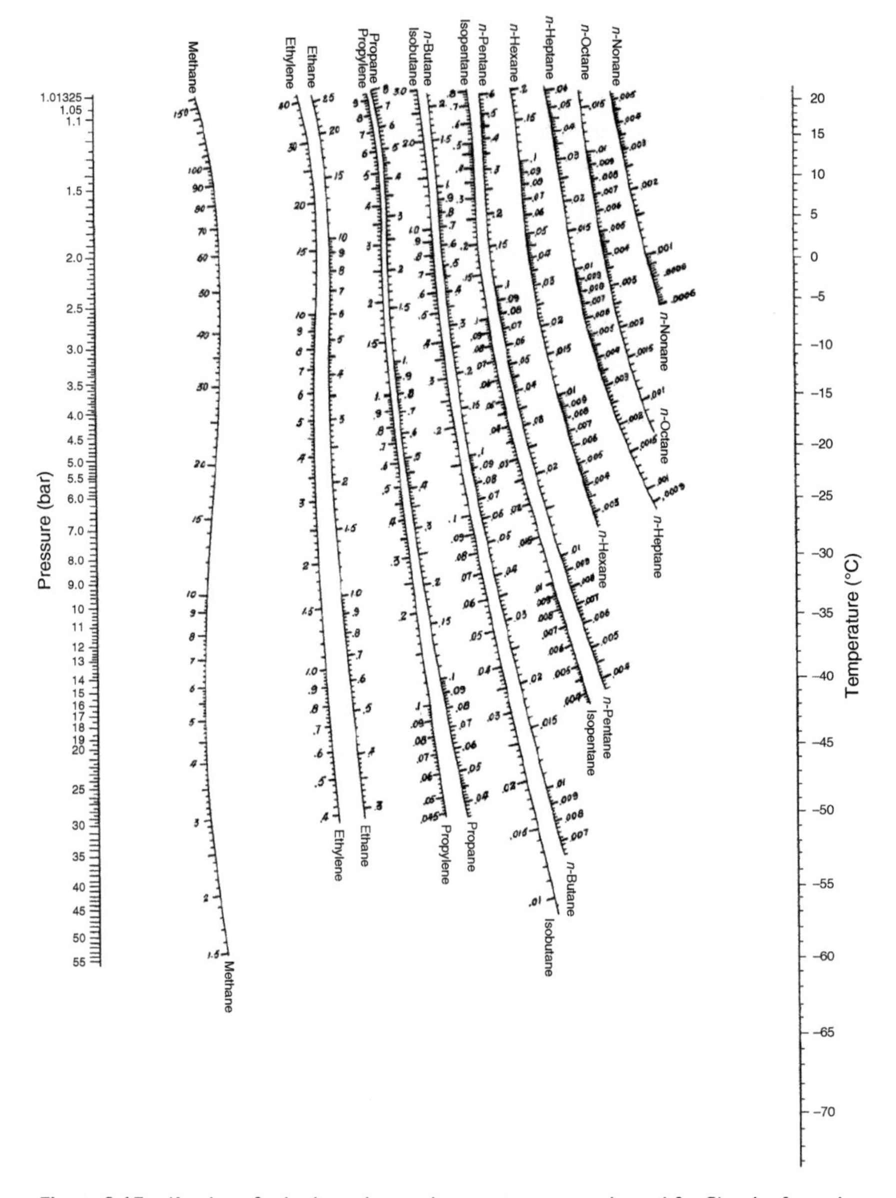

Figure 8.13 *K*-values for hydrocarbon at low temperature adapted for SI units from the original DePriester Nomograms.

Procedure for Dew Point Calculations:

1. Give an estimate of the pressure (or temperature) and liquid composition
2. Calculate densities from EOS for the liquid and vapor
3. Solve the system of equations for a new, P or T, $\{x_i\}$

$$x_i = \frac{y_i}{\hat{\varphi}_i^l/\hat{\varphi}_i^v} \Rightarrow x_i - y_i/K_i = 0 \quad \text{for } i = 1, 2, \ldots, N$$

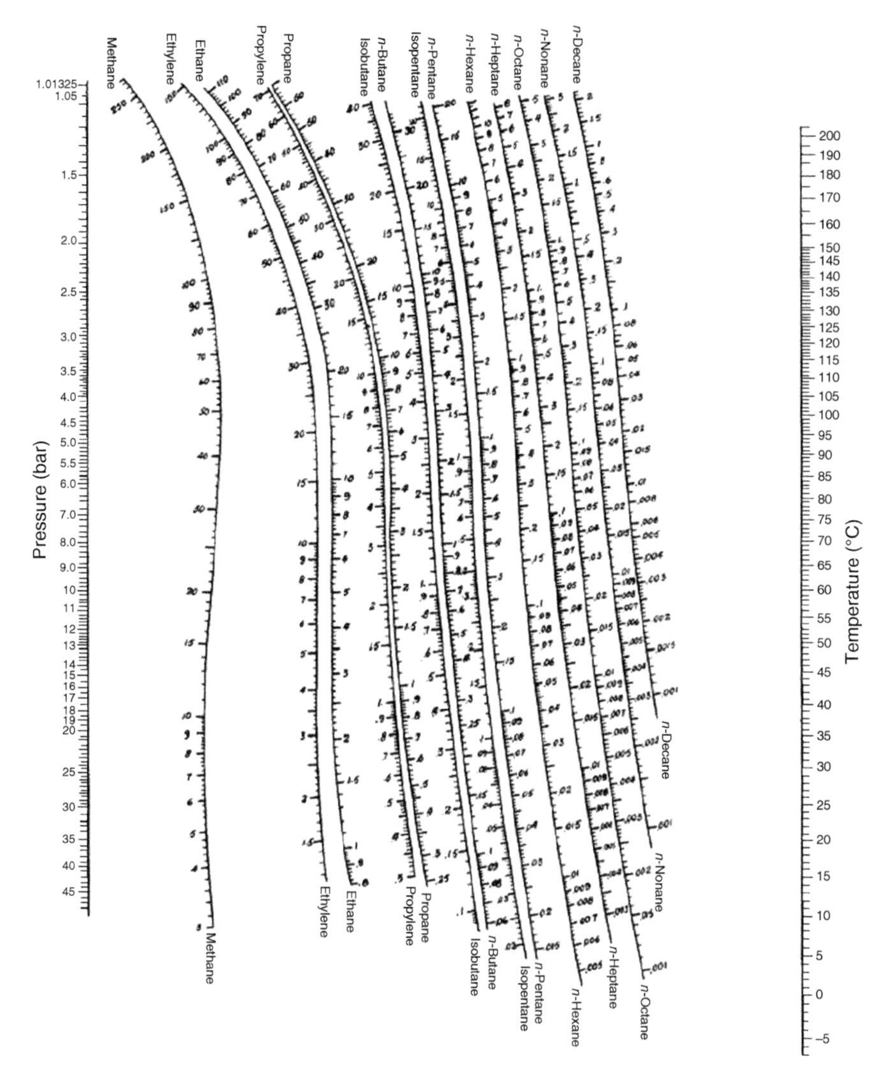

Figure 8.14 K-values for hydrocarbon at high temperature adapted for SI units from the original DePriester Nomograms.

and

$$\sum_{i=1}^{N} y_i = 1 \text{ and } \sum_{i=1}^{N} x_i = 1$$

4. Return to step 2 until values of the pressure or temperature and compositions do not change.

Procedure for the Flash Calculation:

1. Estimate the vapor and liquid compositions and V
2. Calculate densities from the EOS for the liquid and vapor mixture.

3. Select the system of equations and solve for new$\{y_i\}$, $\{x_i\}$, and V

$$f_i = y_i - \frac{\widehat{\varphi}_i^l}{\widehat{\varphi}_i^v} x_i = y_i - K_i x_i = 0 \quad \text{for } i = 1, 2, \ldots, N$$

$$f_{i+N}(\{y_i\}, \{x_i\}, V) = z_i - x_i(1 - V) - y_i V \quad \text{for } i = 1, 2, \ldots, N$$

4. Return to step 2 until the variables do not change appreciably.

In all these calculations, instead of solving the system of equations, we could minimize the objective function:

$$F_{OBJ} = \sqrt{\sum_{i=1}^{nf} f_i^2}$$

DePriester [2] *Nomogram*

Calculations using the DePriester charts (Figures 13 and 14) are simple. To find a K value simply draw a straight line between the pressure (left axis) and the temperature (right axis). The K-value for each substance is the intersection between the straight line and the corresponding curve for the pure substance.

For the bubble point calculations, the working equation is $y_i = K_i x_i$ and the steps are similar as before:

1. Give an estimate of the desired variable: pressure or temperature
2. Calculate the K_i values for each component in the mixture from the DePriester charts at the given temperature and pressure
3. Check if $\sum_{i=1}^{N} y_i = \sum_{i=1}^{N} K_i x_i = 1$. If the summation is not unity, propose a new value of the variable pressure or temperature and return to step 2.
4. Repeat the process until the difference of two consecutive values of the variable is less than a convergence criterion.
   To calculate a new variable, guess two values and use the summation of the compositions for each case. The new variable results from using a straight line:

$$Y^{new} = \frac{Y_{j+1} - Y_j}{\sum_{j+1} - \sum_j} \left(1 - \sum_{j+1}\right) + Y_{j+1} \tag{8.74}$$

   in which Y_{j+1} is a variable nearer to the solution and the summation is over the mole fractions. These iterative procedures can proceed using the two closest values to the solution until reaching convergence.

Dew point calculations follow the above procedure, but the working equation is $x_i = y_i/K_i$, and we must check: $\sum_{i=1}^{N} y_i/K_i = 1$.

For a flash calculation, the K_i values are functions of temperature and pressure, and we solve Eq. (8.70) for V

$$f(V) = \sum_{i=1}^{N} z_i(K_i - 1)/[1 + V(K_i - 1)] = 0$$

Table 8.1 Parameters for Eq. (8.77).

Compound	a_{T1}	a_{T2}	a_{T6}	a_{P1}	a_{P2}	a_{P3}	% error
Methane	−292 860	0	8.2445	−0.8951	59.8465	0	1.66
Ethylene	−600 076.9	0	7.90595	−0.84677	42.94594	0	2.65
Ethane	−687 248.3	0	7.90699	−0.886	49.02654	0	1.95
Propylene	−923 484.7	0	7.71725	−0.87871	47.67624	0	1.90
Propane	−970 688.6	0	7.71725	−0.67984	0	6.90224	2.35
Isobutane	−1 166 846	0	7.72668	−0.92213	0	0	2.52
n-Butane	−1 280 557	0	7.994986	−0.96455	0	0	3.61
Isopentane	−1 481 583	0	7.58071	−0.93159	0	0	4.56
n-Pentane	−1 524 891	0	7.33129	−0.89143	0	0	4.30
n-Hexane	−1 778 901	0	6.96783	−0.84634	0	0	4.90
n-Heptane	−2 013 803	0	6.52914	−0.79543	0	0	6.34
n-Octane	0	−7646.816	12.48457	−0.73152	0	0	7.58
n-Nonane	−255 104	0	5.69313	−0.67818	0	0	9.40
n-Decane	0	−9760.457	12.80354	−0.7147	0	0	5.69

Given two values of the function, it is possible to calculate a value closer to the solution using

$$V^{new} = [(V_{j+1} - V_j)(-f_{j+1})/(f_{j+1} - f_j)] + V_{j+1} \tag{8.75}$$

Continue with this procedure using two closest estimates to the solution until

$$| V_{j+1} - V_j | \leq \varepsilon = 1 \times 10^{-5}$$

Using functions of pressure and temperature instead of nomograms renders the solution as simple as finding the root (temperature or pressure) that satisfies the working equation. Also, some regions of the nomograms and charts have equations. One of the most popular is the Wilson [3] equation,

$$K_i = \frac{P_{C,i}}{P} \exp\left\{5.37(1+\omega_i)\left(1 - \frac{T_{C,i}}{T}\right)\right\} \tag{8.76}$$

This correlation applies for pressures up to 3.5 MPa (500 psia). McWilliams [4] correlates the K values from the DePriester nomographs using

$$\ln K = \frac{a_{T1}}{T^2} + \frac{a_{T2}}{T} + a_{T6} + a_{P1}\ln P + \frac{a_{P2}}{P^2} + \frac{a_{P3}}{P} \tag{8.77}$$

where P is in psia and T is °R. Table 8.1 shows the values of the parameters for Eq. (8.77). This equation is valid for $460 < T < 760$ °R and $14.7 < P < 120$ psia.

Example 8.3

Consider an equimolar hydrocarbon mixture: ethane, propane, and butane. Using the DePriester K-values to find:

(a) The bubble temperature at 10 bar.
(b) The dew temperature at 10 bar.
(c) The equilibrium compositions and the vapor fraction at 10 bar and $T = (2T_B + T_D)/3$

Solution

In the bubble temperature calculation, $z_i = x_i$ and the working equation is $y_i = K_i x_i$. The objective is to find a temperature that makes $\sum_{i=1}^{N} y_i = 1$. Now, select two temperatures, for example, −5 and 0 °C. Using the DePriester Nomogram for low temperatures

		−5		0	
Substance	$z_i = x_i$	K_i	$y_i = K_i x_i$	K_i	$y_i = K_i x_i$
Ethane	0.333333	1.8	0.6	1.95	0.65
n-Propane	0.333333	0.47	0.156667	0.52	0.173333
n-Butane	0.333333	0.11	0.036667	0.131	0.043667
Sum			0.793333		0.867

Now, estimate the following temperature using Eq. (8.74)

$$T^{new} = \frac{-5 - 0}{0.79333 - 0.867}(1 - 0.79333) - 5 = 9.027$$

Because the summation of the y's is 1.062, we use this temperature and 0 °C to calculate a new temperature

$$T^{new} = \frac{0 - 9.027}{0.867 - 1.0616}(1 - 0.867) + 0 = 6.167$$

This temperature brings us closer to the solution, the sum of the y compositions is now 0.97166. Using $T = 6.167$ and $T = 9.027$, we calculate a new temperature as

$$T^{new} = \frac{9.027 - 6.167}{1.061667 - 0.971667}(1 - 1.061667) + 9.027 = 7.068$$

With this new temperature, the summation is almost unity. The bubble point temperature is 7.068 °C. The next table shows the results for each of the temperatures

		9.027		6.167		7.068	
Substance	$z_i = x_i$	K_i	$y_i = K_i x_i$	K_i	$y_i = K_i x_i$	K_i	$y_i = K_i x_i$
Ethane	0.333333	2.33	0.776667	2.14	0.713333	2.2	0.733333
n-Propane	0.333333	0.67	0.223333	0.61	0.203333	0.62	0.206667
n-Butane	0.333333	0.185	0.061667	0.165	0.055	0.17	0.056667
Sum			1.061667		0.971667		0.996667

Now, we repeat the same iterative procedure but use $z_i = y_i$, and the working equation is $x_i = y_i/K_i$. Now, we must find the temperature that makes the summation of the liquid compositions equal to unity. Knowing that the dew temperature should be greater than the bubble temperature, we can choose $T = 40\,°\text{C}$ and $T = 60\,°\text{C}$. The results are (from the high temperature nomogram):

Substance	$z_i = x_i$	40 K_i	$x_i = K_i/y_i$	60 K_i	$x_i = K_i/y_i$
Ethane	0.333333	3.9	0.085470	5.02	0.066401
n-Propane	0.333333	1.49	0.223714	1.98	0.16835
n-Butane	0.333333	0.446	0.747384	0.7	0.47619
Sum			1.056568		0.710942

A new equation comes from the interpolating formula Eq. (8.74)

$$T^{new} = \frac{60 - 40}{0.710942 - 1.056568}(1 - 0.710942) + 40 = 43.273$$

Using this temperature,

Substance	$z_i = x_i$	40 K_i	$x_i = K_i/y_i$
Ethane	0.333333	4.18	0.079745
n-Propane	0.333333	1.47	0.226757
n-Butane	0.333333	0.482	0.691563
Sum			0.998065

The final dew point is 43.273 °C.

For the flash calculation, we have $P = 10\,\text{bar}$ and $T = (2 \times 7.068 + 43.273)/3 = 19.136$. Assuming $V = 0.5$ and 0.2, and checking if

$$f(V) = \sum_{i=1}^{N} z_i(K_i - 1)/[1 + V(K_i - 1)] = 0$$

we find that $f(0.5) = -0.1086$ and $f(0.2) = 0.1317$. Using Eq. (8.75)

$$V_{new} = \frac{0.5 - 0.2}{-0.1086 - 0.1317}(0.1086) + 0.5 = 0.364$$

The function is −0.0028, which is close enough considering the accuracy of the nomogram. The following table contains the results:

Substance	z_i	K_i	$V = 0.5$ $\frac{z_i(K_i-1)}{1+V(K_i-1)}$	$V = 0.2$ $\frac{z_i(K_i-1)}{1+V(K_i-1)}$	$V = 0.364$ $\frac{z_i(K_i-1)}{1+V(K_i-1)}$
Ethane	0.333333	2.9	0.324786	0.458937	0.374399
n-Propane	0.333333	0.9	−0.03509	−0.03401	−0.03459
n-Butane	0.333333	0.252	−0.3983	−0.2932	−0.34262
$f(V)$			−0.1086	0.131728	−0.00281

Example 8.4

Consider a binary mixture of CH_4 and N_2 with an overall mixture composition of 20% CH_4 and 80% N_2. Use the Soave–Redlich–Kwong (SRK) EOS to obtain the bubble pressure at 100 K. Consider a $k_{12} = 0$. Parrish and Hiza [5] give an approximate value for the pressure $P = 6.451$ bar with $y_1 = 0.0236$ at $T = 100$ K and $x_1 = 0.1867$.

Solution

$$T = 100\ \text{K}$$

$$R = 83.14\ \text{cm}^3\ \text{bar}/(\text{mol K})$$

The parameters for the pure substances are

Specie	$T_{c,i}$ (K)	$P_{c,i}$ (bar)	ω_i
CH_4	190.6	45.99	0.012
N_2	126.2	34.0	0.038

The calculated quantities for the pure components are as follows:

Specie	$T_{R,i} = T/T_{c,i}$	α_i Eq. (3.125)	b_i (cm³/mol) Eq. (3.119)	a_i(bar/cm⁶/mol²) Eq. (3.127)
CH_4	0.52465897	1.29395116	29.85299791	3222401.66
N_2	0.79239303	1.12203715	26.73676763	1554910.2

We can calculate the mixture parameters using Eqs. (7.231) and (7.232)

$$a^\alpha = z_1^2 a_1 + 2z_1 z_2 a_{12} + z_2^2 a_2 \text{ with } a_{12} = \sqrt{a_1 a_2}(1 - k_{12})$$

$$b^\alpha = z_1 b_1 + z_2 b_2$$

We use Eqs. (3.151) and (3.152) for mixtures

$$A^\alpha = \frac{a^\alpha P}{(RT)^2} \text{ and } B^\alpha = \frac{Pb^\alpha}{RT}, A^\alpha/B^\alpha = \frac{a^\alpha}{b^\alpha RT} \quad \text{for } \alpha = l \text{ or } v$$

The partial molar properties $\overline{D}_i^{\alpha}$ in each phase (vapor and liquid) for each component using Eq. (7.40) is

$$\overline{D}_1^{\alpha} = (A^{\alpha}/B^{\alpha})\left(\frac{2z_1a_1 + 2z_2a_{12}}{a^{\alpha}} - \frac{b_1}{b^{\alpha}}\right)$$

$$\overline{D}_2^{\alpha} = (A^{\alpha}/B^{\alpha})\left(\frac{2z_2a_2 + 2z_1a_{12}}{a^{\alpha}} - \frac{b_2}{b^{\alpha}}\right) \quad \text{for } \alpha = l \text{ or } v$$

We must calculate the compressibility factor of the liquid and the vapor used for the EOS, Eqs. (3.155) and (3.156), but for mixtures

$$Z_{j+1}^{l} = \frac{Z_j^{2,l} - Z_j^{3,l} + A^l B^l}{(A^l - B^l - B^{2,l})}$$

$$Z_{j+1}^{v} = 1 - (A^v - B^v - B^{2,v})/Z_j^v + A^v B^v/Z_j^{2,v}$$

Using these two equations, the values of the compressibility factors Z^l or Z^v result from iterative calculations. The fugacity coefficient of each component in each phase results from using the equation from example in Section 7.12,

$$\ln \hat{\varphi}_i^{\alpha} = \frac{b_i}{b^{\alpha}}(Z^{\alpha} - 1) - \ln(Z^{\alpha} - B^{\alpha}) - \overline{D}_i^{\alpha} \ln\left(\frac{Z^{\alpha} + B^{\alpha}}{Z^{\alpha}}\right) \quad \text{for } \alpha = l \text{ or } v$$

The objective function to minimize is

$$F_{OBJ} = \sqrt{f_1^2 + f_2^2}$$

with

$$f_1 = y_1 - \frac{\hat{\varphi}_1^l}{\hat{\varphi}_1^v} x_1$$

$$f_2 = y_2 - \frac{\hat{\varphi}_2^l}{\hat{\varphi}_2^v} x_2 = 0$$

Using Solver in Excel, solve for the roots of Z in each phase and then solve for P and y_1 minimizing the objective function. Repeat this procedure until the values of all the functions are less than or equal to a small value. Here are the initial values and the final result:

Initial values

	Liquid	**Vapor**
a^{α} (bar cm^6/mol^2)	1.807796×10^6	1.582095×10^6
b^{α} (cm^3/mol)	27.3600137	26.8103107
A^{α}/B^{α}	7.94736415	7.09775257
B^{α}	0.02122919	0.02080266
$\overline{D}_1^{\alpha}$	11.8730304	11.710112
$\overline{D}_2^{\alpha}$	6.96594759	6.9862699
Z^{α}	0.0304885	0.8566086

(Continued)

	Liquid	Vapor
$\ln \widehat{\varphi}_1^\alpha$	−2.65003182	−0.26128681
$\ln \widehat{\varphi}_2^\alpha$	0.05354205	−0.13127281
$\widehat{\varphi}_1^\alpha$	0.07064896	0.77006003
$\widehat{\varphi}_2^\alpha$	1.05500135	0.87697849

The value of the objective function is 0.00683966. The final values are

	Liquid	Vapor
a^α (bar cm^6/mol^2)	1.807796×10^6	1.575999×10^6
b^α (cm^3/mol)	27.3600137	26.7949372
A^α/B^α	7.94736415	7.07446128
B^α	0.02072987	0.02030173
$\overline{D}_1^\alpha$	11.8730304	11.7049364
$\overline{D}_2^\alpha$	6.96594759	6.98638163
I	0.52810306	0.02330262
Z^α	0.0297966	0.86110961
$\ln \widehat{\varphi}_1^\alpha$	−2.62564635	−0.25410526
$\ln \widehat{\varphi}_2^\alpha$	0.07630259	−0.12799775
$\widehat{\varphi}_1^\alpha$	0.07239295	0.77561016
$\widehat{\varphi}_2^\alpha$	1.07928911	0.87985536

The final results are

$$y_1 = 0.0186,\ y_2 = 1 - y_1 = 0.9814,\ P = 6.2993 \text{ bar}$$

The value of the objective function is 7.0855×10^{-7}.

8.9 Modern Approach to Phase Equilibrium Calculations

Advances in computers and the improvements in software have enabled new methods for phase equilibrium calculations based upon the Gibbs energy. The K-value method converges slowly and can lead to multiple solutions that satisfy the constant fugacity conditions but are at local Gibbs energy minima. It is possible to ensure the global minimum of the Gibbs energy by validating all equilibrium calculations using a phase stability analysis.

First, we examine the Gibbs energy vs the composition diagram shown in Figure 8.15. The first curve shows a mixture in a single phase and the second curve shows the mixture under phase splitting conditions. The equilibrium compositions are those at which a tangent line or plane touches the Gibbs energy curve.

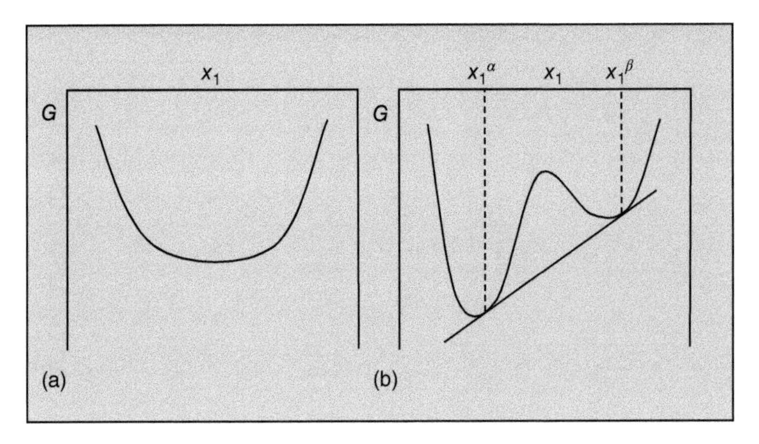

Figure 8.15 Gibbs energy as a function of composition: (a) single phase and (b) two phase.

These points must also satisfy the absolute stability criteria of the mixture. That is, at a certain pressure, temperature, and overall composition at no point should the Gibbs energy surface be below the tangent plane [6] to the surface at the overall composition. Michelsen [7] (1982) suggested these conditions and became the driving force behind many papers developing the tangent plane method for phase equilibrium calculations. This is a powerful but complicated technique.

In 2003, Iglesias-Silva et al. [8] suggested a technique that establishes a set of nonlinear algebraic equations (including material balances and the tangent plane equation). This technique constitutes this section.

8.9.1 Equal Area Rule for Binary Mixtures

For the Gibbs energy for a binary mixture at constant T and P

$$G = x_1\mu_1 + (1 - x_1)\mu_2 \tag{8.78}$$

Taking the derivative with respect to x_1, at constant T and P

$$\left(\frac{\partial G}{\partial x_1}\right)_{T,P} = \mu_1 - \mu_2 \tag{8.79}$$

Substituting this result for the chemical potential of component 2 into Eq. (8.78)

$$G = x_1\mu_1 + (1 - x_1)\left\{\mu_1 - \left(\frac{\partial G}{\partial x_1}\right)_{T,P}\right\} \tag{8.80}$$

The chemical potential of component 1 is

$$\mu_1 = G + (1 - x_1)\left(\frac{\partial G}{\partial x_1}\right)_{T,P} \tag{8.81}$$

For component 2,

$$\mu_2 = \mu_1 - \left(\frac{\partial G}{\partial x_1}\right)_{T,P} = G - x_1\left(\frac{\partial G}{\partial x_1}\right)_{T,P} \tag{8.82}$$

Equilibrium requires equality of chemical potentials in all phases, so

$$\mu_1^\alpha = \mu_1^\beta \Rightarrow G^\alpha + x_2^\alpha \left(\frac{\partial G}{\partial x_1}\right)_{T,P}^\alpha = G^\beta + x_2^\beta \left(\frac{\partial G}{\partial x_1}\right)_{T,P}^\beta \tag{8.83}$$

Under the tangent plane criterion, the derivatives of G with respect to x_1 must equal

$$G^\alpha - G^\beta = \left(\frac{\partial G}{\partial x_1}\right)_{T,P}^\alpha \left(x_2^\beta - x_2^\alpha\right) \Rightarrow G^\alpha - G^\beta = \left(\frac{\partial G}{\partial x_1}\right)_{T,P}^\alpha \left(x_1^\alpha - x_1^\beta\right) \tag{8.84}$$

therefore

$$\left(\frac{\partial G}{\partial x_1}\right)_{T,P}^\beta = \frac{G^\alpha - G^\beta}{x_1^\alpha - x_1^\beta} \tag{8.85}$$

This equation contains the information that a single tangent line intersects the Gibbs energy curve at the phase equilibrium compositions x_1^α and x_1^β. Also, we can write the LHS of the above equations as an integral of the derivative

$$\int_{x_1^\beta}^{x_1^\alpha} \left(\frac{\partial G}{\partial x_1}\right)_{T,P}^\alpha dx_1 = \left(\frac{\partial G}{\partial x_1}\right)_{T,P}^\alpha \left(x_1^\alpha - x_1^\beta\right) \tag{8.86}$$

This equation is similar to the MEAR but, instead of the pressure as the independent variable, it is the derivative of the Gibbs energy. Figure 8.16 demonstrates the equal area rule.

We can follow the same procedure used to find vapor pressures. Rearranging Eq. (8.86)

$$g' \equiv \left[\left(\frac{\partial G}{\partial x_1}\right)_{T,P}^\alpha\right]_{j+1} = \left[\frac{\int_{x_1^\beta}^{x_1^\alpha} \left(\frac{\partial G}{\partial x_1}\right)_{T,P}^\alpha dx_1}{\left(x_1^\alpha - x_1^\beta\right)}\right]_j = \left[\frac{G^\alpha - G^\beta}{x_1^\alpha - x_1^\beta}\right]_j \tag{8.87}$$

Then, the steps to find the equilibrium compositions are

1. Given a temperature and pressure, calculate the fugacity of the pure components
2. Provide two extreme values, e.g. $x_1 = 0.01$ and $y_1 = 0.99$

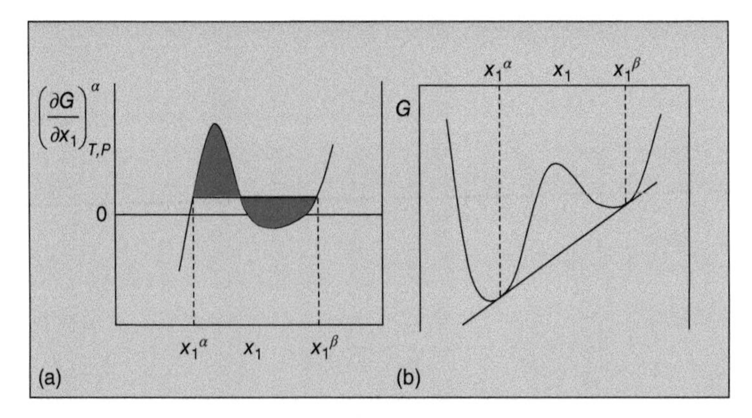

Figure 8.16 Equal area rule for binary mixture two-phase separation: (a) derivative of the Gibbs energy vs. mole fraction, (b) Gibbs energy vs. mole fraction.

3. Calculate a value for g' using Eq. (8.87)
4. Using this value, find the composition roots from the EOS: the smaller one represents one phase and the greater one the other phase.
5. Return to step 3 and repeat the procedure until $\left|g'_{j+1} - g'_{j+1}\right| \leq \varepsilon$. These are the convergence criteria, and generally, $\varepsilon = 1 \times 10^{-3}$. Note: We should calculate the compressibility factor of both phases at every iteration.

Next, we must decide how to calculate the Gibbs energy using an EOS or Excess Gibbs model. Instead of the Gibbs energy, we can use the Gibbs energy of mixing. Either graph should have the same behavior because

$$\Delta_m G \equiv g = G - \sum_{i=1}^{N-1} x_i G_i \text{ because } x_N = 1 - \sum_{i=1}^{N-1} x_i \tag{8.88}$$

Therefore, the value we subtract is a constant. The derivatives are

$$\left(\frac{\partial g}{\partial x_i}\right)_{T,P,x_{j\neq i}} = \left(\frac{\partial G}{\partial x_i}\right)_{T,P,x_{j\neq i}} - (G_i - G_N) \tag{8.89}$$

In each equilibrium phase, the pure component terms are the same, so they cancel. Now, how do we apply the technique if we have the excess Gibbs energy? We simply use the definition of Gibbs energy

$$G^E = G - G^{is} \Rightarrow G^E = g + \sum_{i=1}^{N} x_i G_i - G^{is} \tag{8.90}$$

Then, using Eq. (7.268), replace the ideal solution term to obtain

$$g = G^E + RT \sum_{i=1}^{N} x_i \ln x_i \tag{8.91}$$

When using an EOS, we insert partial molar properties into the Gibbs energy of mixing

$$\Delta_m G \equiv g = \sum_{i=1}^{N} \overline{G}_i x_i - \sum_{i=1}^{N} G_i x_i = \sum_{i=1}^{N} (\overline{G}_i - G_i) x_i \tag{8.92}$$

and from the definition of fugacity coefficient in the mixture, integrate from the pure component to the component i in the mixture

$$\int_{G_i}^{\overline{G}_i} d\overline{G}_i = RT \int_{f_i}^{\hat{f}_i} d\ln \hat{f}_i \Rightarrow \overline{G}_i - G_i = RT \ln\left(\frac{\hat{f}_i}{f_i}\right) \tag{8.93}$$

For convenience, we use fugacity coefficients,

$$\overline{G}_i - G_i = RT \ln\left(\frac{z_i \hat{\varphi}_i}{\varphi_i}\right) \tag{8.94}$$

and

$$g = RT \sum_{i=1}^{N} x_i \ln\left(\frac{x_i \hat{\varphi}_i}{\varphi_i}\right) \tag{8.95}$$

The derivative with respect to x_i is

$$\left(\frac{\partial g}{\partial x_i}\right)_{TPx_{j\neq i}} = RT\ln\left(\frac{z_i\widehat{\varphi}_i}{z_N\widehat{\varphi}_N}\cdot\frac{\varphi_N}{\varphi_i}\right) = RT\ln\left(\frac{\widehat{f}_i f_N}{\widehat{f}_N f_i}\right) \tag{8.96}$$

in which f_i is the pure component fugacity and $\widehat{f}_i$ is the fugacity of component i in the mixture. When using activities, substitute

$$\frac{\widehat{f}_i}{f_i} = x_i\frac{\widehat{\varphi}_i}{\varphi_i} = \gamma_i x_i \Rightarrow \left(\frac{\partial g}{\partial x_i}\right)_{TPx_{j\neq i}} = RT\ln\left(\frac{\gamma_i x_i}{\gamma_N x_N}\right) \tag{8.97}$$

in which γ_i is the activity coefficient of component i, φ_i is the fugacity coefficient of the pure component i at pressure P and temperature T of the mixture, and $\widehat{\varphi}_i$ is the fugacity coefficient of component i in the mixture.

Example 8.5

Consider a binary mixture of CH_4 and C_2H_6. Use the SRK EOS to obtain the equilibrium compositions at 200 K and 13.79 bar. Let $k_{12} = 0$. A.R. Price [9] gives an experimental value for the compositions $x_1 = 0.25$ and $y_1 = 0.814$. T in this experiment is 199.8 K.

Solution

$$T = 200\text{ K}$$

$$R = 83.14\text{ cm}^3\text{ bar/(mol K)}$$

The parameters for the pure substances are as follows:

Specie	$T_{C,i}$ (K)	$P_{C,i}$ (bar)	ω_i
CH_4	190.6	45.99	0.012
C_2H_6	305.3	48.72	0.1

The calculated quantities for the pure components are as follows:

Specie	$T_{R,i} = T/T_{C,i}$	α_i Eq. (3.125)	b_i (cm^3/mol) Eq. (3.119)	a_i (bar cm^6/mol^2) Eq. (3.127)
CH_4	1.04931794	0.97584091	29.85299791	2 277 703.79
C_2H_6	0.79239303	1.25701484	45.13858996	7 105 956.99

Given T and P, we can calculate the parameters for the pure components

$$A_i = \frac{a_i P}{(RT)^2} \text{ and } B_i = \frac{Pb_i}{RT},\ A_i/B_i = \frac{a_i}{b_i RT}$$

We must calculate the compressibility factors of the liquid and vapor using

$$Z_{i,j+1} = \left[\frac{Z_i^2 - Z_i^3 + A_i B_i}{(A_i - B_i - B_i^2)}\right]_j$$

$$Z_{j+1} = \left[1 - (A_i - B_i - B_i^2)/Z_i + A_i B_i / Z_i^2\right]_j$$

respectively. From these two equations, the value of the compressibility factor results from iterative calculations. The fugacity coefficient of each component comes from

$$\ln \varphi_i = (Z_i - 1) - \ln(Z_i - B_i) - (A_i/B_i) \ln\left(\frac{Z_i + B_i}{Z_i}\right)$$

Then,

Quantities	CH_4	C_2H_6
A_i/B_i	4.5884846	9.4674838
B_i	0.0269302	0.0407192
Z_i	0.8972344	0.0537003
$\ln\varphi_i$	−0.0909427	−1.9448148
φ_i	0.9130700	0.1430137

Now, calculate the quantities in each phase

$$g = RT\left\{z_1 \ln\left(\frac{z_1\hat{\varphi}_1}{\varphi_1}\right) + z_2 \ln\left(\frac{z_2\hat{\varphi}_2}{\varphi_2}\right)\right\}$$

and

$$\left(\frac{\partial g}{\partial x_i}\right)_{TPx_{j\neq i}} = RT \ln\left(\frac{z_i\hat{\varphi}_i}{z_N\hat{\varphi}_N} \cdot \frac{\varphi_N}{\varphi_i}\right)$$

Values for the fugacity coefficient in the mixture can be calculated as in the last example. Using Solver in Excel, solve for the roots of the compressibility factor in each phase and then solve for P and y_1 minimizing the objective function. Continue this procedure until the values of all the functions are less than or equal to a specified small value. Here, we present the initial values and the final result:

Initial values

	Liquid	Vapor
a^α (bar cm^6/mol^2)	7.044433×10^6	2.312745×10^6
b^α (cm^3/mol)	44.985734	30.0058538
A^α/B^α	9.41740487	4.63534219
B^α	0.04058131	0.02706807
$\overline{D}_1^\alpha$	4.46044758	4.58845523
$\overline{D}_2^\alpha$	9.46747515	9.27715121

(Continued)

	Liquid	Vapor
Z^α	0.05362877	0.89512589
$\ln \hat{\varphi}_1^\alpha$	1.25481369	−0.09953849
$\ln \hat{\varphi}_2^\alpha$	−1.94188828	−0.29264537
$\hat{\varphi}_1^\alpha$	3.50718488	0.9052551
$\hat{\varphi}_2^\alpha$	0.14343285	0.74628675
g/RT	−0.04309162	−0.04798963
g'/RT	−3.3062595	2.93435465

To find the new value of the derivative, use

$$[g'/RT]_1 = \left[\frac{g^\alpha/RT - g^\beta/RT}{x_1^\alpha - x_1^\beta}\right] = \frac{-0.04309162-(-0.04798963)}{0.01+0.99}$$

$$= -0.004997974$$

With this value, we can find the composition root from the EOS for g'/RT. To find the roots using Solver in Excel, use the function

$$F = \left|\ln\left(\frac{x_1\hat{\varphi}_1^l}{x_2\hat{\varphi}_2^l}\cdot\frac{\varphi_2}{\varphi_1}\right) - [g'/RT]_1\right| + \left|\ln\left(\frac{y_1\hat{\varphi}_1^l}{x_2\hat{\varphi}_2^l}\cdot\frac{\varphi_2}{\varphi_1}\right) - [g'/RT]_1\right|$$

After three iterations, the results are:

	Liquid	Vapor
a^α (bar cm^6/mol^2)	5.585556×10^6	2.824550×10^6
b^α (cm^3/mol)	41.1427441	32.1188675
A^α/B^α	8.16456668	5.2887023
B^α	0.03711458	0.0289742
$\overline{D}_1^\alpha$	4.50328607	4.58285341
$\overline{D}_2^\alpha$	9.46042515	9.34452114
Z^α	0.05232311	0.86294877
$\ln \hat{\varphi}_1^\alpha$	1.08403861	−0.09717659
$\ln \hat{\varphi}_2^\alpha$	−1.92558978	−0.31965167
$\hat{\varphi}_1^\alpha$	2.956596	0.90739575
$\hat{\varphi}_2^\alpha$	0.14578975	0.72640202
g/RT	−0.25317497	−0.18403905
g'/RT	0.11711671	0.1171097

and the equilibrium composition values are

$$y_1 = 0.851764, x_1 = 0.261412$$

The final value of the objective function is 7×10^{-6}.

8.9.2 A General Approach for Multicomponent and Multiphase Systems

Using the definition of a partial molar quantity in the chemical potential

$$\mu_i = G + \left(\frac{\partial G}{\partial x_i}\right)_{TPx_{k\neq i}} - \sum_{j=1}^{N-1} x_j \left(\frac{\partial G}{\partial x_j}\right)_{TPx_{k\neq j}} \tag{8.98}$$

in which $G = G(x_1, x_2, \ldots, x_{N-1})$. The chemical potential of component N according to the last equation is

$$\mu_N = G - \sum_{j=1}^{N-1} x_j \left(\frac{\partial G}{\partial x_j}\right)_{TPx_{k\neq j}} \tag{8.99}$$

If we subtract the last two equations, then

$$\mu_i - \mu_N = \left(\frac{\partial G}{\partial x_i}\right)_{TPx_{k\neq i}} \tag{8.100}$$

At equilibrium, the equality of chemical potentials for π phases is valid, therefore

$$(\mu_i - \mu_N)^\alpha = (\mu_i - \mu_N)^\beta = \cdots = (\mu_i - \mu_N)^\pi \quad \text{for } i = 1,2,\ldots,N-1 \tag{8.101}$$

or

$$\left(\frac{\partial G}{\partial x_i}\right)^\alpha_{TPx_{k\neq i}} = \left(\frac{\partial G}{\partial x_i}\right)^\beta_{TPx_{k\neq i}} = \cdots = \left(\frac{\partial G}{\partial x_i}\right)^\pi_{TPx_{k\neq i}} \quad \text{for } i = 1,2,\ldots,N-1 \tag{8.102}$$

The last expression represents $\pi - 1$ equations that are necessary to have equilibrium for each component i. If we use the reduced Gibbs energy of mixing, the equations are equivalent because

$$\frac{\Delta_m G}{RT} = g(x_1, x_2, \ldots, x_{N-1}) = \frac{G - G_N}{RT} - \sum_{j=1}^{N-1} x_j \frac{(G_j - G_N)}{RT} \tag{8.103}$$

in which G_j is the Gibbs energy of pure component j, and the chemical potential is

$$\frac{\mu_i}{RT} = g + \left(\frac{\partial g}{\partial x_i}\right)_{TPx_{k\neq i}} - \sum_{j=1}^{N-1} x_j \left(\frac{\partial g}{\partial x_j}\right)_{TPx_{k\neq j}} + \frac{G_i}{RT} \tag{8.104}$$

and Eq. (8.100) becomes

$$\frac{\mu_i - \mu_N}{RT} = \left(\frac{\partial g}{\partial x_i}\right)_{TPx_{k\neq i}} + \frac{G_i - G_N}{RT} \quad \text{for } i = 1,2,\ldots,N-1 \tag{8.105}$$

and Eq. (8.101) becomes

$$\left(\frac{\partial g}{\partial x_i}\right)^\alpha_{TPx_{k\neq i}} = \left(\frac{\partial g}{\partial x_i}\right)^\beta_{TPx_{k\neq i}} = \cdots = \left(\frac{\partial g}{\partial x_i}\right)^\pi_{TPx_{k\neq i}} \quad \text{for } i = 1,2,\ldots,N-1 \tag{8.106}$$

So, we must solve $\pi - 1$ equations at equilibrium. Another condition that applies at equilibrium is the minimization of the total Gibbs energy (or the Gibbs energy of mixing). This condition results from finding the tangent plane equation for the Gibbs energy function. Consider the reduced Gibbs energy of mixing and write a function in which the reduced energy of mixing is a variable

$$F \equiv \frac{\Delta_m G}{RT} - g(x_1, x_2, \ldots, x_{N-1}) = F\left(\frac{\Delta_m G}{RT}, x_1, x_2, \ldots, x_{N-1}\right) = 0 \tag{8.107}$$

At an equilibrium point, $\left(g^\alpha, x_1^\alpha, x_2^\alpha, \ldots, x_{N-1}^\alpha\right)$, the tangent plane equation that passes through the point is

$$(g - g^\alpha)\left(\frac{\partial F}{\partial g}\right)^\alpha_{TPx_{k\neq i}} + \sum_{j=1}^{N-1}\left(x_j - x_j^\alpha\right)\left(\frac{\partial F}{\partial x_j}\right)^\alpha_{TPx_{k\neq i}} = 0 \tag{8.108}$$

In terms of g, this equation reduces to

$$(g - g^\alpha) - \sum_{j=1}^{N-1}\left(x_j - x_j^\alpha\right)\left(\frac{\partial g}{\partial x_j}\right)^\alpha_{TPx_{k\neq j}} = 0 \tag{8.109}$$

which is the tangent plane that uses a reference point in phase α. Because this equation must be valid at any equilibrium point, in any phase i

$$(g^i - g^\alpha) - \sum_{j=1}^{N-1}\left(x_j^i - x_j^\alpha\right)\left(\frac{\partial g}{\partial x_j}\right)^\alpha_{TPx_{k\neq j}} = 0 \quad \text{for } i = \beta, \gamma, \ldots, \pi - 1 \tag{8.110}$$

This is a set of $\pi - 1$ equations that must hold at equilibrium. The tangent plane at equilibrium is the equation when the tangent plane becomes the chemical potential of each component, thus satisfying the Gibbs criterion. In this case, we have $(N - 1)\pi$ unknowns and we have $N(\pi - 1)$ equations. This system of equations is valid when the number of phases is greater than or equal to the number of components. If we plot the tangent plane equation,

$$\Lambda = g^\alpha - \sum_{j=1}^{N-1}\left(x_j - x_j^\alpha\right)\left(\frac{\partial g}{\partial x_j}\right)^\alpha_{TPx_{k\neq j}} \tag{8.111}$$

against g_1' by varying the composition x_1, one intersection exists if the mixture splits into two phases. Figure 8.17 shows this behavior. This is equivalent to pure component vapor–liquid equilibrium in P vs G^R/RT. Equations (8.106) and (8.110) are sufficient when the number of phases is greater than or equal to the number of components. If this is not the case, completing the system of equations requires use of material balances, and the system of equations becomes

$$\left(\frac{\partial g}{\partial x_i}\right)^\alpha_{TPx_{k\neq i}} - \left(\frac{\partial g}{\partial x_i}\right)^p_{TPx_{k\neq i}} = 0 \quad \text{for } i = 1, 2, \ldots, N - 1 \text{ and } p = \beta, \gamma, \ldots, \pi \tag{8.112}$$

$$(g^i - g^\alpha) - \sum_{j=1}^{N-1}\left(x_j^i - x_j^\alpha\right)\left(\frac{\partial g}{\partial x_j}\right)^\alpha_{TPx_{k\neq j}} = 0 \quad \text{for } i = \beta, \gamma, \ldots, \pi - 1 \tag{8.110}$$

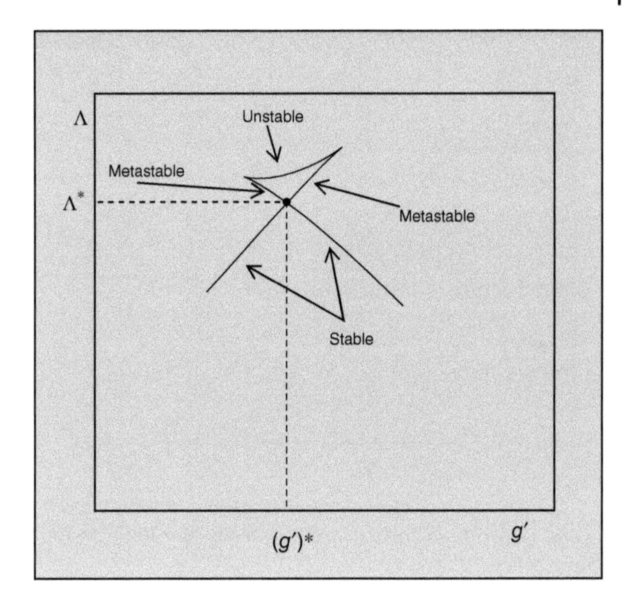

Figure 8.17 Tangent plane equation vs the derivative of the Gibbs energy.

and

$$z_i - \sum_{k=1}^{p} x_i^k \varphi^k = 0 \quad \text{for } i = 1, 2, \ldots, N-1 \tag{8.113}$$

These latter equations complete the system when the number of components is greater than the number of phases, for N and p greater than 3. This completes a system of $(N\pi - 1)$ equations that solves for $\{x_1^{\alpha}, \ldots, x_1^{\pi}, \ldots x_{N-1}^{\alpha}, \ldots, x_{N-1}^{\pi}, \varphi^1, \varphi^2, \ldots, \varphi^{\pi-1}\}$. When the number of phases is 2 and 3, it is not necessary to include the material balance in the form of Eq. (8.113). In two-phase, multicomponent equilibrium, the total derivative of the Gibbs energy of mixing at any equilibrium point must be the same in each phase

$$\left(\frac{dg}{dx_i}\right)^{\alpha} = \left(\frac{dg}{dx_i}\right)^{\beta} \quad \text{for } i = 1, 2, \ldots, N-1 \tag{8.114}$$

This equation is a necessary condition. The total derivative of the Gibbs energy of mixing is

$$\left(\frac{dg}{dx_i}\right) = \sum_{j=1}^{N-1} \left(\frac{\partial g}{\partial x_j}\right)_{TPx_{k\neq i}} \left(\frac{dx_j}{dx_i}\right) \tag{8.115}$$

At equilibrium in a multicomponent, two-phase system, Eqs. (8.106) and (8.114) imply that

$$\left(\frac{dx_j}{dx_i}\right)^{\alpha} = \left(\frac{dx_j}{dx_i}\right)^{\beta} \quad \text{for } i = 1, 2, \ldots, N-1 \text{ and } (j \neq i) \tag{8.116}$$

This derivative is apparent in a ternary system. At equilibrium, we can draw a line through the equilibrium compositions and the overall composition as shown in Figure 8.18.

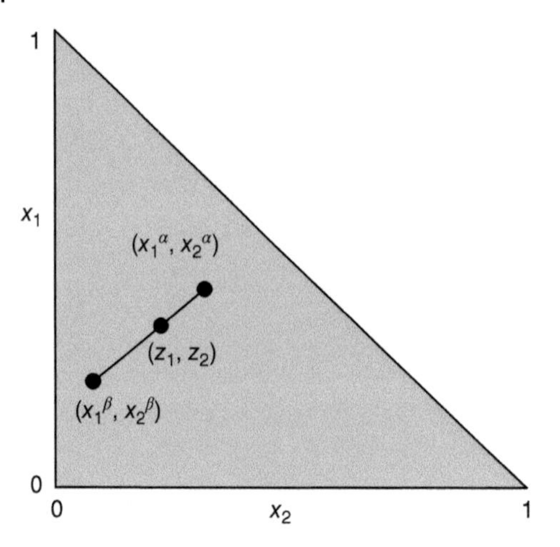

Figure 8.18 Equilibrium compositions in a two-phase, ternary system.

The relative change between two components at the overall composition (z_1, z_2) along a tie-line is

$$x'_{ji} \equiv \left(\frac{dx_j}{dx_i}\right) = \frac{x_j^\beta - x_j^\alpha}{x_i^\beta - x_i^\alpha} = \frac{x_j - z_j}{x_i - z_i} \quad (i \neq j) \tag{8.117}$$

This equality in compositions comes from the lever rule arm and implies the material balance. Equation (8.117) completes the system of $(N-1)\pi$ equations that establish equilibrium for two phases. The same equations apply when $\pi > N$. The sets of equations for different cases are as follows:

Case I: A binary mixture with two phases in equilibrium

$$F_1 = \left(\frac{\partial g}{\partial x_1}\right)^\alpha_{TPx_{k\neq 1}} - \left(\frac{\partial g}{\partial x_1}\right)^\beta_{TPx_{k\neq 1}} = 0 \tag{8.118}$$

$$F_2 = (g^\beta - g^\alpha) - \left(x_1^\beta - x_1^\alpha\right)\left(\frac{\partial g}{\partial x_1}\right)^\alpha_{TPx_{k\neq 1}} = 0 \tag{8.119}$$

This is a system of equations with x_1^α and x_1^β as unknowns given T and P, but they can solve any two variables when keeping the other two fixed. Note that these equations are analogous to the pure component case.

Case II: Binary mixture three-phase equilibrium

In a Λ–g' curve, the equilibrium condition is a single intersection point as shown in Figure 8.15. Other intersections exist, but they correspond to unstable, two-phase equilibrium.

$$F_1 = \left(\frac{\partial g}{\partial x_1}\right)^\alpha_{TPx_{k\neq 1}} - \left(\frac{\partial g}{\partial x_1}\right)^\beta_{TPx_{k\neq 1}} = 0 \tag{8.120}$$

$$F_2 = \left(\frac{\partial g}{\partial x_1}\right)^\alpha_{TPx_{k\neq 1}} - \left(\frac{\partial g}{\partial x_1}\right)^\gamma_{TPx_{k\neq 1}} = 0 \tag{8.121}$$

$$F_3 = (g^\beta - g^\alpha) - \left(x_1^\beta - x_1^\alpha\right)\left(\frac{\partial g}{\partial x_1}\right)^\alpha_{TPx_{k\neq 1}} = 0 \tag{8.122}$$

These three equations have three unknowns: x_1^α, x_1^β, and x_1^γ. Note that from Eqs. (8.110) and (8.113), we have an extra equation. We recommend eliminating one of the tangent plane equations to complete the system of equations.

Case III: Ternary mixture with two phases in equilibrium

$$F_1 = \left(\frac{\partial g}{\partial x_1}\right)^\alpha_{TPx_2} - \left(\frac{\partial g}{\partial x_1}\right)^\beta_{TPx_2} = 0 \tag{8.123}$$

$$F_2 = \left(\frac{\partial g}{\partial x_2}\right)^\alpha_{TPx_1} - \left(\frac{\partial g}{\partial x_2}\right)^\beta_{TPx_1} = 0 \tag{8.124}$$

$$F_3 = (g^\beta - g^\alpha) - \left(x_1^\beta - x_1^\alpha\right)\left(\frac{\partial g}{\partial x_1}\right)^\alpha_{TPx_2} - \left(x_2^\beta - x_2^\alpha\right)\left(\frac{\partial g}{\partial x_2}\right)^\alpha_{TPx_1} = 0 \tag{8.125}$$

and

$$F_4 = \left(x_{21}'\right)^\alpha - \left(x_{21}'\right)^\beta = 0 \tag{8.126}$$

The unknowns are x_1^α, x_1^β, x_2^α, and x_2^β.

Case IV: Ternary mixture, three-phase

The system of equations is

$$F_1 = \left(\frac{\partial g}{\partial x_1}\right)^\alpha_{TPx_2} - \left(\frac{\partial g}{\partial x_1}\right)^\beta_{TPx_2} = 0 \tag{8.127}$$

$$F_2 = \left(\frac{\partial g}{\partial x_2}\right)^\alpha_{TPx_1} - \left(\frac{\partial g}{\partial x_2}\right)^\beta_{TPx_1} = 0 \tag{8.128}$$

$$F_3 = \left(\frac{\partial g}{\partial x_1}\right)^\alpha_{TPx_2} - \left(\frac{\partial g}{\partial x_1}\right)^\gamma_{TPx_2} = 0 \tag{8.129}$$

$$F_4 = \left(\frac{\partial g}{\partial x_2}\right)^\alpha_{TPx_1} - \left(\frac{\partial g}{\partial x_2}\right)^\gamma_{TPx_1} = 0 \tag{8.130}$$

$$F_5 = (g^\beta - g^\alpha) - \left(x_1^\beta - x_1^\alpha\right)\left(\frac{\partial g}{\partial x_1}\right)^\alpha_{TPx_2} - \left(x_2^\beta - x_2^\alpha\right)\left(\frac{\partial g}{\partial x_2}\right)^\alpha_{TPx_1} = 0 \tag{8.131}$$

$$F_6 = (g^\gamma - g^\alpha) - \left(x_1^\gamma - x_1^\alpha\right)\left(\frac{\partial g}{\partial x_1}\right)^\alpha_{TPx_2} - \left(x_2^\gamma - x_2^\alpha\right)\left(\frac{\partial g}{\partial x_2}\right)^\alpha_{TPx_1} = 0 \tag{8.132}$$

The unknowns for this system of equations are x_1^α, x_1^β, x_1^γ, x_2^α, x_2^β and x_2^γ. In this case, we do not need a mass balance to complete the system of equations.

Example 8.6

Esper et al. [10] measured the phase envelope of a near-equimolar mixture of methane–carbon dioxide (52.4% of methane). Perform the following equilibrium calculations using the SRK EOS. Consider $k_{ij} = 0$.

(a) Dew point temperature at $P = 5.2$ bar
(b) Bubble point temperature at $P = 78$ bar
(c) Flash calculation at $T = 225$ K and $P = 50$ bar

Iteration	g'_{new}	x_1^α	x_1^β	g^α	g^β	$\left(\frac{\partial g}{\partial x_1}\right)_{T,P,x_2}^\alpha$	$\left(\frac{\partial g}{\partial x_1}\right)_{T,P,x_2}^\beta$
1	−0.02287	0.023314	0.848621	−0.02078	−0.15653	−0.022870	−0.022872
2	−0.16448	0.01936	0.825575	−0.02042	−0.15432	−0.164478	−0.164478
3	−0.16608	0.01932	0.825288	−0.02042	−0.15427	−0.166077	−0.166077
4	−0.16608	0.01932	0.825288	−0.02042	−0.15427	−0.166078	−0.166078

Use Eqs. (8.118) and (8.119) to perform your calculations.

Solution

Solve the system of equations

$$F_2 = (g^\beta - g^\alpha) - \left(x_1^\beta - x_1^\alpha\right)\left(\frac{\partial g}{\partial x_1}\right)_{TPx_{k\neq 1}}^\alpha = 0$$

From example 8.5, we have the expressions for g and g′. Again, an iterative procedure is necessary because at each step, we must calculate the compressibility factor of the mixture and the compressibility factors of the pure components when needed.

Our objective function to be minimized in Excel using Solver® is

$$F = \sqrt{F_1^2 + F_2^2}$$

We can use central differences to calculate the numerical derivatives. The parameters for the pure substances are as follows:

Specie	$T_{C,i}$ (K)	$P_{C,i}$ (bar)	ω_i
CH_4	190.6	45.99	0.012
CO_2	304.2	73.83	0.224

(a) Dew Temperature Calculation

The calculated quantities for the pure components are as follows:

Specie	$T_{R,i} = T/T_{C,i}$	α_i Eq. (3.125)	b_i (cm^3/mol) Eq. (3.119)	a_i (bar cm^6/mol^2) Eq. (3.127)
CH_4	1.07555089	0.96333898	29.85299791	2248523.12
CO_2	0.67389875	1.31680611	29.67937869	4876892.98

For initial values: $T = 205$ and $x_1 = 0.05$, and using central derivatives in Excel

Quantities	CH_4	CO_2
A_i/B_i	4.41921901	9.641054522
B_i	0.00910809	0.024493969
Z_i	0.96831506	0.03214471
$\ln \varphi_i$	−0.11243788	−1.556022244
φ_i	0.89365286	0.210973606

	Liquid	Vapor
a^α (bar cm^6/mol^2)	4.721606×10^6	3.374295×10^6
b^α (cm^3/mol)	29.6880596	29.7703552
A^α/B^α	9.33134171	6.65020788
B^α	0.00905777	0.00908288
$\overline{D}_1^\alpha$	3.49568943	4.18864354
$\overline{D}_2^\alpha$	9.6384813	9.35999719
Z^α	0.01201888	0.94647733
$\ln \hat{\varphi}_1^\alpha$	2.86523677	−0.02902484
$\ln \hat{\varphi}_2^\alpha$	−0.57931118	−0.07810303
$\hat{\varphi}_1^\alpha$	17.5532089	0.97139233
$\hat{\varphi}_2^\alpha$	0.56028417	0.92486913
g/RT	0.878244	0.05520324
g'/RT	−0.94347539	−1.29843235

After each iteration,

Iteration	F_1	F_2	F	T	x_1
0	0.35495696	−0.20758383	0.41120006	205	0.05
1	−0.01866368	0.00016262	0.01866439	200.5570669	0.029061424
2	−0.00034998	-5.2907×10^{-7}	0.00034998	200.5606093	0.029616194
3	-3.899×10^{-7}	-6.9052×10^{-7}	7.9299×10^{-7}	200.5606127	0.029626605

(b) Bubble temperature calculation

Specie	$T_{R,i} = T/T_{C,i}$	α_i Eq. (3.125)	b_i (cm^3/mol) Eq. (3.119)	a_i (bar cm^6/mol^2) Eq. (3.127)
CH_4	1.31164743	0.90413282	29.85299791	2008045.44
CO_2	0.82182774	1.22649963	29.67937869	4295737.45

Using $T = 220$ K and $y_1 = 0.7$ as initial values

Quantities	CH_4	CO_2
A_i/B_i	3.23620151	6.963585238
B_i	0.11202953	0.024493969
Z_i	0.7563761	0.038534014
$\ln \varphi_i$	0.10625994	−0.121969195
φ_i	1.11211092	0.885175636

	Liquid	Vapor
a^α (bar cm^6/mol^2)	3.239198×10^6	2.816522×10^6
b^α (cm^3/mol)	29.7703552	29.8009121
A^α/B^α	5.94868348	5.16714842
B^α	0.12695386	0.12708417
$\overline{D}_1^\alpha$	3.75946223	3.88253685
$\overline{D}_2^\alpha$	8.35866654	8.16457543
Z^α	0.21673125	0.25991867
$\ln \hat{\varphi}_1^\alpha$	−0.10838781	−0.26821882
$\ln \hat{\varphi}_2^\alpha$	−2.22436462	−1.96842935
$\hat{\varphi}_1^\alpha$	0.89727955	5.16714842
$\hat{\varphi}_2^\alpha$	0.1081361	0.12708417
g/RT	−1.49169227	−1.29643851
g'/RT	0.94394001	1.27939776

After each iteration, the function values are:

Iteration	F_1	F_2	F	T	y_1
0	−0.33545775	−0.02992024	0.33678943	220	0.7
1	−0.13020395	−0.0048838	0.13029551	230.054391	0.600164182
2	−0.06139465	−0.00111903	0.06140485	233.5119093	0.560860259
3	−0.03045445	−0.00027918	0.03045573	235.0452754	0.542449217
4	−0.01542261	-7.1885×10^{-5}	0.01542278	235.782738	0.533364189
5	−0.00784983	-1.8705×10^{-5}	0.00784985	236.140285	0.528778077
6	−0.00404193	-4.9459×10^{-6}	0.00404193	236.1402838	0.5264528
7	−0.0020726	-1.2972×10^{-6}	0.0020726	236.1403061	0.525254422
8	−0.00105757	-3.3649×10^{-7}	0.00105757	236.1403069	0.524639443
9	−0.0005405	-8.7302×10^{-8}	0.0005405	236.1403068	0.524325716
10	−0.00027677	-2.2601×10^{-8}	0.00027677	236.1403061	0.524165344

Iteration	F_1	F_2	F	T	y_1
11	−0.00014074	-5.6992×10^{-9}	0.00014074	236.1403061	0.52408307
12	-7.2709×10^{-5}	-1.4498×10^{-9}	7.2709×10^{-5}	236.1403061	0.524041193

(c) Flash calculation

Specie	$T_{R,i} = T/T_{C,i}$	α_i Eq. (3.125)	b_i (cm^3/mol) Eq. (3.119)	a_i (bar cm^6/mol^2) Eq. (3.127)
CH_4	1.18048269	0.91555863	29.85299791	2136999.33
CO_2	0.73964497	1.24390026	29.67937869	4606880.54

Using $T = 225$ K and $P = 50$ bar

Quantities	CH_4	CO_2
A_i/B_i	3.82669542	8.297735891
B_i	0.07979311	0.024493969
Z_i	0.75147072	0.034334209
$\ln \varphi_i$	0.0434576	−0.812545112
φ_i	1.04441571	0.443727292

Initial values: $y_1 = 0.9$ and $x_1 = 0.1$

	Liquid	Vapor
a^α (bar cm^6/mol^2)	4.317721×10^6	2.341816×10^6
b^α (cm^3/mol)	29.6967406	29.835636
A^α/B^α	7.77236752	4.19589949
B^α	0.07937546	0.07974671
$\overline{D}_1^\alpha$	3.12272827	3.8180852
$\overline{D}_2^\alpha$	8.2889941	7.59622808
Z^α	0.11376924	0.69915497
$\ln \hat{\varphi}_1^\alpha$	0.82622714	−0.23442998
$\ln \hat{\varphi}_2^\alpha$	−1.90292795	−0.64076559
$\hat{\varphi}_1^\alpha$	2.28468267	0.79102162
$\hat{\varphi}_2^\alpha$	0.14913133	0.52688889
g/RT	−1.22815057	−0.55800385
g'/RT	−0.3240722	−0.72789925

After each iteration,

Iteration	F_1	F_2	F	y_1	x_1
0	−2.07162968	−0.72789925	2.19578843	0.1	0.9
1	−0.24490966	−0.00157646	0.24491473	0.797221409	0.325294029
2	−0.08586515	0.00387893	0.08595272	0.800034225	0.400108504
3	−0.03821859	0.0010548	0.03823314	0.802230509	0.430035321
4	−0.0174717	0.00023682	0.0174733	0.802807388	0.442778329
5	−0.00793195	4.9999×10^{-5}	0.0079321	0.802937591	0.448374571
6	−0.00356941	1.0718×10^{-5}	0.00356943	0.802964824	0.450858577
7	−0.00160214	7.4858×10^{-7}	0.00160214	0.802971134	0.451965802
8	−0.0007203	3.2518×10^{-6}	0.00072031	0.802970231	0.452455554
9	−0.00032381	2.1019×10^{-7}	0.00032381	0.802971877	0.452680017
10				0.802970883	0.452777338

8.10 Binary Liquid–Liquid Equilibrium (LLE)

Two liquid phases α and β can coexist in equilibrium. The mathematical conditions are the same as for vapor–liquid equilibria. That is, the equality of chemical potentials or the equations of the Gibbs energy are discussed in Section 8.9. At liquid–liquid equilibrium, the equality of the fugacities for each component in the mixture in each phase must hold

$$\hat{f}_i^{\alpha} = \hat{f}_i^{\beta} \quad \text{for } i = 1, 2, \dots, N \tag{8.133}$$

Then, substituting the definition of the fugacity coefficient in the mixture,

$$x_i^{\alpha}\gamma_i^{\alpha}f_i = x_i^{\beta}\gamma_i^{\beta}f_i \quad \text{for } i = 1, 2, \dots, N \tag{8.134}$$

or

$$x_i^{\alpha}\gamma_i^{\alpha} = x_i^{\beta}\gamma_i^{\beta} \quad \text{for } i = 1, 2, \dots, N \tag{8.135}$$

At a given temperature and pressure, this is a system of N equations in which the unknowns are $2N-2$ compositions. Therefore, completing the system requires material balances

$$z_i = L^{\alpha}x_i^{\alpha} + (1 - L^{\alpha})x_i^{\beta} \quad \text{for } i = 1, 2, \dots, N-1 \tag{8.136}$$

The above equations complete a system of $(2N-1)$ equations that can establish $\{x_1^{\alpha}, \dots, x_1^{\pi}, \dots x_{C-1}^{\alpha}, \dots, x_{C-1}^{\pi}, L^{\alpha}\}$. For a binary system, at constant T and P,

Eq. (8.136) is not necessary because

$$x_1^\alpha \gamma_1^\alpha = x_1^\beta \gamma_1^\beta \tag{8.137}$$

$$\left(1 - x_1^\alpha\right) \gamma_1^\alpha = \left(1 - x_1^\beta\right) \gamma_1^\beta \tag{8.138}$$

The above equations represent a system of two nonlinear equations with two unknowns. *P–x* and *T–x* diagrams appear in Figure 8.19. Drawing a horizontal line at a given pressure or temperature, the equilibrium compositions of the two liquid phases result at the intersections with the curve.

According to the Gibbs phase rule, at equilibrium, $F = 2 + C - P = 2 + 2 - 2 = 2$; therefore, with a fixed pressure, we must specify the temperature to establish a unique equilibrium composition as shown in Figure 8.19a in which the equilibrium line (binodal curve) corresponds to a single temperature. Similarly, in Figure 8.19b, the line corresponds to a single pressure. To the left or the right of the equilibrium curve, only one phase exists, and the pressure, temperature, and overall composition of the mixture come from the phase rule. The highest point (point C) in Figure 8.19b is the "upper critical solution temperature" (UCST), and this is the highest temperature at which a mixture can separate into two liquid phases. On the other hand, Figure 8.19a contains a "lower critical solution pressure" (LCSP) at point C. In this diagram, we see that increasing the pressure corresponds to reducing the solubility; therefore, a separation into two phases is possible. The most common LLE diagrams depict compositions as a function of temperature as in Figure 8.19b. These diagrams are "solubility diagrams."

Figure 8.20 shows three different solubility diagrams. Diagram (a) in Figure 8.20 has a lower critical solution temperature (LCST). At lower temperatures, the liquid components are totally miscible, but at increased temperature, the mixture can separate into two phases. Another type of solubility diagram appears in Figure 8.20b. In this case contains a UCST and a LCST, so the two-phase region is an island. At lower temperatures, a single phase exists, but when increasing the temperature, two phases appear until they disappear at the UCST. Diagram (c) does not have a UCST nor an LCST. The binodal curve can touch other equilibrium curves. If it joins at the freezing line, then only an UCST exists. If it intersects with the vapor–liquid equilibrium curve, then only an LCST exists. When the binodal curve intersects both the freezing and vapor–liquid equilibrium curves, neither UCST nor LCST exist.

To find the equilibrium conditions for a binary mixture, we can use Eqs. (8.118) and (8.119)

$$F_1 = \left(\frac{\partial g}{\partial x_1}\right)^\alpha_{TPx_{k\neq 1}} - \left(\frac{\partial g}{\partial x_1}\right)^\beta_{TPx_{k\neq 1}} = 0 \tag{8.118}$$

$$F_2 = (g^\beta - g^\alpha) - \left(x_1^\beta - x_1^\alpha\right)\left(\frac{\partial g}{\partial x_1}\right)^\alpha_{TPx_{k\neq 1}} = 0 \tag{8.119}$$

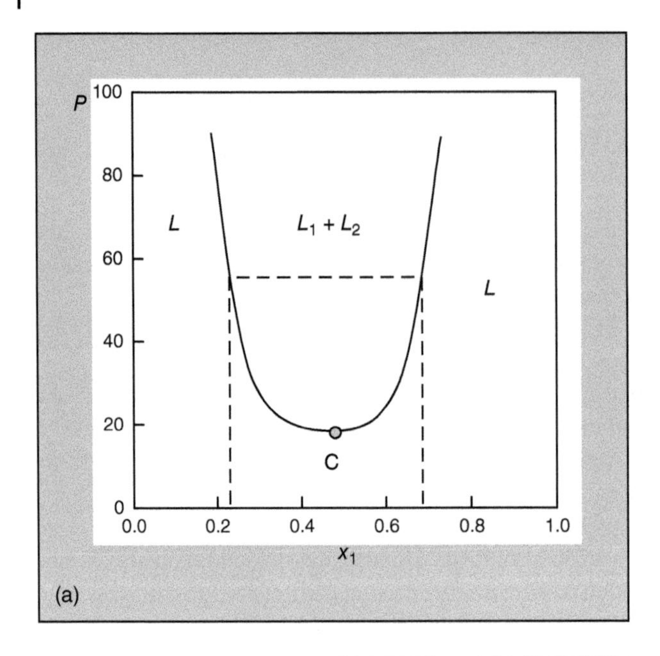

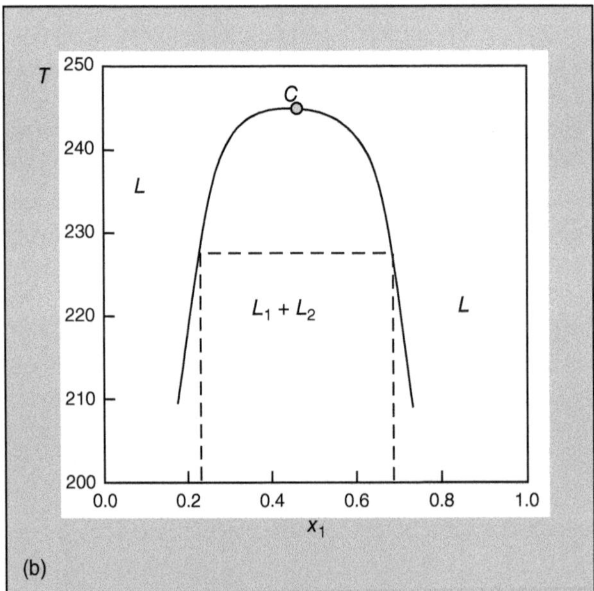

Figure 8.19 Liquid–liquid equilibrium in a P–x and T–x diagram. (a) with Lower Critical Solution Pressure (LCSP). (b) with Upper Critical Solution Temperature (UCST).

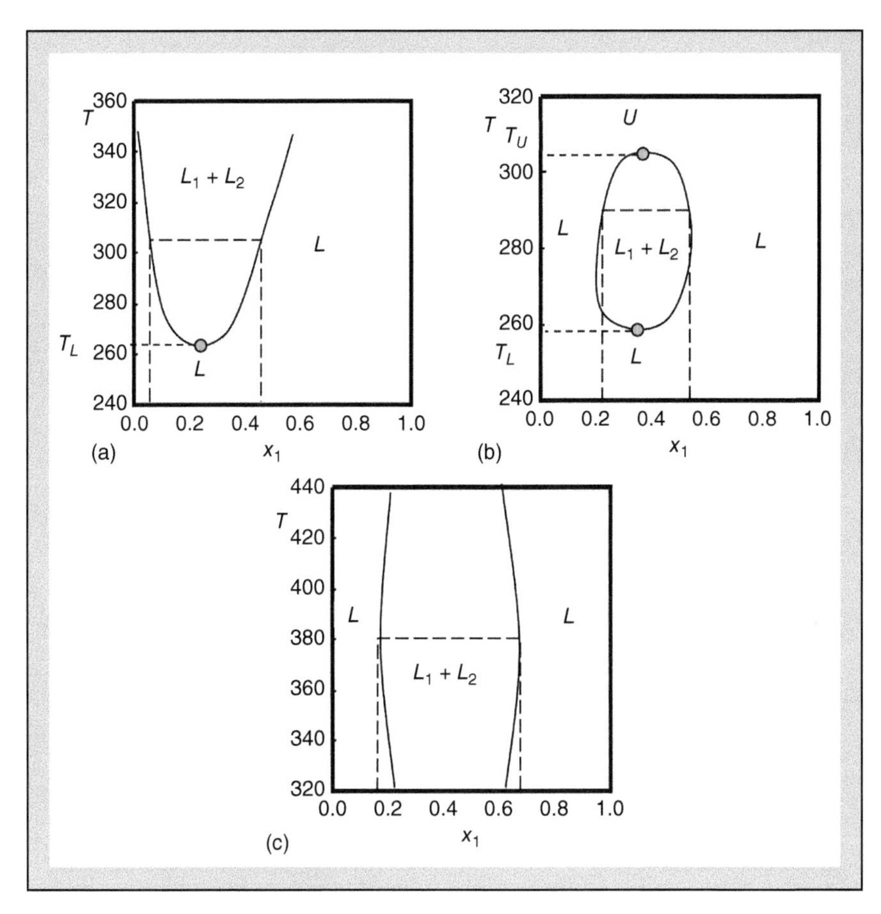

Figure 8.20 Solubility diagrams: (a) with LCST, (b) with LCST and UCSP, (c) without critical solution temperatures.

Now, α and β refer to the liquid phases. Generally, excess Gibbs models are useful for calculating LLE. Any model is satisfactory in Eqs. (8.118) and (8.119). If one uses excess Gibbs models, Eqs. (8.91) and (8.97) apply to the calculations. When using an EOS, Eqs. (8.95) and (8.96) are appropriate.

Example 8.7

Calculate LLE equilibrium compositions for a mixture in which the excess Gibbs energy is

$$\frac{G^E}{RTx_1x_2} = A_{21}x_1 + A_{12}x_2 \text{ with } A_{12} = 4 \text{ and } A_{21} = 1.75$$

Solution

The activity coefficients from the excess Gibbs energy are

$$\ln \gamma_1 = x_2^2[A_{12} + 2(A_{21} - A_{12})x_1]$$

$$\ln \gamma_2 = x_1^2[A_{21} + 2(A_{12} - A_{21})x_2]$$

Then,

$$g = RTx_1x_2[A_{21}x_1 + A_{12}x_2] + RT(x_1 \ln x_1 + x_2 \ln x_2)$$

$$\left(\frac{\partial g}{\partial x_1}\right)_{T,P,x_2} = RT\{(1 - 2x_1)(A_{21}x_1 + A_{12}x_2) + x_1x_2(A_{21} - A_{12}) + \ln x_1 - \ln x_2\}$$

Use the following procedure:

1. Given a value of the derivative, g', find the two roots x_1^α, x_1^β using

$$f(x_1) = \left(\frac{\partial g}{\partial x_1}\right)_{T,P,x_2} - g' = 0$$

or

$$f(x_1) = RT\{(1 - 2x_1)(A_{21}x_1 + A_{12}x_2) + x_1x_2(A_{21} - A_{12}) + \ln x_1 - \ln x_2\} - g' = 0$$

2. Use the new composition values to calculate a new value of the derivative from

$$g'_{new} = \frac{g^\beta - g^\alpha}{x_1^\beta - x_1^\alpha}$$

3. Return to step 2 and continue until the value of the derivative and the values of the compositions do not change appreciably.

Look at the plot of the derivative vs x_1. The maximum of the function occurs at $x_1 = 0.1$ with a value of $g' = 0.62$ (point A in the Figure 8.21), and the minimum value of g' is −0.666, point B at $x_1 = 0.61$. With these two values, we can start the procedure using an average value of g' equal to −0.02287. The following table contains the iterations and the values of the compositions

Iterations	x_1^α	x_1^β	g^α	g^β	g'_{new}	$\left(\frac{\partial g}{\partial x_1}\right)^\alpha_{T,P,x_2}$	$\left(\frac{\partial g}{\partial x_1}\right)^\beta_{T,P,x_2}$
	0.1	0.9	0.0147	−0.14733	−0.02287		
1	0.0233	0.8486	−0.0208	−0.15653	−0.16448	−0.02287	−0.02287
2	0.0194	0.8256	−0.0204	−0.15432	−0.16608	−0.16448	−0.16448
3	0.0193	0.8253	−0.0204	−0.15427	−0.16608	−0.16608	−0.16608

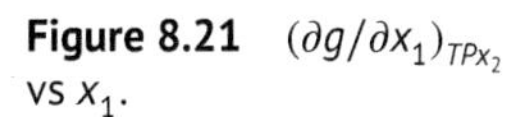

Figure 8.21 $(\partial g/\partial x_1)_{TPx_2}$ vs x_1.

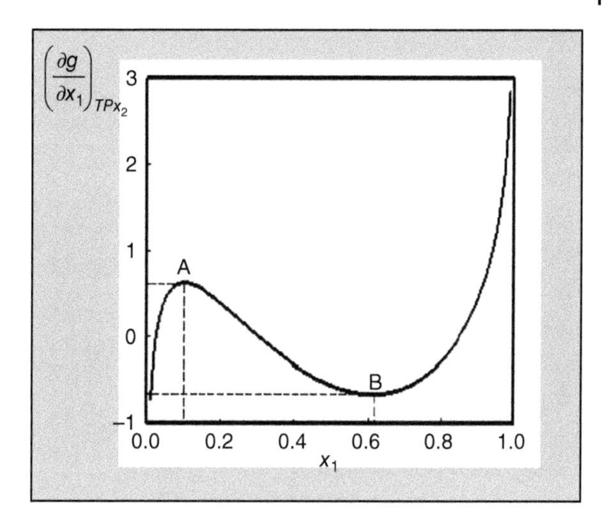

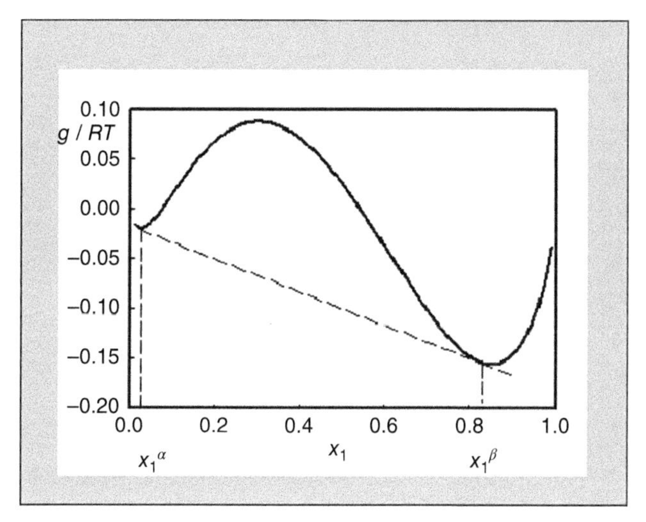

Figure 8.22 $(\partial g/\partial x_1)_{TPx_2}$ vs x_1.

The values of the derivative corresponds to those values at which the function is zero, but the value of the g'_{new} corresponds to the slope of the tangent line that touches the equilibrium compositions as seen in Figure 8.22.

Example 8.8

Aoki and Moriyoshi [11] provide experimental solubility data for 1-butanol + water at 100 kPa. Calculate the temperature dependence for the parameters of the excess Gibbs energy model

$$\frac{G^E}{RTx_1x_2} = A_{21}x_1 + A_{12}x_2, \text{ where } A_{12} = f(T), A_{21} = g(T)$$

Data

T (K)	x_1^{α}	x_1^{β}
302.95	0.0180	0.482
322.75	0.0177	0.459
332.65	0.0180	0.436
342.65	0.0171	0.422
362.65	0.0193	0.359
372.45	0.0249	0.328
380.05	0.0280	0.290
382.95	0.0292	0.281
391.25	0.0396	0.233
395.85	0.0536	0.183
397.05	0.0635	0.165
397.45	0.0739	0.148
397.75	0.0859	0.134

Solution

Expressions for the derivative and the Gibbs energy change on mixing are

$$\left(\frac{\partial g}{\partial x_1}\right) = RT\{(1-2x_1)(A_{21}x_1 + A_{12}x_2) + x_1x_2(A_{21} - A_{12}) + \ln x_1 - \ln x_2\}$$

and

$$g = RTx_1x_2[A_{21}x_1 + A_{12}x_2] + RT(x_1 \ln x_1 + x_2 \ln x_2)$$

Substitute $1 - x_1$ for the composition of the second component. Now, the unknowns are A_{21} and A_{12} at each temperature, and they come from solving the equilibrium Eqs. (8.118) and (8.119). A minimization of the function

$$\Im = \sqrt{F_1^2 + F_2^2}$$

using Solver in Excel produces

T (K)	A_{21}	A_{12}
302.95	2.5192	−4.3317
322.75	2.2315	−5.2536
332.65	1.9001	−6.1744
342.65	1.6371	−7.0853

T (K)	A_{21}	A_{12}
362.65	0.3349	−10.2850
372.45	−0.1469	−10.8005
380.05	−1.1952	−13.1565
382.95	−1.4408	−13.6589
391.25	−2.5726	−15.6595
395.85	−3.9952	−18.4241
397.05	−4.1443	−18.4667
397.45	−4.4449	−18.9752
397.75	−4.5011	−18.9533

Polynomial functions of temperature can represent these results:

$$A_{12} = 653 - 6.0316T + 0.0186T^2 - 1.9315 \times 10^{-5}T^3$$

$$A_{21} = 467 - 4.25T + 0.013T^2 - 1.3312 \times 10^{-5}T^3$$

and they appear in Figure 8.23:

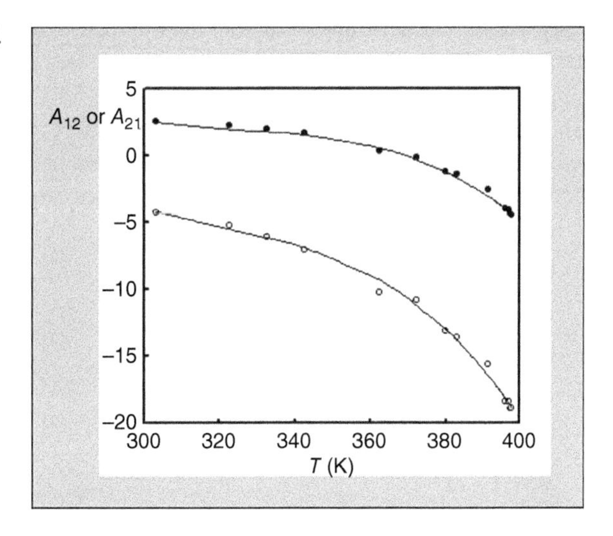

Figure 8.23 A_{12} or A_{21} vs T.

With the temperature expressions for the parameters, we can calculate the equilibrium compositions as in example 8.7. The results are in Figure 8.24 in which the dashed line (-) represents the results using the parameters from the correlation. The Margules equation correlates these data well.

8.11 Binary Vapor–Liquid–Liquid Equilibrium (VLLE)

Vapor–liquid–liquid equilibrium can be a combination of VLE and LLE. If we start at a sufficiently high pressure, VLLE cannot appear, but other possibilities exist: one

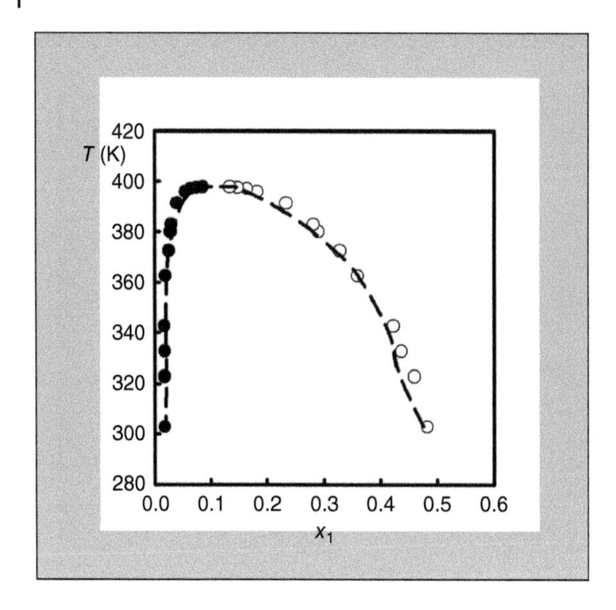

Figure 8.24 T vs x_1.

single phase (vapor or liquid) and two other phases (LL and VL). The formation of each possibility depends on the temperature. At high temperature, a single vapor phase or VLE can form, while at low temperatures, a single liquid phase or LLE can appear. If one lowers the temperature, LLE and VLE intersect to produce VLLE as shown in Figure 8.25. In this figure, it is apparent that at high pressure, the VLE curve is separate from the solubility curve. If the pressure is lower, the VLE curve intersects the solubility curve, and at the intersection, VLLE exists that can also be a heterogeneous azeotrope.

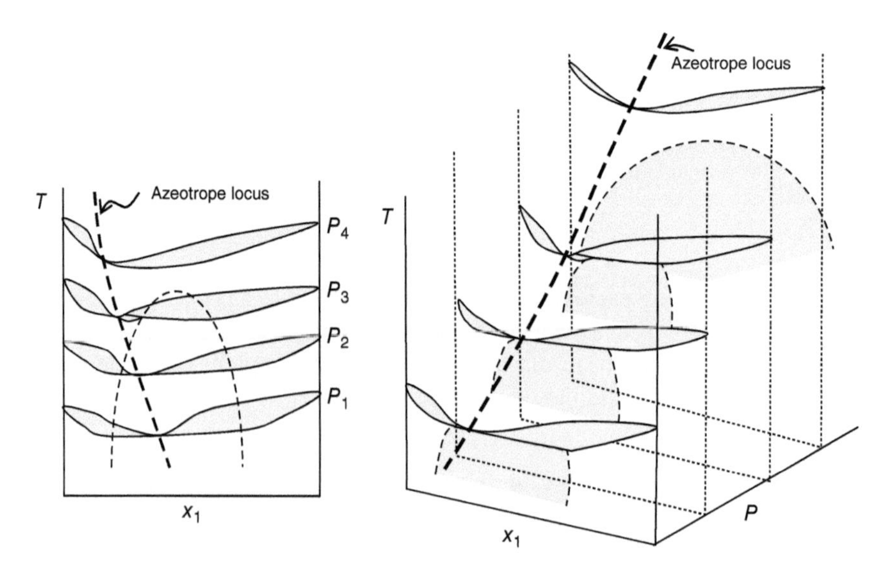

Figure 8.25 Temperature–pressure–composition diagram showing VLE, LLE, and VLLE.

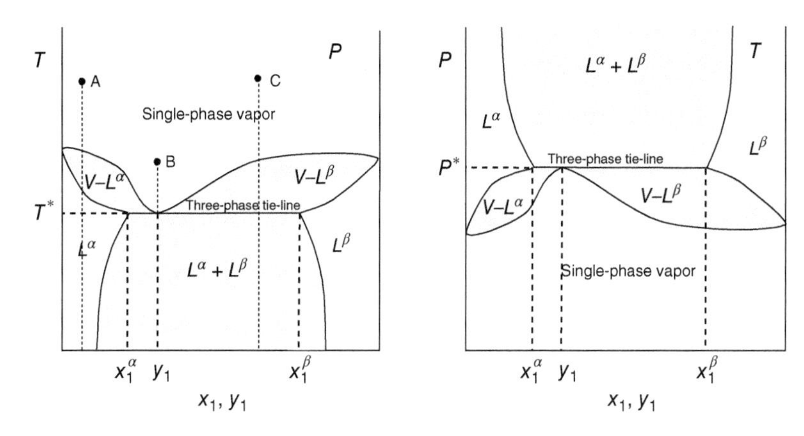

Figure 8.26 Temperature–composition and pressure–composition diagrams for a partially miscible binary mixture.

Now, let us examine a temperature vs composition diagram at a constant pressure, Figure 8.26. If conditions correspond to at point A (a single vapor phase with $y_1 = 0.1$), and we reduce the temperature along a constant composition path, we can reach the following conditions:

- A dew point curve ($y_1 = 0.1$)
- A vapor–liquid (alpha) equilibrium region
- A bubble point curve ($x_1^\alpha = 0.1$)
- A single-liquid region ($x_1 = 0.1$)

Each of these steps requires a reduction in temperature. It is interesting to note that if we move point A to the right, we reach a single-liquid phase region and then reach the LLE curve until we enter the two liquid-phase region. If we start at point B, we do not cross a vapor–liquid region, but we reach the three-phase tie-line at the equilibrium vapor composition. This vapor is in equilibrium with two liquid phases: alpha and beta.

At point C, we reach the following regions or curves by reducing the temperature at constant composition:

- A dew point curve
- A three-phase tie-line (VLLE)
- A two-liquid phase region

A tie-line for VLLE is a consequence of the phase rule because $F = 2 + C - P = 2 + 2 - 3 = 1$. Therefore, fixing the pressure or temperature allows VLLE to occur at a single temperature or pressure. Figure 8.20 shows the pressure composition diagram at a constant temperature. Again, we can pass thorough different phases depending on the initial condition and by increasing the pressure at a constant composition.

The equilibrium conditions for VLLE are the equality of the chemical potentials

$$\mu_i^v = \mu_i^{l,\alpha} = \mu_i^{l,\beta} \text{ or } \hat{f}_i^v = \hat{f}_i^{l,\alpha} = \hat{f}_i^{l,\beta} \quad \text{for } i = 1, 2, \ldots, N \tag{8.139}$$

Then, using the gamma–phi method

$$Py_i\hat{\varphi}_i = x_i^\alpha \gamma_i^\alpha P_i^{sat} \varphi_i^{sat} \quad \text{for } i = 1, 2, \ldots, N \tag{8.140}$$

$$Py_i\hat{\varphi}_i = x_i^\beta \gamma_i^\beta P_i^{sat} \varphi_i^{sat} \quad \text{for } i = 1, 2, \ldots, N \tag{8.141}$$

$$x_i^\alpha \gamma_i^\alpha f_i = x_i^\beta \gamma_i^\beta f_i \quad \text{for } i = 1, 2, \ldots, N \tag{8.142}$$

For a binary mixture at low pressures,

$$Py_i = x_i^\alpha \gamma_i^\alpha P_i^{sat} \quad \text{for } i = 1, 2 \tag{8.143}$$

$$Py_i = x_i^\beta \gamma_i^\beta P_i^{sat} \quad \text{for } i = 1, 2 \tag{8.144}$$

$$x_i^\alpha \gamma_i^\alpha = x_i^\beta \gamma_i^\beta \quad \text{for } i = 1, 2 \tag{8.145}$$

The unknowns are $x_1^\alpha, x_1^\beta, y_1$ for a given temperature and pressure. Six equations exist, but we require only three. In Eqs. (8.140)–(8.145), we ignore the Poynting correction.

We can also consider the case in which the two liquids are completely immiscible. In this case, the vapor phase is a mixture of two components, and two liquid phases exist in which each is a nearly pure component:

Vapor: mixture of components 1 and 2
Liquid Phase α: nearly pure component 1
Liquid Phase β: nearly pure component 2

then

$$x_1^\alpha \cong 1 \text{ and } x_2^\alpha \cong 0$$

Then, if $x_1^\beta \cong 0, x_2^\beta \cong 1$.

Figure 8.27 presents T–x–y and P–x–y diagrams for a binary system of immiscible liquids. Two VLE phases exist, and in each, the vapor is in equilibrium with a pure liquid component. V–L^α contains pure component 1 in the liquid phase, while V–L^β contains pure component 2. LLE exists with two pure liquid components, and the VLLE has a vapor mixture of components 1 and 2 in equilibrium with pure component 1 and pure component 2.

The equilibrium relations in terms of fugacity are Eqs. (8.140)–(8.142):

$$Py_1\hat{\varphi}_1 = x_1^\alpha \gamma_1^\alpha P_1^{sat} \varphi_i^{sat} \quad Py_2\hat{\varphi}_2 = x_2^\alpha \gamma_2^\alpha P_2^{sat} \varphi_2^{sat} \tag{8.146}$$

$$Py_1\hat{\varphi}_1 = x_1^\beta \gamma_1^\beta P_1^{sat} \varphi_1^{sat} \quad Py_2\hat{\varphi}_2 = x_2^\beta \gamma_2^\beta P_2^{sat} \varphi_2^{sat} \tag{8.147}$$

$$x_1^\alpha \gamma_1^\alpha = x_1^\beta \gamma_1^\beta \quad x_2^\alpha \gamma_2^\alpha = x_2^\beta \gamma_2^\beta \tag{8.148}$$

However, some implications exist because we have nearly pure components in the liquid phase:

If $x_1^\alpha \cong 1 \Rightarrow \gamma_1^\alpha \cong 1$ and $x_2^\alpha \cong 0 \Rightarrow \gamma_2^\alpha \cong \gamma_2^{\alpha,\infty}$ (unknown value)
also,
$x_1^\beta \cong 0 \Rightarrow \gamma_1^\beta \cong \gamma_1^{\beta,\infty}$ (unknown value) and $x_2^\beta \cong 1 \Rightarrow \gamma_2^\beta \cong 1$

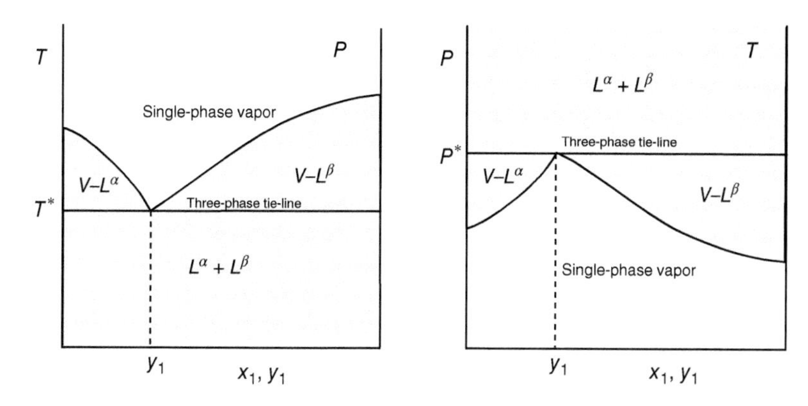

Figure 8.27 Temperature–composition and pressure–composition diagrams for a completely immiscible binary mixture.

From Eqs. (8.146)–(8.148), we select the equations that do not involve unknown quantities

$$Py_1\hat{\varphi}_1 = P_1^{sat}\varphi_i^{sat} \tag{8.149}$$

$$Py_2\hat{\varphi}_2 = P_2^{sat}\varphi_2^{sat} \tag{8.150}$$

Then, the equilibrium pressure of the VLLE is

$$P = P_1^{sat}\varphi_i^{sat}/\hat{\varphi}_1 + P_2^{sat}\varphi_2^{sat}/\hat{\varphi}_2 \tag{8.151}$$

Example 8.9

Consider a mixture of water (1) + benzene (2) + at 200 kPa. At what temperature does VLLE exist? Construct a T–y diagram for the mixture VLE considering that the two components are completely immiscible. The vapor pressure of water from the International Association for the Properties of Water and Steam (IAPWS) is

$$\ln(P^{sat}/P_C) = \left(\frac{T_C}{T}\right)(a_1\tau + a_2\tau^{1.5} + a_3\tau^3 + a_4\tau^{3.5} + a_5\tau^4 + a_6\tau^{7.5})$$

with $\tau = 1 - T/T_C$, $T_C = 647.096$ K, $P_C = 22.064$ MPa, $a_1 = -7.85951783$, $a_2 = 1.84408259$, $a_3 = -11.7866497$, $a_4 = 22.6807411$, $a_5 = -15.9618719$, and $a_6 = 1.80122502$. For benzene,

$$\ln(P^{sat}/P_C) = \left(\frac{T_C}{T}\right)(a_1\tau + a_2\tau^{1.5} + a_3\tau^3 + a_4\tau^5)$$

with $T_C = 562.05$ K, $P_C = 4894$ kPa, $a_1 = -7.0087894$, $a_2 = 1.41137625$, $a_3 = -2.7441613$, and $a_4 = -1.6718406$.

Solution

Assume an ideal gas mixture because the total pressure is only 200 kPa; therefore, Eq. (8.151) becomes

$$P = P_1^{sat} + P_2^{sat}$$

Use both vapor pressure equations to find the temperature that satisfies

$$F(T) = P_1^{sat} + P_2^{sat} - 200 = 0 \tag{A}$$

The next table indicates where the temperature should lie between 362 and 362.5 K.

T (K)	P_1^{sat}	P_2^{sat}	$P = P_1^{sat} + P_2^{sat}$
360	62.195	124.25	186.447
360.5	63.409	126.10	189.505
361	64.644	127.96	192.603
361.5	65.898	129.84	195.743
362	67.173	131.75	198.924
362.5	68.468	133.68	202.148
363	69.784	135.63	205.414
363.5	71.121	137.60	208.723
364	72.479	139.60	29.076
364.5	73.859	141.61	215.473
365	75.260	143.65	218.914

Solving Eq. (A) for temperature yields

$$T = 362.168 \text{ K with } P_1^{sat} = 67.605 \text{ kPa and } P_2^{sat} = 132.395 \text{ kPa}$$

The equilibrium vapor composition is

$$y_1 = \frac{P_1^{sat}}{P} = \frac{67.605}{200} = 0.338$$

To calculate the VLE, we must find the temperature range over which vapor–liquid equilibrium exists. This range lies between the two saturation temperatures at which the vapor pressure is 200 kPa.

For water, $T = 393.361$ K
For benzene, $T = 377.036$ K

Then, the y_1 composition in equilibrium with the liquid in the α-phase is

$$y_1 = \frac{P_1^{sat}}{P} = \frac{P_1^{sat}}{200}$$

The y_2 composition in equilibrium with the liquid β-phase is

$$y_2 = \frac{P_2^{sat}}{P} = \frac{P_2^{sat}}{200} \Rightarrow y_1 = 1 - y_2$$

and the results are

T (K)	P_1^{sat}	y_1	P_2^{sat}	y_2	$y_1 = 1 - y_2$
393.3606	200.000	1.000	301.83		
390	179.642	0.898	278.21		
385	152.518	0.763	245.70		
380	128.852	0.644	216.17		
377.036	116.314	0.582	200.00	1.000	0
375	108.299	0.542	189.44	0.947	0.053
370	90.536	0.453	165.33	0.827	0.173
365	75.260	0.376	143.65	0.718	0.282
362.1676	67.605	0.338	132.40	0.662	0.338

The T–y_1 diagram is shown in Figure 8.28:

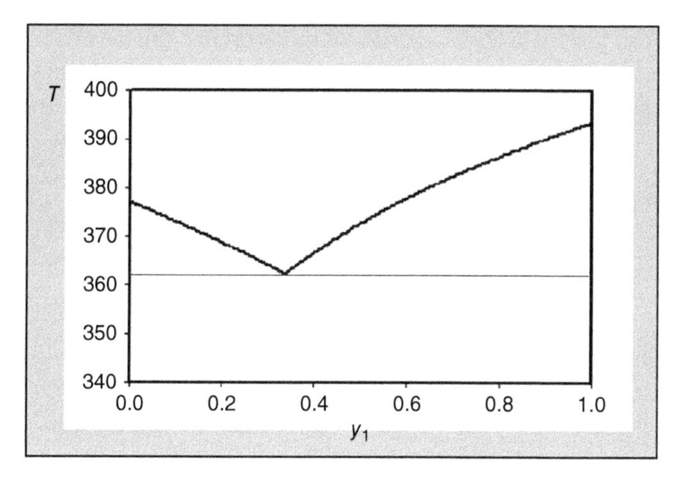

Figure 8.28 T vs y_1.

8.12 Binary Vapor–Solid Equilibrium (VSE)

Vapor–solid equilibrium can exist for a pure component represented as the sublimation curve in a pressure–temperature diagram. In this case, we can use a thermodynamic model valid for gases and solids. If the solid phase in equilibrium is a pure component, then the $\gamma - \varphi$ method simplifies the solution. Again, the solution must satisfy equality of the chemical potentials or equality of fugacities, and if component 1 is the solid phase, only one equilibrium relationship exists

$$\hat{f}_1^v = f_1^s \tag{8.152}$$

We can calculate the fugacity of the solid pure component using an alternative form of Eq. (7.146)

$$f_1^s = \varphi_1^{sat} P_1^{sat} \exp\left[\frac{V_1^s\left(P - P_1^{sat}\right)}{RT}\right] \tag{8.153}$$

in which P_1^{sat} is the vapor–solid saturation pressure at temperature T, and V_1^S is the molar volume of the solid. Using the definition of fugacity coefficient for the vapor phase

$$\hat{f}_1^v = y_1 P \hat{\varphi}_1 \tag{8.154}$$

The vapor composition result if we combine both equations

$$y_1 = \frac{\varphi_1^{sat} P_1^{sat}}{\hat{\varphi}_1 P} \exp\left[\frac{V_1^s\left(P - P_1^{sat}\right)}{RT}\right] \tag{8.155}$$

Generally, the major component in the vapor phase is the "solvent," and the minor component is the "solute." In this treatment, component 1 is the solute and component 2 the solvent. The vapor composition, y_1, is the solubility of the solute in the solvent. At low pressures, ideal gas behavior may be a good assumption, and if the difference between the pressures is sufficiently small, the Poynting correction is negligible. Therefore, Eq. (8.155) becomes

$$y_1 \approx \frac{P_1^{sat}}{P} \tag{8.156}$$

Example 8.10

Find the solubility of benzoic acid (1) in CO_2 (2) at 35 °C using the SRK EOS with $k_{ij} = 0.036$. The sublimation pressure of benzoic acid at 35 °C is 0.36 Pa, and the volume of solid benzoic acid is 92.8 cm^3/mol. Compare the results to the experimental data of Dobbs et al. [12]

Data

P (bar)	y_1
120	1.25×10^{-3}
160	2.19×10^{-3}
200	2.53×10^{-3}
240	2.81×10^{-3}
280	3.03×10^{-3}

Solution

Use Eq. (8.155) to find the vapor composition of benzoic acid:

$$y_1 = \frac{\varphi_1^{sat} P_1^{sat}}{\hat{\varphi}_1 P} \exp\left[\frac{V_1^s\left(P - P_1^{sat}\right)}{RT}\right] \tag{A}$$

Because $\hat{\varphi}_1$ is also a function of y_1, the composition results from finding the root of

$$F(y_1) = \frac{\varphi_1^{sat} P_1^{sat}}{\hat{\varphi}_1 P} \exp\left[\frac{V_1^S\left(P - P_1^{sat}\right)}{RT}\right] - y_1 = 0 \tag{B}$$

Using Eqs. (7.231) and (7.232), we calculate the parameters for the mixture in the vapor phase from

$$a = y_1^2 a_1 + 2y_1 y_2 a_{12} + y_2^2 a_2 \text{ with } a_{12} = \sqrt{a_1 a_2}(1 - k_{12})$$

$$b = y_1 b_1 + y_2 b_2$$

Also, we need

$$A = \frac{aP}{(RT)^2}, B = \frac{Pb}{RT}, \text{and } A/B = \frac{a}{bRT}$$

and the partial molar property $\overline{D}_1$ for the vapor phase:

$$\overline{D}_1 = (A/B)\left(\frac{2z_1 a_1 + 2z_2 a_{12}}{a} - \frac{b_1}{b}\right)$$

The fugacity coefficient of component 1 is

$$\ln \hat{\varphi}_1 = \frac{b_1}{b}(Z - 1) - \ln(Z - B) - \overline{D}_1 \ln\left(\frac{Z + B}{Z}\right) \tag{C}$$

In the above equation, the compressibility factor of the vapor mixture comes from the root of

$$f(Z) = 1 - (A - B - B^2)/Z + AB/Z^2 - Z = 0 \tag{D}$$

Also, we require the fugacity coefficient of the pure component at vapor/solid equilibrium (use P_1^{sat})

$$\ln \varphi_1^{sat} = Z_1^{sat} - 1 - \ln\left(Z_1^{sat} - B_1\right) - (A_1/B_1)\ln\left(\frac{Z_1^{sat} + B_1}{Z_1}\right) \tag{E}$$

In this equation, the compressibility factor is the root of

$$f\left(Z_1^{sat}\right) = 1 - \left(A_1 - B_1 - B_1^2\right)/Z_1^{sat} + A_1 B_1/Z_1^{2,sat} - Z_1^{sat} = 0 \tag{F}$$

The procedure is

1. Calculate Z_1^{sat} that satisfies Eq. (F), then φ_1^{sat} using Eq. (E)
2. Provide an initial guess for y_1
3. Find the compressibility of the mixture (Z) that satisfies Eq. (D)
4. Find the fugacity coefficient of component 1 using Eq. (C)
5. Find a new value for y_1 that satisfies Eq. (B)
6. If $|y_{1,j+1} - y_{1,j}| \leq 10^{-9}$, stop, otherwise return to step 3.

The problem solution using Solver in Excel is

	A	B	C	D	E	F	G	H	I
1		T_c	T_R	ω	α	P_c	b_i	a_i	
2	Benzoic Acid	751	0.41032	0.604	2.22358	44.7	121.0210391	82901296	
3	CO2	304.2	1.01298	0.224	0.98937	73.83	29.67937869	3664195.8	
4									
5	T=	308.15	P=	280			aij=	16801460	
6	R=	83.14		4088					
7									
8	P1Sat	3.60E-06							
9	V1sol	92.8							
10									
11	Mixture					Pure Benzoic Acid			
12	a/bar cm6 mol^{-2}	3756708.423				$(A_1/B_1)^{sat}$	26.73795879		
13	b/ cm3 mol^{-1}	29.99874181				B_1^{sat}=	1.70056E-08		
14	A/B	4.888011917				Z=	0.999999662		
15	B	0.327860336				$\ln\varphi^{sat}$=	-4.37689E-07		
16	$\bar{D}_1$	24.60435278				φ^{sat}=	0.999999562		
17	Z	0.569808817							
18	$\ln\hat{\varphi}_1$	-11.4991435				F(Zsat)	-9.9672E-08		
19	$\hat{\varphi}_1$	1.01388E-05							
20	f(Z)	4.20211E-07							
21									
22	f(y1)=	1.1894E-07							
23									
24	y_1	3.49636E-03	y_2	0.9965036					
25	kij	0.036							

After repeating the same procedure for each pressure,

P (bar)	y_1
120	1.14×10^{-3}
160	1.93×10^{-3}
200	2.57×10^{-3}
240	3.09×10^{-3}
280	3.50×10^{-3}

Figure 8.29 shows the SRK EOS prediction of the solubility of benzoic acid in CO_2.

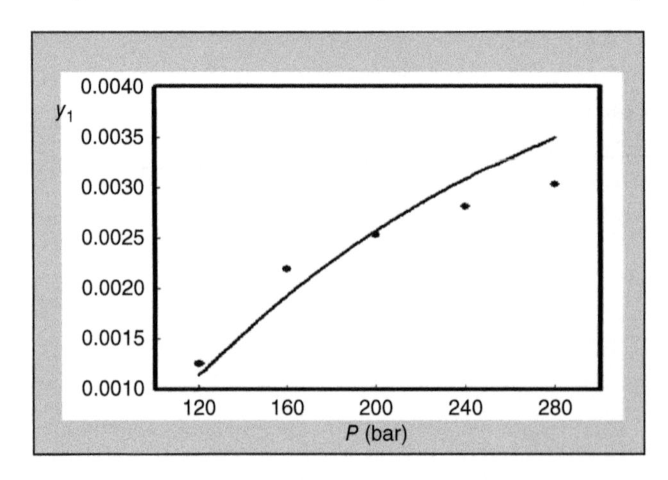

Figure 8.29 y_1 vs *P*.

8.13 Binary Liquid–Solid Equilibrium (LSE)

The separation of a liquid from a solid follows the thermodynamic equilibrium between the phases. In the chemical, crystallization, a common separation process, is liquid–solid equilibrium (LSE). The amounts of solids that exist in equilibrium with a liquid vary enormously. In some cases, such as salt water, the concentration can be high, but it can also be almost negligible. The phase diagrams for LSE are similar to those for vapor–liquid–liquid equilibrium and gas–solid equilibrium.

Classification of the phase equilibrium diagrams involving solids uses the miscibility of the solid component. If the binary components are miscible in the solid phase, the phase diagrams are similar to those found in VLE. Figure 8.30 shows some typical T–x diagrams of this kind. The equilibrium liquid curve is the *liquidus curve*, and the solid equilibrium curve is the *solidus curve* (as shown in Figure 8.30a). As in vapor–liquid equilibrium, some cases have the composition of the solid equal to the composition of the liquid. This state is *solutropes* (in analogy to azeotropes in VLE). Figure 1b,c shows this behavior. Figure 1b illustrates a "minimum melting point," while Figure 1c has two solutropes and two minimum melting points.

Figure 8.30d–f shows the LSE when the liquid components are totally miscible, but the solid components are partially immiscible. In this case, the behavior is analogous to LLE and VLLE. Figure 8.30d,e occur when the melting point is at a high temperature and the solid–solid equilibrium has a maximum similar to the UCST in LLE. The same figure, diagram (e), has a solutrope and two different liquid–solid regions. Another diagram we have seen before is Figure 8.30f. In this case, the mixture melts before the solid–solid equilibrium (SSE) disappears, and the LSE equilibrium curve and the solid–solid equilibrium curve join, as happens in VLLE. We have two LSE curves and a three-phase line where the LSE curve and SSE join. The three-phase line is the *Eutectic* line, and the temperature is the *Eutectic* temperature. In Figure 8.30f, T^* represents this temperature. Situations exist in which three phase equilibria exist, and we can pass from a liquid + solid equilibrium into a single solid phase. This point is the *Peritectic* point, and it appears in diagram (g) as P. The line and the temperature also have the same name. When the liquid components are immiscible, the phase diagram is more complex, and it can form *monotectic* points at which the liquid phase of component 1 can pass to a solid–liquid phase. Finally, when the solids are immiscible, the phase diagrams look like diagrams (h) and (i). Diagram (h) is similar to LLE with two immiscible liquids, and diagram (i) shows that a solid can form two crystalline phases α and β that are in equilibrium with a liquid. They have a point at which the liquid is in equilibrium with the two crystalline forms of one of the components. At equilibrium,

$$\mu_i^l = \mu_i^s \text{ or } \hat{f}_i^l = \hat{f}_i^s \quad \text{for } i = 1, 2, \ldots, N \tag{8.157}$$

The equations in Section 8.9 apply if we have an EOS valid for liquids and solids, or if we have solution models for each phase. If we express the fugacity coefficient of each component in the mixture in terms of the fugacity coefficient

$$x_i \gamma_i^l f_i^l = z_i \gamma_i^s f_i^s \quad \text{for } i = 1, 2, \ldots, N \tag{8.158}$$

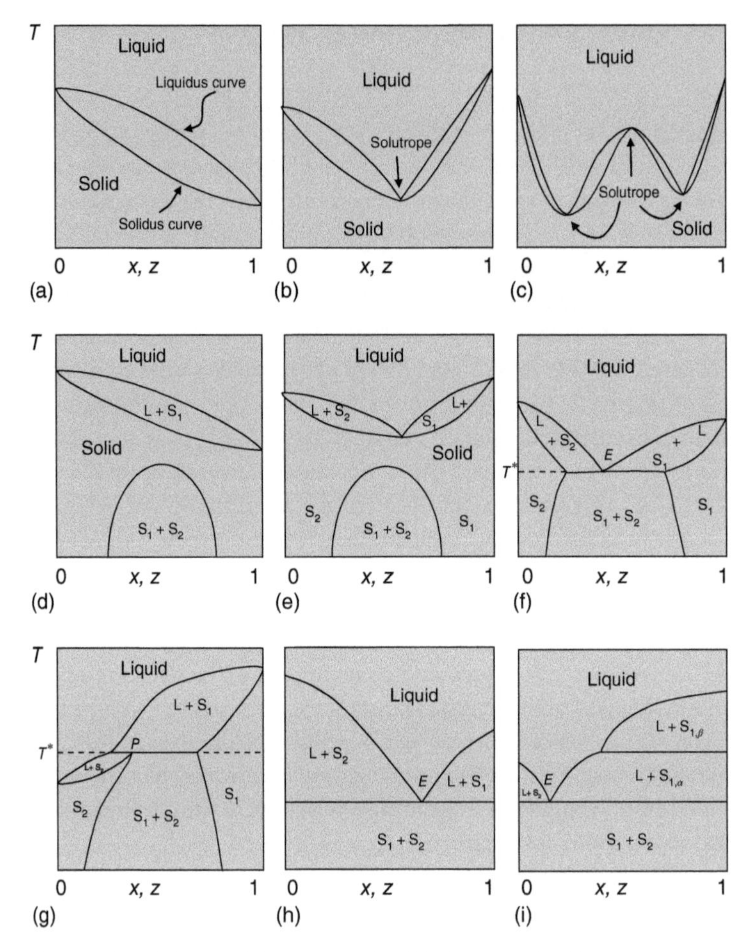

Figure 8.30 Different Liquid–Solid Equilibrium for miscible components (a)–(c), partially miscible solids (d)–(g), and immiscible solids (h)–(i) binary mixtures.

The key point is how to calculate the fugacities of the pure components. This calculation comes from the definition of fugacity

$$\ln\left(\frac{f_i^l(T,P)}{f_i^s(T,P)}\right) = \frac{\Delta G^{ls}(T,P)}{RT} = \frac{G_i^l(T,P) - G_i^s(T,P)}{RT} \tag{8.159}$$

The Gibbs energy is a state function; therefore,

$$\frac{\Delta G^{ls}(T,P)}{RT} = \frac{\Delta G^{ls}(T_{m,i},P)}{RT} + \int_{T_{m,i}}^{T} \left(\frac{\partial \Delta G^{ls}}{\partial(1/T)}\right) d(1/T) \tag{8.160}$$

where $T_{m,i}$ is the melting temperature of component i and $\Delta G^{ls}(T_{m,i}, P) = 0$. Now, the Gibbs–Helmholtz equation can replace the term in the integral

$$\frac{\Delta G^{ls}(T,P)}{RT} = \int_{T_{m,i}}^{T} \frac{\Delta H^{ls}(T,P)}{R} d(1/T) \tag{8.161}$$

Using the definition of the heat capacity at constant pressure for an enthalpy change

$$\left(\frac{\partial \Delta H_i^{ls}}{\partial T}\right)_P = \Delta C_{P,i}^{ls} \tag{8.162}$$

Integrating the above equation from $T_{m,i}$ to T

$$\Delta H^{ls}(T,P) = \Delta H^{ls}(T_{m,i}, P) + \int_{T_{m,i}}^{T} \Delta C_{P,i}^{ls} dT \tag{8.163}$$

in which ΔH_i^{ls} is the enthalpy of fusion and $\Delta C_{P,i}^{ls}$ is the change of the heat capacity from liquid to solid. The evaluation of Eq. (8.163) requires a temperature function, but if we consider it constant,

$$\Delta H^{ls}(T,P) = \Delta H^{ls}(T_{m,i}, P) + \Delta C_{P,i}^{ls}(T - T_{m,i}) \tag{8.164}$$

Substituting this result into Eq. (8.161)

$$\frac{\Delta G^{ls}(T,P)}{RT} = \frac{\Delta H^{ls}(T_{m,i}, P)}{R}\left(\frac{1}{T} - \frac{1}{T_{m,i}}\right) + \frac{\Delta C_{P,i}^{ls}}{R}\left[\ln\left(\frac{T}{T_{m,i}}\right) - \left(\frac{T_{m,i}}{T} - 1\right)\right] \tag{8.165}$$

and finally,

$$\ln\left(\frac{f_i^l(T,P)}{f_i^s(T,P)}\right) = \frac{\Delta H^{ls}(T_{m,i}, P)}{R}\left(\frac{1}{T} - \frac{1}{T_{m,i}}\right) + \frac{\Delta C_{P,i}^{ls}}{R}\left[\ln\left(\frac{T}{T_{m,i}}\right) - \left(\frac{T_{m,i}}{T} - 1\right)\right] \tag{8.166}$$

Now, the equilibrium Eq. (8.158) becomes

$$\ln\left(\frac{z_i\gamma_i^s}{x_i\gamma_i^l}\right) = \ln\left(\frac{f_i^l}{f_i^s}\right) = \frac{\Delta H^{ls}(T_{m,i}, P)}{RT_{m,i}}\left(\frac{T_{m,i}}{T} - 1\right) + \frac{\Delta C_{P,i}^{ls}}{R}\left[\ln\left(\frac{T}{T_{m,i}}\right) - \left(\frac{T_{m,i}}{T} - 1\right)\right] \tag{8.167}$$

This equation considers only one crystalline form, but if more exist, we must consider

$$\ln\left(\frac{f_i^l(T,P)}{f_i^{s,2}(T,P)}\right) = \frac{G_i^l(T,P) - G_i^{s,1}(T,P)}{RT} + \frac{G_i^{s,1}(T,P) - G_i^{s,2}(T,P)}{RT} \tag{8.168}$$

The first change in Gibbs energy is the RHS of Eq. (8.167), and the second change of Gibbs energy comes from

$$\frac{\Delta G^{s1s2}(T,P)}{RT} = \frac{\Delta G^{s1s2}(T_{tr,i},P)}{RT} + \int_{T_{tr,i}}^{T} \left(\frac{\partial \Delta G^{s1s2}}{\partial(1/T)}\right) d(1/T) \tag{8.169}$$

Here, $\Delta G^{s1s2}(T_{tr,i},P) = 0$ because it is an equilibrium condition and $T_{tr,i}$ is the transition temperature. Following the same steps as in Eqs. (8.161)–(8.165)

$$\frac{\Delta G^{s1s2}(T,P)}{RT} = \frac{\Delta H^{s1s2}(T_{tr,i},P)}{RT_{tr,i}}\left(\frac{T_{tr,i}}{T} - 1\right) + \frac{\Delta C_{P,i}^{s1s2}}{R}\left[\ln\left(\frac{T}{T_{tr,i}}\right) - \left(\frac{T_{tr,i}}{T} - 1\right)\right] \tag{8.170}$$

If $\Delta C_{P,i}^{s1s2} \approx 0$, then the equilibrium equation is

$$\ln\left(\frac{z_i\gamma_i^s}{x_i\gamma_i^l}\right) = \frac{\Delta H^{ls}(T_{m,i},P)}{RT_{m,i}}\left(\frac{T_{m,i}}{T} - 1\right) + \frac{\Delta C_{P,i}^{ls}}{R}\left[\ln\left(\frac{T}{T_{m,i}}\right) - \left(\frac{T_{m,i}}{T} - 1\right)\right] + \frac{\Delta H^{s1s2}(T_{tr,i},P)}{RT_{tr,i}}\left(\frac{T_{tr,i}}{T} - 1\right) \tag{8.171}$$

Equation (8.167) can calculate the LSE at constant pressure when the solid has one crystalline form and the heat capacity is constant. Limiting cases depend on the values of the activity coefficient and the miscibility of the solid phase for binary mixtures.

Case I. If $\gamma_i^s = 1$ and $\gamma_i^l = 1$ (ideal solution), the equilibrium equations are

$$z_i = x_i f_i(T) \text{ for } i = 1, 2 \tag{8.172}$$

with

$$f_i(T) = \exp\left\{\frac{\Delta H^{ls}(T_{m,i},P)}{RT_{m,i}}\left(\frac{T_{m,i}}{T} - 1\right) + \frac{\Delta C_{P,i}^{ls}}{R}\left[\ln\left(\frac{T}{T_{m,i}}\right) - \left(\frac{T_{m,i}}{T} - 1\right)\right]\right\} \tag{8.173}$$

In this case, we can solve a system of equations with two unknowns x_1 and z_1:

$$z_1 = x_1 f_1(T) \tag{8.174}$$

$$1 - z_1 = (1 - x_1) f_2(T) \tag{8.175}$$

Solving this system for x_1 and z_1 in terms of f_1 and f_2

$$x_1 = \frac{1 - f_2(T)}{f_1(T) - f_2(T)} \tag{8.176}$$

and

$$z_1 = \frac{1 - f_2(T)}{f_1(T) - f_2(T)} f_1(T) \tag{8.177}$$

This solution provides a diagram similar to the one given by Raoult's law because the equilibrium equations $\left(y_i = x_i\left(P/P_i^{sat}\right)\right)$ are similar. In Figure 8.30a shows this behavior.

Case II. If $\gamma_i^s = 1$, immiscible solid ($z_i = 1$) and ideal solution $\gamma_i^l = 1$, the equilibrium equations are

$$x_1 = 1/f_1(T) \tag{8.178}$$

$$x_2 = 1/f_2(T) \Rightarrow x_1 = 1 - 1/f_2(T) \tag{8.179}$$

These equations are equivalent to those for VLLE. Therefore, Eq. (8.178) provides a liquidus curve S_1–L and Eq. (8.179) gives a second liquidus curve S_2–L. They join at the eutectic point, and the temperature results from

$$1/f_1(T) = 1 - 1/f_2(T) \tag{8.180}$$

The equilibrium curve corresponds to diagram (h) in Figure 8.30.

If the solution is not ideal, Eqs. (8.178) and (8.179) become

$$x_1\gamma_1 = 1/f_1(T) \tag{8.181}$$

$$x_2\gamma_2 = 1/f_2(T) \tag{8.182}$$

and we must solve each of them individually for the composition. The equilibrium curve is also diagram (h).

Example 8.11

Calculate the LSE temperatures for the system indolene + benzene. Assume that the solid components are immiscible and that the nonideality in the phase follows

$$\ln\gamma_1 = 0.71953x_2^2 \text{ and } \ln\gamma_2 = 0.71953x_1^2$$

Yokoyama et al. [13] provide experimental measurements for this system

x_1	T (K)	x_1	T (K)
0	278.73	0.401	279.4
0.0521	275.05	0.501	288.54
0.2	268.31	0.715	303.68
0.251	266.37	0.792	309.88
0.281	269.32	0.902	317.59
0.349	274.58	1	326.26

Find the eutectic temperature if $x_E = 0.2552$ and compare your temperatures to the experimental results if $\Delta H^{ls}(T_{m,1}, P) = 10.9$ kJ/mol, $\Delta H^{ls}(T_{m,2}, P) = 9.6$ kJ/mol, and $\Delta C_{P,i}^{ls} = 0$ for $i = 1, 2$.

Solution

Using Eq. (8.171) with $\Delta C_{P,i}^{ls} = 0$

$$x_1\gamma_1 = \exp\left\{-\frac{\Delta H_{m,2}^{ls}(T_{m,1}, P)}{RT_{m,1}}\left(\frac{T_{m,1}}{T} - 1\right)\right\} \tag{A}$$

$$x_2\gamma_2 = \exp\left\{-\frac{\Delta H^{ls}(T_{m,2}, P)}{RT_{m,2}}\left(\frac{T_{m,2}}{T} - 1\right)\right\} \tag{B}$$

In Eq. (B), we must substitute $x_2 = 1 - x_1$, then γ_2 is a function of x_1. Solving Eqs. (A) and (B) for T involves finding the roots of

$$g_1(T) = x_1 - \exp\left\{-\frac{\Delta H^{ls}(T_{m,1}, P)}{RT_{m,1}}\left(\frac{T_{m,1}}{T} - 1\right)\right\} \Big/ \gamma_1 = 0 \tag{C}$$

$$g_2(T) = (1 - x_1) - \exp\left\{-\frac{\Delta H^{ls}(T_{m,2}, P)}{RT_{m,2}}\left(\frac{T_{m,2}}{T} - 1\right)\right\} \Big/ \gamma_2 = 0 \tag{D}$$

First, we find the eutectic temperature (the point at which Eqs. (A) and (B) intersect), which must satisfy $g_1(T_E) = 0$ and $g_2(T_E) = 0$. Then, use Solver to find the temperature at which

$$F = \sqrt{g_1(T_E)^2 + g_2(T_E)^2} = 0$$

using the following procedure:

1. Select an initial value of T (this can be the experimental value in the table)
2. Calculate the exponential term
3. Use the experimental value x_1 to calculate γ_i using $x_2 = 1 - x_1$ when necessary
4. Calculate the function g_i. Equation (E) is valid for compositions greater than the eutectic composition and Eq. (D) is valid at lower compositions.
5. Go to Solver and assign the cell of function g as your target cell
6. Select as the parameter your initial temperature and solve

Your Excel spreadsheet should be as follows:

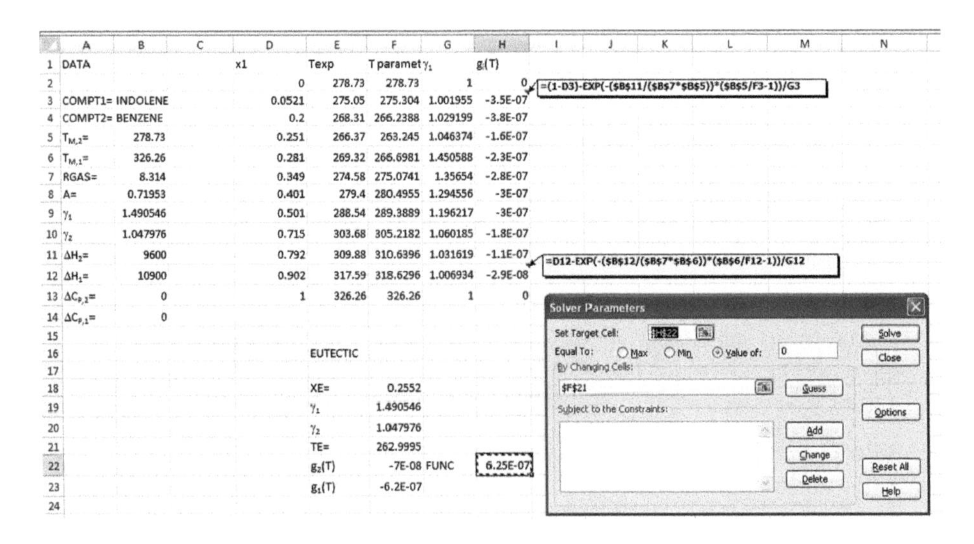

	A	B	C	D	E	F	G	H
1	DATA		x1		Texp	T paramet	γ_1	g(T)
2				0	278.73	278.73	1	0
3	COMPT1=	INDOLENE		0.0521	275.05	275.304	1.001955	-3.5E-07
4	COMPT2=	BENZENE		0.2	268.31	266.2388	1.029199	-3.8E-07
5	$T_{M,2}$=	278.73		0.251	266.37	263.245	1.046374	-1.6E-07
6	$T_{M,1}$=	326.26		0.281	269.32	266.6981	1.450588	-2.3E-07
7	RGAS=	8.314		0.349	274.58	275.0741	1.35654	-2.8E-07
8	A=	0.71953		0.401	279.4	280.4955	1.294556	-3E-07
9	γ_1	1.490546		0.501	288.54	289.3889	1.196217	-3E-07
10	γ_2	1.047976		0.715	303.68	305.2182	1.060185	-1.8E-07
11	ΔH_2=	9600		0.792	309.88	310.6396	1.031619	-1.1E-07
12	ΔH_1=	10900		0.902	317.59	318.6296	1.006934	-2.9E-08
13	$\Delta C_{P,2}$=	0		1	326.26	326.26	1	0
14	$\Delta C_{P,1}$=	0						
15								
16					EUTECTIC			
17								
18					XE=	0.2552		
19					γ_1	1.490546		
20					γ_2	1.047976		
21					TE=	262.9995		
22					g_2(T)	-7E-08	FUNC	6.25E-07
23					g_1(T)	-6.2E-07		
24								

and the equilibrium curve appears in Figure 8.31.

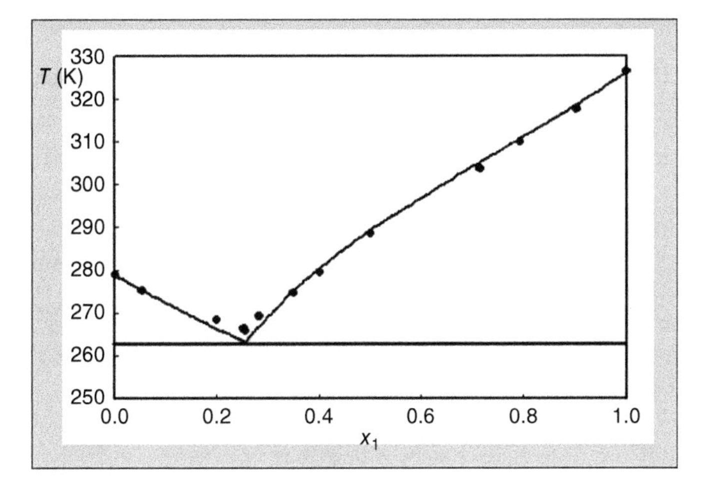

Figure 8.31 T vs x_1.

Problems for Chapter 8

8.1 Vapor–liquid equilibrium for nitrogen and methane occurs at 183 K and 4750 kPa in a container. In the vapor phase, the mixture has 8 kg of nitrogen and 27.5 kg of methane. Find the partial pressures for each component. The molecular weights of nitrogen and methane are 28.014 and 16.043 g/gmol, respectively.

8.2 If the global composition of a mixture has 10 mol of specie A and 20 mol of specie B, what is the vapor composition at the dew point? What is the liquid composition at the bubble point?

8.3 An EOS is

$$P = \frac{RT}{V - b}$$

in which b is a constant. Can this equation represent vapor–liquid equilibrium?

8.4 The vapor pressure of CO_2 at 250 K is 17.85 bar. Calculate the vapor volume using the vdW EOS. Is this volume the saturated vapor volume?
Data

$$T_C = 304.128\text{ K}, P_C = 73.77\text{ bar}$$

8.5 Calculate the vapor pressure and the saturated vapor and liquid volumes of ethane at 150 K using the SRK EOS.

Data

$$T_C = 305.33\text{ K},\ P_C = 48.72\text{ bar},\ \omega = 0.099,\ \text{and}\ V_C = 147.06\ \text{cm}^3/\text{mol}$$

8.6 Substances A and B have vapor pressures of $P_A^{sat} = 1.63$ bar and $P_B^{sat} = 3.59$ bar at 294.6 K. Assuming that mixtures of A and B obey Raoult's law, determine the overall composition for which the mole fraction of A equals 0.437:

a. The range of pressures for which vapor and liquid phases coexist in equilibrium for this mixture at 294.6 K.
b. The fraction of material in the vapor phase and the vapor and liquid compositions at the pressure that is the average of the dew and bubble pressures, all at 294.6 K.

8.7 For a binary mixture of two substances A and B, with $x_A = 0.33$ and vapor pressures for A and B given by the Antoine equation, (T in K and P in kPa), calculate the following, assuming that the mixture obeys Raoult's law.

a. Determine the dew and bubble temperatures (to ±0.1 K) at $P = 186.94$ kPa.
b. Determine the temperature (to ±0.1 K) at which 25% of the material is in the liquid phase at a pressure of 186.94 kPa and the compositions of the liquid and vapor phases.

	a	*b*	*c*
A	13.976	2875	105
B	13.022	3100	137

8.8 The following equation represents the vapor pressures of pure normal butane (n-C_4) and normal pentane (n-C_5) using the parameters given in the table below:

$$\ln P^{sat} = A + \frac{B}{T} + C \ln T + D\frac{P^{sat}}{T^2}$$

The units of T are kelvin and the units of P^{sat} are bar. The parameters are

	A	**B (K)**	**C**	**D (K²/bar)**
n-C_4	41.714	−4065.27	−4.784	2010
n-C_5	46.062	−4827.06	−5.313	2760

Assuming that the mixtures of these materials are described well by Raoult's law,

a. Determine the dew and bubble pressures (to ±0.01 bar) at 350 K for a mixture containing 30 mol% n-C_4 and 70 mol% n-C_5;
b. Determine the amount of vapor (mole fraction vapor) and the vapor and liquid compositions present in a mixture with an overall composition of 30 mol% n-C_4 and 70 mol% n-C_5 at 350 K and 4.45 bar.

8.9 Consider a mixture of two species A and B. The liquid phase for this mixture obeys the van Laar equation for the excess Gibbs free energy as a function of composition. The vapor pressures of species A and B, respectively, are

$$\ln P_A^{sat}\,(\text{kPa}) = 13.603 - \frac{2650}{T\,(\text{K}) - 38.6} \text{ and}$$

$$\ln P_B^{sat}\,(\text{kPa}) = 13.461 - \frac{2825}{T\,(\text{K}) - 29.6}$$

At 350 K, the activity coefficients at infinite dilution of A and B in the liquid phase are 1.876 and 3.136, respectively. The vapor phase may be treated as an ideal gas for this problem.

a. Calculate the bubble point pressure at 350 K for a mixture that contains 45.0 mol% of species A.
b. At 350 K and 154.62 kPa, the equilibrium saturated liquid phase is 35 mol% A. Determine the vapor phase composition and the overall composition that would yield equal numbers of moles in the vapor and liquid phases.

8.10 For a binary mixture of two substances 1 and 2, with $x_1 = 0.617$, vapor pressures for 1 and 2 come from the Antoine equation using the constants given below, and an excess Gibbs function of the van Laar form

$$G^E/RT = x_1 x_2/(0.3860x_1 + 0.7267x_2)$$

Determine the bubble temperature (to ±0.1 K) and the vapor composition at 125.4 kPa, assuming that the vapor phase is an ideal gas.

$$\ln P_i^{sat}\,(\text{kPa}) = a_i - \frac{b_i}{T\,(\text{K}) + c_i}$$

	a	*b*	*c*
1	11.526	2968	51
2	11.083	3096	67

8.11 Two substances, A and B, have vapor pressures of:142 kPa (A) and 96.1 kPa (B) at 55 °C. Two vapor–liquid equilibrium measurements are available for mixtures of A and B at 55 °C:

P (kPa)	x_A	y_A
116	0.250	0.361
136	0.750	0.790

Use these data and the two-suffix Margules equation to calculate the bubble pressure and the vapor composition at the bubble pressure for an equimolar mixture of A and B at 55 °C.

8.12 A mixture of two compounds, A and B, has an azeotrope at $T = 301.4$ K, $P = 136.94$ kPa, and $x_A = y_A = 0.392$. The vapor pressures at this temperature are $P_A^{vap} = 101.57$ kPa and $P_B^{vap} = 86.42$ kPa. To describe the properties of this mixture, use the Margules equation in the form:

$$\frac{G^E}{RTx_Ax_B} = A_{AB}x_A + A_{BA}x_B$$

a. Determine the numerical values of A_{AB} and A_{BA} that describe this mixture.
b. Determine P and $K_A = y_A/x_A$ for the cases where $x_A = 0.221$ and $x_A = 0.634$.

8.13 Determine the composition and pressure of the azeotrope for a mixture that has

$$\frac{G^E}{RT} = -0.215x_Ax_B \quad P_A^{sat} = 81.11\text{ kPa} \quad P_B^{sat} = 70.44\text{ kPa}$$

Assume the vapor phase is an ideal gas.

8.14 A ternary mixture of three substances (A, B, and C) exists as a vapor and a liquid phase in equilibrium at 293.2 K and 74.45 kPa. The mole fractions of A and B in the liquid phase are 0.25 and 0.35, respectively, and in the vapor phase, they are 0.23 and 0.52, respectively. At 293.2 K, the vapor pressures of pure A, B, and C are 89.72, 131.42, and 50.28 kPa, respectively.

a. Calculate the numerical value of G^E/RT for the liquid phase at these conditions based on the Lewis–Randall reference state.
b. Calculate the total pressure and vapor mole fractions that would result if the liquid phase were an ideal solution with the same composition. Compare your results with the actual values.
c. If the mathematical representation of the excess Gibbs energy is $G^E/RT = Ax_ix_jx_k$, derive an expression for the activity coefficient γ_i in terms of A, x_i, x_j, and x_k.

8.15 Consider a binary mixture of compounds A and B. The liquid-phase nonideality is

$$\frac{G^E}{RT} = bx_Ax_B$$

The vapor pressures of the pure substances are 98.79 kPa for substance A and 166.2 kPa for substance B. The pressure in the system is 116.1 kPa when the liquid-phase composition is equimolar ($x_A = x_B$).

a. Calculate the vapor composition that is in equilibrium with the equimolar liquid at 116.1 kPa.
b. Calculate the total pressure and the liquid composition that is in equilibrium with an equimolar phase.

8.16 A mixture containing 12.6 mol% methane, 24.9 mol% ethane, 34.2 mol% propane, and 28.3 mol% n-butane enters a flash drum that operates at 250 K and 9 bar. Use the McWilliams equations to obtain the K-values

a. Determine the fraction of the inlet stream that exits the flash drum as vapor and the compositions of the vapor and liquid streams.
b. Determine the fraction of each species that leaves in the vapor stream.

8.17 Consider a binary mixture of propane (1) and n-pentane (2) at 35 bar and 90 °C. Use the DePriester Nomogram

a. Determine the range of compositions for which two phases (vapor and liquid) exist in equilibrium at these conditions.
b. Determine the vapor, liquid, and overall compositions when 20% of the mixture is in the liquid phase and 80% is in the vapor phase at these conditions.

8.18 A mixture containing 80 mol% ethane, 10 mol% propane, and 10 mol% n-butane has a pressure of 10 bar and a temperature T. If the mole fraction of liquid ethane at equilibrium is 0.525, what is the temperature of the system and the vaporization fraction?

8.19 Two substances, A and B, have the following physical properties:

	P^{sat} (kPa)	V_L^{sat} (cm^3/mol)	B_2 (cm^3/mol)
A	451.2	47.5	−143.6
B	292.4	68.4	−210.5

The interaction second virial coefficient is $B_{2,AB} = -191.4\ \text{cm}^3/\text{mol}$, and the liquid phase follows

$$\frac{G^E}{RT} = -0.9423 x_A x_B$$

For a mixture of A and B with a liquid-phase composition of $x_A = 0.2834$ at 125 °C, determine the pressure of the two-phase mixture and consider the Poynting correction in the fugacity of the liquid. What is the percentage deviation in the pressure using the modified Raoult's law?

8.20 Methane (1) at a pressure of 184 bar is in contact with water (2) at 25 °C in a gas reservoir. Calculate the mole fraction of methane present in the water if Henry's constant for methane in water is 4184 bar at 25 °C. Assume Raoult's law.

8.21 Substance A is dissolved in a liquid substance B. The equilibrium composition of A is 0.035, and the Henry constant is 4500 bar. The vapor pressure of component B is 0.025 bar. Calculate the equilibrium pressure if the nonideality in the liquid is

$$\frac{G^E}{RT} = -0.575x_A x_B$$

Assume ideal solution in the vapor phase.

8.22 Calculate the bubble temperature for a binary system consisting of A and B with equimolar composition in which the vapor pressures are

$$\ln P_A^{sat} = 10 - 2600/T, \quad \ln P_B^{sat} = 12 - 3000/T$$

The temperature is in kelvin and the pressure is in bar. The activity coefficient is

$$\ln \gamma_A = Ax_B^2, \quad \ln \gamma_B = Ax_A^2, \quad A = 2.7 - 0.006T$$

The virial coefficients are

$$B_{2,AA} = -200\ \text{cm}^3/\text{mol}$$

$$B_{2,BB} = -500\ \text{cm}^3/\text{mol}$$

$$B_{2,AB} = -275\ \text{cm}^3/\text{mol}$$

8.23 Calculate the vapor dew point pressure for a binary mixture at 350 K that contains 30 mol% of specie A.

Data

$$\ln P_A^{sat}\ (\text{kPa}) = 16 - \frac{2900}{T\ (\text{K}) - 40}$$

$$\ln P_B^{sat}\ (\text{kPa}) = 15 - \frac{3000}{T\ (\text{K}) - 60}$$

Virial coefficients:

$$B_{2,AA} = -600\ \text{cm}^3/\text{mol}$$

$$B_{2,BB} = -1200\ \text{cm}^3/\text{mol}$$

$$B_{2,AB} = -1100\ \text{cm}^3/\text{mol}$$

The nonideality in the liquid is

$$\ln \gamma_A = 0.529x_B^2 \quad \ln \gamma_B = 0.529x_A^2$$

8.24 A binary mixture exhibits liquid–liquid equilibrium at 300 K. The excess energy is

$$\frac{G^E}{RT} = Ax_1 x_2$$

but the constant A is not the same for both phases. If the equilibrium compositions at 300 K are $x_1^\alpha = 0.4$ and $x_1^\beta = 0.2$. Find the parameter that applies to each phase.

8.25 Calculate LLE equilibrium compositions for a mixture in which the excess Gibbs energy is

$$\frac{G^E}{RTx_1x_2} = A_{21}x_1 + A_{12}x_2 \text{ with } A_{12} = 1 \text{ and } A_{21} = 3$$

Use the equal area rule to obtain the liquid compositions.

8.26 Water (1) and normal heptane (2) (also known as n-C_7) are immiscible in the liquid state. If the pure fluid vapor pressures are

$$\ln P^{sat}_{H_2O}(\text{kPa}) = 16.28861 - 3816.44/T\,(\text{K}) - 46.13$$

and

$$\ln P^{sat}_{n\text{-}C_7}(\text{kPa}) = 13.85871 - 2911.32/T(\text{K}) - 56.51$$

a. Determine the dew and bubble pressures at 425 K for an equimolar mixture of water and normal heptane. What are the compositions of the first liquid drop and of the first vapor bubble?
b. Determine the maximum dew pressure possible at this temperature for binary mixtures of normal heptane and water and the composition at the maximum.

8.27 Estimate the composition of naphthalene in ethylene at 40 bar and 318 K. The second virial of ethylene at 318 K is $-123\ \text{cm}^3/\text{mol}$, and the interaction second virial coefficient virial is $-605\ \text{cm}^3/\text{mol}$. Consider that a truncated second virial equation is valid at this pressure, and the Tsonopolous correlation can estimate the second virial coefficient of naphthalene. The vapor pressure of naphthalene at this temperature is 7.14×10^{-4} bar.
Data

$$T_{C,Naphthalene} = 751.06\ \text{K}, P_{C,Naphthalene} = 41.14\ \text{bar}, \text{ and } \omega_{Naphthalene} = 0.306$$

References

1 Hall, K.R. and Eubank, P.T. (1995). Equal area rule and algorithm for determining phase compositions. *AIChE J.* 41: 924–927.
2 DePriester, C.L. (1953). Light-hydrocarbon vapor liquid distribution coefficients. Pressure–temperature-composition charts and pressure–temperature nomographs. In: *Chemical Engineering Progress Symposium Series*, vol. 49, 1–43. American Institute of Chemical Engineers, New York, United States.
3 Wilson, G.A. (1968). A modified Redlich–Kwong EOS. Application physical data calculation. Paper 15C presented at the 65th AICHE National Meeting, Cleveland, Ohio (May, 1968).

4 McWilliams, M.L. (1973). An equation to relate *K*-factors to pressure and temperature. *Chem. Eng.* 80: 138–140.

5 Parrish, W.R. and Hiza, M.J. (1974). Liquid–vapor equilibria in the nitrogen-methane system between 95 and 120 K. *Adv. Cryog. Eng.* 19: 300.

6 Baker, L.E., Pierce, A.C., and Luks, K.D. (1981). Gibbs energy analysis of phase equilibria. SPE/DOE Second Joint Symposium on Enhanced Oil Recovery, Tulsa, Oklahoma, United States.

7 Michelsen, M.L. (1982). The isothermal flash problem. Part I. Stability. *Fluid Phase Equilib.* 9: 1–19.

8 Iglesias-Silva, G.A., Bonilla-Petriciolet, A., Eubank, P.T. et al. (2003). An algebraic method that includes Gibbs minimization for performing phase equilibria calculations for any number of components or phases. *Fluid Phase Equilib.* 210: 229–245.

9 Price, A. R. (1957). PhD thesis. Low Temperature Vapor–Liquid Equilibrium in Light Hydrocarbon Mixtures: Methane - Ethane - Propane Systems. Rice University, Houston.

10 Esper, G.J., Bailey, D.M., Holste, J.C., and Hall, K.R. (1989). Volumetric behavior of near-equimolar mixtures for $CO_2 + CH_4$ and $CO_2 + N_2$. *Fluid Phase Equilib.* 49: 35–47.

11 Aoki, Y. and Moriyoshi, M. (1978). Mutual solubility of *n*-butanol + water under high pressures. *J. Chem. Thermodyn.* 10: 1173–1179.

12 Dobbs, J.M., Wong, J.M., Lahiere, R.J., and Johnston, K.P. (1987). Modification of supercritical fluid phase behavior using polar cosolvents. *Ind. Eng. Chem. Res.* 26: 56–65.

13 Yokoyama, C., Ebina, T., and Takahashi, S. (1993). Solid–liquid equilibria of six binary mixtures containing indole. *J. Chem. Eng. Data* 38 (4): 583–586.

9

Chemical Reaction Equilibria

9.1 Introduction

The chemical industry has many processes that transform a raw material into an intermediate or a final product. Many of these products finally result in materials that make our lives more comfortable. For example, we can produce ethylene via a chemical reaction and then use it to produce polyethylene, gasoline, etc. Most reactions are not equilibrium events, so why do we need to study chemical reaction equilibria? Well, it can provide information about maximum conversion and help with decisions as to whether the process is feasible or not. Also, it can provide information about operating conditions. For example, if the reaction has an equilibrium conversion of 70% at a given temperature and pressure (T_{eq} and P_{eq}), then we can use this conversion value as our target in an industrial process. On the other hand, we can also change T_{eq} and P_{eq} and add a catalyst to improve the yield in the equilibrium conversion. If the equilibrium conversion is too low, the chances of having an economically feasible process decline substantially.

Time is not a variable in chemical equilibrium; therefore, the result of the conversion after a certain time is unknown. Also, the equilibrium condition represents a single value in the time domain. The behavior of the reaction with time results from the chemical kinetics of the reaction. A chemical engineer also studies reaction engineering, in which the design of the reactor (size, type, residence time, etc.) is the main goal.

In a chemical reaction, individual reactants and products may be present as gas, liquid, or solid. Often different states are present in a reaction at equilibrium, such as

$$H_{2(g)} + \frac{1}{2}O_{2(g)} \leftrightarrow H_2O_{(l)} \tag{9.1}$$

Here, gaseous hydrogen reacts with gaseous oxygen. At equilibrium, liquid water coexists with a gaseous mixture of hydrogen and oxygen. The fundamental equations covering any reaction and the ensuing chemical equilibrium are within the domain of thermodynamics as are some applications.

Thermodynamics for Chemical Engineers, First Edition. Kenneth R. Hall and Gustavo A. Iglesias-Silva.

9.2 Nature of Reactions

A chemical reaction, such as Eq. (9.1), may start with only the reactants present – either in stoichiometric ratio (e.g. 2 mol of H_2 for every mole of O_2) or not. The reaction proceeds over time along some thermodynamic path. The reaction may be isothermal, adiabatic, isobaric, or isochoric and occur in either a closed system or a flow system. Eventually, equilibrium may occur, as in an insulated closed system. A uniform temperature, T_{eq}, and a uniform pressure, P_{eq}, exist throughout an equilibrium system. Also, the chemical potential of each component, $\mu_{i,eq}$, is uniform throughout the system, even though multiple phases may coexist.

A nonzero amount of each component, whether a reactant or a product, must exist uniformly in each phase (α, β, …, π) satisfying

$$\mu_{i,eq}^{\alpha} = \mu_{i,eq}^{\beta} = \dots = \mu_{i,eq}^{\pi}$$

If none of component i is present at equilibrium, then the value of its chemical potential would be $-\infty$. Of course, very little of a given product or reactant may be present at equilibrium. For example, the reaction given by Eq. (9.1) performed at room temperature and pressure with stoichiometric amounts of hydrogen and oxygen proceeds nearly to 100 mass percent water so that only minute percentages of hydrogen and oxygen remain at chemical equilibrium. From the knowledge of T_{eq} and P_{eq}, and the original amounts of the reactants (and products, if present), thermodynamics allows calculation of equilibrium amounts of each component in each phase.

As usual, thermodynamics fails to give information concerning time, i.e. how long it takes to reach equilibrium, and can only predict equilibrium yield, temperature, and pressure. Chemical kinetics describes reaction rates that together with diffusion rates of mass transfer control the overall speed of the reaction. Extremely fast reactions, often idealized as instantaneous, are diffusion controlled. Many industrial reactions, however, are sufficiently slow that equilibrium does not occur (nor even closely approach) at the end of a flow reactor or completion of a batch reaction. Increasing the flow reactor length or decreasing the velocity of flow may not be economically feasible, nor may a long reaction period for a batch reaction in a closed system. Thermodynamics can only provide valuable boundary conditions for reactor design, e.g. the maximum obtainable yield for an infinite reaction time.

9.3 Chemical Reaction Stoichiometry

Chemical reactions conserve atoms, providing that the reactions are not nuclear. That is, *mass is conserved as energy, but the total number of moles is not*. For example, look at the ammonia synthesis reaction,

$$3H_{2(g)} + N_{2(g)} \leftrightarrow 2NH_{3(g)} \tag{9.2}$$

This reaction produces 1 mol of ammonia for every 2 mol of the reactants, hydrogen and nitrogen. If the conversion is 100%, then initially 4 mol exist, while at the end of the reaction, only 2 mol remain. If M_i denotes the chemical symbols of the species involved in the reaction (reactants and products) and ν_i their corresponding coefficients, then

$$|\nu_1|M_1 + |\nu_2|M_2 \leftrightarrow |\nu_3|M_3 \tag{9.3}$$

Also, the above reaction as an equation is

$$0 = \nu_1 M_1 + \nu_2 M_2 + \nu_2 M_2 \tag{9.4}$$

in which $\nu_1 = -3$, $\nu_2 = -1$, and $\nu_3 = 2$ are the stoichiometric coefficients of the reaction (9.2), and $M_1 = H_2$, $M_2 = N_2$, and $M_3 = NH_3$. Notice that moving the reactants to the RHS of the equation causes the stoichiometric coefficients for the *reactants to be negative*, whereas those from the *products remain positive.* In general, for any single reaction, Eq. (9.4) is

$$\sum_{i=1}^{N} \nu_i M_i = 0 \tag{9.5}$$

in which N is the total number of components (both reactants and products).

9.4 Extent of Reaction

Suppose that $n_{H_2,0}$ moles of hydrogen, $n_{N_2,0}$ moles of nitrogen, and $n_{NH_3,0}$ moles of ammonia are present at the start of the ammonia synthesis reaction. After any given reaction time, the moles present become

$$n_{H_2} = n_{H_2,0} + \Delta n_{H_2} \tag{9.6}$$

$$n_{N_2} = n_{N_2,0} + \Delta n_{N_2} \tag{9.7}$$

$$n_{NH_3} = n_{NH_3,0} + \Delta n_{NH_3} \tag{9.8}$$

in which Δn_i is the number of moles consumed or produced and may be either positive or negative depending on whether the reaction has proceeded forward or backward. Inspection of Eq. (9.2) reveals that $\Delta n_{H_2} = 3\Delta n_{N_2}$ or 3 mol of hydrogen are consumed or produced for every mole of nitrogen consumed or produced. Likewise, $\Delta n_{H_2} = -(3/2)\Delta n_{NH_3}$. In general,

$$\frac{\Delta n_1}{\nu_1} = \frac{\Delta n_2}{\nu_2} = \frac{\Delta n_3}{\nu_3} = \cdots = \frac{\Delta n_i}{\nu_i} = \xi \tag{9.9}$$

with

$$\frac{\Delta n_i}{\nu_i} = \frac{n_i - n_{i,0}}{\nu_i} = \xi \text{ for any component } i \tag{9.10}$$

ξ is the *extent of reaction*. It is independent of component i, but it depends upon time.

The extent of the reaction is positive for a forward reaction and negative for a backward reaction, and it depends on the *size* of the reaction. For example, for the reaction of Eq. (9.1), if 1 mol of hydrogen and 0.5 mol of oxygen react to produce 1 mol of water, then the extent of the reaction equals unity, $\xi = 1$, using Eq. (9.10). However, the complete conversion of 2 mol of hydrogen and 1 mol of oxygen yields $\xi = 2$. If the reaction is

$$2H_{2(g)} + O_{2(g)} \leftrightarrow 2H_2O_{(l)} \tag{9.11}$$

starting with 1 mol of hydrogen and 1.5 mol of oxygen, then $\xi = 1/2$ for complete conversion. Equation (9.10) in differential form is

$$dn_i = \nu_i d\xi \tag{9.12}$$

Equations (9.6)–(9.8) written in terms of the extent of the reaction using Eq. (9.10) or (9.12) are

$$n_i = n_{i,0} + \nu_i \xi \tag{9.13}$$

and

$$n_T = \sum_{i=1}^{N} n_i = \sum_{i=1}^{N} n_{i,0} + \xi \sum_{i=1}^{N} \nu_i = n_{T,0} + \xi \sum_{i=1}^{N} \nu_i \tag{9.14}$$

where n_T is the total number of moles. The mole fraction from the above equations is

$$z_i = \frac{n_i}{n_T} = [n_{i,0} + \nu_i \xi] \Big/ \left[n_{T,0} + \xi \sum_{i=1}^{N} \nu_i \right] \tag{9.15}$$

in which $z_i = y_i$ or $z_i = x_i$ if the component i is vapor or liquid, respectively.

Example 9.1

If the initial moles are 2, 3, and 1 of A, B, and C, respectively, what are the equilibrium compositions in terms of the extent of the reaction for the following reaction?

$$A_{(l)} + 3B_{(l)} \leftrightarrow 2C_{(l)}$$

Solution

$$n_A = n_{A,0} - \xi = 2 - \xi$$

$$n_B = n_{B,0} - 3\xi = 3 - 3\xi$$

$$n_C = n_{C,0} + 2\xi = 1 + 2\xi$$

The total number of moles is $n_T = 6 - 2\xi$, and the compositions are

$$y_A = \frac{2 - \xi}{6 - 2\xi}, y_B = \frac{3 - 3\xi}{6 - 2\xi}, \text{and } y_C = \frac{1 + 2\xi}{6 - 2\xi}$$

9.5 Phase Rule for Reacting Systems

The phase rule,

$$F = N - P + 2 \tag{9.16}$$

requires adjustment for reacting systems. R independent reactions exist at equilibrium, each of the form $\sum_{i=1}^{N} \nu_i M_i = 0$, which reduces the number of independent state variables by 1, so the amended phase rule becomes

$$F = (N - R) - P + 2 \tag{9.17}$$

Think of $(N - R)$ as the number of independent components because this is the minimum number of components required in the laboratory to produce all N species via the R independent reactions.

Sometimes, finite, inherent stoichiometric (composition) constraints diminish the number of composition variables by an equal number in the system. The existence of these constraints is not always readily apparent, but Eq. (9.17) must account for them. For example, a vapor–liquid azeotrope contains a special constraint: the compositions in both phases are equal, $y = x$. If we designate this constraint as S, then

$$F = (N - R) - P + 2 - S \tag{9.18}$$

Finding the number of independent reactions from a system with a particular number of species requires the following:

1. Write the formation reaction for each of the chemical species
2. Combine these equations to eliminate all the elements not present in the system from the set of equations

The resulting equations are a set of R independent reactions.

Example 9.2

Determine the number of degrees of freedom for a system formed by $CH_3OH_{(g)}$, $CH_{4(g)}$, $CO_{(g)}$ and $H_{2(g)}$, and $H_2O_{(g)}$.

Solution

First, write the reaction of formation for all the components

$$C + \tfrac{1}{2}O_2 + 2H_2 = CH_3OH \tag{A}$$

$$C + 2H_2 = CH_4 \tag{B}$$

$$H_2 + \tfrac{1}{2}O_2 = H_2O \tag{C}$$

$$C + \tfrac{1}{2}O_2 = CO \tag{D}$$

Because oxygen and carbon are not in the system, we eliminate them. Substituting Eq. (D) into Eqs. (A) and (B) eliminates carbon

$$CO + 2H_2 = CH_3OH$$

$$CO - \tfrac{1}{2}O_2 + 2H_2 = CH_4 \tag{E}$$

Now, substitute Eq. (C) into Eq. (E) to eliminate oxygen

$$CO - H_2O + 3H_2 = CH_4$$

Two independent equations remain

$$CO + 2H_2 = CH_3OH$$

$$CO + 3H_2 = CH_4 + H_2O$$

So, the number of degrees of freedom (with $N = 5, R = 2, P = 1, S = 0$) is

$$F = 5 - 2 - 1 + 2 - 0 = 4$$

Thus, one can specify, for example, T, P and two mole fractions.

9.6 Principles of Reaction Equilibria

Besides the conservation of mass and energy, the remaining principle used to solve reaction equilibria problems is that the total Gibbs function of the reacting system is a minimum at equilibrium. That is, the reaction proceeds spontaneously in a direction (forward or backward) to decrease the total Gibbs energy of the system. This concept, one of the most basic in thermodynamics, also appeared in the Chapter 8. Indeed, reaction equilibria involving more than one phase requires utilizing all the available knowledge of phase equilibria because the phases and the reaction are in equilibrium.

Consider the ammonia synthesis reaction, Eq. (9.2). The total Gibbs energy of the system is

$$nG = n_1\mu_1 + n_2\mu_2 + n_3\mu_3 \tag{9.19}$$

or in general

$$nG = \sum_{i=1}^{N} n_i\mu_i \tag{9.20}$$

Because $\mu_i = \mu_i^0 + RT\ln\left(\hat{f}_i/\hat{f}_i^0\right)$ Eq. (9.20) becomes

$$nG = \sum_{i=1}^{N} n_i\mu_i^0 + RT\ln\left[\prod_{i=1}^{N}\left(\hat{f}_i/\hat{f}_i^0\right)^{n_i}\right] \tag{9.21}$$

Here, μ_i^0, the standard state chemical potential, is only a function of T that generally varies as the reaction progresses. The fugacity $\hat{f}$ is that of component i in the mixture, whereas $\hat{f}_i^0$ is the fugacity of component i in the mixture at the same standard state as μ_i^0. Because the standard state is a pure species, i, Eq. (9.21) becomes

$$nG = \sum_{i=1}^{N} n_i\mu_i^0 + RT\ln\left[\prod_{i=1}^{N}\left(\hat{f}_i/f_i^0\right)^{n_i}\right]$$

The fundamental equation for $d(nG)$

$$dnG = -nSdT + nVdP + \sum_{i=1}^{N}\mu_i dn_i$$

combined with Eq. (9.12) yields

$$dnG = -nSdT + nVdP + \left(\sum_{i=1}^{N}\mu_i\nu_i\right)d\xi \tag{9.22}$$

At a constant temperature and pressure

$$\mathcal{A} \equiv \left(\frac{\partial nG}{\partial \xi}\right)_{T,P} = \sum_{i=1}^{N}\nu_i\mu_i \tag{9.23}$$

in which $\mathcal{A}$ is the affinity of the reaction. When the derivative is negative, the reaction proceeds forward as indicated by Eq. (9.10) with ν_i positive for the products and negative for the reactants. The reaction follows a path that causes nG to decrease with time attaining equilibrium only when nG is a minimum with respect to each of the independent variables. Thus,

$$\sum_{i=1}^{N} \nu_i \mu_i = 0 \tag{9.24}$$

at equilibrium regardless of the path of the reaction process, i.e. the process need not be isothermal, $T = T_{eq}$, nor isobaric $P = P_{eq}$. The above equation describes a single reaction, but it is valid regardless of the number of phases present at equilibrium. In terms of fugacities,

$$\sum_{i=1}^{N} \nu_i \mu_i^0 = -RT \ln \left[\prod_{i=1}^{N} \left(\hat{f}_i / f_i^0 \right)^{\nu_i} \right] \tag{9.25}$$

This equation is the *fundamental equation of reaction equilibria*. The quantity on the LHS is ΔG^0, the standard change of the Gibbs energy

$$\Delta G^0 = \sum_{i=1}^{N} \nu_i \mu_i^0 = -RT \ln \left[\prod_{i=1}^{N} \left(\hat{f}_i / f_i^0 \right)^{\nu_i} \right] \tag{9.26}$$

or

$$\Delta G^0 = \sum_{i=1}^{N} \nu_i \mu_i^0 = -RT \ln \left[\prod_{i=1}^{N} a_i^{\nu_i} \right] \tag{9.27}$$

in which a_i is the activity defined as the ratio of the fugacities, $\hat{f}_i / \hat{f}_i^0$.

ΔG^0 is only a function of temperature for a given reaction and is on a per mole basis as seen from the RHS of Eq. (9.26). ΔG^0 is the change in the Gibbs energy of a system when ν_j moles of a given product j forms from a complete reaction of reactants in exact stoichiometric ratios to ν_j at the start of the reaction. This hypothetical reaction occurs at a constant temperature, T, the equilibrium temperature, with each reactant as a pure component. Both the reactants and products are unmixed in the initial and final states.

As noted in Section 9.4, the values of the stoichiometric coefficients depend upon the direction of the reaction, but the ratio of stoichiometric coefficients is constant for a given reaction whether written forward or backward. For example, consider the following three statements of the same reaction:

$$3H_{2(g)} + N_{2(g)} \leftrightarrow 2NH_{3(g)} \tag{A}$$

$$\tfrac{3}{2}H_{2(g)} + \tfrac{1}{2}N_{2(g)} \leftrightarrow NH_{3(g)} \tag{B}$$

and

$$NH_{3(g)} \leftrightarrow \tfrac{3}{2}H_{2(g)} + \tfrac{1}{2}N_{2(g)} \tag{C}$$

Reaction (A) is the ammonia synthesis reaction, Eq. (9.2). Reaction (B) has stoichiometric coefficients one half of (A), and reaction (C) is the decomposition of ammonia, the exact reverse of (B). For the three reactions, ν_{NH_3} is 2, 1, and −1, respectively. However, the ratio (ν_{NH_3}/ν_{H_2}) is (−2/3) in each case. ΔG^0 is

−16 450 J/mol for Eq. (B), i.e. joule per mole of NH_3 formed. ΔG^0 is 2(−16 450) J/mol for Eq. (A), i.e. joule per mole of N_2 reacted. ΔG^0 is −16 450 J/mol for Eq. (C), i.e. joule per mole of NH_3 decomposed. Thus, the values of ΔG^0 depend upon the particular form of the reaction equation. Examination of Eq. (9.26) shows that for given values of T_{eq} and P_{eq}, the values of $\hat{f}_i$ remain unchanged, so the effect results solely from the ν_i's, which appear on both sides of Eq. (9.25).

To calculate ΔG^0, or any other property at the standard state, requires an equation similar to the one used for the enthalpy of reaction,

$$\Delta G_f^0 \equiv (1)G_{compound}^0 - \sum_{\substack{constituent\\elements}} n_k G_k^0$$

Here, ΔG_f^0 is the standard Gibbs energy of formation of 1 mol of a compound, $G_{compound}^0$ is the Gibbs energy of the compound, and G_k^0 is the Gibbs energy of a constituent element of the compound in its standard state. Because G_k^0 equals zero at standard conditions

$$\Delta G_f^0 = G_{compound} \tag{9.28}$$

and we can substitute this value into the fundamental equation to obtain

$$\Delta G^0 = \sum_{i=1}^{N} \nu_i \Delta G_{f,i} \tag{9.29}$$

Equation (9.29) is a difference of the product of the standard Gibbs energy of formation and the stoichiometric coefficient of the products less the product of the standard Gibbs energy times the stoichiometric coefficient of the reactants. ΔG^0 is the standard Gibbs energy change of the reaction. Generally, the standard temperature is 25 °C.

Example 9.3

Calculate the standard Gibbs energy change of the reaction for

$$C_3H_{8(g)} + 5O_{2(g)} \leftrightarrow 3CO_{2(g)} + 4H_2O_{(g)}$$

$\Delta G_{f,C_3H_8} = -24\,290$ J/mol	$\nu_{C_3H_8} = -1$
$\Delta G_{f,O_2} = 0$ J/mol	$\nu_{O_2} = -5$
$\Delta G_{f,CO_2} = -394\,359$ J/mol	$\nu_{CO_2} = 3$
$\Delta G_{f,H_2O} = -228\,572$ J/mol	$\nu_{H_2O} = 4$

Solution

Using Eq. (9.29)

$$\Delta G^0 = \sum_{i=1}^{N} \nu_i \Delta G_{f,i} = \nu_{CO_2}\Delta G_{f,CO_2} + \nu_{H_2O}\Delta G_{f,H_2O} + \nu_{C_3H_8}\Delta G_{f,C_3H_8} + \nu_{O_2}\Delta G_{f,O_2}$$

$$\Delta G^0 = 3 \cdot (-394\,359) + 4 \cdot (-228\,572) - 1 \cdot (-24\,290) - 5 \cdot 0 = -2\,073\,075 \text{ J/mol}$$

9.7 Understanding the Reaction Equilibria

This section provides the reader with a better physical understanding of the basic principles and equations from Section 9.6. Consider the reaction

$$3A + B \leftrightarrow 2C + 2D$$

occurring in a single gas phase, in which all mixtures of the four components are perfect gas mixtures (PGM). Starting with 3 mol of A and 1 mol of B, unmixed, each at a temperature T and a pressure P, mix these perfect gases at a constant T and P:

$$\Delta(nG_{mix}) = \sum_{i=1}^{2} n_i \left(\overline{G}_i - \overline{G}_i^{ig}\right) = \sum_{i=1}^{2} n_i \left(\mu_i - \mu_i^{ig}\right) \tag{9.30}$$

$$= RT \sum n_i \ln y_i$$

$$= -RT[3\ln 3 - 4\ln 4] = nG_m - nG_I \tag{9.31}$$

where nG_I is the initial state of the unmixed gases

$$nG_I = 3\mu_A^{ig} + \mu_B^{ig} \tag{9.32}$$

and nG_m is the mixed state at T and P immediately before the start of the reaction, which is isothermal and, hence, isobaric (moles are observed in this case and we assume PGM).

Along the reaction path, from Eq. (9.10), the material balance is

$$\xi = \frac{3 - n_A}{3} = 1 - n_B = \frac{n_C}{2} = \frac{n_D}{2} \tag{9.33}$$

and

$$nG = \sum_{i=1}^{4} n_i \mu_i \tag{9.34}$$

$$nG = \sum_{i=1}^{4} n_i \left[\mu_i^0 + RT \ln\left(\frac{Py_i}{P^0}\right)\right] \tag{9.35}$$

for a PGM in which μ_i^0 is the chemical potential of pure i in its standard state as a perfect gas at P^0. If $P = P^0$, then $\mu_i^{ig} = \mu_i^0$; otherwise,

$$\mu_i^{ig} - \mu_i^0 = RT \ln\left(\frac{P}{P^0}\right) \tag{9.36}$$

and Eq. (9.35) becomes

$$nG = \sum_{i=1}^{4} n_i \mu_i^0 + RT \left[n_T \ln\left(\frac{P}{P^0}\right) + \sum n_i \ln\left(\frac{n_i}{n_T}\right)\right] \tag{9.37}$$

in which n_T is the total of moles present (here, $n_T = 4$). Then,

$$nG = 3(1-\xi)\mu_A^0 + (1-\xi)\mu_B^0 + 2\xi\left(\mu_C^0 + \mu_D^0\right)$$
$$+ RT\left[4\ln\frac{P}{P^0} + 3(1-\xi)\ln\frac{3(1-\xi)}{4} + (1-\xi)\ln\frac{(1-\xi)}{4} + 4\xi\ln\frac{\xi}{2}\right] \tag{9.38}$$

Knowing that compounds A, B, C, and D, μ_A^0, μ_B^0, μ_C^0, and μ_D^0 appear from tables (such as the Thermodynamics Research Center (TRC) tables) for the given temperature. As the reaction proceeds at a constant T and P, the only independent variable on the RHS of Eq. (9.38) is ξ. To find the minimum value of nG corresponding to the equilibrium extent of reaction, ξ_e, differentiate Eq. (9.38)

$$\left(\frac{\partial nG}{\partial \xi}\right)_{T,P} = -3\mu_A^0 - \mu_B^0 + 2\left(\mu_C^0 + \mu_D^0\right) + RT\left[4\ln\frac{2\xi}{1-\xi} - 3\ln 3\right]$$

$$= \sum_{i=1}^{4} \nu_i\mu_i^0 + RT\left[4\ln\frac{2\xi}{1-\xi} - 3\ln 3\right] \tag{9.39}$$

At equilibrium, $\left(\frac{\partial nG}{\partial \xi}\right)_{T,P} = 0$, or

$$\sum_{i=1}^{4} \nu_i\mu_i^0 = -RT\left[4\ln\frac{2\xi}{1-\xi} - 3\ln 3\right] \tag{9.40}$$

Note the analogy between these specific equations and the general equations of Section 9.6 – Eq. (9.39) is analogous to Eq. (9.25) because

$$\prod_{i=1}^{4}\left(\hat{f}_i/f_i^0\right)^{\nu_i} = \prod_{i=1}^{4}(p_i/P^0)^{\nu_i} = \prod_{i=1}^{4} y_i^{\nu_i} = \prod_{i=1}^{4} n_i^{\nu_i}$$

$$= \frac{(2\xi_e)^2(2\xi_e)^2}{[3(1-\xi_e)]^3(1-\xi_e)} = \left[\frac{2\xi_e}{1-\xi_e}\right]^4 /27$$

and thus

$$\ln\left[\prod_{i=1}^{4}\left(\hat{f}_i/f_i^0\right)^{\nu_i}\right] = \left[4\ln\frac{2\xi}{1-\xi} - 3\ln 3\right]$$

Calculating $nG(\xi)$ from Eq. (9.38) produces a graph similar to Figure 9.1. nG decreases from point I to M because of mixing of the reactants. As the reaction proceeds, nG further decreases to the equilibrium point, e, because of the additional change in Gibbs energy of mixing caused by the appearance of the products, C and D, in the reaction mixture. Roughly, each new component brought into a mixture via chemical reaction causes a further decrease in G because

$$G = H - TS$$

and

$$\Delta_m G = \Delta_m H - T\Delta_m S \cong -T\Delta_m S$$

Thus, the positive entropy change upon mixing (or increased randomness) is the reason why complete conversions are never possible. Some part of each component is always present at the equilibrium; point e cannot exactly coincide with either M or N. The latter point corresponds to complete conversion, $\xi = 1$, for our reaction, but it could be the starting point for the reaction

$$2C + 2D \rightarrow 3A + B$$

which would proceed from point N to e after mixing pure C and pure D at point Q.

Finally, let us examine the terms $\sum_{i=1}^{4}\nu_i\mu_i$, $\sum_{i=1}^{4}\nu_i\mu_i^{ig}$, and $\sum_{i=1}^{4}\nu_i\mu_i^0$. The first sum is zero at equilibrium, Eq. (9.24). The second term is the total change in nG for

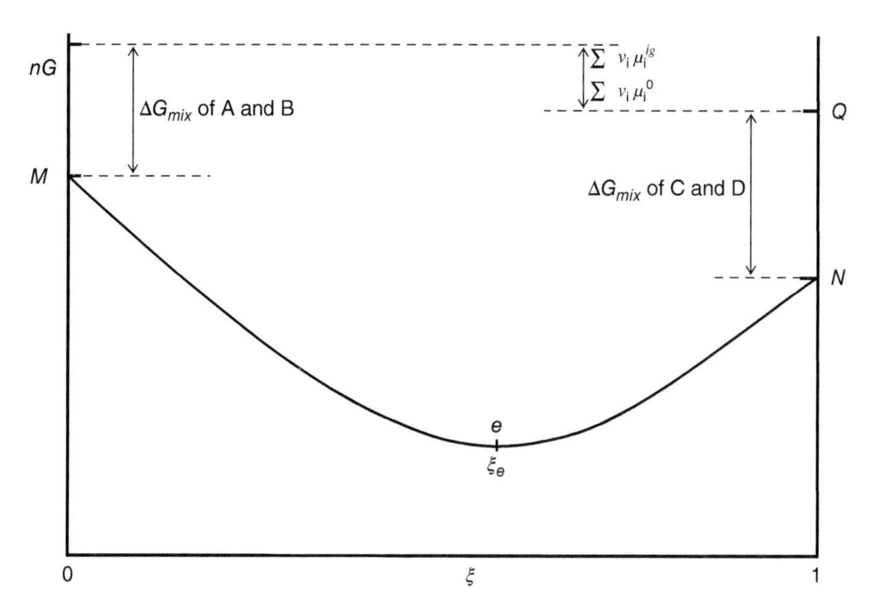

Figure 9.1 Gibbs energy vs the extent of the reaction.

complete conversion of the pure reactants at T_I and P_I to de-mixed products at T_Q and P_Q in Figure 9.1 or

$$nG_Q - nG_I = \sum_{i=1}^{4} \nu_i \mu_i^{ig} \tag{9.41}$$

De-mixing of the products at N yields

$$nG_Q - nG_N = -RT\sum_{i=1}^{2} n_i \ln y_i = 4RT \ln 2 \tag{9.42}$$

in which $nG_Q = 2\mu_C^{ig} + 2\mu_D^{ig}$. The third sum, which appears in the fundamental Eq. (9.25), resembles the second sum

$$\sum \nu_i \mu_i^{ig} = \sum \nu_i \mu_i^0 + RT \ln \prod \left(\frac{\hat{f}_i^{ig}}{f_i^0} \right)^{\nu_i}$$

If each pure gas, i, is perfect, then

$$\sum \nu_i \left(\mu_i^{ig} - \mu_i^0 \right) = RT \ln \prod \left(\frac{P}{P^0} \right)^{\nu_i} = RT \ln \left(\frac{P}{P^0} \right)^{\sum \nu_i}$$

In the reaction, $\sum \nu_i = 0$, so $\sum \nu_i \mu_i^{ig} = \sum \nu_i \mu_i^0$, although $\mu_i^{ig} = \mu_i^0$, only when $P = P^0$.

9.8 Equilibrium Constant

The fundamental equation of reaction equilibria, Eq. (9.25), could also be

$$\Delta G^0 = \sum_{i=1}^{N} \nu_i \mu_i^0 = -RT \ln K_f \tag{9.43}$$

in which

$$K_f \equiv \prod_{i=1}^{N} \left(\hat{f}_i / f_i^0\right)^{\nu_i} \tag{9.44}$$

K_f is the equilibrium constant. In terms of activities, the equilibrium constant is

$$K_f \equiv \prod_{i=1}^{N} a_i^{\nu_i} \tag{9.45}$$

Now, we can calculate the equilibrium constant for reactions in different phases. The trick is to use the correct form of the fugacity coefficient of component i in the mixture and the correct standard state. The general form of the fugacity coefficient is

$$\hat{f}_i = P\hat{\varphi}_i z_i = p_i \hat{\varphi}_i = P\gamma_i \varphi_i z_i = \gamma_i \varphi_i p_i \tag{9.46}$$

Here, we must include the definition of activity coefficient in terms of fugacity coefficients, $\gamma_i = \hat{\varphi}_i / \varphi_i$, and z_i represents the composition, $z_i = y_i$ for a vapor phase and $z_i = x_i$ for a liquid phase. If we include our expression into the definition of equilibrium constant

$$K_f = K_{\hat{\varphi}} K_z \prod_{i=1}^{N} (P/f_i^0)^{\nu_i} = K_\gamma K_\varphi K_z \prod_{i=1}^{N} (P/f_i^0)^{\nu_i} \tag{9.47}$$

in which

$$K_z \equiv \prod_{i=1}^{N} z_i^{\nu_i} \tag{9.48}$$

$$K_\gamma \equiv \prod_{i=1}^{N} \gamma_i^{\nu_i} \tag{9.49}$$

$$K_\varphi \equiv \prod_{i=1}^{N} \varphi_i^{\nu_i} \tag{9.50}$$

and

$$K_{\hat{\varphi}} \equiv \prod_{i=1}^{N} \hat{\varphi}_i^{\nu_i} \tag{9.51}$$

Using Eqs. (9.43) and (9.47)

$$\sum_{i=1}^{N} \nu_i \mu_i^0 = -RT \ln \left\{ K_{\hat{\varphi}} K_z \prod_{i=1}^{N} (P/f_i^0)^{\nu_i} \right\} = -RT \ln \left\{ K_\gamma K_\varphi K_z \prod_{i=1}^{N} (P/f_i^0)^{\nu_i} \right\} \tag{9.52}$$

While K_γ, K_φ, K_z, and $K_{\hat{\varphi}}$ are all equilibrium constants, for activity coefficients, fugacity coefficients, composition, and fugacity coefficients in the mixture, none are as basic as K_f, as ΔG^0 can only vary with T and the form of the reaction equation.

Example 9.4

Find the value of the equilibrium constant at 25 °C for the following reaction:

$$C_3H_{8(g)} + 5O_{2(g)} \leftrightarrow 3CO_{2(g)} + 4H_2O_{(g)}$$

Solution

We calculated the standard Gibbs energy change of the reaction at 25 °C in Section 9.6. Now, using Eq. (9.43) and the result from example 9.3,

$$-RT \ln K_f = -2\,073\,075$$

$$K_f = \exp\left[\frac{2\,073\,075}{RT}\right] = \exp\left[\frac{2\,073\,075}{8.314 \cdot 298.15}\right] = \exp[836.32] = 1.619 \times 10^{363}$$

9.9 Temperature Dependence of the Equilibrium Constant

Because K_f is independent of pressure, the temperature dependence results from the thermodynamic relationship between ΔG^0 and the enthalpy and entropy. The thermodynamic relationships of the Gibbs energy of i at the standard state with these two properties considering that they are independent of pressure are

$$G_i^0 = H_i^0 - TS_i^0 \tag{9.53}$$

$$dH_i^0 = C_{P,i}^0 dT \tag{9.54}$$

$$dS_i^0 = C_{P,i}^0 \frac{dT}{T} \tag{9.55}$$

and

$$H_i^0 = -RT^2 \frac{d\left(G_i^0/RT\right)}{dT} \tag{9.56}$$

Multiplying both sides by the corresponding stoichiometric coefficient, ν_i, and summing over all species

$$\Delta G^0 = \Delta H^0 - T\Delta S^0 \tag{9.57}$$

$$d\Delta H^0 = \Delta C_P^0 dT \tag{9.58}$$

$$d\Delta S^0 = \Delta C_P^0 \frac{dT}{T} \tag{9.59}$$

and

$$\Delta H^0 = -RT^2 \frac{d(\Delta G^0/RT)}{dT} \tag{9.60}$$

in which ΔH^0, ΔS^0, and ΔC_P^0 are the standard enthalpy change of the reaction (standard heat of the reaction), standard entropy change of the reaction, and standard heat capacity change of the reaction with

$$\Delta H^0 = \sum_{i=1}^{N} \nu_i H_i^0 \tag{9.61}$$

$$\Delta S^0 = \sum_{i=1}^{N} \nu_i S_i^0 \tag{9.62}$$

and

$$\Delta C_P^0 = \sum_{i=1}^{N} \nu_i C_{P,i}^0 \tag{9.63}$$

Equation (9.60) expressed in terms of the equilibrium constant is

$$\frac{d\ln K_f}{dT} = \frac{\Delta H^0}{RT^2} \tag{9.64}$$

This equation is the *van't Hoff equation*, and it relates the effect of T on the equilibrium constant. According to the *van't Hoff equation*, the sign of the slope of ln K_f with respect to temperature depends upon the sign of the heat of reaction. If the reaction is endothermic, ΔH^0 is positive, the slope is positive and ln K_f increases as the temperature increases, and therefore, the equilibrium constant increases. On the other hand, if the reaction is exothermic, ΔH^0 is negative, and the equilibrium constant decreases as the temperature increases. We can also write Eq. (9.64) as

$$\frac{d\ln K_f}{d(1/T)} = -\frac{\Delta H^0}{R} \tag{9.65}$$

If we consider ΔH^0 a constant, we can integrate over two arbitrary temperatures obtaining

$$\ln\frac{K_f}{K_1} = -\frac{\Delta H^0}{R}\left(\frac{1}{T} - \frac{1}{T_1}\right) \tag{9.66}$$

This equation is useful for extrapolation purposes over short intervals of temperature, and it implies that a plot of the ln K_f vs $1/T$ is a straight line.

Now, we can obtain the temperature dependence of ln K_f using Eqs. (9.57)–(9.59). First, integrate Eqs. (9.58) and (9.59) from T_0 to T,

$$\Delta H_T^0 = \Delta H_{T_0}^0 + \int_{T_0}^{T} \Delta C_P^0 dT \tag{9.67}$$

and

$$\Delta S_T^0 = \Delta S_{T_0}^0 + \int_{T_0}^{T} \Delta C_P^0 \frac{dT}{T} \tag{9.68}$$

in which $\Delta H_{T_0}^0$ and $\Delta S_{T_0}^0$ are the standard enthalpy change of the reaction and standard entropy change of the reaction at the given temperature T_0. Substituting both equations into Eq. (9.57)

$$\Delta G_T^0 = \Delta H_{T_0}^0 + \int_{T_0}^{T} \Delta C_P^0 dT - T\Delta S_{T_0}^0 - T\int_{T_0}^{T} \Delta C_P^0 \frac{dT}{T} \tag{9.69}$$

It is more convenient to introduce the $\Delta G_{T_0}^0$ using Eq. (9.57) evaluated at T_0

$$\Delta S_{T_0}^0 = \frac{\Delta H_{T_0}^0 - \Delta G_{T_0}^0}{T_0} \tag{9.70}$$

and then

$$\Delta G_T^0 = \Delta G_{T_0}^0(T/T_0) + \Delta H_{T_0}^0(1 - T/T_0) + \int_{T_0}^{T} \Delta C_P^0 dT - T\int_{T_0}^{T} \Delta C_P^0 \frac{dT}{T} \tag{9.71}$$

Division by RT yields

$$\frac{\Delta G_T^0}{RT} = -\ln K_f = \frac{\Delta G_{T_0}^0}{RT_0} + \frac{\Delta H_{T_0}^0}{RT}(1 - T/T_0) + \frac{1}{RT}\int_{T_0}^{T} \Delta C_P^0 dT - \frac{1}{R}\int_{T_0}^{T} \Delta C_P^0 \frac{dT}{T} \tag{9.72}$$

If we know the value of an equilibrium constant at T_1, the equilibrium constant at a different temperature becomes

$$-\ln\frac{K_f}{K_{f,1}} = \frac{\Delta H^0_{T_0}}{R}\left(\frac{1}{T}-\frac{1}{T_1}\right)+\frac{1}{RT}\int_{T_0}^{T}\Delta C_P^0 dT-\frac{1}{RT_1}\int_{T_0}^{T_1}\Delta C_P^0 dT$$
$$-\frac{1}{R}\int_{T_0}^{T_1}\Delta C_P^0\frac{dT}{T} \tag{9.73}$$

Notice that the first term resembles the term obtained from the *van't Hoff equation* when the heat of reaction is constant. We usually have expressions for the heat capacity as a function of temperature in the form

$$C^0_{P,i}/R = A_i + B_i T + C_i T^2 + D_i/T^2 \tag{9.74}$$

The integrals are

$$\frac{1}{RT}\int_{T_0}^{T}\Delta C_P^0 dT = \frac{1}{T}\left[\Delta A(T-T_0)+\frac{\Delta B}{2}\left(T^2-{T_0}^2\right)\right.$$
$$\left.+\frac{\Delta C}{3}\left(T^3-{T_0}^3\right)-\Delta D\left(\frac{1}{T}-\frac{1}{T_0}\right)\right] \tag{9.75}$$

and

$$\frac{1}{R}\int_{T_0}^{T}\Delta C_P^0\frac{dT}{T} = \Delta A\ln\left(\frac{T}{T_0}\right)+\Delta B(T-T_0)+\frac{\Delta C}{2}\left(T^2-{T_0}^2\right)$$
$$-\frac{\Delta D}{2}\left(\frac{1}{T^2}-\frac{1}{{T_0}^2}\right) \tag{9.76}$$

with $\Delta A = \sum_{i=1}^{N} v_i A_i$, $\Delta B = \sum_{i=1}^{N} v_i B_i$, $\Delta C = \sum_{i=1}^{N} v_i C_i$, and $\Delta D = \sum_{i=1}^{N} v_i D_i$.

Example 9.5

Find the equilibrium constant at 500 °C for the following reaction:

$$C_3H_{8(g)} + 5O_{2(g)} \leftrightarrow 3CO_{2(g)} + 4H_2O_{(g)}$$

Solution

Use Eq. (9.72) and select the standard temperature as 25 °C = 298.15 K, therefore

$$\frac{\Delta G^0_{T_0}}{RT_0} = \frac{\Delta G^0_{298.15}}{RT_0} = 836.32$$

and

$$\frac{\Delta H^0_{T_0}}{RT} = \frac{\sum_{i=1}^{N} v_i \Delta H_{f,i}}{RT} = \frac{3\cdot(-393\,509)+4\cdot(-241\,818)-1\cdot(-104\,680)-5\cdot 0}{8.314\cdot 773.15}$$
$$= -317.85$$

Now calculate $\Delta C_P^0/R$

Species	ν_i	A_i	$\nu_i \cdot A_i$	B_i	$\nu_i \cdot B_i$	C_i	$\nu_i \cdot C_i$	D_i	$\nu_i \cdot D_i$
CO_2	3	5.457	16.371	1.05×10^{-3}	3.1350×10^{-3}	0	0	-1.16×10^{5}	-3.471×10^{5}
H_2O	4	3.470	13.88	1.45×10^{-3}	5.8000×10^{-3}	0	0	1.21×10^{4}	4.840×10^{4}
C_3H_8	−1	1.213	−1.213	2.88×10^{-2}	-2.8765×10^{-2}	-8.82×10^{-6}	8.824×10^{-6}	0	0
O_2	−5	3.639	−18.195	5.06×10^{-4}	-2.5300×10^{-3}	0	0	-2.27×10^{4}	1.135×10^{5}
Sum			10.843		−0.02236		8.824×10^{-6}		-1.852×10^{5}

The integrals (9.75) and (9.76) are

$$\frac{1}{RT}\int_{T_0}^{T}\Delta C_P^0 dT = \frac{1}{773.15}\left[10.843(773.15-298.15) - \frac{0.02236}{2}(773.15^2-298.15^2)\right.$$
$$+\frac{8.824\times10^{-6}}{3}(773.15^3-298.15^3)$$
$$\left.+1.852\times10^5\left(\frac{1}{773.15}-\frac{1}{298.15}\right)\right] = 0.4670$$

and

$$\frac{1}{R}\int_{T_0}^{T}\Delta C_P^0\frac{dT}{T} = 10.843\ln\left(\frac{773.15}{298.15}\right) - 0.02236(773.15-298.15) + \frac{8.824\times10^{-6}}{2}$$
$$\times(773.15^2-298.15^2) + \frac{1.852\times10^5}{2}\left(\frac{1}{773.15^2}-\frac{1}{298.15^2}\right)$$
$$= 1.0694$$

and the final value of the equilibrium constant is

$$-\ln K_f = -836.32 - 317.85\left(1-\frac{773.15}{298.15}\right) + 0.4670 - 1.0694 = -330.54$$

or

$$K_f = \exp[330.54] = 3.55\times10^{143}$$

9.10 Standard States

The standard state of the species plays an important role in the calculation of the equilibrium constant. Generally, for gases, the standard state of each species is the ideal gas at 1 atm or 1 bar. For liquids, the standard state usually is the pure liquid at a standard pressure, P^0. This pressure can be the vapor pressure evaluated at the equilibrium temperature, T. For solids, the standard state is the pure solid at the equilibrium temperature.

9.11 Applications to Different Types of Reactions

We have obtained the equilibrium constant at different temperatures. Now, we can calculate the equilibrium extent of reaction at a given temperature and pressure.

However, first we must know the phases of the species, and the conditions at which the reaction occurs. As seen in Eq. (9.47), the expression for the equilibrium constant in terms of composition, number of moles, and extent of the reaction uses the expression for the fugacity of component i in the mixture. Now, we only need to select the expression or value for the standard state.

9.11.1 Reactions in Single-Phase Systems

This section provides the expression of the equilibrium constant for reactions in which only one phase exists (gas, liquid, or solid).

9.11.1.1 Gas-Phase Reactions

For the gas phase, the standard state is the pure component as a ideal gas then

$$f_i^0 = f_i^{ig} = P^0 = 1 \tag{9.77}$$

the equilibrium constant can be

$$K_f = K_p K_{\hat{\varphi}} = K_p K_\gamma K_\varphi \tag{9.78}$$

in which

$$K_p \equiv \prod_{i=1}^{N} (p_i/P^0)^{\nu_i} = (P/P^0)^{\Sigma \nu_i} \prod_{i=1}^{N} y_i^{\nu_i} \equiv (P/P^0)^{\Sigma \nu_i} K_y \tag{9.79}$$

This is the partial pressure equilibrium constant. Using Eqs. (9.52) and (9.79)

$$\sum_{i=1}^{N} \nu_i G_i^0 = -RT \ln K_\gamma K_\varphi K_p = -RT \ln K_\gamma K_\varphi (P/P^0)^{\Sigma \nu_i} K_y = -RT \ln K_{\hat{\varphi}} K_p \tag{9.80}$$

For real gases

$$K_f = K_{\hat{\varphi}} K_y (P/P^0)^{\Sigma \nu_i} = (P/P^0)^{\Sigma \nu_i} \prod_{i=1}^{N} (\hat{\varphi}_i y_i)^{\nu_i} \tag{9.81}$$

and to calculate the fugacity coefficients of each component i in the mixture, use an EOS such as the virial equation. Note that the fugacity coefficient depends upon the composition; therefore, Eq. (9.47) is nonlinear in the composition and extent of reaction. If we consider that the gas mixture behaves as an *ideal solution*, the equilibrium constant becomes

$$K_f = K_\varphi K_p = K_\varphi (P/P^0)^{\Sigma \nu_i} K_y = (P/P^0)^{\Sigma \nu_i} \prod_{i=1}^{N} (\varphi_i y_i) \text{ (IS)} \tag{9.82}$$

The advantage of using this equation over the preceding one is that now we calculate the fugacity coefficients of pure components that do not depend upon composition. Hydrocarbon mixtures tend to behave as *ideal solutions*. Finally, considering a PGM

$$K_f = K_p = (P/P^0)^{\Sigma \nu_i} K_y = (P/P^0)^{\Sigma \nu_i} \prod_{i=1}^{N} (y_i)^{\nu_i} \text{ (PGM)} \tag{9.83}$$

which comes from the fundamental equation, Eq. (9.43).

For a PGM that has a large value for the equilibrium constant, the reaction goes to the RHS, which means a large conversion. On the other hand, for a small equilibrium constant, the contrary applies.

9.11.1.2 Liquid-Phase Reactions

Because the standard state is the pure component as a liquid at T and P^* (vapor pressure at T), then using Eq. (9.47)

$$K_f = K_\gamma K_\varphi K_x \prod_{i=1}^{N} (P/f_i^*)^{\nu_i} \tag{9.84}$$

Using the definition of fugacity coefficient in Eq. (9.49)

$$K_f = K_\gamma K_x \prod_{i=1}^{N} (f_i/P)^{\nu_i} \prod_{i=1}^{N} (P/f_i^*)^{\nu_i} = K_\gamma K_x \prod_{i=1}^{N} (f_i/f_i^*)^{\nu_i} \tag{9.85}$$

The last term of the above equation is a difference of Gibbs energy for pure components at different conditions. Therefore, using the definition of fugacity and the fundamental equation for the excess Gibbs energy at a constant temperature

$$G_i - G_i^* = RT \ln\left(\frac{f_i}{f_i^*}\right) = \int_{P^*}^{P} V_i dP \tag{9.86}$$

without pressure dependence in the volume

$$\left(\frac{f_i}{f_i^*}\right) \approx \exp\left[\frac{V_i(P - P^*)}{RT}\right] \tag{9.87}$$

Raising the above equation to the power of ν_i and using the properties of exponentials

$$\left(\frac{f_i}{f_i^*}\right)^{\nu_i} = \exp\left[\frac{\nu_i V_i(P - P^*)}{RT}\right] \tag{9.88}$$

and

$$\prod_{i=1}^{N}\left(\frac{f_i}{f_i^{ig}}\right)^{\nu_i} = e^{[\nu_1 V_1 (P-P^*)/RT]} \cdot e^{[\nu_2 V_2 (P-P^*)/RT]} \ldots\ldots e^{[\nu_N V_N (P-P^*)/RT]}$$

$$= \exp\left[\left(\sum_{i=1}^{N} \nu_i V_i\right) \frac{(P - P^*)}{RT}\right] \tag{9.89}$$

P^* can be the vapor pressure at T. This term is different from unity only at high pressures, so generally, it is omitted. Finally, the expression for the equilibrium constant is

$$K_f = \exp\left[\left(\sum_{i=1}^{N} \nu_i V_i\right) \frac{(P - P^*)}{RT}\right] K_x K_\gamma \tag{9.90}$$

or

$$K_f = \exp\left[\left(\sum_{i=1}^{N} \nu_i V_i\right) \frac{(P - P^*)}{RT}\right] \prod_{i=1}^{N} (x_i \gamma_i)^{\nu_i} \tag{9.91}$$

As mentioned before, at low to medium pressures

$$K_f = K_x K_\gamma = \prod_{i=1}^{N} (x_i \gamma_i)^{\nu_i} \text{ (low–medium pressures)} \tag{9.92}$$

and considering an ideal solution, the equilibrium constant depends upon the composition

$$K_f = K_x = \prod_{i=1}^{N} (x_i)^{\nu_i} \text{ (IS and low–medium pressures) } \textit{law of mass action} \tag{9.93}$$

In the liquid phase, we must calculate the activity coefficient using a solution model for the excess Gibbs energy. Examples of such models appear in Chapter 7.

Sometimes, it is more convenient to start with the definitions of the fugacities to develop the expression for the equilibrium constants. For example, there are cases in which equilibrium exists at a temperature higher than the critical temperature of one of the pure components in the mixture. The vapor pressure of the pure component does not exist, so we need to use another standard state such as Henry's law. In this case,

$$\hat{f}_i = x_i \gamma_i^{\dagger} H_i$$

Here, $\gamma_i^{\dagger}$ is the asymmetric activity coefficient equal to $\gamma_i^{\dagger} = \gamma_i / \gamma_i^{\infty}$ and H_i is the Henry constant at T and at P. If we choose the standard state as a hypothetical pure i whose fugacity equals Henry's law at T and an appropriate pressure, P_i^0

$$f_i^0 = H_i^0 \tag{9.94}$$

Now, to correct Henry's law for pressure

$$\left(\frac{\partial \ln f_i^0}{\partial P} \right)_{T,x} = \frac{\overline{V}_i^0}{RT}$$

and in the limit as $x_i \to \infty$

$$\left(\frac{\partial \ln H_i}{\partial P} \right)_{T,x} = \frac{\overline{V}_i^{\infty}}{RT} \tag{9.95}$$

Integrating between P_i^0 and P

$$\frac{H_i}{H_i^0} = \exp \left[\int_{P_i^0}^{P} \frac{\overline{V}_i^{\infty}}{RT} dP \right] \tag{9.96}$$

and the equilibrium constant becomes

$$K_f = \prod_{i=1}^{N} \left(\frac{x_i \gamma_i^{\dagger} H_i}{H_i^0} \right)^{\nu_i} = \exp \left[\sum \nu_i \int_{P_i^0}^{P} \frac{\overline{V}_i^{\infty}}{RT} dP \right] \prod_{i=1}^{N} \left(x_i \gamma_i^{\dagger} \right)^{\nu_i} \tag{9.97}$$

or

$$K_f = \exp \left[\sum \nu_i \int_{P_i^0}^{P} \frac{\overline{V}_i^{\infty}}{RT} dP \right] (\gamma^{\infty})^{-\sum \nu_i} K_x K_\gamma \tag{9.98}$$

A special case of Eq. (9.97) is when the species exist at low concentration in an aqueous solution. In this case, the concentration in terms of molality is $m\left[=\frac{\text{moles of solute}}{\text{kg of solvent}}\right]$. We use the same hypothetical standard state in which the Henry constant is at T and P of the mixture. At these low concentrations, the asymmetric activity coefficient equals unity, $\gamma^{\dagger}=1$, and Eq. (9.97) becomes

$$K_f = \prod_{i=1}^{N}(m_i)^{\nu_i} \tag{9.99}$$

with $\hat{f}_i/f_i^0 = m_i$.

9.11.1.3 Solid Reactions

The standard state is the pure solid at the equilibrium temperature T and 1 bar, and the standard fugacity is

$$f_i^0 = f_i^s \tag{9.100}$$

This is the case of the liquid when the equilibrium constant is Eq. (9.90) or (9.91), but replacing P^* by 1,

$$K_f = \exp\left[\left(\sum_{i=1}^{N}\nu_i V_i^S\right)\frac{(P-1)}{RT}\right]K_x K_\gamma \tag{9.101}$$

or

$$K_f = \exp\left[\left(\sum_{i=1}^{N}\nu_i V_i^S\right)\frac{(P-1)}{RT}\right]\prod_{i=1}^{N}(x_i\gamma_i)^{\nu_i} \tag{9.102}$$

In all these cases, the LHS is at the equilibrium temperature and the RHS of the equation is a function of the extent of the reaction that generally is unknown.

Example 9.6

Find the extent of the reaction at 25 °C and 2 bar for the synthesis of methanol

$$CO_{(g)} + 2H_{2(g)} \leftrightarrow CH_3OH_{(g)}$$

Assume a stoichiometric feed of 1 mol of carbon monoxide, 2 mol of hydrogen, and consider the mixture to be a perfect gas.

Solution

First, calculate the equilibrium constant from the Gibbs energy of formation

$\Delta G_{f,C_3OH} = -161\,960$ J/mol	$\nu_{C_3OH} = 1$
$\Delta G_{f,H_2} = 0$ J/mol	$\nu_{H_2} = -2$
$\Delta G_{f,CO} = -137\,169$ J/mol	$\nu_{CO} = 1$

Using Eq. (9.29)

$$\Delta G^0 = \sum_{i=1}^{N}\nu_i\Delta G_{f,i} = \nu_{C_3OH}\Delta G_{f,C_3OH} + \nu_{H_2}\Delta G_{f,H_2} + \nu_{CO}\Delta G_{f,CO}$$

$$\Delta G^0 = 1\cdot(-161\,960) - 2\cdot 0 - 1\cdot(-137\,169) = -24\,791 \text{ J/mol}$$

$$-RT \ln K_f = -24\,791$$

$$K_f = \exp\left[\frac{2\,073\,075}{RT}\right] = \exp\left[\frac{24\,791}{8.314 \cdot 298.15}\right] = \exp[10.001] = 1.2048 \times 10^5$$

Now, the material balance is

$$n_{H_2} = 2 - 2\xi$$

$$n_{CO_2} = 1 - \xi$$

$$n_{CH_3OH} = +\xi$$

$$n_T = 3 - 2\xi$$

Considering a PGM, use Eq. (9.83)

$$K_f = (P/P^0)^{\Sigma \nu_i} \prod_{i=1}^{N} (y_i)^{\nu_i} = (2/1)^{1-1-2} \frac{\xi/(3-2\xi)}{\left(\frac{2-2\xi}{3-2\xi}\right)^2 \left(\frac{1-\xi}{3-2\xi}\right)} = \left(\frac{1}{4}\right) \frac{\xi(3-2\xi)^2}{(2-2\xi)^2(1-\xi)}$$

This is a cubic equation of the form

$$4 \cdot K_f (2-2\xi)^2 (1-\xi) = \xi(3-2\xi)^2 \Rightarrow 16 \cdot K_f (1-\xi)^3 - \xi(3-2\xi)^2 = 0$$

Solving the equation using a root-finding routine

$$\xi = 0.9919$$

The equilibrium compositions are

$$y_{H_2} = \frac{2-2\xi}{3-2\xi} = \frac{2 - 2 \cdot 0.9919}{3 - 2 \cdot 0.9919} = 0.016$$

$$y_{CO} = \frac{1-\xi}{3-2\xi} = \frac{1 - 0.9919}{3 - 2 \cdot 0.9919} = 0.008$$

$$y_{CH_3OH} = \frac{\xi}{3-2\xi} = \frac{0.9919}{3 - 2 \cdot 0.9919} = 0.976$$

Simplifying the cubic equation by neglecting terms divided by the equilibrium constant yields

$$(1-\xi)^3 = 0 \Rightarrow \xi = 1$$

This gives a 100% conversion and an error of 3.4%.

9.11.2 Heterogeneous Reactions (Different Phase Systems)

So far, we have considered only a single phase in the reaction, but a gas component, A, can react with a liquid component, B, to form a liquid component C. Because we have a chemical reaction in equilibrium, we must consider the physical equilibrium among the reactants and the products. Several methods can solve these problems. The first problem is to express the equilibrium constant in terms of the compositions. This is not a difficult task; we simply use Eq. (9.44) and the correct value for the fugacity coefficient in the mixture. For example

$$K_f = \frac{(x_C \gamma_C)}{(y_A P \hat{\varphi}_A / P^0)(x_B \gamma_B)} \tag{9.103}$$

This equation does not consider corrections at high pressure for the liquids. The next question is: do we have a value for the equilibrium constant because it also depends upon the standard states? If we have a value for K_f, then we must solve for the extent of the reaction in Eq. (9.103) and satisfy $\hat{f}_i^v = \hat{f}_i^l$ and the material balances: $\sum_{i=1}^{N} x_i = 1$ and $\sum_{i=1}^{N} y_i = 1$. Sometimes, it is easier to find the equilibrium constant at a common standard state, either gas or liquid. If this is the case, we can consider the reaction as a gas or liquid reaction and use the physical equilibrium conditions and the material balance to solve for the extent of reaction. First, considering a gas phase reaction, the equilibrium constant becomes

$$K_f = \frac{\hat{f}_C^v}{\hat{f}_A^v \hat{f}_B^v} P^0 = \frac{y_C \hat{\varphi}_C}{(y_A \hat{\varphi}_A)(y_B \hat{\varphi}_B)} \frac{P^0}{P} \tag{9.104}$$

Substituting the equilibrium conditions $\hat{f}_B^v = \hat{f}_B^l, \hat{f}_C^v = \hat{f}_C^l$,

$$K_f = \frac{\hat{f}_C^l}{\hat{f}_A^v \hat{f}_B^l} P^0 = \frac{x_C \gamma_C f_C}{(y_A \hat{\varphi}_A P)(x_B \gamma_B f_B)} P^0 \tag{9.105}$$

and the only remaining equation to consider is the mass balance.

Example 9.7

The hydrogenation of propylene to form isopropanol occurs in two phases at $T = 423.15$ K and 10 bar. Estimate the vapor and liquid compositions at these conditions. Assume no other reactions and that the solubility of propylene is negligible. Consider that the nonideality in the vapor is $\hat{\varphi}_i = \varphi_i$ (ideal solution)

Data

$$C_3H_6 + H_2O \rightarrow C_3H_8O$$

$$\ln K_f = -17.578 + 6099 \cdot 1/T \text{ for reaction in the gas phase}$$

Activity coefficients for isopropanol + water system

$$\ln \gamma_{iso} = x_w^{\ 2}[A_{12} + 2(A_{21} - A_{12})x_{iso}]$$

$$\ln \gamma_w = x_{iso}^{\ 2}[A_{21} + 2(A_{12} - A_{21})x_w]$$

$$A_{12} = -0.4$$

$$A_{21} = 1.5$$

Virial coefficients

$$B_{2,isopropanol} = -558 \text{ mol/cm}^3$$

$$B_{2,water} = -287 \text{ mol/cm}^3$$

$$B_{2,propylene} = -164 \text{ mol/cm}^3$$

Vapor pressures

$$P^{sat}_{isopropanol} = 868.8 \text{ kPa}$$

$$P^{sat}_{water} = 476.16 \text{ kPa}$$

Solution

Use Eq. (9.105)

$$K_f = \frac{\hat{f}_C^l}{\hat{f}_A^v \hat{f}_B^l} P^0 = \frac{x_C \gamma_C f_C}{(y_A \hat{\varphi}_A P)(x_B \gamma_B f_B)} P^0$$

with A = propylene, B = water, and C = isopropanol. The fugacity of liquids B and C are

$$f_C = \phi_C^{sat} P_C^{sat}, f_B = \phi_B^{sat} P_B^{sat}$$

and the equation is

$$K_f = \frac{\hat{f}_C^l}{\hat{f}_A^v \hat{f}_B^l} P^0 = \frac{x_C \gamma_C P_C^{sat} \phi_C^{sat}}{(y_A \hat{\varphi}_A P)\left(x_B \gamma_B P_B^{sat} \phi_B^{sat}\right)} P^0 \tag{A}$$

We know that

$$y_A + y_B + y_C = 1 \Rightarrow y_A = 1 - y_B - y_C$$

and if we use the equilibrium condition for B and C

$$y_B = \frac{x_B \gamma_B P_B^{sat} \varphi_B^{sat}}{P \varphi_B}$$

$$y_C = \frac{x_C \gamma_C P_C^{sat} \varphi_C^{sat}}{P \varphi_C}$$

then Eq. (A) is only a function of the liquid compositions x_B and x_C

$$\left(1 - \frac{x_B \gamma_B P_B^{sat} \varphi_B^{sat}}{P \varphi_B} - \frac{x_C \gamma_C P_C^{sat} \varphi_C^{sat}}{P \varphi_C}\right) K_f (\hat{\varphi}_A P)\left(x_B \gamma_B P_B^{sat} \phi_B^{sat}\right) = x_C \gamma_C P_C^{sat} \phi_C^{sat} P^0$$

The material balance on the liquid side is

$$x_B + x_C = 1$$

Substituting for the liquid composition of water in Eq. (A), we have only one unknown composition of isopropanol. So, we can solve for the composition of isopropanol

$$\left(1 - \frac{x_B \gamma_B P_B^{sat} \varphi_B^{sat}}{P \varphi_B} - \frac{x_C \gamma_C P_C^{sat} \varphi_C^{sat}}{P \varphi_C}\right) K_f (\varphi_A P)\left(x_B \gamma_B P_B^{sat} \phi_C^{sat}\right) - x_C \gamma_C P_C^{sat} \phi_C^{sat} P^0 = 0 \tag{B}$$

With

$$\ln \gamma_C = {x_B}^2 [A_{12} + 2(A_{21} - A_{12}) x_C]$$

$$\ln \gamma_B = {x_C}^2 [A_{21} + 2(A_{12} - A_{21}) x_B]$$

$$A_{12} = -0.4$$

$$A_{21} = 1.5$$

The fugacity coefficients are

$$\ln \varphi_A = \frac{PB_{2,AA}}{RT} = \frac{1\ \text{MPa} \times (-164)\ \text{cm}^3/\text{mol}}{8.314\ \text{cm}^3/\text{mol MPa/K} \times 423.15\ \text{K}}$$
$$= -0.0466 \Rightarrow \varphi_A = 0.954453$$

$$\ln \varphi_B = \frac{PB_{2,BB}}{RT} = \frac{1\ \text{MPa} \times (-287)\ \text{cm}^3/\text{mol}}{8.314\ \text{cm}^3/\text{mol MPa/K} \times 423.15\ \text{K}}$$
$$= -0.0816 \Rightarrow \varphi_B = 0.92166$$

$$\ln \varphi_C = \frac{PB_{2,CC}}{RT} = \frac{1\ \text{MPa} \times (-558)\ \text{cm}^3/\text{mol}}{8.314\ \text{cm}^3/\text{mol MPa/K} \times 423.15\ \text{K}}$$
$$= -0.1586 \Rightarrow \varphi_C = 0.8533$$

$$\ln \varphi_B^{sat} = \frac{P_B^{sat} B_{2,BB}}{RT} = \frac{0.47616\ \text{MPa} \times (-287)\ \text{cm}^3/\text{mol}}{8.314\ \text{cm}^3/\text{mol MPa/K} \times 423.15\ \text{K}}$$
$$= -0.03884 \Rightarrow \varphi_B^{sat} = 0.9619$$

$$\ln \varphi_C^{sat} = \frac{P_C^{sat} B_{2,CC}}{RT} = \frac{0.8688\ \text{MPa} \times (-558)\ \text{cm}^3/\text{mol}}{8.314\ \text{cm}^3/\text{mol MPa/K} \times 423.15\ \text{K}}$$
$$= -0.1378 \Rightarrow \varphi_C^{sat} = 0.87127$$

Substituting the values into Eq. (B), the final equation is

$$f(x_C) = (1 - 0.496949 x_B \gamma_B - 0.887069 x_C \gamma_C) K_f (0.43716)(x_B \gamma_B) - 0.075696 \times x_C \gamma_C = 0 \quad \text{(C)}$$

with

$$\ln \gamma_C = x_B^2[-0.4 + 3.8 x_C]$$
$$\ln \gamma_B = x_C^2[1.5 - 3.8 x_B]$$

and

$$x_B = 1 - x_C$$

Using Solver, the composition for isopropanol is

$$x_C = 0.0891$$

One can do a trial-and-error solution, assuming two values of x_C and then extrapolating linearly using

$$x_{C,j+2} = \frac{x_{C,j+1} - x_{C,j}}{f(x_{C,j+1}) - f(x_{C,j+1})}(0 - f(x_{C,j+1})) + x_{C,j}$$

The next table shows the iterations

j	x_C	$f(x_C)$
	0.5	−0.05419
	0.3	−0.02977
	0.056201	0.00328
	0.080399	0.000927
	0.089929	-8.4×10^{-5}

The final values are (A = propylene, B = water, and C = isopropanol):

$x_A = 0.0$	$y_A = 0.4792$
$x_B = 0.9709$	$y_B = 0.4456$
$x_C = 0.0891$	$y_C = 0.0752$

9.12 Multi-reactions

For the case of multiple reactions, we have an extent of reaction for each of the independent reactions; therefore, Eq. (9.10) applies to each reaction and the total change of moles is the sum of all the contributions of each reaction

$$\Delta n_i = \sum_{j=1}^{R} \nu_{i,j}\xi_j \text{ for any } i \text{ component} \tag{9.106}$$

In differential form

$$dn_i = \sum_{j=1}^{R} \nu_{i,j} d\xi_j \tag{9.107}$$

The total number of moles after a certain time is

$$n = \sum_{i=1}^{N} n_{i,0} + \sum_{i=1}^{N}\sum_{j=1}^{R} \nu_{i,j}\xi_j = n_{T,0} + \sum_{i=1}^{N}\sum_{j=1}^{R} \nu_{i,j}\xi_j \tag{9.108}$$

and the mole fraction is

$$y_i = \frac{n_{i,0} + \sum_{i=1}^{R} \nu_{i,j}\xi_j}{n_{T,0} + \sum_{i=1}^{N}\sum_{j=1}^{R} \nu_{i,j}\xi_j} \tag{9.109}$$

Consider the following two independent reactions

$$A + 2B \rightarrow C$$

$$C + 3D \rightarrow 2E$$

The number of moles present after a certain time from Eq. (9.108) is

$$n_A = n_{A,0} - \xi_1$$

$$n_B = n_{B,0} - 2\xi_1$$

$$n_C = n_{C,0} + \xi_1 - \xi_2$$

$$n_D = n_{D,0} - 3\xi_2$$

and the total number of moles is

$$n_T = n_{T,0} - 2\xi_1 - 4\xi_2$$

in which $n_T = n_{A,0} + n_{B,0} + n_{C,0} + n_{D,0}$. We can calculate the composition using Eq. (9.109)

$$y_A = \frac{n_{A,0} - \xi_1}{n_{T,0} - 2\xi_1 - 4\xi_2}$$

$$y_B = \frac{n_{B,0} - 2\xi_1}{n_{T,0} - 2\xi_1 - 4\xi_2}$$

$$y_C = \frac{n_{C,0} + \xi_1 - \xi_2}{n_{T,0} - 2\xi_1 - 4\xi_2}$$

and

$$y_D = \frac{n_{D,0} - 3\xi_2}{n_{T,0} - 2\xi_1 - 4\xi_2}$$

Now, for the calculation of the equilibrium constants, we have one for each reaction and

$$K_{f,j} \equiv \prod_{i=1}^{N} \left(\hat{f}_i / f_i^0\right)^{\nu_{i,j}} \text{ for } j = 1,2, \ldots, R \tag{9.110}$$

and we have a system of equations formed by (9.110) in which the unknowns are the extents of the reaction.

Example 9.8

Consider the production of component C by two independent reactions at 1.5 bar

$$A + B \rightarrow C$$

$$D + E \rightarrow C$$

If the equilibrium constants are $K_{f,1} = 0.26$ and $K_{f,2} = 0.005$ at 800 K, what is the product composition when the reactions reach equilibrium? Is this process feasible?

Solution

Using the procedure in section 9.11 considering a stoichiometric feed and Eq. (9.109)

$$y_A = \frac{1 - \xi_1}{4 - \xi_1 - \xi_2}, y_B = \frac{1 - \xi_1}{4 - \xi_1 - \xi_2}, y_C = \frac{\xi_1 + \xi_2}{4 - \xi_1 - \xi_2},$$

$$y_D = \frac{1 - \xi_2}{4 - \xi_1 - \xi_2}, \text{ and } y_E = \frac{1 - \xi_2}{4 - \xi_1 - \xi_2}$$

At the reaction conditions, considering PGM and for both reactions

$$\sum_{i=1}^{N} \nu_{i,j} = 1 - 1 - 1 = -1$$

and

$$K_{f,1} = (1.5)^{-1} \frac{\left(\frac{\xi_1 + \xi_2}{4 - \xi_1 - \xi_2}\right)}{\left(\frac{1 - \xi_1}{4 - \xi_1 - \xi_2}\right)\left(\frac{1 - \xi_1}{4 - \xi_1 - \xi_2}\right)} = (1.5)^{-1} \frac{(\xi_1 + \xi_2)(4 - \xi_1 - \xi_2)}{(1 - \xi_1)(1 - \xi_1)} \tag{A}$$

$$K_{f,2} = (1.5)^{-1} \frac{\left(\frac{\xi_1 + \xi_2}{4 - \xi_1 - \xi_2}\right)}{\left(\frac{1 - \xi_2}{4 - \xi_1 - \xi_2}\right)\left(\frac{1 - \xi_2}{4 - \xi_1 - \xi_2}\right)} = (1.5)^{-1} \frac{(\xi_1 + \xi_2)(4 - \xi_1 - \xi_2)}{(1 - \xi_2)(1 - \xi_2)} \tag{B}$$

Dividing Eqs. (A) and (B)

$$\frac{(1-\xi_1)(1-\xi_1)}{(1-\xi_2)(1-\xi_2)} = \frac{(1-\xi_1)^2}{(1-\xi_2)^2} = \left[\frac{1-\xi_1}{1-\xi_2}\right]^2 = \frac{K_{f,1}}{K_{f,2}} = K$$

or

$$\frac{1-\xi_1}{1-\xi_2} = K^{0.5} \Rightarrow \xi_1 = 1 - K^{0.5}(1-\xi_2) \tag{C}$$

Substituting this result into Eq. (A) provides

$$(3 - 2K^{0.5} - K - 1.5K_1K) + (2 + 6K^{0.5} + 3K_{f,1}K)\xi_2 - (1 - K + 1.5K_{f,1}K)\xi_2{}^2 = 0$$

The roots are $\xi_2 = 0.7655$ and $\xi_2 = -3.55934$. Substituting these results into Eq. (C)

$$\xi_1 = 1 - 52^{0.5}(1 - 0.7655) = -0.691$$

$$\xi_1 = 1 - 52^{0.5}(1 + 2.6918) = -31.87787$$

Selecting the second root gives negative compositions, which is physically impossible! Then, our results are $\xi_2 = 0.7655$ and $\xi_1 = -0.691$ and

$$y_A = y_B = 0.05974, y_C = 0.01898, \text{and } y_D = y_E = 0.43077$$

The process is not feasible at the conditions of the reaction because the product composition is too small.

9.13 Nonstoichiometric Solution

As mentioned before, the total Gibbs function of the reacting system is a minimum at equilibrium. Then, we can write Eq. (9.20) as

$$\min(nG) = \min \sum_{i=1}^{N} n_i\mu_i \tag{9.111}$$

This equation is subject to the conservation of mass in such way that the number of moles n_i can vary as long as the total number of atoms remain constant. This mass constraint is

$$\text{subject to} \sum_{i=1}^{N} a_{ki}n_i - b_k = 0 \quad k = 1,2,3,\ldots,n_{elem} \tag{9.112}$$

in which b_k is the total number of atoms, a_{ki} is the number of atoms of element k in one molecule of species i, and n_{elem} is the total number of elements. For example, assume that we have methane, CH_4, carbon, C, and hydrogen H_2, then, we have three species and two elements (C and H). The total number of elements is

$$\sum_{i=1}^{N} a_{Ci}n_i = n_{CH_4}\cdot 1 + n_C\cdot 1 + n_{H_2}\cdot 0$$

$$\sum_{i=1}^{N} a_{Hi}n_i = n_{CH_4}\cdot 4 + n_C\cdot 0 + n_{H_2}\cdot 2$$

The value of b_k comes from Eq. (9.112) using the initial number of moles, that is

$$b_C = n_{CH_4,0} \cdot 1 + n_{C,0} \cdot 1 + n_{H_2,0} \cdot 0$$

$$b_H = n_{CH_4,0} \cdot 4 + n_{C,0} \cdot 0 + n_{H_2,0} \cdot 2$$

Solving Eqs. (9.111) and (9.112) is a nonlinear programming problem (NLP) using computer codes found in the literature [1]. Here, we use Lagrange multipliers to convert the optimization problem into a problem of solving a system of nonlinear equations. We cannot solve Eq. (9.111) by differentiation with respect to n_i because they depend upon the constraint (9.112). First, we multiply the constraint by a Lagrange multiplier. We need as many Lagrange multipliers as equality constraints

$$\left(\sum_{i=1}^{N} a_{ki} n_i - b_k \right) \lambda_k = 0 \quad k = 1, 2, 3, \ldots, n_{elem} \tag{9.113}$$

Because this set of equations equals zero, we can substitute them into Eq. (9.111) without altering the solution,

$$\min \sum_{i=1}^{N} n_i \mu_i + \sum_{k=1}^{n_{elem}} \left(\sum_{i=1}^{N} a_{ki} n_i - b_k \right) \lambda_k \tag{9.114}$$

The problem is now to minimize Eq. (9.114) when the variables are the number of moles and the set of Lagrange multipliers. We can differentiate this equation and set it equal to zero because the constraints are gone

$$\sum_{i=1}^{N} \mu_i dn_i + \sum_{k=1}^{n_{elem}} \left(\sum_{i=1}^{N} a_{ki} dn_i \right) \lambda_k + \sum_{k=1}^{n_{elem}} \left(\sum_{i=1}^{N} a_{ki} n_i - b_k \right) d\lambda_k = 0 \tag{9.115}$$

Rearranging

$$\sum_{i=1}^{N} \left(\mu_i + \sum_{k=1}^{n_{elem}} a_{ki} \lambda_k \right) dn_i + \sum_{k=1}^{n_{elem}} \left(\sum_{i=1}^{N} a_{ki} n_i - b_k \right) d\lambda_k = 0 \tag{9.116}$$

Because dn_i and $d\lambda_k$ can have any values, Eq. (9.116) can be zero only if the terms in parentheses are zero

$$\mu_i + \sum_{k=1}^{n_{elem}} a_{ki} \lambda_k = 0 \quad \text{for } i = 1, 2, \ldots, N \tag{9.117}$$

$$\sum_{i=1}^{N} a_{ki} n_i - b_k = 0 \quad \text{for } k = 1, 2, 3, \ldots, n_{elem} \tag{9.118}$$

Equations (9.117) and (9.118) are a $N + n_{elem}$ system of equations whose unknowns are N mole fractions (or N unknowns) and n_{elem} Lagrange multipliers. Introducing the expression for chemical potential into Eq. (9.117)

$$\mu_i^0 + RT \ln \left(\hat{f}_i / f_i^0 \right) + \sum_{k=1}^{n_{elem}} a_{ki} \lambda_k = 0 \quad \text{for } i = 1, 2, \ldots, N \tag{9.119}$$

The fugacity coefficient in the mixture i and the fugacity at the standard state come from Sections 9.11 and 9.12 and $\mu_i^0 = \Delta G_{f,i}^0$. The system becomes

$$\Delta G_{f,i}^0 + RT\ln\left(\hat{f}_i/f_i^0\right) + \sum_{k=1}^{n_{elem}} a_{ki}\lambda_k = 0 \quad \text{for } i = 1, 2, \ldots, N \tag{9.120}$$

$$\sum_{i=1}^{N} a_{ki}n_i - b_k = 0 \quad \text{for } k = 1, 2, 3, \ldots, n_{elem} \; (9.118)$$

Remember that for a PGM, $\hat{f}_i/f_i^0 = Py_i/P^0$

Example 9.9
Calculate the equilibrium composition of the reaction

$$CH_{4(g)} + H_2O_{(g)} \rightarrow CO_{(g)} + 3H_{2(g)}$$

at 1000 K and 1 bar using the nonstoichiometric method. Consider that initially, we have 2 mol of CH_4 and 2 mol of H_2O.

Data

$$\Delta G_{f,H_2O}^0 = -192\,420, \Delta G_{f,CO}^0 = -200\,240, \Delta G_{f,CH_4}^0 = 19\,720$$

Solution
We have four species CH_4, H_2O, CO, and H_2 and three elements C, O, and H. First, calculate the total number of atoms

$$b_C = n_{CH_4,0}\cdot 1 + n_{CO,0}\cdot 1 + n_{H_2,0}\cdot 0 + n_{H_2O,0}\cdot 0 = 2 + 0 + 0 + 0 = 2$$

$$b_O = n_{CH_4,0}\cdot 0 + n_{CO,0}\cdot 1 + n_{H_2,0}\cdot 0 + n_{H_2O,0}\cdot 1 = 0 + 0 + 0 + 2\cdot 1 = 2$$

$$b_H = n_{CH_4,0}\cdot 4 + n_{CO,0}\cdot 0 + n_{H_2,0}\cdot 2 + n_{H_2O,0}\cdot 2 = 2\cdot 4 + 0 + 0 + 2\cdot 2 = 12$$

The constraints, Eq. (9.118), are

$$n_{CH_4} + n_{CO} - 2 = 0$$

$$n_{CO} + n_{H_2O} - 2 = 0$$

$$n_{CH_4}\cdot 4 + n_{H_2}\cdot 2 + n_{H_2O}\cdot 2 - 12 = 0$$

and the material balance is

$$n_{CH_4} + n_{H_2} + n_{H_2O} + n_{CO} = n_T$$

The rest of the equations come from Eq. (9.120). Assume a PGM because the temperature is sufficiently high and the pressure is sufficiently low. For a PGM, $\hat{f}_i/f_i^0 = Py_i/P^0$, and

$$\frac{\Delta G_{f,i}^0}{RT} + \ln y_i + \sum_{k=1}^{n_{elem}} a_{ki}\frac{\lambda_k}{RT} = 0 \quad \text{for } i = 1, 2, \ldots, N$$

Now

$$\frac{\Delta G^0_{f,CH_4}}{RT} + \ln y_{CH_4} + \sum_{k=1}^{n_{elem}} a_{k,CH_4}\frac{\lambda_k}{RT} = 0$$

$$\frac{\Delta G^0_{f,CO}}{RT} + \ln y_{CO} + \sum_{k=1}^{n_{elem}} a_{k,CO}\frac{\lambda_k}{RT} = 0$$

$$\frac{\Delta G^0_{f,H_2O}}{RT} + \ln y_{H_2O} + \sum_{k=1}^{n_{elem}} a_{k,H_2O}\frac{\lambda_k}{RT} = 0$$

$$\frac{\Delta G^0_{f,H_2}}{RT} + RT\ln y_{H_2} + \sum_{k=1}^{n_{elem}} a_{k,H_2}\frac{\lambda_k}{RT} = 0$$

If we use $\lambda_i^* = \lambda_i/RT$ and replace the values, then

$$2.3719 + \ln y_{CH_4} + 1\cdot\lambda_C^* + 4\cdot\lambda_H^* = 0$$

$$-24.085 + \ln y_{CO} + 1\cdot\lambda_C^* + 1\cdot\lambda_O^* = 0$$

$$23.1441 + \ln y_{H_2O} + 2\cdot\lambda_H^* + 1\cdot\lambda_O^* = 0$$

$$\ln y_{H_2} + 2\cdot\lambda_H^* = 0$$

$$y_{CH_4} + y_{CO} - 2/n_T = 0$$

$$y_{CO} + y_{H_2O} - 2/n_T = 0$$

$$y_{CH_4}\cdot 4 + y_{H_2}\cdot 2 + y_{H_2O}\cdot 2 - 12/n_T = 0$$

$$y_{CH_4} + y_{H_2} + y_{H_2O} + y_{CO} = 1$$

We have a system of eight nonlinear equations with eight unknowns y_{CH_4}, y_{H_2}, y_{H_2O}, y_{CO}, n_T, λ_C^*, λ_H^*, and λ_O^*. Simultaneous computer solution produces

$$y_{CH_4} = 0.0502, y_{H_2} = 0.6747, y_{H_2O} = 0.0502, y_{CO} = 0.2249, n_T = 7.2705,$$
$$\lambda_C^* = -0.1664, \lambda_H^* = 0.1967, \text{and } \lambda_O^* = 25.743$$

9.14 Equal Area Rule for Reactive Thermodynamic Equilibrium Calculations

In Section 8.9, we learned of a new technique for performing phase equilibrium calculations. Here, we can modify that technique to address issues such as reactive distillation. Ung and Doherty [2, 3] suggested using transformed mole fractions to address this issue. Iglesias-Silva et al. [4] inserted this concept into the equal area procedure from Section 8.9 for phase equilibrium.

We start with the total Gibbs energy for a mixture with N components and R reactions:

$$G = \sum_{i=1}^{N} n_i \mu_i \tag{9.121}$$

with

$$\sum_{i=1}^{N} v_{i,j} \mu_i = 0 \tag{9.122}$$

In these equations, n_i is the number of moles of component i and $v_{i,j}$ is the stoichiometric coefficient of component i in reaction j. Thus, $N - R$ mole fractions are independent variables. Ung and Doherty transform the mole fractions using R reference components to obtain an unconstrained system of $N - R$ equations:

$$X_i = \frac{\widehat{n}_i}{\widehat{n}_T} \equiv \frac{\left[n_i - v_i^T (V_{ref})^{-1} n_{ref}\right]}{\left[n_T - v_T^T (V_{ref})^{-1} n_{ref}\right]} = \frac{\left[x_i - v_i^T (V_{ref})^{-1} x_{ref}\right]}{\left[1 - v_T^T (V_{ref})^{-1} x_{ref}\right]} \quad i = 1, 2, \ldots, C - N \tag{9.123}$$

in which

$$v_i^T = (v_{i,1}, v_{i,2}, \ldots, v_{i,R}) \tag{9.124}$$

$$v_T^T = (v_{T,1}, v_{T,2}, \ldots, v_{T,R}) \quad \text{with } v_{T,i} = \sum_{k=1}^{N} v_{k,i} \tag{9.125}$$

$$n_{ref} = (n_{N-R+1}, n_{N-R+1}, \ldots, n_N)^T \tag{9.126}$$

$$x_{ref} = (x_{N-R+1}, x_{N-R+1}, \ldots, x_N)^T \tag{9.127}$$

$$\sum_{i=1}^{N-R} X_i = 1 \tag{9.128}$$

and

$$V_{ref} = \begin{bmatrix} v_{N-R+1,1} & v_{N-R+1,1} & \cdots & v_{N-R+1,R} \\ v_{N-R+2,1} & v_{N-R+2,2} & \cdots & v_{N-R+2,R} \\ \vdots & \vdots & \cdots & \vdots \\ v_{N,1} & v_{N,2} & \cdots & v_{N,R} \end{bmatrix} \tag{9.129}$$

In these equations, n_T is the total number of moles, x_i is the mole fraction of component i, n_{ref} and x_{ref} are row vectors of dimension R denoting the number of moles and mole fractions of the R reference components. Equation (9.123) depends upon the equilibrium constants for each reaction. Equations (9.123)–(9.129) provide the unconstrained Gibbs energy:

$$G = G(T, P, \widehat{n}_1, \widehat{n}_2, \ldots, \widehat{n}_{N-R}) \tag{9.130}$$

The transformed, unconstrained molar Gibbs energy is

$$\widehat{g} = \frac{G}{\widehat{n}_T} = \frac{G/n_T}{\left[1 - v_T^T (V_{ref})^{-1} x_{ref}\right]} = \frac{g}{\left[1 - v_T^T (V_{ref})^{-1} x_{ref}\right]} \tag{9.131}$$

in which g is the molar Gibbs energy. Applying normal thermodynamics relationships to a binary system using the transformed mole fractions with $N - R = 2$

$$\hat{g} = X_1\hat{\mu}_1 + X_2\hat{\mu}_2 \tag{9.132}$$

and

$$\hat{\mu}_i = \left(\frac{\partial \hat{n}_T\hat{g}}{\partial \hat{n}_i}\right)_{T,P,\hat{n}_{j=1}} \tag{9.133}$$

allows us to determine the phase compositions at a constant T and P. In these equations, $\hat{\mu}_i$ is the transformed chemical potential. The total differential is

$$d\hat{g} = \hat{\mu}_1 dX_1 + \hat{\mu}_2 dX_2 \tag{9.134}$$

whose derivative with respect to X_1 is

$$\left(\frac{\partial \hat{g}}{\partial X_1}\right) = \hat{\mu}_1 + \hat{\mu}_2\frac{dX_2}{dX_1} = \hat{\mu}_1 - \hat{\mu}_2 \tag{9.135}$$

Then

$$X_1\left(\frac{\partial \hat{g}}{\partial X_1}\right) = X_1\hat{\mu}_1 + X_1\hat{\mu}_2 = X_1\hat{\mu}_1 - (1 - X_2)\hat{\mu}_2$$
$$= X_1\hat{\mu}_1 + X_2\hat{\mu}_2 - \hat{\mu}_2 = \hat{g} - \hat{\mu}_2 \tag{9.136}$$

At equilibrium, the transformed chemical potential is equal in each phase, so

$$\hat{g}^\alpha - X_1^\alpha\left(\frac{\partial \hat{g}}{\partial X_1}\right)_{T,P}^\alpha = \hat{g}^\beta - X_1^\beta\left(\frac{\partial \hat{g}}{\partial X_1}\right)_{T,P}^\beta \tag{9.137}$$

but from Eq. (9.135), the orthogonal derivative is equal in each phase at equilibrium, so a tangent line exists that touches $\hat{g}$ vs X_1 at X_1^α and X_1^β; thus, Eq. (9.137) becomes

$$\hat{g}^\alpha - \hat{g}^\beta - \left(\frac{\partial \hat{g}}{\partial X_1}\right)_{T,P}^\alpha \left(X_1^\alpha - X_1^\beta\right) = 0 \tag{9.138}$$

which becomes

$$\int_{X_1^\beta}^{X_1^\alpha}\left(\frac{\partial \hat{g}}{\partial X_1}\right)_{T,P} dX_1 - \left(\frac{\partial \hat{g}}{\partial X_1}\right)_{T,P}^\alpha \left(X_1^\alpha - X_1^\beta\right) = 0 \tag{9.139}$$

This is an equal area rule that determines the equilibrium-transformed compositions for a reactive system with $N - R = 2$.

To use the equal area algorithm, we must guess an initial value for the orthogonal derivative and find the transformed equilibrium compositions. Then, we can adjust the slope using

$$\left(\frac{\partial \hat{g}}{\partial X_1}\right)_{k+1} = \frac{\left[\int_{X_1^\beta}^{X_1^\alpha} (\partial\hat{g}/\partial X_1)_{T,P} dX_1\right]_k}{\left[X_1^\alpha - X_1^\beta\right]_k} \tag{9.140}$$

This new slope allows calculation of a set of transformed equilibrium compositions. The two transformed mole fractions are functions of the reference mole fraction from Eq. (9.123) and the reference mole fractions in any phase calculated from the equilibrium constants for each reaction. This requires solving a system of R non-linear equations:

$$K_j(T) - \prod_{i=1}^{N} a_i^{\nu_{i,j}} = 0 \quad \text{for} \quad j = 1, 2, \ldots, R \tag{9.141}$$

in which K is the equilibrium constant and a_i is the activity coefficient of component i. We can terminate the algorithm when two successive values of the derivative are within a predetermined tolerance.

Generally, we can use the transformed molar Gibbs energy of mixing for the transformed molar Gibbs energy

$$\frac{\Delta_m \hat{g}}{RT} = \frac{\hat{g}}{RT} - \frac{1}{RT}\sum_{i=1}^{N-R} X_i g_i = \frac{1}{RT}\sum_{i=1}^{N-R} X_i[\mu_i - g_i] \tag{9.142}$$

in which μ_i and g_i are the molar values for the pure component. Then, the transformed Gibbs energy of mixing is

$$\frac{\Delta_m \hat{g}}{RT} = \frac{1}{RT}\sum_{i=1}^{N-R} X_i[g_i + RT\ln x_i + RT\ln\gamma_i - g_i] = \sum_{i=1}^{N-R} X_i \ln(x_i\gamma_i) \tag{9.143}$$

whose orthogonal derivative for $N - R = 2$ is

$$\left(\frac{\partial \Delta_m \hat{g}/RT}{\partial X_1}\right)_{T,P} = (\ln x_1\gamma_1 - \ln x_2\gamma_2) \tag{9.144}$$

Example 9.10

Assume a mixture undergoing the reversible reaction:

$$A_1 + A_2 \leftrightarrow A_3$$

at a given temperature and a pressure such that the system is liquid but exhibits phase splitting during the reaction. Use component 3 as the reference component. Also, assume that the nonideality in the liquid is given by a regular solution for the excess energy (which it is equivalent to a Margules equation at a constant temperature)

$$G^E/RT = x_1x_2A_{12} + x_1x_3A_{13} + x_2x_3A_{23}$$

With $A_{12} = 2, A_{13} = 3$, and $A_{23} = 2$ and the equilibrium constant K_f is equal to 4 Calculate the reactive equilibrium compositions

Solution

Here, we have that $N = 3$ and $R = 1$, then using Eqs. (9.124), (9.125), (9.129),

$$V_{ref} = \nu_{3,1} = 1$$
$$V_{ref}^{-1} = 1$$
$$\nu_1^T = \nu_1 = -1$$
$$\nu_2^T = \nu_2 = -1$$
$$\nu_T^T = \nu_1 + \nu_2 + \nu_3 = -1$$

The transformed mole fractions are from Eq. (9.123)

$$X_1 = \frac{x_1 + x_3}{1 + x_3}$$

$$X_2 = \frac{x_2 + x_3}{1 + x_3}$$

in each phase at equilibrium.

Now, assuming a regular solution for the excess energy

$$G^E/RT = x_1x_2A_{12} + x_1x_3A_{13} + x_2x_3A_{23}$$

The activity coefficients are

$$\frac{nG^E}{RT} = A_{12}\frac{n_1n_2}{n} + A_{13}\frac{n_1n_3}{n} + A_{23}\frac{n_2n_3}{n}$$

$$\ln\gamma_1 = \left[\frac{\partial\left(\frac{nG^E}{RT}\right)}{\partial n_1}\right]_{T,P,n_2,n_3} = A_{12}n_2\left\{\frac{1}{n} - \frac{n_1}{n^2}\right\} + A_{13}n_3\left\{\frac{1}{n} - \frac{n_1}{n^2}\right\} - A_{23}\frac{n_2n_3}{n^2}$$

$$\ln\gamma_1 = A_{12}\{x_2 - x_1x_2\} + A_{13}\{x_3 - x_1x_3\} - A_{23}x_2x_3$$

$$\ln\gamma_1 = A_{12}x_2\{1 - x_1\} + A_{13}x_3\{1 - x_1\} - A_{23}x_2x_3$$

but $1 - x_1 = x_2 + x_3$, then

$$\ln\gamma_1 = A_{12}x_2\{x_2 + x_3\} + A_{13}x_3\{x_2 + x_3\} - A_{23}x_2x_3$$

$$\begin{aligned}\ln\gamma_1 &= A_{12}{x_2}^2 + A_{12}x_2x_3 + A_{13}x_2x_3 + A_{13}{x_3}^2 - A_{23}x_2x_3\\ &= A_{12}{x_2}^2 + A_{13}{x_3}^2 + (A_{12} + A_{13} - A_{23})x_2x_3\end{aligned}$$

Similarly,

$$\ln\gamma_2 = A_{12}{x_1}^2 + (A_{12} + A_{23} - A_{13})x_1x_3 + A_{23}{x_3}^2$$

$$\ln\gamma_3 = A_{13}{x_1}^2 + (A_{13} + A_{23} - A_{12})x_1x_2 + A_{23}{x_2}^2$$

Using Eq. (9.45), the equilibrium constant is

$$K_f = {a_1}^{-1}{a_2}^{-1}a_3 = \frac{a_3}{a_1a_2}$$

Considering the nonideal solution using Eq. (9.92),

$$K_f = {a_1}^{-1}{a_2}^{-1}a_3 = \frac{x_3\gamma_3}{(x_1\gamma_1)(x_2\gamma_2)} \Rightarrow K_f(x_1\gamma_1)(x_2\gamma_2) - x_3\gamma_3 = 0$$

Now expressing the last equation in terms of the transform compositions and x_3

$$K_f[(\{1 + x_3\}X_1 - x_3)\gamma_1][(\{1 + x_3\}X_2 - x_3)\gamma_2] - x_3\gamma_3 = 0 \tag{A}$$

We use a root finding technique to solve for x_3. We use the transformed Gibbs energy of mixing instead of the transformed Gibbs energy, then the orthogonal derivative is

$$\left(\frac{\partial\Delta_m\hat{g}/RT}{\partial X_1}\right)_{T,P} = (\ln x_1\gamma_1 - \ln x_2\gamma_2) \tag{B}$$

To create a plot of the orthogonal derivative, $[\partial(\Delta_m\hat{g}/RT)/\partial X_1]$ vs X_1, we fix a value of X_1 and then we can solve for x_3 using Eq. (A). For $X_1 = 0.5$, we obtained $x_3 = 0.875958$. The shape of the orthogonal derivative is shown below.

The plot contains two extrema at P. These points $\left\{X_1, \left(\frac{\partial\Delta_m\hat{g}/RT}{\partial X_1}\right)_{T,P}\right\}$ are approximately: (0.55, 1.92633169) and (0.75, 0.97738696). We can use the average of those values to obtain the initial values for X_1^α and X_1^β. Our initial values is $\left(\frac{\partial\Delta_m\hat{g}/RT}{\partial X_1}\right)_{T,P} =$ 1.45186. The procedure is as follows:

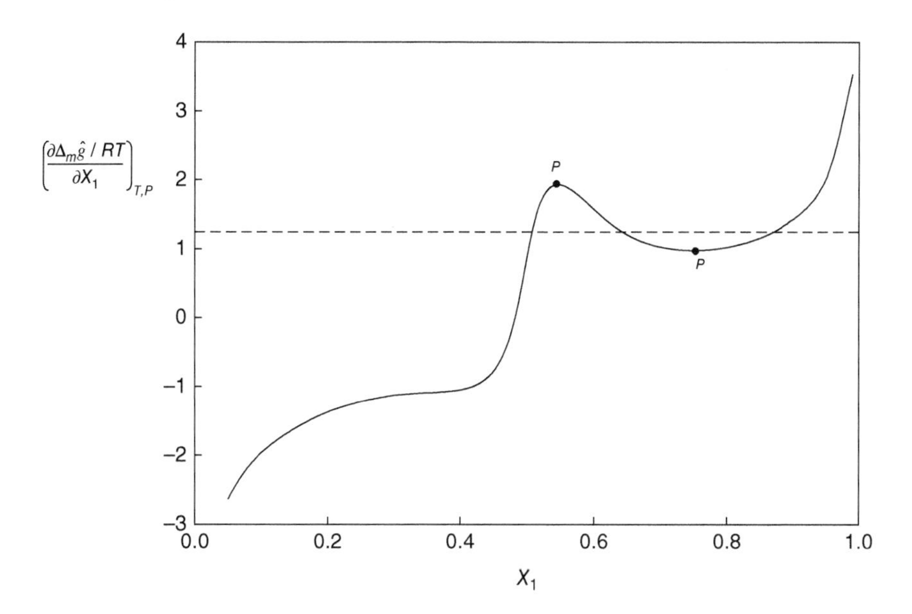

1. Use the initial value of the orthogonal derivative to find X_1^α and X_1^β; unfortunately, the procedure is complicated because it is necessary to calculate X_1 and x_3 simultaneously at each phase because X_1 depends on x_3, and x_3 is required for the calculation of the activity coefficients. Eqs. (A) and (B) are solved simultaneously.
2. After the values of X_1 and x_3 are found for each phase, we apply Eq. (9.140) for the transformed Gibbs energy of mixing

$$\left(\frac{\partial\Delta_m\hat{g}/RT}{\partial X_1}\right)_{k+1} = \frac{\left[\int_{x_1^\beta}^{x_1^\alpha}(\partial\Delta_m\hat{g}/RT/\partial X_1)_{T,P}dX_1\right]_k}{\left[X_1^\alpha - X_1^\beta\right]_k}$$

The integral is

$$\int_{x_1^\beta}^{x_1^\alpha}(\partial\Delta_m\hat{g}/RT/\partial X_1)_{T,P}dX_1 = \sum_{i=1}^{2}X_i^\beta \ln(x_i\gamma_i)^\beta - \sum_{i=1}^{2}X_i^\alpha \ln(x_i\gamma_i)^\alpha$$

Four iterations sufficiently determine the orthogonal derivative to be 1.281118 and the equilibrium compositions are:

$$x_1^\alpha = 0.082500$$

$$x_2^\alpha = 0.051180$$

$$x_1^\beta = 0.866320$$

$$x_2^\beta = 0.051180$$

The table below shows the iterations and the values of the iterative variables.

Iteration		x_3^α	X_1^α	x_3^β	X_1^β
0	1.451830	0.857270	0.512741	0.063237	0.902571
1	1.285510	0.866135	0.508493	0.081890	0.877306
2	1.281122	0.866320	0.508391	0.082499	0.876509
3	1.281118	0.866320	0.508391	0.082500	0.876508
4	1.281118	0.866320	0.508391	0.082500	0.876508

Problems for Chapter 9

9.1 Methane is burned with air. Calculate the moles of nitrogen, if the combustion reaction is

$$CH_4 + 2O_2 \rightarrow CO_2 + 2H_2O$$

9.2 In the following reaction, air supplies the oxygen. If the system contains 1.5 mol of O_2 for each 4 mol of HCl, what is the percent of excess air? Assume that the nitrogen is inert. What is the number of moles present at the end of the reaction?

$$4HCl + O_2 \rightarrow 2H_2O + 2Cl_2$$

9.3 A system contains initially 2 mol of A and 3 mol of B. For the following reaction

$$3A_{(g)} + 2B_{(g)} \rightarrow C_{(g)} + 5D_{(g)}$$

determine the mole fractions of the species as a function of the extent of the reaction.

9.4 Estimate the equilibrium constant of a reaction at 1000 K, if the standard heat of reaction is independent of temperature and equals 5000 J/mol. The equilibrium constant at 300 K is 1.5.

9.5 Determine the standard Gibbs energy of formation for NO_2 at $T = 1667$ K and $P = 1$ bar.
Data

Species	A_i	B_i	C_i	D_i	$\Delta G_{f,i}$	$\Delta H_{f,i}$
N_2	3.280	5.900×10^{-4}	0	4.00×10^{3}	0	0
O_2	3.639	5.060×10^{-4}	0	-2.27×10^{4}	0	0
NO_2	4.982	1.195×10^{-3}	0	-7.92×10^{4}	51 310	33 180

Ideal gas heat capacity

$$C_P^{ig}/R = A + BT + CT^2 + D/T^2$$

9.6 Calculate the standard Gibbs energy of formation for benzene as a liquid at 95 °C and 1 bar based on free element standard states of hydrogen as an ideal gas and carbon as solid graphite, each at 95 °C and 1 bar.
Data

Species	A_i	B_i	C_i	D_i	$\Delta G_{f,i}$	$\Delta H_{f,i}$
C	1.77100	7.71000×10^{-4}	0	-8.67×10^{4}	0	0
H_2	3.24900	4.22000×10^{-4}	0	8.30×10^{3}	0	0
C_6H_6	−0.20595	3.90864×10^{-2}	-1.3301×10^{-5}	0	124 520	49 080

Ideal gas heat capacity

$$C_P^{ig}/R = A + BT + CT^2 + D/T^2$$

9.7 Calculate the equilibrium composition for the reaction $N_2 + C_2H_2 \rightarrow 2HCN$ at 650 °C and 1.53 bar when the reactor feed composition is 75 mol% nitrogen and 25 mol% acetylene. The standard Gibbs energy of reaction for this reaction at 650 °C and 1 bar is 32.42 kJ/mol. All substances are in the gas phase. You may assume ideal gas behavior for this calculation.

9.8 For the following reaction at 40 °C and 2 atm, n_o

$$\mathrm{A} \rightarrow 2\mathrm{B} + \mathrm{C}$$

the fractional dissociation of A is 0.4 at equilibrium. If the initial number of moles of A is n_o:

a. Calculate the chemical equilibrium constant for the gas-phase reaction.

b. The chemical equilibrium constants for the reaction have the values of 0.9 and 0.4 at the temperatures 320 and 298 K, respectively. Find an expression for the heat of reaction as a function of temperature. If

$$C_{P_i}^{ig} = a + bT + cT^{1.5}$$

with

Substance	a	$b \times 10^3$	$c \times 10^5$
A	5.1	1.2	0
B	10.1	2.25	0
C	3.1	1.45	3.1

9.9 Propylene may be produced by dehydrogenation of propane according to $C_3H_8 \rightarrow C_3H_6 + H_2$. Calculate the equilibrium compositions resulting from a pure propane feed, at 298.15 and 910 K, at 2 atm pressure. (The free energy of the reaction, $\Delta_r G°$, at 910 K is 4.94 kJ/mol.) How will the various concentrations change if the pressure increases at a constant temperature? Data

Substance	$\Delta G_{f,i}$ (J/mol)
Propane	−24 300
Propylene	62 200
Hydrogen	0

T (K)	Mole fractions		
	C_3H_8	C_3H_6	H_2
298.15	1.000	1.86×10^{-8}	1.86×10^{-8}
900.00	0.379	0.311	0.311

9.10 One concept proposed for cooling aircraft that operates at very high speed is to use an endothermic reaction such as the dehydrogenation of cyclohexane to form benzene, or of methylcyclohexane to form toluene. Such endothermic reactions are capable of absorbing relatively large quantities of energy at elevated temperatures. Determine the equilibrium conversion and the amount of energy absorbed per mole of cyclohexane fed for the reaction $C_6H_{12} \rightarrow C_6H_6 + 3H_2$ if the reaction occurs in the gas phase at 1.5 atm and 650 K. At this temperature and 1 atm, the standard free energy change of the reaction and the standard heat of the reaction are $\Delta_r G° = -37\,367$ J/mol and $\Delta_r H° = 219\,700$ J/mol, respectively. Will the ability to absorb energy as heat increase or decrease as the temperature rises? Explain how you reached your conclusion.

9.11 The gas-phase reaction, $A + 2B \rightarrow AB_2$, occurs in a laboratory with the following results: at 25.00 °C and 1.24 bar, the equilibrium composition is 0.25A, 0.05B, and $0.70AB_2$, while at 50.00 °C and 1.43 bar, the equilibrium composition is 0.20A, 0.11B, and $0.69AB_2$.

a. Using a standard state pressure of 1 bar, determine the free energies of reaction at 25 °C and 50 °C for this reaction.
b. Calculate the heat of reaction in the temperature range 25–50 °C for this reaction. Is this reaction endothermic or exothermic?

9.12 Consider the reaction $A + B \rightarrow C$ that occurs in the vapor phase at 350 °C. The equilibrium constant, K_f, for this reaction at this temperature is 0.876. Calculate the equilibrium composition and the fraction of species A destroyed for each of the following cases.

a. The reaction occurs at 1.25 bar, beginning with an equimolar mixture of A and B.
b. The reaction occurs at 1.25 bar beginning with a mixture of composition $y_A = 0.2$, $y_B = 0.2$, $y_C = 0.0$, and $y_D = 0.6$. Specie D is inert and does not participate in the reaction.

9.13 Consider the reaction: $A + 2B \rightarrow 2C$. At 450 K and 220 bar, the equilibrium compositions are $y_A = 0.268$, $y_B = 0.435$, and $y_C = 0.297$. Calculate the free energy of reaction at 450 K relative to perfect gas standard states at 450 K. Assume the mixture behaves as an ideal solution, but not as an ideal gas. Pertinent properties for substances A, B, and C appear in Table P9.13.1.

Table P9.13.1 Critical constants and acentric factors.

	T_C (K)	P_C (bar)	V_C (cm^3/mol)	Z_C	ω
A	346.2	44.0	182.5	0.279	0.151
B	173.0	36.7	114.4	0.292	0.020
C	428.6	73.3	133.2	0.274	0.261

9.14 Consider a reaction $A + 3B \rightarrow 2C$ that occurs in the liquid phase. The liquid phase is highly nonideal, and the Wilson equation describes its behavior well by using the parameters given in Table P9.14.1. If the equilibrium compositions at 450 K and 2 bar are $x_A = 0.268$, $x_B = 0.435$, and $x_C = 0.297$, calculate

Table P9.14.1 Wilson equation parameters for the liquid phase.

$\Lambda_{AA} = 1.0000$	$\Lambda_{AB} = 0.1273$	$\Lambda_{AC} = 1.1258$
$\Lambda_{BA} = 0.7038$	$\Lambda_{BB} = 1.0000$	$\Lambda_{BC} = 1.7562$
$\Lambda_{CA} = 1.5563$	$\Lambda_{CB} = 1.4230$	$\Lambda_{CC} = 1.0000$

the free energy of reaction at 450 K relative to pure liquid standard states at 450 K and 1 bar. (You may omit the Poynting correction.)

9.15 Determine the equilibrium composition at 1000 K and 1.8 bar for a feed of pure methane that decomposes into carbon, which precipitates as a solid, and hydrogen. Also, formulate a set of equations that can determine the equilibrium composition if the feed contains 10% oxygen and 90% methane. To work this problem, you must decide which substances can be formed as products in this reactor.

Data

Species	A_i	B_i	C_i	D_i	$\Delta G_{f,i}$	$\Delta H_{f,i}$
CH_4	1.7025	9.0861×10^{-3}	-2.16528×10^{-6}	0	−50 460	−74 520
C	1.7710	7.710×10^{-4}	0	-8.670×10^{4}	0	0
H_2	3.2490	4.220×10^{-4}	0	8.3000×10^{3}	0	0

Ideal gas heat capacity

$$C_P^{ig}/R = A + BT + CT^2 + D/T^2$$

9.16 Solid ammonium chloride (NH_4Cl) decomposes upon heating to form ammonia (NH_3) and hydrogen chloride (HCl), both of which are gases at the conditions of decomposition. Neglect the Poynting corrections when necessary.

a. Calculate the partial pressures of ammonia and hydrogen chloride over solid NH_4Cl at 25 °C. For solid ammonium chloride at this temperature, the standard Gibbs energy of formation is −202.870 kJ/mol, and the standard enthalpy of formation is −314.430 kJ/mol.
b. If solid ammonium chloride is placed into a sealed container, and all the air in the container is evacuated before heating, determine the total pressure in the container at 300 °C if some solid still remains in the container. The standard Gibbs energy of reaction for the decomposition reaction at 300 °C is 9.334 kJ/mol at 300 °C.

Data

At 298.15 K

Species	$\Delta G_{f,i}$	$\Delta H_{f,i}$
NH_4Cl	−202 870	−314 430
HCl	−95 299	−92 307
NH_3	−16 450	−46 110

9.17 Consider a reaction $A_{(g)} + B_{(l)} \rightarrow 2C_{(l)}$. The liquid mixture that contains only A and B is not an ideal solution and the excess Gibbs function for the liquid is

$$\frac{G^E}{RT} = 0.7x_B x_C$$

At 150 °C, at equilibrium, the liquid compositions are $x_B = 0.357$ and $x_C = 0.643$ at a total pressure of 65.7 kPa. Calculate the free energy of reaction for this reaction based on the following standard states: ideal gas at 1 bar for A and pure liquid at 1 bar for B and C. Consider the vapor pressures of B and C to be 0.637 and 0.321 kPa. You may neglect the Poynting corrections in this calculation.

9.18 Consider the reaction: $A + B \rightarrow C$. This reaction has a free energy of reaction of 17.67 kJ/mol at 413.2 K and 1 bar based on the conventional choice of standard states. A reactor vessel containing A, B, and C has both liquid and vapor phases present when it is at 413.2 K and 0.856 bar. Calculate the composition of the vapor and liquid phases given the following information. The critical temperature of substance A is 352.87 K; the vapor pressures of substances B and C at 413.2 K are 7.32×10^{-3} and 0.564×10^{-3} bar, respectively; Henry's constant for substance A in the liquid phase is 7895 bar, and the excess Gibbs energy for binary liquid mixtures of B and C is

$$\frac{G^E}{RT} = -1.524x_B x_C$$

9.19 Upon heating, calcium carbonate converts to calcium oxide and carbon dioxide according to the following reaction:

$$CaCO_{3(s)} \rightarrow CaO_{(s)} + CO_{2(g)}$$

Solid calcium carbonate in the amount of 5 g is placed into a vessel with a volume of 0.5 l. The vessel is evacuated, sealed, and heated to high temperatures. What is the pressure in the vessel at 1000 K and how much mass (in grams) of each substance (calcium carbonate, calcium oxide, and carbon dioxide) are present in the vessel? Assume that CO_2 behaves as an ideal gas.

For this problem, calculate the Gibbs energy of reaction at the reaction temperature from the values at 298.15 K. Assume that the standard heat of reaction, $\Delta_r H°$, and the standard entropy of reaction, $\Delta_r S°$, are independent of temperature. The standard heat of reaction at 298.15 is 178 kJ/mol and the Gibbs energy of reaction is 130.5 kJ/mol.

Molecular weights: carbon dioxide: 44.01; calcium carbonate: 100.09; and calcium oxide: 56.08

9.20 Methanol (CH_3OH) may react with oxygen (O_2) to produce formaldehyde (CH_2O) at 800 K. Also, a thermal decomposition of methanol can occur. The outlet stream of a reactor with a feed composition containing 2 mol of methanol to 1 mol of oxygen (no formaldehyde, hydrogen, or water in the feed) contains CH_2O, CH_3OH, H_2O, H_2, and O_2. Write the system of equations that must be solved to determine the equilibrium composition at 800 K and 2 bar using the method of reaction coordinates. You may assume ideal gas behavior.

9.21 Consider the decomposition of a compound according to $AC_2 \rightarrow AC + C$ and $AC \rightarrow E + C$ at 25 °C in the liquid phase. The free energies of the reaction at 25 °C are 11.416 and 13.135 kJ/mol for the first and second reactions, respectively. These free energies have standard states as liquids at 1 bar pressure. Assume that the liquid mixture is an ideal solution and the difference between the system and standard state pressures are sufficiently small that all Poynting corrections are negligible.

a. Calculate the free energy of formation for AC_2 based on the same choice of standard states, where A and C are the constituent elements.
b. Calculate the mole fraction of each species (A, C, AC, and AC_2) in the liquid phase at 25 °C if the starting condition is pure AC_2.

9.22 Calculate the equilibrium concentrations of NO and NO_2 in air that has been heated to 2000 K at 1 bar. Consider the two reactions:

$$\frac{1}{2}N_2 + \frac{1}{2}O_2 \rightarrow NO \text{ and } NO + \frac{1}{2}O_2 \rightarrow NO_2$$

The equilibrium constants are 3.027×10^{-3} and 1.887×10^{-2}, respectively.

a. Formulate the two equilibrium equations in terms of the variables $X_1 = \varepsilon_1 / n_T{}^\circ$ and $X_2 = \varepsilon_2 / n_T{}^\circ$
b. Solve for the equilibrium compositions, assuming that $X_2 \ll X_1$. Do the results satisfy this assumption?

9.23 Consider two gas-phase reactions at 1500 K

$$A + B \rightarrow C + D \quad (1)$$

$$A + C \rightarrow 2E \quad (2)$$

The free energies of the reaction at the reaction temperature are
(1) $\Delta G^0 = -4078$ J/mol
(2) $\Delta G^0 = -4835$ J/mol.
The pressure in the reactor is 10 atm, and the feed consists of 2 mol of A and 1 mol of B.
Calculate the composition of the reaction mixture if both reactions reach equilibrium. Assume that the standard state pressure is 1 atm.

9.24 Consider the production of component C by two independent reactions at 1.5 bar

$$A + B \rightarrow C$$

$$D + E \rightarrow C$$

If the chemical equilibrium constants are $K_1 = 0.26$ and $K_2 = 0.005$ at 800 K, what is the product composition when the reactions reach equilibrium? Is this process feasible?

9.25 Methanol (CH_3OH) may react with oxygen (O_2) to produce formaldehyde (CH_2O) at 800 K. A second reaction forms formaldehyde by decomposing methanol into formaldehyde and water (H_2O). The outlet stream of a reactor with a feed composition containing 2 mol of methanol to 1 mol of oxygen (no formaldehyde, hydrogen, or water) contains CH_2O, CH_3OH, H_2O, H_2, and O_2. The free energies of formation of the compounds relative to the usual free element standard states at 800 K are −88.33 kJ/mol (CH_3OH), −95.32 kJ/mol (CH_2O), and −203.65 kJ/mol (H_2O). Write the system of equations that must be solved to determine the equilibrium composition by minimizing the Gibbs free energy. The only variables in these equations should be $\{yi\}$ and $\{\lambda^*\}$.

References

1 Schittkowski, K. (2015). Nonlinear programming software for members and students of academic institutions. http://klaus-schittkowski.de/SOFTWARE_A.pdf (accessed April 14, 2022).

2 Ung, S. and Doherty, M.F. (1995). Vapor–liquid equilibrium in systems with multiple reactions. *Chem. Eng. Sci.* 50: 23–48.

3 Ung, S. and Doherty, M.F. (1995). Theory of phase equilibria in multireaction systems. *Chem. Eng. Sci.* 50: 3201–3216.

4 Iglesias-Silva, G.A., Bonilla-Petriciolet, A., and Hall, K.R. (2006). An algebraic formulation for an equal area rule to determine phase compositions in simple reactive systems. *Fluid Phase Equilib.* 241: 25–30.

A

Appendices

A.1 Instructions to Add an Add-In Your Computer

Windows:

- Download the file and put it in a folder of your choice
- Click the File tab, click **Options**, and then click the **add-Ins** category.
- In the Manage box, click **Excel add-Ins** and then click **Go**. The **add-Ins** dialog box appears.
- In the **add-Ins** available box, select the check box next to the **add-In** that you want to activate and then click OK.
- If the add-In is not there, click **Add** and then locate the add-In and check the box next to the **add-In**.

Macintosh (Mac):

- Download the file at any folder.
- Open Excel.
- From the top-level Mac menu bar, click the **Tools** menu and select Excel **add-Ins** to open the **add-Ins** dialog.
- If the **Add-In** is listed, simply check it in the list.
- If the **Add-In** is not listed, use the **Select bottom** to browse and select it and click Open. The add-In file should now appear in the list of available add-Ins in the **add-Ins** dialog.
- Press OK.

A.2 Excel® LK CALC Add-In

The add-In LK CALC calculates compressibility factors, residual enthalpy, and residual entropy using the Lee–Kesler equation of state (EOS) [1]. The valid ranges are

Thermodynamics for Chemical Engineers, First Edition. Kenneth R. Hall and Gustavo A. Iglesias-Silva.

$0.3 \leq T_R \leq 4$ and $0.01 \leq P_R \leq 10$. Once the add-In is in the computer, using the above instructions from Appendix A.1, the following formulas should appear in excel:

z_0(tr,pr)	Calculates the z_0 compressibility factor using the reduced temperature and pressure.
z_1(tr,pr)	Calculates the z_1 compressibility factor using the reduced temperature and pressure.
hres_0(tr,pr)	Calculates the reduced residual enthalpy, $\left[\frac{H^R}{RT_C}\right]^0$, using the reduced temperature and pressure.
hres_1(tr,pr)	Calculates the reduced residual enthalpy, $\left[\frac{H^R}{RT_C}\right]^1$, using the reduced temperature and pressure.
sres_0(tr,pr)	Calculates the reduced residual entropy $\left[\frac{S^R}{R}\right]^0$ using the reduced temperature and pressure.
sres_1(tr,pr)	Calculates the reduced residual enthalpy $\left[\frac{S^R}{R}\right]^1$ using the reduced temperature and pressure.
fug_0(tr,pr)	Calculates the fugacity coefficient φ^0 using the reduced temperature and pressure.
fug_1(tr,pr)	Calculates the fugacity coefficient φ^1 using the reduced temperature and pressure.

To use and access these functions, simply type = in any cell, and the function name and the values of the reduced temperature and pressure appear separated by a comma. For example, to calculate $z^{(0)}$, go to any cell for A1, type =z_0(1,1) and the value appears in your cell.

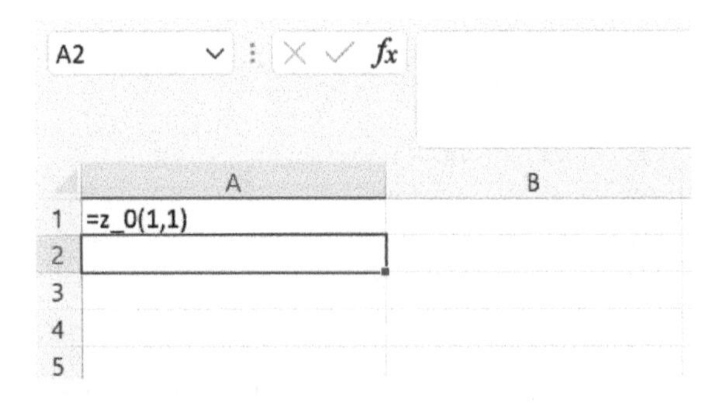

Your values can be in different cells, for example, if the reduced temperature is in cell B1 and the reduced pressure is in B2, use

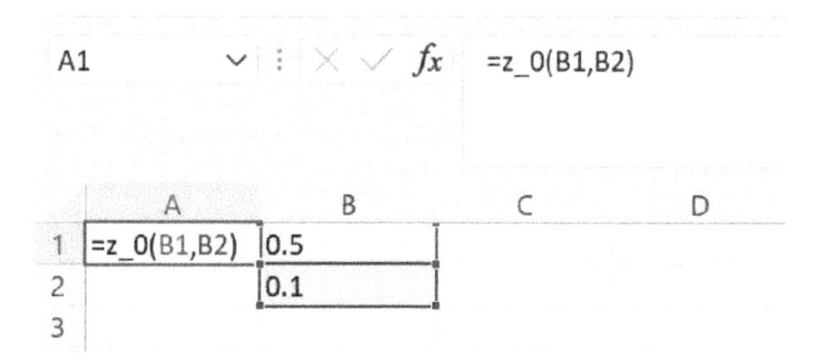

The same procedure applies for any function.

A.3 Excel® STEAM CALC Add-In

The instructions to install the STEAM CALC Add-In appear in Appendix A.1, and you use the same approach as in Appendix A.2. Once the add-In is in your computer, it appears when you open an excel sheet using the credits

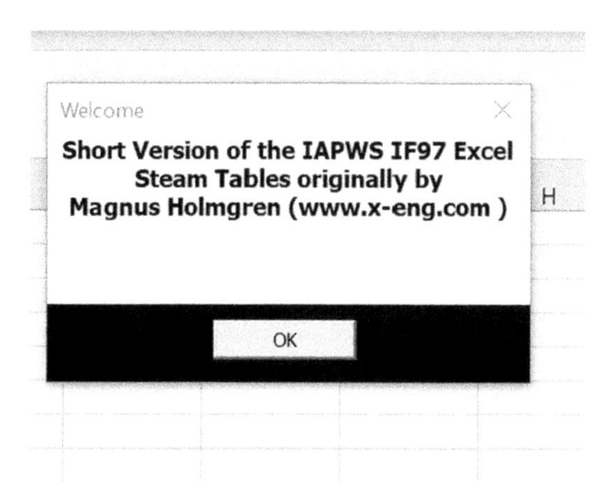

Click OK and continue working in your excel program.

The add-In Steam is a short version of the steam tables developed by IAPWS IF97 Excel Steam Tables originally by Magnus Holmgren (www.x-eng.com). This version is only to solve the problems in the text Thermodynamics for Chemical Engineering.

The pressure and temperature should appear in kPa and °C. The volume units are cm^3/g and the units for the enthalpy, internal energy, and entropy are kJ/kg, kJ/kg, and kJ/(kg °C).

The add-In contains the following functions

Temperature Functions

Tsat_p(p)	Calculates the temperature at saturation at a given pressure
Tsat_s(s)	Calculates the temperature at saturation at a given entropy
T_hs(h,s)	Calculates the temperature at a given enthalpy and entropy
T_ph(p,h)	Calculates the temperature at a given pressure and enthalpy
T_ps(p,s)	Calculates the temperature at a given pressure and entropy
T_pv(p,v)	Calculates the temperature at a given pressure and volume

Pressure Functions

psat_T(T)	Calculates the vapor pressure at a given temperature
psat_s(s)	Calculates the vapor pressure at a given entropy
p_hs(h,s)	Calculates the pressure given the enthalpy and entropy

Enthalpy Functions

hVsat_p(p)	Calculates the vapor enthalpy at saturation given a pressure
hLsat_p(p)	Calculates the liquid enthalpy at saturation given a pressure
hVsat_T(T)	Calculates the vapor enthalpy at saturation given a temperature
hLsat_T(T)	Calculates the liquid enthalpy at saturation given a temperature
h_ps(p,s)	Calculates the enthalpy at a given pressure and entropy
h_pT(p,T)	Calculates the enthalpy at a given pressure and temperature
h_pv(p,v)	Calculates the enthalpy at a given pressure and volume

Volume Functions

vVsat_p(p)	Calculates the saturated vapor volume at a given pressure
vLsat_p(p)	Calculates the saturated liquid volume at a given pressure
vLsat_T(T)	Calculates the saturated liquid volume at a given temperature
vVsat_T(T)	Calculates the saturated vapor volume at a given temperature
v_pT(p,T)	Calculates the volume at a given pressure and temperature
v_ph(p,h)	Calculates the volume at a given pressure and enthalpy
v_ps(p,s)	Calculates the volume at a given pressure and entropy

Entropy Functions

sV**sat_p(p)**	Calculates the vapor entropy at saturation given a pressure
sLsat_p(p)	Calculates the liquid entropy at saturation given a pressure
sVsat_T(T)	Calculates the vapor entropy at saturation given a temperature
sLsat_T(T)	Calculates the liquid entropy at saturation given a temperature
s_ph(p,h)	Calculates the entropy at a given pressure and enthalpy
s_**pT(p,T)**	Calculates the entropy at a given pressure and temperature
s_pv(p,v)	Calculates the enthalpy at a given pressure and volume

Internal Energy Functions

uVsat_p(p)	Calculates the vapor internal energy at saturation given a pressure
uLsat_p(p)	Calculates the liquid internal energy at saturation given a pressure
uVsat_T(T)	Calculates the vapor internal energy at saturation given a temperature
uLsat_T(T)	Calculates the liquid internal energy at saturation given a temperature
u_ph(p,h)	Calculates the internal energy at a given pressure and enthalpy
u_ps(p,s)	Calculates the internal energy at a given pressure and entropy
u_pT(p,T)	Calculates the internal energy at a given pressure and temperature
u_pv(p,v)	Calculates the internal energy at a given pressure and volume

To access the functions, type = plus the function name and between parenthesis two properties separated by a comma. For example, in cell B1 type the pressure 100 kPa and in cell B2, the temperature 100 °C, then you can calculate the enthalpy, entropy, specific volume, and internal energy because they use as arguments the pressure and temperature as shown below

	A	B	C
1	=h_pT(B1,B2)	100	
2	=s_pT(B1,B2)	100	
3	=v_pT(B1,B2)		
4	=u_pT(B1,B2)		
5			
6			
7			

and the results are

	A	B	C
1	2675.76737	100	
2	7.36099921	100	
3	1.69595941		
4	2506.17143		
5			
6			
7			

A.4 Heat Capacity Equations for an Ideal Gas

The excel file named Ideal Heat Capacities.xls contains the constants A, B, C, D, and C' for the equation

$$\frac{C_P^{ig}}{R} = A + BT + CT^2 + DT^3 + C'/T^2$$

Depending on the numerical value of the constants, the equation reduces to

$$\frac{C_P^{ig}}{R} = A + BT + CT^2 + DT^3$$

$$\frac{C_P^{ig}}{R} = A + BT + CT^2 + C'/T^2$$

or

$$\frac{C_P^{ig}}{R} = A + BT + C'/T^2$$

The accuracy is related to the number of parameters. The maximum temperature is 1500 K. The constants come from Refs. [2, 3].

A.5 Antoine Equation Constants

The excel file Antoine Constants.xlsx contains the parameters used in the Antoine vapor pressure equation,

$$\ln P^{sat} = A - \frac{B}{T + C}$$

where P^{sat} is in MPa and T is in K. The constants have been taken from Yaws and Yang [4].

A.6 Heat Capacity Equations for liquids

The excel file Liquid Heat Capacities.xlsx contains the constants A, B, C, and D for the equation

$$\frac{C_P^L}{R} = A + BT + CT^2 + DT^3$$

The constants come from Coker [5].

A.7 Iterative Procedures for the Calculation of Vapor Liquid Equilibrium

A.7.1 Bubble Point Calculations

A.7.1.1 Bubble Pressure Calculation

Specified: T, $\{x_i\}$	**unknown: P, $\{y_i\}$**

Working equation for calculation of bubble pressure:

$$\sum_{i=1}^{N} y_i = 1 \Rightarrow \sum_{i=1}^{N} \frac{x_i \gamma_i P_i^{sat}}{\Phi_i P} = 1 \Rightarrow P = \sum_{i=1}^{N} \frac{x_i \gamma_i P_i^{sat}}{\Phi_i}$$

One-time calculations: γ_i, P_i^{sat}
Iterative calculations: y_i, ΦP_i^{sat}

Calculation Procedure for Bubble Pressure:

1. Calculate the vapor pressure for each substance $\{P_i^{sat}\}$ using the specified temperature
2. Calculate the activity coefficient for each substance $\{\gamma_i\}$ using the specified temperature and liquid-phase composition
3. Note: Steps 1 and 2 need only be calculated once because the temperature and liquid compositions are known.
4. Assume $\{\Phi_i\} = 1$, then calculate P, $\{y_i\}$ using

$$P = \sum \frac{x_i \gamma_i P_i^{sat}}{\Phi_i} \text{ and } y_i = \frac{x_i \gamma_i P_i^{sat}}{\Phi_i P}$$

5. Use the values of P, $\{y_i\}$ obtained in step 4 to calculate new values for $\{\Phi_i\}$ using:

$$\Phi_i = \exp\left[\frac{B_{ii}\left(P - P_i^{sat}\right) + \frac{1}{2}\sum_{j=1}^{N}\sum_{k=1}^{N} y_i y_k (2\delta_{ij} - \delta_{jk})}{RT}\right]$$

Repeat steps 3 and 4 until

$$|\delta P| < \varepsilon \text{ where } \delta P = P_{n+1} - P_n$$

and ε is the convergence criterion.

***Note*: The convergence criterion should be at least 1 order of magnitude smaller than the desired accuracy *and* 1 order of magnitude larger than the ultimate precision of the calculating device (computer).**

A.7.1.2 Bubble Temperature Calculation

Specified: $\{x_i\}$, P	**Determined: $\{y_i\}$, T**

The working equation for calculation of bubble temperature is

$$\sum_{i=1}^{N} y_i = 1 \Rightarrow \sum_{i=1}^{N} \frac{x_i \gamma_i P_i^{sat}}{\Phi_i P} = 1 \Rightarrow P = \sum_{i=1}^{N} \frac{x_i \gamma_i P_i^{sat}}{\Phi_i}$$

One-time calculations: γ_i
Iterative calculations: y_i, Φ_i, P_i^{sat}

Calculation Procedure for Bubble Temperature:

1. Initial estimate of temperature: $T = \sum x_i T_i^{sat}$ in which (for the Antoine equation):

$$T_i^{sat} = \frac{B_i}{A_i - \ln P} - C_i$$

2. Calculate new vapor pressures $\{P_i^{sat}\}$ for each substance using:

$$P_i^{sat} = \exp\left[A_i - \frac{B_i}{T + C_i}\right]$$

3. Calculate an activity coefficient for each substance $\{\gamma_i\}$

4. Choose one component arbitrarily, designate it as species J (in the initial iteration $\Phi_i = 1$)

$$P = \sum_{i=1}^{N} \frac{x_i \gamma_i P_i^{sat}}{\Phi_i} = \sum_{i=1}^{N} \frac{x_i \gamma_i P_i^{sat}}{\Phi_i} \left(\frac{P_J^{sat}}{P_J^{sat}} \right) = P_J^{sat} \sum_{i=1}^{N} \frac{x_i \gamma_i}{\Phi_i} \left(\frac{P_i^{sat}}{P_J^{sat}} \right)$$

Then:

$$P_J^{sat} = \frac{P}{\sum_{i=1}^{N} \frac{x_i \gamma_i}{\Phi_i} \left(\frac{P_i^{sat}}{P_J^{sat}} \right)}$$

5. Calculate a new temperature estimate, T, from:

$$T = \frac{B_J}{A_J - \ln P_J^{sat}} - C_J$$

6. Calculate new vapor pressures $\{P_i^{sat}\}$ using the latest temperature estimate in

$$P_i^{sat} = \exp\left[A_i - \frac{B_i}{T + C_i}\right]$$

7. Calculate new estimates for the vapor composition $\{y_i\}$ using:

$$y_i = \frac{x_i \gamma_i P_i^{sat}}{\Phi_i P}$$

8. Use P, $\{y_i\}$ to calculate new estimates for $\{\Phi_i\}$, $\{\gamma_i\}$.
9. Calculate a new estimate for P_J^{sat} (see step 4)
10. Calculate a new estimate for T (see step 5)

Repeat steps 6–10 until convergence.

A.7.2 Dew Point Calculations

A.7.2.1 Dew Pressure Calculation

Specified: T, $\{y_i\}$	**Unknown: P, $\{x_i\}$**

Working equation for calculation of dew pressure:

$$\sum x_i = 1 \Rightarrow \sum \frac{y_i}{K_i} = 1 \Rightarrow \sum_{i=1}^{N} \frac{y_i \Phi_i P}{\gamma_i^L P_i^{sat}} = P \sum_{i=1}^{N} \frac{y_i \Phi_i}{\gamma_i^L P_i^{sat}} = 1$$

One-time calculations: P_i^{sat}
Iterative calculations: P, x_i, Φ_i, γ_i

Calculation Procedure for Dew Pressure:

1. Calculate P_i^{sat} (once only because T is specified)
2. Set $\{\Phi_i\} = 1$ and $\{\gamma_i\} = 1$.
3. Calculate P using

$$P = \frac{1}{\sum \frac{\Phi_i y_i}{\gamma_i P_i^{sat}}}$$

4. Calculate $\{x_i\}$ using

$$x_i = \frac{y_i \Phi_i P}{\gamma_i P_i^{sat}}$$

5. Calculate $\{\Phi_i\}$ using $\{y_i\}$ and P.
6. Calculate $\{\gamma_i\}$ using $\{x_i\}$, T and P.

Repeat steps 3–6 until $|\delta P| < \varepsilon$.
***Note*: An internal loop could normalize $\{x_i\}$ before recalculating P, but the procedure used here has much the same effect.**

A.7.2.2 Dew Temperature Calculation

Known: $\{y_i\}$, P	**Determine** $\{x_i\}$, T

Working equation for calculation of dew temperature:

$$\sum x_i = 1 \Rightarrow \sum \frac{y_i}{K_i} = 1 \Rightarrow \sum_{i=1}^{N} \frac{y_i \Phi_i P}{\gamma_i^L P_i^{sat}} = P \sum_{i=1}^{N} \frac{y_i \Phi_i}{\gamma_i^L P_i^{sat}} = 1$$

One-time calculations: None
Iterative calculations: $T, x_i, \Phi_i, \gamma_i, P_i^{sat}$

Calculation Procedure for Dew Temperature:

1. Calculate the initial estimate of temperature using:

$$T = \sum y_i T_i^{sat} = \sum y_i \left[\frac{B_i}{A_i - \ln P} - C_i \right]$$

2. Calculate $\left\{ P_i^{sat} \right\}$ using:

$$P_i^{sat} = \exp \left[A_i - \frac{B_i}{T + C_i} \right]$$

3. Choose one component arbitrarily, designate it as species J, then multiply both sides of the working equation (step 1) by P_J^{sat} to obtain (in the initial iteration $\Phi_i = 1$):

$$P_J^{sat} = P \sum_{i=1}^{N} \frac{y_i \Phi_i}{\gamma_i} \left(\frac{P_J^{sat}}{P_i^{sat}} \right)$$

4. Calculate T using:

$$T = \frac{B_J}{A_J - \ln P_J^{sat}} - C_J$$

5. Calculate $\left\{ P_i^{sat} \right\}$ using:

$$P_i^{sat} = \exp \left[A_i - \frac{B_i}{T + C_i} \right]$$

6. Calculate $\{\Phi_i\}$ using the specified $\{y_i\}$, P and the newly calculated values for T, $\left\{ P_i^{sat} \right\}$.

7. Calculate $\{x_i\}$ using

$$x_i = \frac{y_i \Phi_i P}{\gamma_i P_i^{sat}}$$

8. Calculate $\{\gamma_i\}$ using $T, P, \{x_i\}$

Repeat steps 7 and 8 until $\{\gamma_i\}$ and $\{x_i\}$ converge.
Repeat steps 3–8 until convergence: $(|\delta T| < \varepsilon)$

A.7.3 Flash Calculation

Specified: $T, P, \{z_i\}$	**To be determined**: $V, L, \{x_i\}, \{y_i\}$

Material balance:

$$z_i = y_i V + x_i L \Rightarrow \begin{cases} y_i = \dfrac{z_i K_i}{1 + V(K_i - 1)} \\ x_i = \dfrac{z_i}{1 + V(K_i - 1)} \end{cases}$$

Working equation choices:

$$\sum_{i=1}^{N} x_i = 1 \Rightarrow F_x = \sum_{i=1}^{N} \frac{z_i}{1 + V(K_i - 1)} - 1 = 0$$

$$\sum_{i=1}^{N} y_i = 1 \Rightarrow F_y = \sum_{i=1}^{N} \frac{z_i K_i}{1 + V(K_i - 1)} - 1 = 0$$

$$F = F_y - F_x = \sum_{i=1}^{N} \frac{K_i z_i}{1 + V(K_i - 1)} - \sum_{i=1}^{N} \frac{z_i}{1 + V(K_i - 1)} = \sum_{i=1}^{N} \frac{(K_i - 1) z_i}{1 + V(K_i - 1)} = 0$$

The third alternative is the best behaved mathematically because it has no spurious roots

$$F = \sum_{i=1}^{N} \frac{z_i (K_i - 1)}{1 + V(K_i - 1)}$$

$$\frac{dF}{dV} = -\sum_{i=1}^{N} \frac{z_i (K_i - 1)^2}{[1 + V(K_i - 1)]^2}$$

Newton's method works well here because there is no spurious root in this working equation:

$$V_{j+1} = V_j - \frac{F_j}{\left(\frac{dF}{dV}\right)_j}$$

The $\{K_i\}$ come from:

$$K_i = \frac{y_i}{x_i} = \frac{\gamma_i P_i^{sat}}{\Phi_i P}$$

Flash Calculation Procedure:

Specified: $T, P, \{z_i\}$
Determine: $V, L, \{x_i\}, \{y_i\}$

Procedure:

1. Calculate P_{DP}, $\{x_i\}$ at T.
2. Calculate P_{BP}, $\{y_i\}$ at T.
3. Proceed with flash calculation only if

$$P_{DP} < P < P_{BP}$$

(If this condition is not satisfied, two phases do not exist, so the flash calculation is meaningless.)

4. Calculate initial estimates using:

$$\frac{\gamma_i - \gamma_i^{DP}}{\gamma_i^{BP} - \gamma_i^{DP}} = \frac{P - P_{DP}}{P_{BP} - P_{DP}}$$

$$\frac{\Phi_i - \Phi_i^{DP}}{\Phi_i^{BP} - \Phi_i^{DP}} = \frac{P - P_{DP}}{P_{BP} - P_{DP}}$$

$$V = \frac{P_{BP} - P}{P_{BP} - P_{DP}}$$

5. Calculate $\{K_i\}$ using

$$K_i = \frac{y_i}{x_i} = \frac{\gamma_i P_i^{sat}}{\Phi_i P}$$

6. Calculate F, dF/dV using

$$F = \sum_{i=1}^{N} \frac{z_i(K_i - 1)}{1 + V(K_i - 1)}$$

$$\frac{dF}{dV} = -\sum_{i=1}^{N} \frac{z_i(K_i - 1)^2}{[1 + V(K_i - 1)]^2}$$

Calculate a new estimate for the vapor fraction using:

$$V_{j+1} = V_j - \frac{F_j}{\left(\frac{dF}{dV}\right)_j}$$

Iterate step 6 until V converges to a solution.

7. Calculate $\{x_i\}$ from:

$$x_i = \frac{z_i}{1 + V(K_i - 1)}$$

8. Calculate $\{\gamma_i\}$ using the $\{x_i\}$ calculated in step 7.
9. Calculate $\{y_i\}$ using:

$$y_i = K_i^{old} x_i \frac{\gamma_i^{new}}{\gamma_i^{old}}$$

10. Calculate new $\{\Phi_i\}$ using the $\{y_i\}$ calculated in step 9.

Repeat steps 5–10 until convergence is achieved for V, $\{x_i\}$, and $\{y_i\}$.

References

1 Lee, B.I. and Kesler, M.G. (1976). A generalized thermodynamic correlation based on three-parameter corresponding states. *AlChE J.* 21: 510–527.
2 Spencer, H.M. (1948). Empirical heat capacity equations of gases and graphite. *Ind. Eng. Chem.* 40: 2152–2154.
3 Pankratz, L.B. (1982). Thermodynamic properties of elements and oxides with a section on process applications by R.V. Mrazek. *U.S. Bur. Mines Bull.* 672, 1–509.
4 Yaws, C.L. and Yang, H.C. (1989). To estimate vapor pressure easily. *Hydrocarbon Process.* 68 (10): 65–68.
5 Coker, A. *Ludwig's Applied Process Design for Chemical and Petrochemical Plants*, Distillation, Packed Towers, Petroleum Fractionation, Gas Processing and Dehydration, 4e, vol. 2.

Index

Thermodynamics for Chemical Engineers, First Edition. Kenneth R. Hall and Gustavo A. Iglesias-Silva.

s

t

u

v

w

z